1991 SOLAR WORLD CONGRESS

VOLUME 2, PART II

Proceedings of the Biennial Congress of
the International Solar Energy Society,
Denver, Colorado, USA, 19-23 August 1991

Edited by

M.E. ARDEN
SUSAN M.A. BURLEY
MARTHA COLEMAN

American Solar Energy Society, Inc.,
Boulder, Colorado, USA

PERGAMON PRESS

OXFORD • NEW YORK • BEIJING •
FRANKFURT • SEOUL • SYDNEY • TOKYO

Pergamon Press Offices:

U.K.	Pergamon Press plc, Headington Hill Hall, Oxford OX3 0BW, England
U.S.A.	Pergamon Press, Inc., Maxwell House, Fairview Park, Elmsford, New York 10523, U.S.A.
PEOPLE'S REPUBLIC OF CHINA	Maxwell Pergamon China, Beijing Exhibition Centre, Xizhimenwai Dajie, Beijing 100044, People's Republic of China
GERMANY	Pergamon Press GmbH, Hammerweg 6, D-6242 Kronberg, Germany
KOREA	Pergamon Press Korea, KPO Box 315, Seoul 110-603, Korea
AUSTRALIA	Maxwell Macmillan Pergamon Publishing Australia Pty Ltd., Lakes Business Park, 2 Lord Street, Botany, NSW 2019, Australia
JAPAN	Pergamon Press, 8th Floor, Matsuoka Central Building, 1-7-1 Nishi-Shinjuku, Shinjuku-ku, Tokyo 160, Japan

First Edition 1991

Library of Congress Cataloging-in-Publication Data

ISBN: 0-08-041690-X

NOTICE

Neither the International Solar Energy Society®, the United States Section of the International Solar Energy Society, nor any of the cosponsors of this Congress make any warranty, expressed or implied, to accept any legal liability or any information, apparatus, product, or process disclosed, or represent that its use would not infringe privately on rights of others. The contents of the proceedings express the opinion of the authors and are not necessarily endorsed by the International Solar Energy Society®, the United States Section of the International Solar Energy Society, or any of the cosponsors of this Congress.

Printed in the United States of America

The paper used in this publication meets the minimum requirements of American National Standard for Information Sciences -- Permanence of Paper for Printed Library Materials, ANSI Z39.48-1984

INTERNATIONAL SOLAR ENERGY SOCIETY

The International Solar Energy Society is a worldwide nonprofit organization dedicated to the advancement of the utilization of solar energy. Its interests embrace all aspects of solar energy, including characteristics, effects and methods of use, and it provides a common meeting ground for all those concerned with the nature and utilization of this renewable non-polluting resource.

Founded in 1954, the Society has expanded over the years into a truly international organization with members in more than 90 of the world's countries. It has been accepted by the United Nations as a nongovernmental organization in consultative status, and it is widely regarded as the premier body of its type operating in the solar energy field.

The Society is interdisciplinary in nature and numbers among its members most of the world's leading figures in solar energy research and development, as well as many with an interest in renewable energy and its practical use. High academic attainments are not a prerequisite for membership, only a special interest in this particular field.

Organization

Day-to-day administration is provided by the Society's headquarters office, which since 1970 has been located in Australia. The headquarters house the Secretary-Treasurer and the Administrative Secretary, together with members of their supporting staff.

In countries and regions in which sufficient interest exists, Sections of the Society have been established. These Sections, which are largely autonomous, organize meetings and other local activities and in some cases produce their own publications. All Society members are eligible to belong to their respective national or regional Sections, although in some cases this may involve the payment of an additional Sectional fee. In recent years the number of Sections has increased slowly but steadily.

Activities of the Society are:

1. Publications of *Solar Energy*, a monthly scientific journal of an archival nature, containing scientific and technical papers on solar energy and its utilization, reviews, technical notes and other items of interest to those working in the field of solar energy.
2. Publication of a less technical magazine, *SunWorld*.
3. Publication of a newsletter for members, *ISES News*.
4. Organization of major International Congresses on solar energy at which numerous scientific and technical papers are presented and discussed. These Congresses are held every two years in different countries, normally in conjunction with equipment exhibitions, and are widely attended.
5. Publication of the Proceedings of each International Congress. Whereas copies of the Society's three periodicals (items 1-3 above) are supplied to all members as part of their membership, copies of Congress Proceedings are available (from the publisher) only on special order and at an additional cost. Special pre-publication prices are normally available to Society members.
6. More recently ISES has become increasingly involved with other major Non-Governmental Organizations in matters relating to the application of renewable energy and other global environmental problems, and is currently preparing its contribution for presentation at the United National Conference on Environment and Development (UNCED - or popularly referred to as ECO 92).

Headquarters:

International Solar Energy Society
PO Box 124
Caulfied East, Vic. 3145
AUSTRALIA

Telephone: 61 3 571 7557
Fax: 61 3 563 5173
Telex: AA 154 087 CITVIC

AMERICAN SOLAR ENERGY SOCIETY

The American Solar Energy Society (ASES) is the United States Section of ISES and presently has over 4,000 members.

ASES seeks to promote the widespread near-term and long-term use of solar energy. To achieve that goal, ASES:
• Fosters the use of science and technology in the application of solar energy;
• Encourages basic and applied research and development in solar energy;
• Promotes education in fields related to solar energy; and
• Provides information relating to all aspects of solar energy.

Activities:

• ASES conducts the National Solar Energy Conference as a annual forum for exchange of information about advances in solar energy technologies, programs, and concepts. The conference features speakers who are national leaders in their technical and professional fields. Workshops, exhibits and tours of solar applications highlight this annual event, which is attended by more than 450 solar energy enthusiasts from throughout the country.

• ASES publishes *Solar Today*, a bi-monthly magazine. Each issue highlights practical applications of solar energy, presents the latest results of solar energy research, covers developments in the nation's solar energy industry, and includes member discussion of solar-related issues.

• Each year, ASES sponsors a Roundtable in Washington, DC, bringing together energy decision-makers in a highly visible public forum. Each Roundtable addresses an issue of critical importance to ASES members and the nation.

• To ensure worldwide dissemination of information about solar energy developments, ASES annually publishes *Advances in Solar Energy*. This compendium of the latest R&D developments is authored by ASES members who are nationally recognized experts on their respective topics.

• Technical, regulatory and educational issues are addressed in periodic White Papers, which present critical analyses of important solar energy topics.

• ASES educates the public and energy decision-makers on the benefits of solar energy through a public relations campaign and information materials.

• ASES has 16 state and regional chapters, which are independently incorporated organizations providing services to their members appropriate to the local areas. Typical activities include newsletters, technical meetings, public outreach activities, and government relations.

Headquarters
2400 Central Avenue, Suite B-1
Boulder, CO 80301
Telephone   : 303-443-3130
Fax   : 303-443-3212

Table of Contents

Volume 1: Solar Electricity, Biofuels, Renewable Resources

Volume 2: Active Solar and Solar Heat

Volume 3: Passive Solar, Socio-Economic, Education

Volume 4: Plenaries, State-of-the-Art, Farrington Daniels Lecture

Contents of Volume 2

x

2.17 Posters: Active II

A SYSTEM FOR RECUPERATING SOLAR ENERGY
FALLING ON A HORIZONTAL SLAB

M. ADJ, Y. SFAXI*, A. GIRARDEY and M. GRIGNON*

Laboratoire d'Energétique Appliquée, ENSUT   B.P 5085
Dakar (Sénégal).

*Laboratoire d'Energie Solaire d'Evry, France

ABSTRACT

In Senegal, we observe that the surface temperature of a thick concrete
slab subjected to solar radiation can, without any particular precautions,
reach between 50° and 60 °C. Within a depth of 5 cm we still find 40 -
45°C. A simple calculation also shows that a 15 cm thick slab, at certain
hours, stores energy of about 1,2 Kwh per m2. For comparison purposes,
about 2,3 Kwh is needed to raise the temperature of 100 litres of water
from 25 to 45°C. We have studied the possibility of recuperating part of
the energy stored in the slab for the production of hot sanitary water.
This article presents the theoretical and experimental study of the system.

KEYWORDS

Recuperation; Solar energy; Hot sanitary water; Concrete slab.

DESCRIPTION OF THE SYSTEM AND THE MEASURING SET

The experiment   prototype appears in the form of a horizontal 3m by 3m and
15cm thick concrete slab. In this slab is embedded , during casting, a 40m
long 14/16m diameter copper tube in which water is allowed to flow. The
copper tube is connected to the water supply network. The following
equipment is used for the instrumentation of the structure's different
elements and for measuring climatic data: thermocouples installed at
different spots to measure temperatures, Eppley pyranometer , a flowrate
meter, an automatic data acquisition and processing system composed of a
data acquisition station driven by a microcomputer.

ANALYTICAL APPROACH TO THE PROBLEM

FIELD TEMPERATURE IN THE SLAB
The surface of the slab is subjected to external thermal solicitations such
as the global solar flux and the ambient air temperature. These two

variables are assumed to be periodic functions according to the following simplified hypothesis:

$$Ta(t) = Tm + To.cos(\Omega.t - \phi) \qquad (1)$$

$$Eng(t) = Engm + Eo.cos(\Omega.t) \qquad (2)$$

The slab and the underlying ground can be assimilated to a homogeneous semi-infinite environment with constant thermophysical properties. The surface (x=0) exchanges verify Newton's law.
Introducing a functon F(t) defined as environment function, it can easily be shown that F(t) can be written as follows:

$$F(t) = (T_m + \frac{\alpha}{h}Eng_m) + (\frac{\alpha^2.Eo^2}{h^2} + To^2 + \frac{2.\alpha.Eo.To.cos(\phi)}{h})cos(\Omega t - \Psi) \qquad (3)$$

$$tg\Psi = - \frac{To.sin\phi}{\frac{\alpha.Eo}{h} + To.cos\phi}$$

the form $F(t) = F_m + F_0 cos(\Omega t - \psi)$.

By introducing F(t) into the general heat equation, the expression for the steady state solution is found (CARSLAW and JAEGER, 1947)
for $k = (\Omega/2.a)^{1/2}$ and $\delta = arctg(k/((h/\lambda)+k)$

$$T(x,t) = (T_m + \frac{\alpha}{h}Eng_m) + Fo\frac{h/\lambda}{\{((h/\lambda)+k)^2 + k^2\}^{1/2}}exp(-kx).cos(\Omega t - kx - \delta - \Psi) \qquad (4)$$

From equation (4) the temperature in different points of the slab and the delay time between the maxima of the temperature curves and the sunlight are calculated. This approach is interesting as it makes it possible to examine periodic non sinusoidal ambient functions by using Fourier's serial alteration of curves obtained from a "meteo" card-index.

FLUID TEMPERATURE : ONE TEMPERATURE MODEL

We formulate the already verified hypothesis that the thermal field in the slab is slightly affected by the recuperation tube and that, consequently, the latter is in contact with a medium whose temperature is Tp(t) = T(x,t), X being the average depth of the tube in the slab.
The thermal balance applied to the fluid and the tube which are assumed to be at the same temperature, gives:

$$q_m.C_p.\frac{\partial T_f}{\partial y} + \frac{M.C_p}{L}.\frac{\partial T_f}{\partial t} = - K.(T_f - T_p) \qquad (5)$$

in which K is the global linear transfer coefficient between the tube and the concrete slab (w/m.k) and M the equivalent mass of the fluid in the tube and the tube L is the length of the tube.

This first order partial differential equation solved by charateristic
lines method gives

$$\frac{y - yo}{L} = \frac{1}{M} \int_{to}^{t} q_m(u) \, . du \tag{6}$$

$$T(y,t) = T(yo,t) \, . \exp(-\frac{t-to}{\tau}) + \frac{1}{\tau} \exp(-\frac{t}{\tau}) \int_{to}^{t} T_p(u) \, . \exp(\frac{u}{\tau}) \, du \tag{7}$$

with $\tau = M.C_p/K.L$ and $m = K/q_m.C_p$

Equations (6) and (7) make it possible to calculate the heat conveying
fluid output temperature for different conditions (steady state, flowrate
step response, periodic function Tp)

NUMERICAL APPROACH BY THE MODAL ANALYSIS METHOD

The thermal field in the slab may be geometrically schematised according to
figure 1.

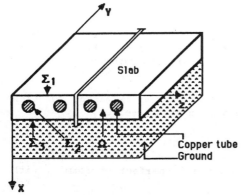

Σ_1 : the boundary between the slab
and the ambiant surroundings,

Σ_2 : the boundary between the slab
and the tube,

Σ_3 : the boundary between the slab
and the ground,

Ω : the domain inside the slab.

Fig. 1. bloc diagram of the slab

$T = T(M,t)$, the slab temperature at point M and at instant t is determined
by the resolution of the heat equation considering the initial and boundary
conditions respectively on Σ_1, Σ_2 and Σ_3 :

$$-\lambda(\partial T/\partial n) = h \, . (T - Ta) + \alpha.Eng \tag{8}$$

$$-\lambda(\partial T/\partial n) = K \, . (T - T_f) \tag{9}$$

$$\lambda(\partial T/\partial n) = \lambda_{sol}(\partial T/\partial n) \tag{10}$$

where λ_{sol} is the thermal conductivity of the ground.

The numerical resolution method we have used is that of the modal analysis
(BACOT P., 1984 ; PEUBE J.L.,1984).

It has the advantage of being able to closely estimate the initial model by using a reduced model, this curtailing the calculation time considerably. After separating only the space variables, we obtain a set of equations the linear structure of which makes it possible to regroup them in the form of differential matrix equation of the first order:

$$C(dT/dt) = A.T(t) + B.E(t) \tag{11}$$

A,B and C are matrices
T(t) is the temperature vector
E(t) is the solicitation vector

After base changement and passage into the proper mode space, system 11 is written as follows:

$$dX/dt = D.X(t) + F.E(t) \tag{12}$$

$$Y = Z.X + G.E(t) \tag{13}$$

where D is the diagonal matrix of proper values, Y is the vector of observable scales (in this case, the various temperatures) and X the vector of temperatures in the proper base. X is linked to the vector T by the relation $X = P^{-1}.T$ with P as passage matrix.
By adopting a sampling pitch Δt we obtain :

$$X_{((n+1)\Delta t)} = e^{D.\Delta t}X_{(n.\Delta t)} - D^{-1}(I-e^{D.\Delta t}).F.E(K.\Delta t) \tag{14}$$

$$Y_{((n+1)\Delta t)} = Z.X_{((n+1)\Delta t)} + G.E_{((n+1)\Delta t)} \tag{15}$$

The reduction of the initial system is carried out by retaining only unstable and slow modes and by neglecting rapid modes. The classification in decreasing order of the n modes of the initial system enables us to fix the number m of the modes to be retained (PEUBE J.L.,1984).

The matrix G is determined in such a way that the permanent functions of the initial system and of the reduced system are preserved.

$$G = G_0 + (Z_m.D_m^{-1}.F_m) \tag{16}$$

This method   obtains results which are in perfect conformity with the measurements.

RESULTS

To validate the analytical approach,it is verified if the experimental curves   obey the exponential law (Fig.2). Thus for several flowrates the curves $Log(T^*) = Log((Tp-Te)/(Tp-T_u))$ are drawn . It is noticed that the curves are linear with a slope m which when determined, permits to find K and τ.

The measurements are taking on the set-up according to the following experimental protocols:

a) continuous functioning with the constant fluid flow rate(Fig 3)
b) intermittent functioning with varying fluid flow rate (Fig 4).

The analytical model retained gives the temperatures and the delays which values approach the values gotten experimentally.To be more specific, the results obtained in Dakar, on 27th, March 1989 are shown in table 1.

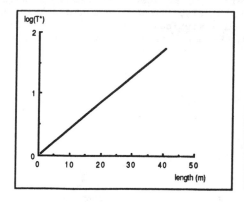

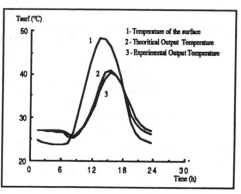

Fig. 2 . Experimental curves of Log(T*)with respect to the length of the tube, for m = 0.05.

Fig. 3 . Output temperature, on 13th, march 1991 for Te=26°C and qm=150l/h.

Analytical and experimental results on 13th, March, 1991 are obtained for:
λ = 1.65 w/m.k ; Ω = 7,272 .10-5rd/s ; Engmax = 1kW/m2 ; To = 2°C
a = 5.6.10-7m2/s ; h = 17 W/m2.k; Tm = 28 °C; α = 0.65.

The output temperature curves of the heat conveying fluid are continuously monitored, the temperature levels attained are shown in Fig. 3 and are of the order of 40 °C, with neither a transparent cover nor a sun rays concentrating device on the slab. These results give an idea on the system's performance. It is observed that when sunlight is at its peak at about 01 H p.m., the instantaneous recovery rate is about 12 to 15%.

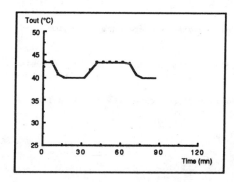

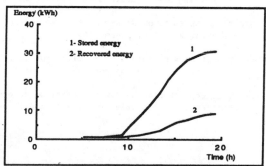

Fig. 4 . Response of the système to a jump of the fluid flow rate.

Fig. 5 . Evolution of the stored and recovered energies during the day 13 march, 1991.

Fig. 5 shows the evolution of the values of the stored energy and the recovered energy.

CONCLUSION

Our study shows that it is absolutely possible to produce hot water in a very economical manner by recuperating solar energy stored in a wall; a solution which should interest, particularly, the developing countries with sufficient sunshine. The numerical and analytical models provide results in total conformity with the measurements and can be used to optimise the system.

NOMENCLATURE

a   : diffusivity (m^2/s)
Cp : specific heat capacity $(J/kg.^\circ C)$
Eo : instantaneous solar energy mean (W/m^2)
Eng : instantaneous solar energy (W/m^2)
h   : superficial heat transfer coefficient $(W/m.^\circ C)$
L   : length
A   : matrix of thermal conductances
C   : matrix of thermal capacities in the new base

qm: mass flow rate (kg/hr)
Te: input temperature $(^\circ C)$
Tf: output temperature $(^\circ C)$
Tm: ambiant temperature mean $(^\circ C)$
To: ambiant temperature amplitude
Ts: surface temperature $(^\circ C)$
α : absorption coefficient
B : matrix of solicitations
Z : matrix of output temperatures in the new base

REFERENCES

Carslaw,H.S. and J.C.Jaeger (1947). Conduction of heat in solids. Clarendon Press, Oxford.

D. TRAORE, M. ADJ & A. GIRARDEY. Recuperation of solar energy stored in a concrete wall for producing hot sanitary water. Proceedings First World Renewable Energy Congress, Vol 2,p1025-1029, Reading, UK 23-28 Sept 1990.

Traore, D. (1989). Contribution à l'étude théorique et expérimentale d'un dispositif de récupération de l'énergie solaire stockée dans une dalle en béton. Thèse de docteur-ingénieur, Université Cheihk Anta Diop de Dakar.

BACOT P., NEVEU A. et SICARD J. Analyse modale des phénomènes thermiques en régime variable. Rev. Gén. Therm. 1984.

PEUBE J. L. Une technique de modélisation thermique des régimes instationnaires conduisant à une représentation d'ordre réduit- Journées d'Etudes ASME-GUT-ENSMA de Poitier, 1984

SFAXI Y. Contribution à l'étude numérique et expérimentale du comportement des dalles chauffantes. Thèse Univ. Paris XII -1987.

A SUPERINSULATED PASSIVE/ACTIVE SOLAR HOUSE
DESIGN, PERFORMANCE, AND ECONOMICS

A. F. Burke

Idaho National Engineering Laboratory
EG&G Idaho, Inc.
P.O. Box 1625
Idaho Falls, Idaho 83415

ABSTRACT

The design and construction of an active/passive solar house, which utilizes solar-air systems to heat the house and domestic hot water and to cool the house in the summer, are described. The house was built in Idaho Falls, Idaho in 1989 and has been lived in since completion. The solar systems include a sunspace, a solar collector using a fiber mesh absorber and evacuated glass tube glazing, a basement rock bed for heating and cooling house air, and a preheat water unit for year-round use. Records of electricity and wood pellet use over the 1 1/2 year of occupancy show that the solar fraction for the winter months is 40-45% and for the entire year it is 65%. The alternative energy fraction (solar plus wood pellets) is 65% for the winter months and 78% for the year. Based on the results of FCHART calculations of life-cycle costs, the economics of the complete system would be favorable in locations where the cost of electricity was 10 cents/kWh or higher.

KEYWORDS

Solar-air systems; fiber-mesh, evacuated tube glazing collector; rock bed thermal storage; active/passive solar house

DESIGN AND CONSTRUCTION

The intention was to design a solar house that could be replicated with minor exceptions anywhere in the northern United States by an experienced builder, and his subcontractors, using locally available building materials and construction techniques. The objective of the solar design was to meet as large a fraction of the heating, domestic hot water, and cooling demands of the home year-round as is practical and consistent with the climate (Idaho Department of Water Resources) of the Idaho Falls Area which is cold (8000 deg-days and many nights at sub-zero temperature) in the winter and hot and dry in the summer. The solar insolation is inconsistent in the winter with extended periods of near-perfect sun followed by extended periods of cloudiness not uncommon. The insolation from May through October is consistent with good sun most days. The climatic data for the Idaho Falls Area is summarized in Table 1.

The solar systems are all solar thermal with no use of photovoltaic panels. All the solar units use air as the heat transfer medium, and the thermal storage is done using rock beds and masonry. The solar air systems were selected to minimize problems with low, sub-zero temperatures in the winter and rock beds and masonry were used for thermal storage, because such systems are readily compatible with the solar air approach and could be constructed at reasonable cost by the local subcontractors. The passive portion of the solar system consists of a sunspace and south-facing glass which heats the interior of the house primarily by natural convection on sunny days. The sunspace is closed off from the main part of the house at night during the winter to minimize heat loss from the house. A roof-mounted solar collector (4'high by 30' long) utilizes a fiber-mesh as the absorber and evacuated tubes as the glazing. The solar heated air is used to heat water all year round and to heat the house directly or indirectly via the rock bed storage in the winter. The rock bed thermal storage is used to cool the house in the summer. The summer nights in Idaho Falls are consistently cool (temperatures in the low fifties or high forties) making it easy to maintain the rock bed at about 60 F throughout the summer.

All the solar systems are designed to operate automatically based on the collector, outdoor, rock bed, and water temperatures. The airflow rate through the collector is modulated by a solid-state electronic controller based on the collector temperature. The solar system is reconfigured manually in the spring and fall by repositioning several air-dampers which direct the solar air to the rock bed in the winter or by pass it in the spring and summer.

Construction of the house was started on May 1, 1989 and it was completed on October 13, 1989 (the move-in day). All construction was done by an experienced local builder and his subcontractors except for the solar collector, which was built on-site by Duggan Solar Systems, Schenectady, New York. The solar features of the house were specified by drawings/sketches and/or on-site instruction to the craftsmen working on the house.

The layout of the house is shown in Figure 1. It is a one-floor plan having about 2700 ft^2 including the sunspace. A photograph of the completed house is given in Figure 2. The detailed characteristics of the house and the solar units are summarized in Table 2 (also see Figures 2-8). In-depth descriptions and the performance of the various components are given in (Burke, 1991). Only the roof-mounted solar collector will be discussed in some detail in this paper.

The Solar Collector

The primary component of the active solar system is a 4'x 30' roof-mounted solar collector. The collector utilizes fiber mesh as the absorber and evacuated glass tubes as the glazing. Air is pulled through the collector and the associated duct system (see Figure 9) by fans. The air is heated through contact with the fiber mesh. The body of the collector and all inlet and outlet manifolds are constructed of 1 1/2" Thermax. The collector was built on-site by Duggan Solar Systems of Schenectady, New York. It is tilted at a 20^0 angle with respect to the vertical rather than a more optimum 45^0, because there had been no previous experience with this type of collector other than in the vertical orientation. Tilting the collector at 20^0 presented challenging enough problems relative to sealing against wind, rain, and snow without going to the more optimum 45^0 slope. No problems have been encountered with the collector through two Idaho winters. Photographs of the collector at several stages of construction are shown in Figures 7-8. The complete fabrication, roof mounting of the collector, and initial operation spanned a period of less than one week. The collector has operated continuously from July 3, 1989 until the present time-February 1991-without any difficulties.

On sunny winter days the collector begins operation at about 9 AM in the morning with an output air temperature of 90-95 F and continues to operate until between 4-5 PM when the output temperature falls to 85-90 F. The maximum output air temperature of 140-150 F occurs around solar noon (12:30 PM in Idaho Falls) with an input temperature of 50-55 F from the basement rock bed. The output temperature slowly increases and then decreases as the sun moves across the southern sky. The airflow (CFM) through the collector is automatically varied from a minimum of about 150 CFM (STP) to maximum of 400 CFM as a function of the temperature of the solar heated air. On partly cloudy days when the insolation is intermittent and/or mostly diffuse, the collector still operates, but the output air temperature is reduced and the collector turns off and on periodically as the insolation fluctuates. The collector has a low thermal mass so the output air temperature approaches the maximum value after only 1-2 minutes of high insolation. The collector produces useful air temperatures of 100-105 F at diffuse insolation levels of only 30 Btu/hr ft2 at low ambient temperatures (10-20 F).

From May to mid-October, the collector operates most days from about 9 AM to 5 PM as that part of the year is very sunny in Idaho Falls. The input air to the collector is drawn from the attic as the basement rock bed is decoupled from the system. The maximum output air temperature during the summer is 180-185 F since the input air from the attic is 90-95 F. In the summer all the solar heated air is used to heat domestic hot water to temperature as high as 140 F. After passing through the hot water preheat unit, the air is vented through the roof. The collector is turned-off automatically in the summer when the output air temperature falls below the temperature of the water in the hot water unit.

The efficiency of the collector has been measured at field installations in both Schenectady, New York and Idaho Falls. The results of the efficiency measurements are given in Figure 10. There is some scatter in the data, but the Y-intercept and the slope of the efficiency line can be determined with reasonable confidence. The efficiency values obtained outdoors on large collectors are in good agreement with those obtained in the laboratory on much smaller collectors using artificial radiation (Herrick, 1983/1; Herrick, 1983/2; Twerdok, 1984). The response of the collector to a sudden change in airflow was also determined from data taken on large collectors. The thermal time constant of 1.1 minutes was estimated from the data shown in Figure 11. The short time constant explains the fast response of the collector to intermittent insolation on partly cloudy days in the winter.

PERFORMANCE SUMMARY OF THE HOUSE

Electrical energy and wood use records have been used to calculate the solar and alternative energy fractions for the house. As shown in Table 3, the electrical energy use is essentially constant from March-October and increases significantly in November-February during the winter, heating months. The details of the solar fraction calculations are given in Table 4. The energy requirements for heating the house were determined using a calculated UA value of 350 Btu/h °F and the average winter degree-days for each month in Idaho Falls (Table 1). The energy requirements for domestic hot water were calculated for a water heater set point of 125 F and an average ground water temperature of 50 F. A water usage of 40 gallons per day (two people in the house) was assumed. This resulted in a total energy requirement of 76.4 million Btu for heating and domestic hot water. The data given in Table 5 show that the solar fraction for domestic hot water was 33% for the winter months and 65% for the year. It was assumed that the additional electrical energy use in the winter months was used for heating and that 90% of the wood was used during November-February. The separate contributions of electrical energy and wood to meeting the energy requirements of the house are shown in Table 4. The solar contribution is calculated as the difference between the requirement and the purchased electrical and wood energies. It was found that for the winter months (November-February) the solar fraction (passive plus active) was 41% and the renewable energy fraction (solar plus wood) was 65%. For the year, the solar fraction was calculated to be 63% and the renewable energy fraction- 78%. This performance was obtained before the solar system was altered to permit solar heated air to be ducted directly into the house. It is estimated that this improvement in the system will increase the solar energy fraction to at least 50% in the winter months and to 70% for the year. Both wood and electrical energy use will be reduced using the improved system. During the summer all the cooling requirements are met using the diurnal cycle (night time cooling of the rock bed), so the solar fraction for cooling is essentially 100%. In a climate having more consistent winter insolation and less degree-days than Idaho Falls, significantly higher solar fractions would be expected using the same solar design (see Table 6).

CALCULATED PERFORMANCE USING FCHART

Calculations were made using both the passive direct gain and active collector-pebble bed options of the FCHART program for Pocatello, Idaho (near Idaho Falls) and Denver, Colorado. Although it is only an approximation, the solar fractions obtained for the separate passive and active calculations were added to get the total solar fraction for the present passive-active solar house design. The results of the FCHART calculations are given in Table 6. They indicate an annual solar fraction of .81 for Pocatello and .98 for Denver. The value of .81 for Pocatello is significantly higher than the actual solar fraction of 70% projected for Idaho Falls based on experience with the house in 1989-91. It should be noted that Idaho Falls is colder than Pocatello (8000 deg-days compared to about 7000 deg-days) and that Pocatello is slightly sunnier than Idaho Falls. Another explanation of this difference is that most of the passive south-facing glass in the house is in an enclosured sunspace having high thermal mass which traps some of the solar heat before it is distributed to the main part of the house.

The FCHART calculations predict relatively high solar fractions for the present solar house even though the areas of the roof collector and south-facing glass are modest for the size (one-story - 2700 ft^2) of the house. Calculations were also made for the same solar systems on smaller houses. This was done by reducing the UA of the house to 260 (2000 ft^2) and 195 (1500 ft^2). The results of those calculations are also shown in Table 6. In the smaller houses the contribution of the active system is larger and the total solar fraction approaches 1.0 even in Idaho, where the insolation in the winter months is less consistent than in Colorado. Application of the solar systems used in the present house is thus attractive in houses of all sizes.

ECONOMIC CONSIDERATIONS

Detailed cost information is available on the house and the costs of solar related components are known accurately. A list of these components and their costs are given in Table 7. Interpretation of these costs as they relate to the economic attractiveness or lack of attractiveness of solar designs for meeting heating, domestic hot water, and cooling requirements depend on what comparable non-solar house is used for comparison and what fuel is used in that house. All the subsequent economic analyzes will be done assuming the non-solar house uses electricity at a cost of 6.5-10 cents/kWh. It is known at the outset that on a strict economic basis (payback period or life cycle costs), solar designs are not attractive if the fuel used in the non-solar house is natural gas or oil at about \$1/gal. Most of the solar component costs shown in Table 7 are offset by related costs in a non-solar house of comparable quality. It is assumed that both the solar and non-solar houses have equivalent insulation characteristics and have central heating and air-conditioning. The non-solar house is also assumed to have two fireplaces which is common for high quality (expensive) houses in Idaho Falls. For the solar house, it is assumed that the sunspace is one of the most attractive functional rooms in the house and the costs associated with it are not considered as solar costs. The costs of the thermal chimney and interior brick are credited against the cost of the fireplaces in the non-solar house. The cost of the basement rock bed is credited against the cost of air-conditioning in the non-solar house. Using the rock bed for summertime cooling works very well in Idaho Falls, but it likely would not be satisfactory in high humidity climates. The electric furnace in the non-solar house would be much larger than in the solar house, but that difference and the cost of any fireplace inserts are used to offset the cost of the pellet wood stove in the solar house. Hence based on the above arguments, the only special solar costs are those associated with the active solar collector and the water preheat unit.

The costs of these active solar components are not small. However, some of the costs would undoubtedly be somewhat lower if the collector had not been hand built on-site and the subcontractors had been more familiar with solar projects. Part of the cost was also due to the desire to instrument the system to track its performance and the need to provide for some conservatism in a new design. (After all, this was an "engineering project" funded by the author.) For the purposes of the present economic analysis of the active solar system, the cost of the collector was taken to be \$30/ft2 with an area independent cost of \$4000 to cover the water preheat unit, extra duct work, wiring, and plumbing, and system electronic control. The origin of these cost inputs is given in Table 8. The projected future cost of an active solar system like that in the present house would be \$7600 rather than the actual cost of \$8300. FCHART runs were made for a number of passive/active solar houses with UA values between 195 and 350. The payback period and life-cycle cost results for Idaho and Colorado are summarized in Table 9. In all cases, the economics of the passive system is more favorable than the active system, but both systems are needed to achieve high solar fractions especially in locations with inconsistent winter insolation. Life cycle cost savings result for an electricity cost of 10 cents/kWh, but not for 6.5 cents/kWh.

REFERENCES

Idaho Solar and Weather Information, Idaho Department of Water Resources, (Boise, Idaho)

Burke, A.F.(1991) A Superinsulated Passive/Active Solar House- Design, Performance, and Economics (EG&G Report)

Herrick, C.(1983/1) An Air-Cooled Solar Collector Using All-Cylindrical Elements in a Low-Loss Body (Solar Energy) Vol 30, No.2

Herrick, C. (1983/2) Optical Measurements on a Solar Collector Cover of Cylindrical Glass Tubes (Solar Energy) Vol 30, No.3

Twerdok, J., Design (1984) Construction, and Testing of a Solar Air Heating Facility (Master's Thesis, Union College, Dept. of Mechanical Engineering)

SOLAR ASSISTED ADS AND RSS RUBBER DRYING

Mohd Yusof Othman and Kamaruzzaman Sopian

Solar Energy Research Group
Universiti Kebangsaan Malaysia
43600 Bangi, Malaysia

ABSTRACT

This paper describes the design of both solar assisted smokehouse and drier for RSS and ADS drying respectively. In addition, RSS requires the introduction of smoke during the process while ADS requires only drying. The conventional energy source used in producing RSS is rubberwood and ADS is diesel. The introduction of solar assisted smokehouse and drier promotes energy conservation in this sector.

KEYWORDS

Energy conservation, ribbed smoke sheet (RSS), air dried sheet (ADS), smoke-house, drying tunnel.

INTRODUCTION

Malaysia is the world largest producer of natural rubber (havea brasilliensis). The annual production of natural rubber in Malaysia increases from 1.540×10^6 tonnes in 1980 to 1.580×10^6 tonnes in 1987 [Ministry of Primary Commodities, 1987]. There are at least six types of rubber produced for marketing purposes, namely latex concentrate, rubber ribbed smoke sheet (RSS), superior processing rubber (SPR), Standard Malaysian Rubber (SMR - block rubber), rubber air dried sheet (ADS) and crepe rubber.

Study on the status of drying of agricultural produced in Malaysia has been conducted on paddy, cocoa, coffee, pepper, tobacco, tea, banana, anchovies, and rubber [Othman et al, 1987]. The summary of the report was publised in the Solar World Congress in Japan [Othman et al, 1989] which indicated that the immediate potential market for solar drying is rubber both for RSS and ADS.

One of the important processes involved in producing both RSS and ADS is drying. In addition, RSS requires the introduction of smoke whereby ADS requires only drying during the process. The rubberwood that were used as source of heat for producing RSS requires approximately one kg of rubberwood to dry one kg of RSS rubber [Sethu, without date]. Hence, the ratio of rubberwood to dry rubber is typically 1:1.

Rubberwood was once a cheap source of heat. Presently, it is one of the expensive energy source. The price of rubberwood increased fivefold within five years from M$30.00 per 6x6x6 cu. ft. in 1980 to M$150.00 in the early 1986 [Othman et al, 1987]. This is due to the fact that when chemically treated, it can be used as furniture.

The increase in demand of rubberwood and its availability has pressured the rubber industry in Malaysia to change the source of heat to other sources. The initial work conducted by Rubber Research Institute of Malaysia (RRIM) shows that it is more economical (the saving of 66% of rubberwood) and the drying is more faster using solar assisted smokehouse than conventional rubberwood smokehouse [Muniandy, 1989]. Futhermore, the high solar radiation intensity enjoyed by Malaysia, average between 500 - 700 W/m² [Chuah and Lee, 1982], making it more attractive to work with.

RUBBER SHEETS PROCESSING

Figure 1 shows the art of making rubber sheets in Malaysia. The process of collecting latex until dripping is similar for both RSS and ADS. The wet rubber sheets are hung in air for 24 hours for dripping. Next, the trolly loaded with hung sheet is transfered into standard RRIM tunnel type smokehouse for drying and smoking.

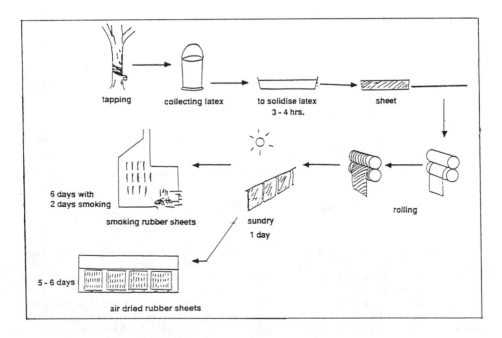

Fig. 1. The process of making rubber sheets.

Conventional method of producing RSS needs 4 - 5 days drying around 50°C - 60°C, depending on weather, follows by 2 days smoking in the smokehouse. Normally rubberwood is used as heat source for smoking and drying. In producing ADS, diesel is considered as appropriate energy source since it can be used continuously for 5 - 6 days at 58°C (140°F). At the moment, the source is considered economically fea sible method of producing ADS.

AIR DRYING OF RUBBER SHEET

Drying of rubber sheet has been discussed in detail by [Gale, 1962] and the drying curves for a typical rubber sheet is shown in Figure 2. Considerable amount of water is removed if the sheet is hung overnight. As soon as the temperature reaches 50°C, the water from the sheet exudes under humid condition through a process known as synthesis. About 80% of drying time is required to remove the last 10 % of the moisture.

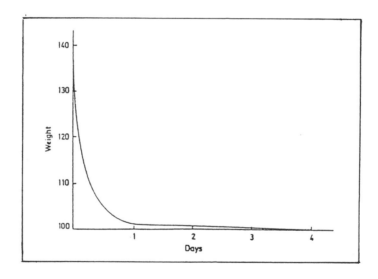

Fig. 2. Drying curve for rubber sheets.

Natural drying takes a longer period because drying occur only during the day. Humidity in Malaysia is very high so that at night, when the temperature falls, the humidity rises to near saturation point. Moreover, the lower the humidity the more water it can absorb. Furthermore, the hotter the air the more moisture can be absorbed before it become saturated. Air, which is saturated at one temperature, can still be made to absorb moisture if its temperature is raised. In brief, most moisture is taken up when the air temperature is maintained at the highest possible figure that can be used without damage to the rubber. A further advantage of a high drying temperature is because of the properties of wet rubber; the water held inside the rubber is assisted to diffuse to the surface. Hence, the design of drying equipment takes into account the above factors.

SOLAR ASSISTED SMOKEHOUSE FOR RSS RUBBER DRYING

Solar Assisted Smokehouse

An experimental force convective solar assisted smokehouse was constructed by
RRIM with the size of drying chamber is 1800 mm long by 1500 mm wide by 2200 mm
high [Mundiandy, 1986]. It is fitted with two levels of beroties for drying
three sheets of rubber weighing 1 kg each. Hence, the chamber is estimated to
contain 1500 kg of dry rubber. Access to the sheet hanging space is through two
doors on the eastern and western sides of the chamber. A "portable smokehouse
furnace" is fitted on the southern sides while the solar panels are attached to
the northern sides. The temperature is maintained at about 50°C - 55°C during
effective sunshine hours. During the night the smokehouse is operated by a single stoking of firewood in the evenings. The detailed performance of the system
is reported by Muniandy et al (1986). The RSS rubber produced using the solar
assisted smokehouse is comparable to the one produced conventionally.

Based on the above experimental trial result, an improved large scale solar
assisted smokehouse with capacity of 1500 kg was constructed as shown in Figure
3 with the drying performance is shown in Figure 4. It was shown by Muniandy
(1989) the payback period for the designed solar assisted smokehouse is about 38
months for full capacity operation of 7500 kg per month.

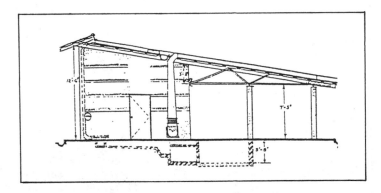

Fig. 3. Side view of 1500 kg RRIM's solar assisted smokehouse.

SOLAR ASSISTED DRIER FOR ADS RUBBER DRYING

The process of producing ADS has been described above. Moreover, summary of the
drying specifications are shown in Table 1. Figure 5 shows the existing drying
tunnel located in a 2400 acres plantation. The plantation has four drying tunnel of which two are in operation. The production is about 3.5 ton per day and
consumed about 140 litre per mton of dry sheet. Hence, the fuel consumption is
about 399 litre/day of diesel in continous operation for 5 to 6 days. At the
present price of diesel about M$0.53 per litre, the monthly fuel bill is about
M$7791. Hence, there is potential for a solar assisted drier for ADS rubber.
Conceptually, a collector area of 20 m² and a fan less than 2 hp will be installed. It is expected that the fuel savings will be in between 20% to 30%.
This is an ongoing project whereby proposal have been submitted to the respecta-
tive plantation management.

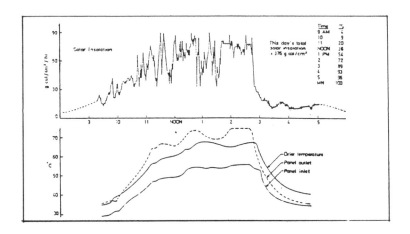

Fig, 4. Measured solar insolation and corresponding smokehouse
temperature.

TABLE 1. Specification of drying tunnel for ADS rubber sheet.

Dripping	24 hours
Number of tunnel/plantation	4 (2 in operation)
Heat source	Diesel
Tunnel capacity	12 trolleys
Trolley capacity	1000 kg
Number of dried rubber sheets	3 trolleys/day
Tunnel temperature	140°F (58°C)
Drying period	5 - 6 days
Energy consumption	140 litre/tonnes day sheet (140x3.5=399 litre/day)
Cost	3.5x140x0.53=M$259.70 per day ≈ M$7791.00 per month

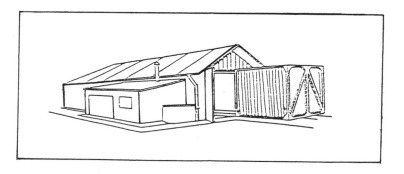

Fig. 5. A single drying tunnel for ADS rubber.

CONCLUSION

The use of solar energy in assisting operation of a smokehouse has been effective and comparable to conventional smokehouse. Furthermore, the quality of rubber sheets are also as comparable to the fully smoked sheets. A similar situation is envisaged for the solar assisted drier for ADS drying. The use of solar energy will enhanced energy conservation in drying of rubber sheet. The uncertainty of price of conventional fuel makes solar energy more favourable. However, the following factors have to be considered when solar energy is going to be utilised for rubber drying,

- Non-availability of direct solar heat throughout the drying cycle.

- To qualify as a RSS rubber, it needs to have the appearance of smoked sheet and smell of smoke.

- The need to provide heating facilities during the night, prolonged rainy weather and cloudy period since ADS rubber needs constant temperature of 140 F.

REFERENCES

Chuah. D.G.S and Lee. S.L., 1981. Solar Radiation Estimates In Malaysia, Solar Energy, Vol 26. pp 33-40.

Gale, R.S., 1962. Drying of Rubber Sheet in the Falling Rate Period. Trans. Inst. Rubber. Ind. 38 T19 1962.

Ministry of Agriculture of Malaysia, 1988. Agricultural, Livestock and Fisheries Statistics for Management 1986-87. Agriculture Economic Division,

Ministry of Primary Industries, Malaysia, 1988. Statistic on Commodities.

Mohd Yusof Hj Othman, M. Noh Dalimin, Baharudin Yatim and Muhamad Mat Saleh, 1987. Status of Drying Technologies of Agricultural Produces in Peninsular Malaysia. Tech. Report, FGFS, Universiti Kebangsaan Malaysia 2(2), pp 188-212 (In Malay).

Muniandy. V, 1989.Utilization of Solar Energy For Rubber Sheet Drying. Proc. of Seminar Penggunaan Tenaga Suria Dalam Pengeringan, Universiti Kebangsaan Malaysia, October. pp 65-73.

Muniandy V., Graham D.J., Rama Rao P.S., 1986. Solar Power Boosted Smokehouse. Rubber Growers Conference 1986, Ipoh, RRIM.

University of Malaya Agricultural Graduates Alumni and Pantai Maju, 1986. Malaysia Agricultural Index.

An Experimental Study of Basin-Type Integral Solar Regenerator/Dehumidifier

W.J. Yan and L.C. Chen

Energy and Resources Laboratories, ITRI
Chutung, Hsinchu, Taiwan, R.O.C.

ABSTRACT

In this study, a basin-type integral solar regenerator/ dehumidifier with a collecting area of $1.9m^2$ was built and tested. During the day, the weak lithium chloride solution is concentrated by solar heating. The evaporated water vapor is taken out by an air stream blowing over the surface of solution. During the night, the humid air is flowed over the desiccant surface. The water vapor in the air is absorbed by strong solution due to vapor pressure difference. The experimental results showed that the solar regeneration efficiency can be as high as 31 %. For a typical summer night in Taiwan, the average rate of dehumidification is 0.112 kg H_2O/hr. When water cooling coils were used in the system, the dehumidification rate can be further increased to an average value of 0.147 kg H_2O/hr.

KEYWORDS

Integral solar regenerator/dehumidifier; liquid desiccant; dehumidification rate; lithium chloride; regeneration efficiency.

INTRODUCTION

A basin-type solar distillation process with air flow through the still was found more efficient than without flowing air (Yeh, 1985). If the sea water in the basin is replaced by liquid desiccant such as lithium chloride, lithium bromide or calcium chloride, then it can be used for dehumidification purpose at night. The weak desiccant solution could be regenerated by solar energy during the day. The goal of this paper is to investigate the performance of such a basin-type integral solar regenerator/ dehumidifier operating in intermittent cycle.

EXPERIMENTAL DESCRIPTION

The schematic diagram of integral solar regenerator/dehumidifier in this experiment is shown in Fig.1. The solar regenerator/ dehumidifier has a collecting area of $1.9m^2$. It is made from wood coated with a water proof black plastic film as absorber. The basin is separated by a baffle-wall to form an air circulation loop. A small fan supplies air flow at an average speed of 2 m/s. The distance between absorber and glass cover is 8cm. Lithium chloride solution is the working fluid and is put in the basin. Air humidity was measured by a Humicap sensor with digital indicator. Four T-type thermocouples were used to measure input and output temperatures of both desiccant solution and air. The concentration of LiCl solution was determined by measuring the refractive index of solution sample. There is a linear relation between refractive index and concentration of the solution. The solar radiation was measured by a pyranometer. The water cooling coils consist of two copper tubes both are 3 m long and 1 cm in diameter. They were used to remove the evolved heat of absorption during the night.

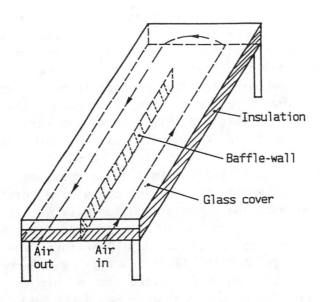

Fig.1.   Schematic   diagram   of   a   basin-type   integral solar regenerator/dehumidifier

RESULTS AND DISCUSSION

The experiments were performed for several days. The regenerator/ dehumidifier was tested to obtain its regeneration efficiency. Study was also made to investigate the effect of cooling water on the dehumidification capability. Daily experimental results of four days are shown in Table 1. These experiements were done with water cooling coils in the basin except for that of Aug.16. The regeneration efficiency is positively proportional to solar insolation. The dehumidification rate with cooling coils is better than that without it. Here, the daily regeneration efficiency (η) and average condensation rate of the regenerator/ dehumidifier (D_c, kg H_2O/hr) are defined as

$$\eta = \frac{D_e \times \lambda}{A \times I_o} \tag{1}$$

$$D_c = \frac{M_1 \times (\frac{C_1}{C_2} - 1)}{\Delta t} \tag{2}$$

Where D_e is the total evaporation rate in kg, λ is the latent heat of water in kJ/kg, A is the area of regenerator in m^2, I_o is the solar irradiance in W/m^2, M_1 is the initial weight of solution in dehumidification process in kg, Δt is the time interval of dehumidification process in hr, C_1 and C_2 are the initial and final concentrations of LiCl during the dehumidification process in weight %. Fig.2 shows the variations of inlet and outlet temperatures (T_{ai} and T_{ao}) of air, solution temperature (T_s), solution concentration (C_s), and solar irradiance (I_o) with time on Aug.16, 1990. Fig.3 shows the same information on Sep.5, 1990 on which the cooling coils were integrated in the basin to enhance the dehumidification rate at night. Comparing the two figures, from solar irradiance data, Aug.16 was a clear day and Sep.5 was a cloudy day. Therefore, the variation of concentration on Aug.16 (35% to 41%) was larger than Sep.5 (33.5% to 36%) which resulted in a higher regeneration efficiency (31.70%) on Aug.16 than on Sep.5 (25.88%). The dehumidification rate can be calculated from equation (2), and the value on Sep.5 (0.140 kg H_2O/hr) is higher than on Aug.16 (0.112 kg H_2O/hr). It was due to the enhancement by the water cooling coils which maintain the solution in a lower temperature and

TABLE 1. Daily results of the experiments

Date	Average Solar Irradiance (W/m²)	Average Concentration (%)		Average Temperature (°C)		Regeneration Efficiency (%)	Average Rate (kg H₂O/hr)
		Reg. Process	Deh. Process	Reg. Process	Deh. Process		
8.16	780	38.491	40.117	60.0	40.2	31.70	0.112
8.28	706	36.682	38.509	56.1	31.9	27.78	0.164
9.5	489	34.964	35.252	47.2	31.7	25.88	0.140
9.11	480	34.914	35.156	49.2	30.3	25.43	0.138

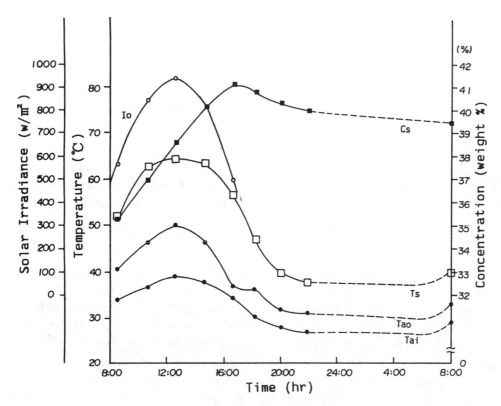

Fig.2. Experimental results on Aug.16, 1990

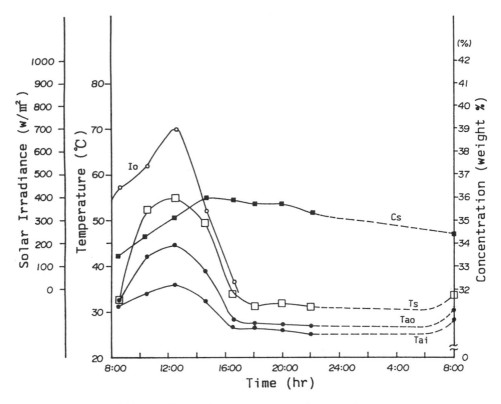

Fig.3, Experimental results on Sep. 5, 1990

result in a larger dehumidification capability. The temperatures on Figs.2 and 3 show the same trend. The variations of temperatures for air and LiCl solution positively depend on the solar irradiance.

CONCLUSIONS

A novel idea of using basin-type solar still as solar regenerator/ dehumidifier is given and proven in this study. It can regenerate liquid desiccant during the day and dehumidify air in the night. A prototype regenerator/dehumidifier was built and tested. Experimental results showed that its regeneration efficiency is better than that of the regenerator/collector of a solar liquid desiccant cooling system. Its dehumidification capability can be enhanced by integrating cooling coils in the basin.

REFERENCE

Yeh, H. M., and L.C. Chen (1985). Basin-type solar distillation with air flow through the Still. Energy, 10, 1237-1241.

EXPERIMENTS AND SIMULATION RESULTS ON THE
THERMAL PERFORMANCE OF A SOLAR TEA DRYER IN INDONESIA

Kamaruddin Abdullah* and G. Brouwer**

*Agricultural Engineering Department, Fateta, IPB
PO Box 122, Bogor, Indonesia
**Van Heugten Consulting Engineers,
Solar Energy Department
PO Box 305, 6500 AH Nijmegen, the Netherlands

ABSTRACT

Tests have been conducted to study the performances of a CITAS-EES flat plate collector tea drying. The tests included the determination of collector efficiency, collector temperature, hot air temperature distribution, the drying efficiency and the overall efficiency of the system. In addition to this it was also found necessary to determine several drying parameters of the commodity, namely the drying constant, the equillibrium moisture content which were not available during the test period. The share of solar energy was 55% of the total energy required to accomplish the drying proses.

KEYWORDS

Thermal performance; solar tea dryer; drying parameters

INTRODUCTION

The aim of solar energy use in tea drying factories is to reduce the energy costs without changing the drying condition. At present the drying process is accomplished by means of artificial drying with fossil fuel as its energy source. By making use of solar energy during bright sunny days or even during cloudy periods the fuel oil could be conserved and where sunshine period in a year is high the amount of oil saving could reach a significant amount. It is expected that the cumulative reduction of fuel cost could surpass the cost incurred by installing the solar thermal conversion device for drying.

The objective of the study was to obtain the performance of the solar drying system, determine the drying parameters such as the Me, and K values and evaluate the amount of fuel savings.

DESCRIPTION OF THE SOLAR DRYER

The solar drying system designed in this study consisted of three main components: a). the heat generating solar collector, b). the drying chamber and c). the auxilliary heating unit.

a). Heat generating solar collector

The collector was a flat plate collector type tilted 10 degrees from horizontal facing North where 16 copper tubes with 8 mm ID were welded to a chrome-oxide spectral selective heat aluminium absorber. The collector was provided with single glazed cover and 4 cm backside insulation. The total surface area of the collector was 26,1 m². The entire collector was supported by aluminium casing. Water was circulated through the collector tubes, and the collector heat was passed through a water-air heat exchanger, to heat the drying air. (see Fig. 1).

b). The drying chamber

The drying chamber was made of 0.3 mm steel plate with inlet air was introduced from the side of the drying chamber and distributed to each rack where drying trays were placed one above another (Fig. 1). Variable direction louvers were placed near the air duct inlet within the drying chamber to enable the drying air to flow in parallel direction across each rack and then left the chamber through the chimney.

c). Auxilliary unit

In case of no sunshine, additional power for heating could be supplied from a water tank with 25 KW electric heater. The hot water from the system was pumped through the water-air heat exchange placed within the main air duct.

ENERGY AND MASS BALANCES

1. Energy balance in the collector

Energy coming from the sun was transmitted to the collector heat absorber after passing through sunlit glass cover. This shortwave length solar ray of between 2 - 4 microns was then absorbed by the selected absorber plate, covered with thin layer coating which could admit short wave radiaton but opaque to long wave radiaton. With this arrangement the plate temperature could be increased to closely 95°C. The absorbed heat then was transferred the working fluid, in this case water, which in turn dissipated all the energy to the drying air in the heat exchanger.

The amount of useful heat collected within the heat collector was computed using the following relation (Duffie, 1974).

$$q_u = A_c \, F_R \, [G_t \, (\tau\alpha) - U_L \, (T_i - T_a)] \qquad \dots\dots\dots\dots\dots \quad (1)$$

here F_R is the heat removal factor and U_L is the overall heat loss coefficient of the collector.

Since the optical parameter ($\tau\alpha$) varies according to the angle of incidence, therefore, to obtain a more general expression for arbitrary direction of incoming solar rays, the following equation was used.

$$(\tau\alpha) = K \, (\tau\alpha)_n \qquad \dots\dots\dots\dots\dots \quad (2)$$

where K is given by

$$K = 1 + bo \left(\frac{1}{\cos \theta} - 1 \right) \qquad \dots\dots\dots\dots\dots \quad (3)$$

From eqs.(1), (2) and (3) the collector efficiency, η_c, defined as the ratio between the useful energy qu and the incoming solar energy Gt was obtained. Hence

$$\eta_c = F_R \ (\tau\alpha)_n \ [\ 1 + bo(\ \frac{1}{\cos \ \theta} - 1] - F_R \ U_L \ (\frac{Ti - Ta}{G_t}) \ \ \dots \ (4)$$

The amount of heat transferred the drying air within the water-air heat exchanger is given by

$$m_a \ Cp_a \ (To_a - Ti_a) = \eta_c \ G_t \ Ac \qquad\qquad \dots\dots\dots (5)$$

or
$$To_a = \frac{\eta_c \ G_t \ Ac}{m_a \ Cp_a} + Ti_a \qquad\qquad \dots\dots\dots (6)$$

2. The Drying Process

The parallel trays containing tea were considered to form hypothetical thin layers in which the drying front started from hot air inlet and moves from right to left. Each layer could be further subdivided into several sections or widths which continuosly exchange heat and mass with the drying air. The resulting change of moisture content within each layer could be estimated from the simple formula (Henderson, 1979),

$$\frac{M - Me}{Mo - Me} = A \ exp \ (\ - k \ t \ ) \qquad\qquad \dots\dots\dots (7)$$

The amount of hot air necessary to accomplish the drying process was calculated from the following relation

$$m_a \ Cp_a \ dTx = W_i \ H_{fg} \ (\ dM/dt \ ) \qquad\qquad \dots\dots\dots (8)$$

where from eq.(7) $dM/dt = - k \ (\ M - Me \ )$ $\qquad\qquad \dots\dots\dots (9)$

EXPERIMENTAL PROCEDURES

Fig. 1 shows the schematic diagram of the entire experimental set-up and measurement sensors were placed at critical points to obtain the required data. The solar collector was supplied by the P.T. CITAS-Engineering and the EES International B.V. the Netherlands.

Data collected during the experimental runs were :
        a. global solar irradiation (Eppley pyranometer)
        b. water and air temperatures at inlet and outlet of the collectors, heat exchangers and the drying chamber (cc-thermocouples and data logger)
        c. water and air flow rates (water flow meter and thermistor type anemometer)
        d. electric power (boiler and blower)
        e. moisture change of tea (oven)

Test samples were gathered from a tea factory in Cibadak, Sukabumi. The tea samples had already passed the withering process and were ready for drying. During each test run a small sample of tea was collected to measure the initial moisture content by means of a vacuum oven model DPA-30 (Yamato) and after that about 33.6 kg of tea leaves were spread over each tray to make a layer of about 3 - 4 cm.

RESULTS AND DISCUSSION

Typical solar radiation available during experiments and the resulting collector temperature are given in Table 1. The table shows the varying outlet water temperature during the day of the experiment. From these data and other pertinent data it was also possible to measure the amount of useful heat Qu and the coefficient of heat losses UL from the collector. During this test, as can be seen in the Table 1, the water temperature exceeded 75 C but still considered inadequate for tea drying. The weather in Bogor seems to be unfit for solar drying due to high cloudiness and rains through out the year. Despite these difficulties, however, the current study had provided the authors with very valuable data for further design improvement and application of the solar system. The following are some further results of the tests.

1. Efficiency of the Solar Collector

Solar collector efficiency was calculated for different incident angles of incoming solar rays. One of the typical result for less then 30 degrees is shown in Fig. 2 which shows the linear relationship between efficiency η_c and the value of (Tw-Ta)/Gt.

The general expression of the collector efficiency was found to be as follow

$$\eta_c = 0.728 \; K - 5.58 \; (\; Ti - Ta \;)/G_t \qquad \ldots\ldots\ldots\ldots\ldots \quad (10)$$

where $\qquad K = [\; 1 - 0.223 \; (\; 1/\cos \theta \;) - 1 \;] \qquad \ldots\ldots\ldots\ldots\ldots \quad (11)$

2. The Drying Performances

Whether condition in Dramaga, Bogor, during the test were not good enough to obtain the proper operating condition for the current solar drying system. In addition to this several improvements of the current system are still required, for example, on proper insulation at the main air duct and the efficiency of the heat exchanger to prevent significant temperature drop. In addition to this some improvement in the auxilliary heating system which could be used as energy heat storage could also be further improved. For example, although the collector temperature could reach as high as 90 C, the maximum drying temperature was at 54 C. This had prevented proper drying process in which a minimum temperature of 70 C is required. Because of this reason it took 6 hours of drying time to reduce the initial moisture content from 193.3% db to 5.7% db.

Since during this study the basic drying data for tea leaves were not available to the authors, therefore, it was decided to determine these parameters using their apparent values. Expanding eq.(7) using the Taylor's series following the algorithm developed earlier (Kamaruddin, 1980, Harris, 1980) each value of A, K and Me were determined as follows

$$A = 0.99456 \quad \text{with} \quad SD \; +/- \; 0.051 \qquad \ldots\ldots\ldots\ldots\ldots \quad (12)$$

$$k = \exp [0.27998 - 2754.86 \; /Td] \qquad \ldots\ldots\ldots\ldots\ldots \quad (13)$$

$$Me = 4.44292 - 0.33968 \; (Td - Tw) + 0.00846 \; (Td - Tw)^2 \qquad \ldots\ldots\ldots \quad (14)$$

Using the above value of the drying parameters a simulation study was conducted. Table 2 shows one of the typical results which the initial moisture content of the leaves was at 155.7% db and maximum air temperature available was 50 C. Comparison between the experimental data and those predicted from the theory indicated a fair agreement, although the theory tended to underestimate the rate of drying at the tray located at the inlet portion of the drying chamber.

1810

c. Oil Saving

The amount of energy required for the drying operation was not the same during each operating condition. For the case of experimental run on December 16, 1989, the amount of energy received by the collector was 233530.4 kJ or equivalent to 5.7 liter of fuel oil, 185328 kJ for electric heating and 11185.5 kJ to operate the blower. Under this condition the share of solar energy for heating then was about 55.7%.

REFERENCES

Duffie, J.A. and W.A. Beckman. 1974. Solar Energy Thermal Process, John Wiley & Sons, New York.

Gani Bratawidjaja, 1990. Graduation Thesis. Dept. of Agricultural Engineering, Fateta, IPB

Henderson, S.M. and Perry R.L. 1979. Agricultural Process Engineering, AVI Publication Co.

Kamaruddin Abdullah, 1980. Lecture notes, the Graduate School, IPB.

Sandra Harris, 1980. Graduation thesis. Dept. Agricultural Engineering, IPB

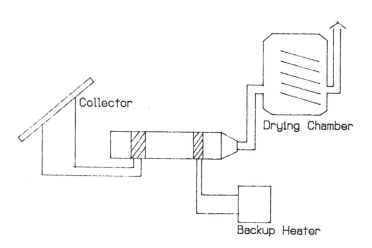

Fig. 1. The solar drying set up

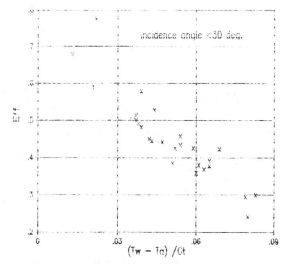

Fig. 2. Collector efficiency

Table 1. Coefficient of Heat Loss (U_L)

Data: September 18, 1989

Local Time	twin (C)	two (C)	ta (C)	Qu (kJ/sec)	UL (W/m2 K)
08:00-08:30	39.4	41.8	29.4	7643.4	3.23
08:30-09:00	49.7	53.3	30.5	10545.5	3.39
09:00-09:30	60.1	64.6	31.3	10173.7	3.51
09:30-10:00	63.1	68.7	32.8	14935.1	3.52
10:00-10:30	65.3	71.0	34.4	15481.1	3.54
10:30-11:00	69.8	75.8	35.0	16078.1	3.58
11:00-11:30	74.6	77.9	36.1	9027.6	3.59
11:30-12:00	68.8	75.8	36.4	18860.9	3.57
12:00-12:30	71.4	77.5	40.5	16648.9	3.58
12:30-13:00	70.8	75.9	42.5	13943.3	3.57
13:00-13:30	68.5	73.0	42.4	12063.4	3.54

Table 2. Drying performance

Data: December 16, 1989

Local Time	Dryer inlet			Dryer outlet		
	td(C)	RH(%)	M(%db)	td(C)	RH(%)	M(%db)
10:45	49.2	21.0	155.7	49.2	30.0	155.7
10:55	48.5	24.7	133.5	46.5	28.6	135.8
11:04	49.4	23.7	114.6	47.7	27.0	117.9
11:15	48.9	24.0	98.7	47.4	26.7	102.5
11:25	49.7	21.5	84.9	48.4	23.8	88.6
11:45	50.1	18.2	63.3	49.2	19.6	66.3
12:04	50.0	17.7	47.4	49.3	18.7	49.8
13:04	45.3	11.4	22.2	45.1	11.7	23.4
14:04	50.5	25.8	10.3	50.4	25.4	12.2

MONITORING OF TEN WATER HEATING SOLAR SYSTEMS

M.J. Carvalho*, M. Collares-Pereira**, J. Cruz Costa*, J. Oliveira**

*Dpto de Energias Renováveis – LNETI; Est. do Paço do Lumiar; 1600 LISBOA – PORTUGAL
**Centro para a Conservação de Energia; Est. de Alfragide, Praceta 1; 2700 AMADORA PORTUGAL

ABSTRACT

The results obtained through the monitoring of different types of water heating systems are presented. To check system performance it is shown that a simplified, utilizability based calculation method, based on the average radiation and temperature values of the measuring period, can be used, even for short monitoring periods (lower than one month), as long as the thermal characteristics of the system, specially the collector efficiency curve and heat exchanger effectiveness (when present) are known.

KEYWORDS

Solar Water Heating System, Monitoring, Modeling, System Performance

INTRODUCTION

This work presents the results obtained through the monitoring of different types of water heating systems. The programme includes the monitoring of ten systems. At the time of this writing only four systems have been monitored.

The work presented here is organized as follows: a brief description of the four systems and of the monitoring equipment is presented; the calculation method adopted for comparison with the measured values is also briefly described; the measured results obtained are presented and compared with calculated results and, finally, some conclusions are presented. At the time of the Conference in Denver an analysis of the main problems encountered in all of the ten systems, the lessons to be learned by the solar collector industry, installing companies, etc, will be summarized and presented.

SYSTEM CHARACTERIZATION

Three of the systems monitored are conventional solar water heating systems and a fourth system is a heat pump using an external flat plate evaporator. In Table 1a) their principal characteristics are summarized. In Table 1b), the dimensions of each system, and the thermal characteristics of their main components are listed.

Some comments on Table 1b) are in order. The storage thermal losses of systems A, B and D were estimated (for the measured conditions) using storage tank dimensions and the thermal properties of the insulation, considering that thermal losses from tank wall to ambient were caused by natural convection (storage tank is indoors). In the case of system C, thermal losses of the storage tank were directly measured. The tank was left with hot water for several hours, and the temperature of the water in the tank was measured using two temperature sensors placed through the lateral wall, near the top and the bottom of the tank. The ambient temperature was also measured during this period. The global heat loss coeficient of the tank was taken to be given by:

$$(UA)_{tank} = \frac{MC_p}{\Delta t} \ln \frac{T_i - \overline{T}_{amb}}{T_f - \overline{T}_{amb}} \qquad (1)$$

where C_p is the water specific heat, M is the mass of water in the tank, Δt is the time the tank was left with hot water to cool, T_i and T_f are the initial and final temperatures of the tank, $\overline{T}_{amb}$ is the average ambient temperature during time Δt. The error estimated for this measurement is 24% of the measured value, but the influence of this error in the estimate of the total delivered energy is only of $\pm 2\%$.

TABLE 1a) - SYSTEM TYPE, BACKUP AND LOAD

SYSTEM	LOCATION	SYSTEM TYPE	BACKUP	LOAD
A	Camping (Lisbon, $\lambda=38.7°$)	Forced Circulation External Heat Exchanger	Electrical (in series with Storage)	Showers day/night
B	Thermal Baths (Longroiva, $\lambda=41°$)	Forced Circulation Heat Exchanger, incorporated in Storage (Jackets type)	Wood Furnace (in paralel with Storage)	Showers day/night
C	Hospital (Lisbon, $\lambda=38.7°$)	Forced Circulation Heat Exchanger, incorporated in Storage (Jackets type)	Electrical (in series with Storage)	Showers day/night
D	Soccer field (Lagos, $\lambda=37.1°$)	Heat Pump	Electrical	Showers day

TABLE 1b) - SYSTEM DIMENSIONS AND THERMAL CHARACTERISTICS

SYSTEM	Collector Area (m^2)	Collector Efficiency $F'\eta_0$[1] $F'U_L$ $(W/°C.m^2)$	Azimuth and Tilt α β	Storage Volume (m^3)	Storage Thermal Losses $(W/°C)$	Heat Exchanger Area (m^2)	Effectiveness
A	39.6	0.73 ± 0.01 6.5 ± 0.3	$0°$ $40°$	3	6.8	0.45	0.63 ± 0.06
B	45.4	0.81 ± 0.02 8.5 ± 0.3	$-14°$ a $-22°$ $25°$ a $35°$	4	13.4	6.5	0.48 ± 0.03
C	35.6	0.81 ± 0.02 8.5 ± 0.3	$0°$ $28°$	2	16 ± 4	7.3	0.35 ± 0.06
D	20.0	----- -----	$0°$ $82°$	3	24.1	-----	-----

[1]Measurements done at LNETI-DER as part of its regular collector testing activity.

The effectiveness of the heat exchangers was determined experimentally. For system A (external heat exchanger - counterflow type) the effectiveness was evaluated using the equation (Kreith, F. and Kreider, J.F., 1978):

$$\epsilon = \frac{(\dot{m}C_p)_h \ (T_{h1} - T_{h2})}{(\dot{m}C_p)_{min} \ (T_{h1} - T_{c1})} \tag{2}$$

where $(\dot{m}C_p)_h$ is the flow capacitance of the hot fluid; $(\dot{m}C_p)_{min}$ is the minimum flow capacitance of the hot and cold fluid; T_{h1}, T_{h2} are, respectively, the temperature of the hot fluid entering and exiting the heat exchanger; T_{c1} is the temperature of the cold fluid entering the heat exchanger. Instantaneous measurements were made of T_{h1}, T_{h2} and T_{c1}, as well as the measurement of $(\dot{m}C_p)$ for the hot and cold fluid. For systems B and C (Jacket type heat exchanger) the equation used for the effectiveness was:

$$\epsilon = \frac{(T_{h1} - T_{h2})}{(T_{h1} - \overline{T}_{tank})} \tag{3}$$

where T_{h1} and T_{h2} are, respectively, the temperature of the hot fluid entering and exiting the heat exchanger and $\overline{T}_{tank}$, the average temperature of the water in the tank. Instantaneous measurements of T_{h1}, T_{h2} and $\overline{T}_{tank}$ gave the value of ϵ in Table 1b. However, the uncertainty associated with these measurements is large due to the variation of T_{h1}, T_{h2}, T_{c1} and $\overline{T}_{tank}$ during the day and lack of the appropriate temperature sensors for greater accuracy.

MONITORING OF SYSTEM PERFORMANCE

Long term system performance was evaluated through the continuous measurement of the following quantities: daily radiation incident on the collectors, daily energy collected and daily energy delivered to load[2]. The daily solar radiation was measured with a pyranometer and a solar integrator used to register the daily values of radiation. The daily energy values were measured with an energy totalizer, formed by a flowmeter and two RTD temperature sensors. These energy totalizers have only a digital indication of the total volume of water circulated (in m^3) and of the total energy delivered (in KWh). The daily totals were obtained by daily reading of the instruments. The value of ambient temperature was periodically registered through the reading of a mercurythermometer, when available. When this was not the case, the ambient temperature values of the day were obtained from the nearest meteorological station, by request to the National Institute of Meteorology. The measurement period was roughly 30 days in each case, from July to December 1990.

The fourth system monitored is not a solar system. It is a heat pump with an external flat plate evaporator. The monitoring of this system consisted of the measurement of solar radiation, ambient temperature, energy delivery to load and consumption of electrical energy by the heat pump compressor.

ESTIMATING THE SYSTEM PERFORMANCE

Given the instrumentation used, sophisticated calculation methods, like detailed simulation, are not appropriate. The calculation method chosen to compare with measured energy delivered to the load is a simplified method usually used as a design method (Gordon,J.M. and Zarmi,Y., 1985). This method is valid for systems working at low energy thresholds, as is the case of domestic hot water systems. This simplified method permits the consideration of three different load profiles:

i) daytime constant load (daytime being approximately equal to the solar captation period)
ii) nighttime constant load
iii) 24 hours constant load

The method considers radiation values to be constant during the day and the system to be working at a

---

[2]The energy delivered to load will only be considered as the energy produced by the solar system. It does not include the energy produced by the backup system.

constant threshold, $\bar{I}_c$. The collectible energy is given by:

$$\overline{Q}_{col} = A_c \, (F\eta_0)_t \, \overline{K} \, \overline{H}_{col} \, \phi(\bar{I}_c) \tag{4}$$

where A_c is the collector aperture area, $(F\eta_0)_t$ is a modified $F_R\eta_0$ taking into account pipe losses and heat exchanger effectiveness (as will be detailed below), $\overline{K}$ is the average incidence angle modifier (an energy weighted average of the variation of the glass cover transmission coefficient with incidence angle (Collares Pereira,M. and Rabl,A., 1979)), $\overline{H}_{col}$ is the average daily radiation incident on the collector and $\phi(\bar{I}_c)$ is the utilizability (fraction of solar radiation incident on the collector aperture above threshold $\bar{I}_c$).

For comparison with the measured collectible energy, $\overline{Q}_{col}$ is calculated using the measured value of $\overline{H}_{col}$. $\bar{I}_c$ is determined considering the measured value of daily average ambient temperature, $\overline{T}_{amb}$, and the value of mains temperature, T_c, as well as the thermal characteristic of the system, according to Gordon,J.M. and Zarmi,Y. (1985). $\phi(\bar{I}_c)$ is calculated using the simplified functional form for utilizabilty given in reference (Carvalho,M.J. and Collares Pereira,M., 1990).

The flexibility of this calculation consists in the possibility of introducing different correction factors for the usual coefficients $F_R\eta_0$ and F_RU_L which take into account the pipe losses between collector and storage tank (Beckman, W.A., 1978) and the penalty imposed on the solar system by the existence of a heat exchanger with a determined effectiveness (de Winter,F., 1975). The coefficients $(F\eta_0)_t$ and $(FU_L)_t$ are:

$$(F\eta_0)_t = (F_R\eta_0)' \times F_x \quad \text{and} \quad (FU_L)_t = (F_RU_L)' \times F_x \tag{5}$$

where:

$$(F_R\eta_0)' = (F_R\eta_0)\left(1 + \frac{(UA)_{out}}{\dot{m}_cC_p}\right)^{-1} \tag{6}$$

$$(F_RU_L)' = (F_RU_L)\left(1 - \frac{(UA)_{in}}{\dot{m}_cC_p} + \frac{(UA)_{in} + (UA)_{out}}{A_cF_RU_L}\right)\left(1 + \frac{(UA)_{out}}{\dot{m}_cC_p}\right)^{-1} \tag{7}$$

and:

$$F_x = \left(1 - \frac{A_cF_RU_L}{\dot{m}_cC_p} + \frac{A_cF_RU_L}{\epsilon\,(\dot{m}C_p)_{min}}\right)^{-1} \tag{8}$$

$(UA)_{in}$ and $(UA)_{out}$ being, respectively, the global pipe losses coeficient for the inlet pipes to collector and outlet pipes from collector to storage tank, $\dot{m}_cC_p$ the fluid flow capacitance between collector & storage tank, $(\dot{m}C_p)_{min}$ de minimum fluid flow capacitance in the heat exchanger and ϵ the heat exchanger effectiveness.

The energy delivered to load can be determined by:

$$\overline{Q}_u = \overline{Q}_{col} - \overline{Q}_p \tag{9}$$

where $\overline{Q}_p$ is the energy lost by the storage tank and is determined considering an average daytime temperature, $\overline{T}_{day}$ and an average nightime temperature, $\overline{T}_{night}$, in the tank:

$$\overline{Q}_p = (UA)_{tank}\,[(\overline{T}_{day} - T_{amb})\Delta t_{day} - (\overline{T}_{night} - T_{amb})\Delta t_{night} \tag{10}$$

Δt_{day} and Δt_{night} are, respectively the daytime and nightime duration.

COMPARISON BETWEEN MEASURED AND ESTIMATED VALUES

The average daily measured values of radiation incident on the collectors, $\overline{H}_{col}$, ambient temperature, $\overline{T}_{amb}$, mains water temperature, $\overline{T}_c$, daily demand of hot water, $\overline{V}$, collected energy, $\overline{Q}_{col}$, and delivered energy, $\overline{Q}_u$, as well as the estimated values of collected and delivered energy, $\overline{Q}'_{col}$ and $\overline{Q}'_u$, respectively, are listed in Table 2, for systems A,B and C. For each system the monitoring period is indicated. The difference between measured and estimated data is also indicated and is defined as:

$$\delta_{col} = \frac{\overline{Q}_{col} - \overline{Q}'_{col}}{\overline{Q}_{col}} \times 100 \; (\%) \qquad \delta_u = \frac{\overline{Q}_u - \overline{Q}'_u}{\overline{Q}_u} \times 100 \; (\%) \tag{11}$$

TABLE 2

System	Monitoring Period	$\overline{H}_{col}$ MJ/m^2	$\overline{T}_{amb}$ °C	$\overline{T}_c$ °C	$\overline{V}$ m^3	$\overline{Q}_{col}$ MJ	$\overline{Q}_u$ MJ	$\overline{Q'}_{col}$ MJ	$\overline{Q'}_u$ MJ	δ_{col} %	δ_u %
A	14/7 - 13/8	22.4	25.3	21.0	11.9	390	393	406	403	-4.1	-2.5
B	26/8 - 2/9	22.9	24.3	35.0	3.5	326	280	320	288	1.6	-2.8
	3/9 - 17/9	21.0	23.6	35.0	4.0	299	362[3]	298	268	0.3	----
C	8/11 - 2/12	14.3	13.8	15.0	4.4	201	----	195	180	2.9	----
	13 - 26/12	10.6	10.0	15.0	4.0	141	117	132	117	6.2	-0.1

System A:

Hot water demand was, during the measuring period, greater than expected, since the storage volume, as indicated in Table 1b), is 3 m^3 [4]. For calculation of $\overline{Q'}_{col}$ and $\overline{Q'}_u$, the load profile adopted was constant load 24 hours per day. The choice of such a load profile takes into account that a percentage of the load will occur before sunrise and mostly after sunset, and that, during daytime, the load is greater than the storage volume but lower than the total demand of hot water.

The measured value of $\overline{Q}_u$ is about 1% larger than the measured value of $\overline{Q}_{col}$. This discrepancy is small, since it is of the same order of magnitude of the experimental errors associated with $\overline{Q}_{col}$ and $\overline{Q}_u$ and also of the same order of magnitude of the daily average tank losses estimated $(\overline{Q'}_{col} - \overline{Q'}_u)$. The measured and calculated values are in good agreement. The calculated values slightly overestimate the measured values.

System B:

The monitoring period for system B was, for comparison purposes, divided into two periods because, on the first period, no auxiliary energy was used, while in the second period the furnace had to be used. As the furnace is in parallel with the storage tank, it is difficult to evaluate what fraction of the $\overline{Q}_u$ measured corresponds to the solar system; it is also difficult to measure the energy delivered by the furnace to the load, given the specific nature of the furnace, its location, maintenance, etc. Although the hot water demand was almost equal to the storage volume (Table 1b), the load profile considered was, constant load 24 hours per day, due to the large consumption of hot water early in the morning (before sunrise).

The measured and calculated values are in good agreement even considering the short periods of time for which the results were obtained (9 and 15 days). The difference between calculated and measured values of delivered energy can only be analysed for the first period. The difference observed is greater for the energy delivered than for the collectible energy. A longer data acquisition period (greater than one month) will tend to eliminate these discrepancies. A lesson learned was that more accuracy can be obtained with a more appropriate choice of data acquisition routine, based on an anticipated knowledge of the daily load profile.

---

[3]Auxiliary on.

[4]Solar system design usually considers the storage volume equal to the demand of hot water.

System C:

For system C it was necessary to consider two functioning periods. There were problems with the monitoring equipment which resulted in no values for $\overline{Q}_u$ during the first measuring period. The daily demand of hot water was almost twice that of the storage volume. A detailed monitoring of the demand profile indicates that the most adequate load profile for the calculation is, in this case, constant load 24 hours per day.

A good agreement is observed between measured and estimated values. The collectible energy is slightly under estimated, while the measured and estimated values for the delivered energy are identical. As in the case of system B, the reason for this lies mostly in the mismatch between the actual time at which of the "hand" measurements were made, and the proper time at which these measurements should have been done.

System D:

This system (Axergie, France) is of a very different nature. The heat pump may have higher COP when the sun shines on the evaporator, but it is not really a solar system in the conventional sense. It has an evaporator consisting of unglazed metal plates, with browning color finish (uncoated copper), laid in several rows. Measurements were made during the period of 1/11 to 23/11/90.

The data are still being analysed and will be presented at the time of the Conference. However, some conclusions can already be drawn: 1) daily load variability seems to have the strongest influence on the variation of daily COP; 2) when daily load was equal to the storage volume, COP varied between 2 and 3 (with an average value of 2.6); 3) for those few days during which daily load was between 1.5 and 2 times the storage volume, COP went to values around 4; 4) there seems to be also a COP dependence on average daily temperature and daily radiation, however, this is certainly a second effect in comparison with the load effect refered above; 5) these results are certainly below the ones quoted by the manufacturer.

CONCLUSIONS

The results obtained for the three solar systems show that the monitoring of these systems can be done with simple instrumentation. System performance is well described by a simplified method (Gordon,J.M. and Zarmi,Y., 1985), even when it is used for a relatively short period of measurement days. The results depend strongly on the knowledge of the thermal characteristics of the system, specially collectors efficiency curve and heat exchanger effectiveness. The uncertainty in the determination of the average effectiveness must be clarified with a detailed study of daily variation of heat exchanger efectiveness, when used in solar systems.

At the time of the Conference in Denver, the authors will present a summary of practical considerations involving system design and functioning problems, which result from the complete set of observations made.

REFERENCES

Beckman, W.A. (1978). Solar Energy, 21, page 531
Carvalho, M.J. and Collares Pereira, M. (1990). "Utilizability Correlations". EUFRAT/FD/04
Collares Pereira, M. and Rabl, A. (1979). Solar Energy, 23, page 223
de Winter, F. (1975). Solar Energy, 17, page 335
Gordon, J.M. and Zarmi, Y. (1985). Solar Energy, 35, page 35
Kreith, F. and Kreider, J.F. (1978). Principles of Solar Engineering. Hemisphere Publishing Corporation.

COMPUTER CONTROL AND MONITORING OF A SOLAR SPACE HEATING SYSTEM

S. U. Chaudhry and L. F. Jesch

**Solar Energy Laboratory
Uiversity of Birmigham
Birmingham B15 2TT, UK**

ABSTRACT

The design of the control system has a strong influence on the success of any solar space heating system. Minimizing the use of auxiliary energy and providing the required comfort are the main objectives of most control strategies. The analysis of an operating solar space heating system can only determine the thermal performance of the building if the monitoring conditions have been set to enable such analysis. Both the control and the monitoring functions can be well served by the same computer system.

An improved control/monitoring system for a solar house operating in Birmingham, UK is described in the paper. A BBC microcomputer equipped with A/D converters and optoisolation records the data received from the sensors and operates the active solar system. The monitored data is transferred to an IBM AT computer for further sorting and analysis.

The paper emphasizes factors associated with the computer control and monitoring of the solar space heating system and describes some of the difficulties experienced in using the control/monitoring system and the various solutions tried which were aimed at overcoming the operational difficulties. The performance results of the system based on monitored data are also presented.

KEYWORDS

Active solar, Space heating, Computer control, Data acquistion, Monitoring

INTRODUCTION

The successful operation of an active solar energy system is very much dependent upon the design of the control system, which must be accurate, reliable, safe, and capable of utilizing solar energy most effectively and efficiently. In some cases, the response and benefit of a specific operational mode or control strategy is evident, like using auxiliary energy at night when it is cheaper. In others, a comprehensive analysis of the system components over longer periods will be required. However, the accuracy of the analysis will depend on the extent to which the system components that carry out the thermal processes are instrumented and monitored, so that the performance of the system can be analyzed most precisely.

The Bournville solar demonstration house in Birmingham is part of a Solar Village of some three hundred housing units. This demonstration house uses both active and passive solar technologies. It provides the opportunity to study many aspects of an active solar space heating system in several experimental modes and hence allows for the analysis of the system behavior. The original control and monitoring system (1) served during the first phase of the project from 1985 to 1988.

In the following year alterations were made in control and monitoring system. These alterations made use of the experience gained during the first phase and resulted in a new control system.

The new system with data acquisition facility became operational in January, 1990. During the first few weeks the emphasis was placed on the adjustment and fault correction in the solar collection subsystems (pumps, valves and auxiliary thermostat failures). This was followed by almost uninterrupted system operation and data acquisition starting from 27 January until 21 August. However, some data are missing due to either hardware or software failure. Since the data gaps are random, they are not believed to influence the overall performance evaluation of the system.

SYSTEM DESCRIPTION AND CONTROL STRATEGY

The solar space heating system consists of flat plate solar collectors, underfloor and ceiling embedded heating coils, two phase change energy storage tanks, (2) electrical immersion heaters for auxiliary supply, a heat rejection tank for dumping the excess heat and microcomputer control system. The flat plate solar collectors are made of stainless steel, selectively coated and having a total area of 15.68 m^2. They are mounted on the roof facing South with a tilt of 35 degrees. The storage tanks situated one on each floor are packed with vertical tubes containing phase change material and the rest of these tanks are filled with the heating fluid circulating in the system.

The control of the system is based on the concept of the critical level of solar radiation rather than temperature. The idea of using radiation control is to utilize the source of heating energy, which is radiation, instead of using temperature which is a consequence of it. The critical level of radiation is defined as one below which no useful energy can be collected due to the heat losses from the collector and parameters related to collector efficiency.

The heating fluid is circulated through the collector circuit, whenever the total solar radiation is above the critical level of radiation. The energy collected by the solar collectors is preferably supplied to the heating zones of the house. The reason for this is that the massive concrete floors of the house themselves serve as heat storage and distribute the heat evenly to the space.

If the house does not demand energy and the solar supply is still available, the energy is stored in the storage tanks. The tank at ground floor is preferably charged with solar energy. The first floor tank is charged when the ground floor tank has reached the upper limit of temperature.

If the house heating demand is zero and both tanks are also fully charged, but the collector pump is still operating to protect the collectors from any damage caused by overheating, the heat is diverted to the dump tank, from where domestic hot water supply makes use of it.

There is one auxiliary unit in the system having 3kW and 6kW electrical immersion heaters. At night, during the time when electricity is cheaper, the auxiliary heaters are employed. The cheap rate electricity is considered to be available between 2 am and 7 am. The auxiliary heater power is switched in steps of 3, 6 and 9 kW depending upon the difference between the ambient air and room temperature.

If neither the collector nor the auxiliary heaters are supplying any useful energy and the house heating demand is non zero, then storage tanks satisfy the load. The storage tank at the ground floor is discharged preferably. If the temperature of this tank is not high enough to satisfy the load, then the first floor tank is employed.

The heating space is divided into three separate zones, upper floor zone, lower floor zone and lower ceiling zone. The three zones are operated individually or together depending upon the heating requirement of the house. The number of active heating zones is determined in a similar manner to the auxiliary heater power.

COMPUTER CONTROL

As mentioned earlier, the design of the control system is based on the concept of critical level of radiation rather than temperature. Since the temperature differential controller or even a microprocessor controller is unable to perform complicated mathematical calculations, therefore a microcomputer is used to implement the radiation control. The space heating system is controlled by a program running on a BBC micro based EUROBEEB computer.

The control and monitoring system information flow is shown in Fig.1. The information from all the sensors reades the analogue-to-digital converter of the EUROBEEB computer in millivolt signals.

The signals are processed by the converter and their equivalent digital outputs are produced and read by the computer. The automatic control program evaluates the status of the system elements on the basis of the measurements taken from the sensors and determines the operating mode of the system. The executive devices are operated accordingly so as to extract the heat from solar collectors or auxiliary, deliver it to the loads upon demand or store it for later use, if the demand does not exist at that time.

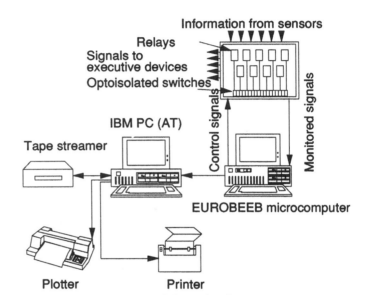

Fig. 1. *Control and monitoring information flow*

DATA ACQUISITION AND PROCESSING

The data acquisition system is developed to record each operating mode with a complete set of measurements, so that the performance of the system during each mode can be analyzed. Continuous measurements including conditions within the building, climatic conditions, and all other parameters affecting the performance of the system are scanned at every eight seconds time step. These measurements are averaged and recorded on floppy diskettes after each quarter of an hour.

The EUROBEEB computer used to control the space heating system is a BBC micro based computer with BBC BASIC interpreter. It has been developed for the purpose of process control. Due to the limitations imposed by the EUROBEEB computer and the basic interpreter, it has been a very problematic job to control the system and satisfy all the requirements of data acquisition and processing with the same computer.

The data collected by the EUROBEEB computer on floppy diskettes can not be read by IBM computers, due to the incompatibility of the two systems. The data files are therefore transferred from the EUROBEEB to an IBM computer by connecting the two systems via serial ports and using the communication software 'Kermit' installed in the IBM machine. An Everex tape streamer is connected to the IBM, which is used for permanent storage of the data. The output of the data files is produced either in tabular or graphic form by the help of a printer and a plotter, which are also connected with the IBM computer.

ANALYSIS AND DISCUSSION

The performance of a system can be analysed from several different perspectives, chief among which are the short term and long term performance analyses. (3) Short term analysis of instantaneous data provide useful information on the system's dynamic behaviour, such as the collector and storage dynamics, status of executive devices and operational modes. Whereas long term analysis provide the information on the overall performance of the system.

As an example, Fig.2a. shows instantaneous values of the total solar radiation, Fig.2b. shows the room temperature, storage tank temperature, and ambient air temperature, plotted as functions of time for a period of one week in April. While the ambient air temperature had 10 to 12°C swings, the room temperatures had swings of about 3 to 5°C during the same period due to the solar and auxiliary power supplied to the house. The fluctuation of room temperature is relatively small, which illustrates the thermal inertia of the house. The charging and discharging cycle of the storage tank is fairly regular.

Additionally, it has been noticed that the room air temperature was consistently higher at the lower floor than at the upper floor. This was a consequence of the sunspace attached to the lower floor, which provided relatively higher passive solar gains during the day and lower heat losses during the night.

The monthly distribution of heating load based on degree day calculations is shown in Fig.3. Note, that only positive values of temperature difference (T_{room} - T_{amb}) have been considered. As the number of days in each month, for which the data are available are different, therefore, care must be taken while comparing the results. The heating load for the entire

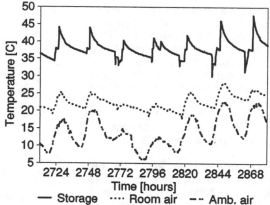

Fig. 2a. *Various temperatures as functions of time*

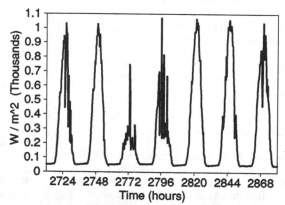

Fig. 2b. *Total solar radiation as a function of time*

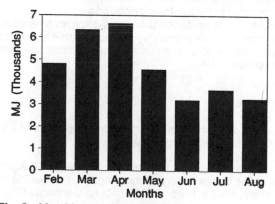

Fig. 3. *Monthly space heating load*

1822

season was measured to be 32.5 GJ.

Fig.4 shows the monthly totals of incident solar radiation for the entire period (I_t), incident solar radiation when the collectors were operating (I_{col}) and useful heat delivered to the system (Q_{col}). Fig.5 shows the relative contribution of solar and auxiliary supplies to the system. About 32 percent of the total load was supplied by the solar energy by active means, which is 10,292 MJ. Electric heaters accounted for 4,928 MJ, total.

As the storage tanks are inside the house, heat losses from the storage tanks contribute to the supply of heat for the building. The calculation of heat losses from the storage tanks prove that a considerable amount of energy, namely 1,875 MJ was delivered to the house as a result of storage tank heat losses. Additionally, it was found that during the month of August, when the house heating load was small, the heating demand was entirely satisfied by the storage losses and the passive solar gains.

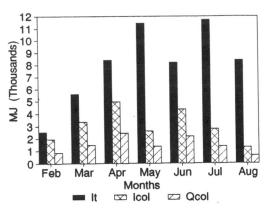

Fig. 4. *Monthly totals of solar radiation incident upon collectors for the entrie periode (It), solar radiation incident upon collectors when collectors were operating (Icol) and useful heat delivered to the system (Qcol)*

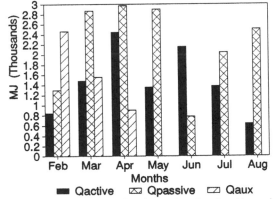

Fig. 5. *Relative contribution to heating load by solar and auxiliary*

CONCLUSIONS

The use of a microcomputer for the control and monitoring of a solar space heating system has been demonstrated. It is an intelligent, optimal, flexible and programmable controller, which far exceeds thermostatic and microprocessor controllers in performance and provides total comfort control. The system works in all operational modes which are aimed to ensure the minimum use of auxiliary and optimal use of solar energy and enables the user to evaluate the performance. Both the control and the monitoring functions can be well served with the same computer system.

REFERENCES

Jankovic, L.: *Solar Energy Monitoring, Control and Analysis in Buildings*, Ph.D. Thesis, The University of Birmingham, 1988.

Laing, D., Jesch, L.F., Jankovic, L., Fellague, A.A.,: *Simulation of Phase Change Energy Storage for Solar Space heating*, Procedings of the Biennial Congress of the International Solar Energy Society, Vol.2 p.1192-1196. Hamburg, Germany, 1987

Duffie, J.A. and Beckman, W.A.: *Solar Engineering of Thermal Processes*, John Wiley and Sons, New York, 1980.

STUDY IN THE SELECTIVE TRANSPARENT MATERIALS OF SOLAR SPECTRUM

Jiao Xiao-huan, Zhang Shu-xia, Hu Wen-xu

Shaanxi Physics Institute,

Shaanxi Teachers University,Xi'an, China

ABSTRACT

This paper is to introduce the experiment of the films of controlling solar radiation in improving the efficiency of energy saving buildings. The films are made by vacuum coating to coat SiOx/Ag or SiOx/Al, etc. on glass or plastic basis. Our experiments have proved that the films have a better selectivity of transmitting and reflecting solar radiation. These films are suitable to be used in energy saving buildings.

KEYWORDS

Film of controlling solar radiation, Selective transparency, Selective reflection

TEXT

The technique of using vacuum coating to coat SiOx/Ag or SiOx/Al etc. on glass or plastic basis to form a film controlling solar radiation and using these films to improve the performance of energy saving buildings is a developing technology in recent years.

There are a lot of technology methods to make these films of

1824

controlling solar radiation, such as vacuum deposition, vacuum
impregnation, vacuum splash and chemistry deposition etc.. Using
one of the above methods can coat the SiOx/Ag or SiOx/Al etc. on
glass or plastic basis.

By comparison among the above methods, we've found the vacuum
coating method is a better one. It has a simple technology, low
cost and suitably light and thermal properties. It is suitable
for spreading in developing countries.

According to the climatic feature in China, in the south of
China it is mainly to save energy consumption for cooling, in
the north of China is mainly to save energy consumption for
heating and in the Midland to save energy consumption for
heating in winter and for cooling in summer.

For the above three regions, we have tested three films of
controlling solar radiation:
A. ZnS/Ag Film of Controlling Solar Radiation: This film has the
   feature of middle reflection and middle transparence for
   short wave radiation (0.35—0.7 micrometer), and high
   reflection for long wave radiation to the outdoor heat
   radiation (see Fig.1). Therefore this film is suitable for
   the south of China.

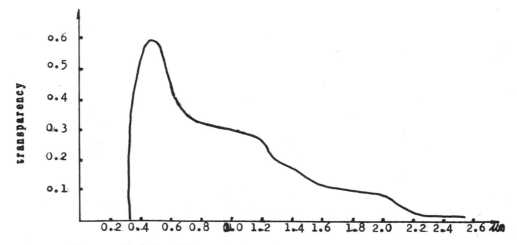

Fig.1. ZnS/Ag Film of Controlling Solar Radiation

B. In O :Sn/Ag Film of controlling Solar Radiation: This film has the feature of high transparence for outdoor solar radiation (0.35~2.5micrometer) and high reflection for indoor heat radiation (see Fig. 2). Therefore this film is suitable for the north of China.

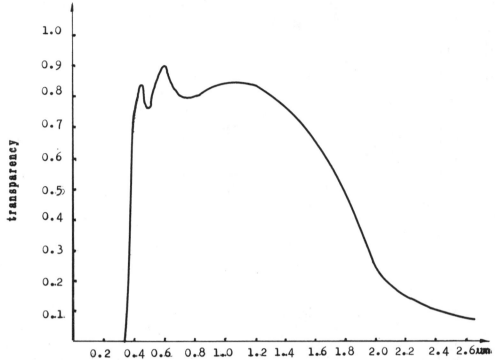

Fig. 2 In_2O_3:Sn/Ag Film of Controlling Solar Radiation

C. SiOx/Ag and SiOx/Al Films of Controlling Solar Radiation:As we can see from Fig. 1, when the wave of film is 0.35~0.7 micrometer, the transparence is quite low, which is not suitable to the middle part of China. As we experimented, and analysed , the effect would be suitable if the transparence is 70% for the wave of 0.35~0.7 micrometer. We therefore used the medium film of SiOx to improve the transparence(see Fig. 3, 4).It has proved that the medium film of SiOx cannot only increase the transparence but also strengthen the chemical and physical stabilities. This stability can propect metal film from oxidization and raise its durability.

1826

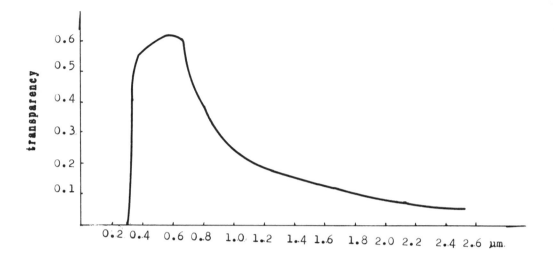

Fig.3 SiOx/Ag Film of Controlling Sloar Radiation

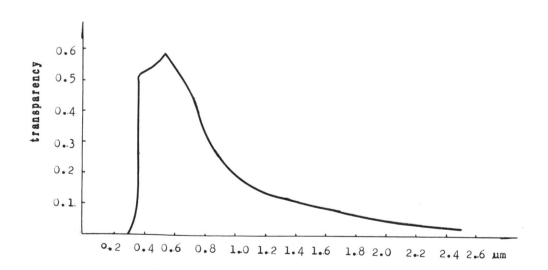

Fig.4 SiOx/Al Film of controlling Solar Radiation

CONCLUSION

A. The feature of the film of controlling solar radiation should be different in different climatic regions. In fact it is a kind of film with the selective spectrum for controlling transparence and reflection in accordance with different demands;

B. The technique to make the film of controlling solar radiation with vacuum coating is simple, low cost and is easy to use. It is especially suitable for developing countries;

C. The film of controlling solar radiation with SiOx/Ag or SiOx /Al on glass basis is suitable for new energy saving buildings. (About 500 million square meter building spaces of these buildings will be built each year in China.). And the film on plastic basis can be stuck on the window glazing of the existing buildings for energy saving (There are about more than 10 billion square meters of building spaces of existing buildings which should be reconstructed fo energy saving.)

REFERNCES

W.D.Munz and S.R.Reineck, in R.I.Seddon (Ed).
   Proc.SPIE, Vol 325; Optical Thin Films, (1982) 65.
P.Nath and R.F.Bunshah. Thin Solid Films, 69 (1980) 63.

THE PREPARING PROCESS AND PROPERTIES ANALYSIS OF SOLAR SELECTIVE ABSORBING SURFACE ON MILD STEEL

WAN Taixin*, ZHAO Qinhua*, CHEN Hong* and CHENG Xiaoxi**

*Gansu Natural Energy Research Institute (GNERI),
Gansu Academy of Science, Lanzhou, Gansu, P. R. China
**The Research Institute of Modern Physics,
Chinese Academy of Sciences, Lanzhou, Gansu, P. R. China

ABSTRACT

This paper emphatically introduces the solar selective absorbing surface processed by alternating current oxidation method on cold-rolled milled steel plate. The performance of the surface, such as solar absorptance α, thermal infrared emittance ε, durability aging and humidity resistance was tested and evaluated. Furthermore, the micrograph and constitution of solar absorbing surface were observed and analyzed by Transmission of Electron Microscope (TEM), Auger Electron Spectroscopy (AES), and Mössbaur Spectrum. The results showed that the alternating current oxidation method can prepare solar selective absorbing surfaces that basically meet the requirements for use in low temperature solar applications.

KEYWORDS

Alternating current oxidation; selective absorbing surface; absorptance; infrared emittance; constitution; micrograph.

INTRODUCTION

In China, a black paint coating was often used for absorber plates used for domestic solar thermal collectors. This painted coating was not a selective surface, and the solar absorptance α and the thermal emittance ε were almost equal. An ideal selective absorbing surface would be one that completely absorbed the incident solar energy with wavelength <2µm (α=1). It should not emit radiation in the infrared region (wavelength>2µm) and the total hemispherical emittance at the operating temperature would equal zero ε=0 [1.2]. Of course, such an ideal surface coating was not achievable, but it was possible to produce selective surfaces having high solar absorptance and low long-wave emittance [3.4]. In conventional practice one attempted to maximize α, minimize ε, and obtained a high α/ε ratio. However, in the practical systems, it showed that the collector efficiency at relatively low temperature was generally more sensitive to α

than to ε, which was not only to enhance α but also reduced ε. The prepared process to achieve these selective surfaces was expensive and complicated. This work presented an alternating current oxidation method for preparing selective surface, the procedure was simple and the cost was low. Yet the results showing this basic preparation would meet the needs for low temperature solar utilization.

PREPARATION OF THE SOLAR SELECTIVE ABSORBING SURFACES

The substrate was cold-rolled milled steel plate with thickness of 0.75 mm and size of 70x50 (mm²). The substrate surface was cleaned with acetone or trichlorethylene to remove any oil contamination on the sheets. The samples were then rinsed with water, and put them into dilute hydrochloric acid for four minutes to scale the oxide. After the treatment, the samples were rinsed again with water and transferred immediately to the oxidation bath.

The alternative current oxidation was carried out in a medium (named as WZS) with concentration of 8-10%, and PH<1. The following operating conditions were used:

-- AC supply with V < 5V, I=12-14A/dm² (Both samples were anode and cathode alternatively).
-- Temperature: 25-40°C.
-- Time: 40-50 minutes.

After oxidation the samples were rinsed immediately in clear water and then put them into soapsuds at 90°C for about 20 minutes. They were then moved from the bath and air dried.

TESTING THE PROPERTIES OF THE ABSORBING SURFACE

The thickness of the surface oxidation layer was measured using a Model Qcc Magnetic Thickness Tester. The total reflective measurements were performed using a Model SSR-E solar spectrum reflector in the range of 0.25-2.5 µm. The total hemispherical emittance was measured by a Model AE radiator at 90°C in the range of 2.5-10 µm. The samples were also exposed to a Model WEL-SUR-DC all autosunlight aging chamber for 240 hours to observe the change in appearance of the surface and to measure α and ε. The conditions in the aging chamber were: carbon arc lamp with radiation intensity of ultraviolet light (< 350 nµ) of 540 w/m², temperature of 25°C, and zero humidity.

The samples were then heat treated in a constant temperature drying chamber at 100°C for 10 hours and then in a freezing chest at -30°C for 14 hours. They were subjected to this temperature cycling for 30 periods; total testing time of 720 hours. After this exposure, the change of surface films was observed and the absorptance α and the emittance ε were measured.

The samples were tested for 30 additional cycle periods for an additional 720 hours. These cycles consisted of placing the samples in a WS/08-01 Humidistat for 10 hours under conditions of 64°C dry, 63°C absolute humidity and 96±2% relative humidity

followed by 14 hours at room temperature with 75% relative humidity. After being cycled, the change in appearance of the surface films was observed.

MICROGRAPHIC OBSERVATION AND CONSTITUTION ANALYSES OF THE SOLAR ABSORBING SURFACE

After the surface film was replicated, the microstructure was
observed using a H-600 TEM with 20,000 magnification.

The constitution analysis was carried out using a model CEMS
Mössbaur. The internal Conversion Elections Mössbaur Spectrum
(CEMS) at room temperature was observed on a constant acceleration drive spectrometer using a 15 mci 57co(pd) radioactive
source and graduating by α-Fe of thickness 25 µm. The CEMS measurement was performed with a gas-flow proportional counter, with
the flow about 4-5 bubbles/minutes. The working gas was a mixture
of 95% helium and 5% methane. The range of channels was 512. The
spectrum was analyzed by the AWMI (algorithm without matrix
inversion) program, on an IBM PC/XT microcomputer.

The chemical composition of the surface film was measured with a
Model PHI & SIMS 3500 system. AES depth profiling was made using
a deferentially pumped ion gun with an Ar^+ ion beam at 3 Kev and
0.5 A/cm² current density in an Argon pressure of 2.5×10^{-8} Torr.
The system pressure was 1×10^{-10} Torr.

RESULTS AND DISCUSSION

Any oil contamination and rust on the substrate can cause dis-
coloration or reduce adhesiveness of the oxide layer to the
substrate surface. Therefore, it was important to clean the
substrate thoroughly before the oxidation step.

The Table showed that α was increased by increasing the surface
roughness in the range of 5-10 µm, but it did not cause a significant difference in ε. If the roughness was decreased, it
would be difficult to obtain the result in blackening. Therefore,
the cold-rolled substrate surface of the steel can meet the needs
of the oxidized blackening, not perform to finish treatment
again.

TABLE 1. Affect of Substrate surface roughness on α and ε

Roughness	α	ε
Cold-rolled (about 10 µm)	0.93	0.31
Grinding surface on 180# paper (<10 µm)	0.90	0.32
Grinding surface on 240# paper (<6 µm)	0.87	0.32

The affect of the oxidation medium concentration, temperature, current density, and time on performance of the oxide surface was determined by the experiments previously described.

It was difficult to obtain a black surface through oxidation when the medium concentration was less than 8%. But if it was over 12%, the adhesiveness of the oxide to the substrate surface would be less.

The experimental results showed that the temperature had a large effect on the performance of the oxide surface. For example, at the room temperature, it was possible to obtain an oxidized surface with $\alpha=0.94$ and $\varepsilon=0.31$. But when the temperature was below 20°C, the oxidizing process needed longer time and the surface blackening was not easy. If the temperature was up to 50°C, the absorptance α equal 0.92 which was the same as the situation at the room temperature, but the thermal emittance $\varepsilon=0.60$ was sharply increasing. The rising temperature can accelerate the rate of metallic ions oxidation, so as to make oxidized surface thick and rough, which led to the emittance ε rising at last.

The oxidizing process required a longer time when current density was below 10 A/dm². However, for a density over 15 A/dm², it led to a rise in bath temperature because of electrode heating. The result was a poor oxide surface.

If the oxidizing time was less than 30 minutes, the oxide surface had a brown appearance and the absorptance α was lower. The oxide film thickness was about 2.0 μm and it did not increase even though the oxidizing time was continued for 50 minutes. There was no further affect of optical performance of the oxide surface.

The samples for oxidation were switched alternately as anode and cathode. Therefore, the desired oxide surface could be achieved on both samples simultaneously. It was aggressive and lost metallic ions when the sample was placed to anode. And vice versa, when it was converted to cathode, the metallic ions were oxidized and deposited on the surface of samples to form oxide film. While the sample was switched to anode again, the next corrosion circulations started in the thinnest part of the oxide film. By repeating the above cycle, there was dense and rough oxide film formed on both substrate surface. The resulting film was able to absorb solar radiation efficiently. The efficient absorption resulted from the multi-reflected and scattered sunlight which came from the rough features of the oxidized surface. The oxidation speed slowed and the film growth stopped when film thickness reached about 2.0 μm. The whole procedure took about 40 minutes. The oxide film appeared black brown and the thickness was about 2.0 μm with $\alpha=0.92-0.94$, $\varepsilon=0.30-0.32$.

As shown in the figure on the next page the constitution of the film was observed under high magnification with the Transmission Electron Micrograph (TEM). A spherical particle structure was apparent with particles about 0.25 μm in diameter, compactly and uniformly distributed on the surface. The Auger Electron Spectroscopy (AES) and Mössbaur spectrum (CEMS) analysis showed that the structure was 46.93% Fe and 47.22% O_2. The main phase included $FeOOH$, $r-Fe_2O_3$, Fe_3O_4 and Ferrite.

FIG.1.   TEM replice micrograph
of film constitution X 20,000

The TEM analysis showed the quantity of particle in the oxide film constitution was of the same order of magnitude as the solar spectral wavelength. This will increase reflection of the solar radiation and increase the absorption of the solar energy. In addition, the structure, which consisted of uniform size and rank compactly spherical particles, made the longer wavelength to be easy transmitted and the thermal infrared emission was decreased as a result.

The experimental results showed that the oxide surface still appeared as black brown after ultraviolet light accelerated aging for 240 hours, the absorptance α was in the range of 0.92-0.93, and the emittance ε was in the range of 0.30-0.32. The results got near those that were not aged. After undergoing the temperature aging treatment (-30°C to +100°C for 720 hours) the absorptance only degraded from $\alpha=0.93$ to $\alpha=0.91$. However, the emittance dropped from $\varepsilon=0.40$ to 0.26. It should be noted that if the emittance could be reduced, the ratio of absorption α and emittance ε will increase, which is favorable to solar heat utilization. This oxide surface has such an excellent aging-resistance that may be well related to their morphology with spherical particles on the oxidized surface. The structure in thermodynamics and energy distribution was well placed to balance and stability.

The oxide surface did not corrode even after exposed to a humidity experiment for 720 hours. The Mössbaur spectrum analysis showed that the oxide surface was made up of Fe_3O_4, Fe_2O_3, Ferrite and FeOOH. These substances, which were insoluble in water, would protect the substrate. The compacted structure of uniform spherical particles could act as a protecting barrier from corrosion for the surface. This would increase the moisture resistance of the oxide surface.

CONCLUSION

The selective solar absorbing surface can be prepared on cold-rolled steel plate by an alternating current oxidation method in a WZS medium. This oxide surface has an excellent aging resistance and humidity resistance. The TEM and CEMS analysis showed that the optical properties and durability of the oxide surface were related to their constitution and morphology. In addition, the principal advantage over electrode deposited, sputtering, CVD and spray pyrolysis and so on was the simple production and the low cost. The resulting surface can basically meet all requirements for domestic low temperature solar applications.

ACKNOWLEDGEMENTS

The author gratefully acknowledges Ms. Fu Liandi and Engineer Cao Quixun for performing the optical testing and TEM analysis done in support of this work.

REFERENCES

1. Mattox, D. M. (1976). J.Vac.Sci.Technol., Vol.13, No.1, pp.128.
2. Erben, E., A. Muenhlratzer (1983). Solar Energy Materials. Vol.9, pp.281.
3. Turisson, J., R. E. Peterson, and H. Y. B. Mar (1975). J.Vac.Sci.Technol., Vol.12, No.5, (Sept./Oct.1975) pp.1011.
4. Mattox, D. M. (1975). J.Vac.Sci.Technol., Vol.12, No.5, pp.1023.

ACTIVE HOT-WATER SUPPLY IN COMBINATION WITH THE SEASONAL STORAGE FOR HEATING

Bart Jan van den Brink and Henk Buijs

De Realiteit 17

1316 TA Almere

The Netherlands

ABSTRACT

A large vessel for the (seasonal) storage of solar energy is fed by water collectors by means of a discharge system. In the cold season the heat stored is used to heat the building.
Domestic warm tap water is heated by a heat exchanger that has been incorporated in the heat storage vessel.
In the winter season, when the water temperature in the vessel is below 50 $^{\circ}$C, the hot tapwater is heated additionally, where necessary, by a small electrical boiler or geyser.
To attain an optimal result it must also be possible to (pre-)heat the cold water for the shower and the bath by the solar heat storage vessel.
Starting point for the construction of the system is that in winter the temperature of the water in the solar heat storage vessel will drop to 30 $^{\circ}$C.
By means of a 10 litre electrical boiler filled with water to 90 $^{\circ}$C, in this way 50 litres of water can be heated to the desired 40 $^{\circ}$C, sufficient for one shower.

Fig.1 : The dwelling "Meerzicht", Almere, The Netherlands

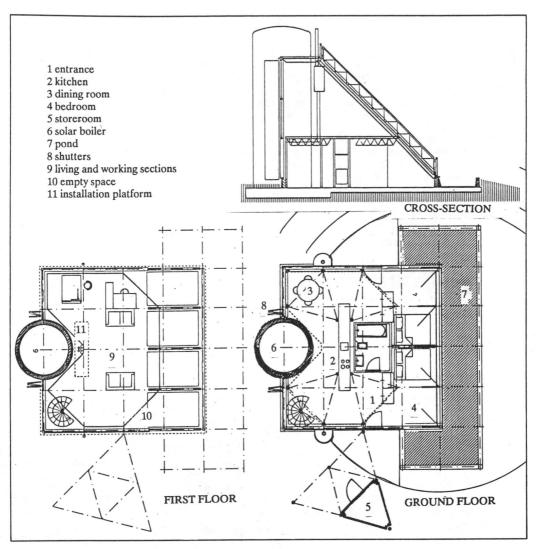

1 entrance
2 kitchen
3 dining room
4 bedroom
5 storeroom
6 solar boiler
7 pond
8 shutters
9 living and working sections
10 empty space
11 installation platform

CROSS-SECTION

FIRST FLOOR

GROUND FLOOR

Fig.2 . Ground-plan and sectional drawings of "Meerzicht"

DESCRIPTION OF THE DWELLING AND THE SOLAR ENERGY SYSTEM

The house "Meerzicht" (see figs. 1 and 2) is being constructed in Almere, The Netherlands and has been lived in since 1988. The solar-energy system has functioned since May 1988, but has not yet been completed.
The living has 100 m^2 of living space , outer walls of glass, a collector roof of 70 m^2 area and has been built around a vessel of 40,000 litre of water for heat storage.

The solar energy system is a discharge system. The solar energy collection is based on a open water system. The only heat exchanger in the system is needed for domestic hot water. For security and efficiency a discharge system is essential. The solar collectors are selected for optimal result in winter season. So the absorbers are spectral selected painted and fitted with two layers of glass.
When the sun cannot contribute positively towards the heating of the water in the storage vessel any longer, than the 80 litre of water are discharged from the collectors and stored in the vessel.

DESCRIPTION OF THE TAPWATER SYSTEM

Warm tapwater for kitchen, washbasin, shower and bath is heated via a heat exchanger that has been incorporated in the storage vessel. In the winter season , when the water temperature in the collector vessel is below 45 °C the tap water for shower and bath is heated additionally by a 10 litre electrical boiler. The fact that the water for the other warm tapwater consumers is less hot, is no problem. The dishwasher itself takes care of the additional (electrical) heating of the washing-water up to the appropriate temperature.

In the winter season, when additional heating is necessary , the supply of cold water is conveyed to the storage vessel via the heat exchanger. Then mixing of this tepid water with additionally heated tepid water can provide appropriate service temperature.

In the winter season the temperature of the storage vessel ultimately drops to 30 °C. Then the tap water from the electrical boiler, which is additionally heated to 90 °C will provide a 40 litre shower of sufficient temperature , after having been mixed with the tepid water available.

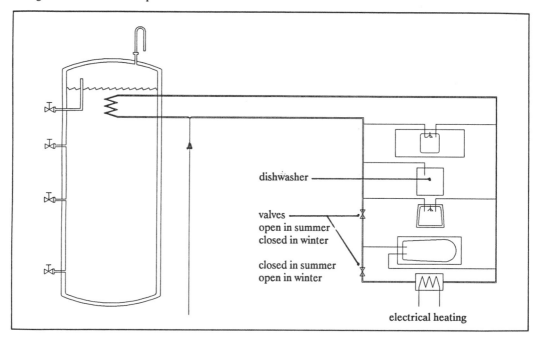

Fig.3 . The tap water system

THE CONTROL SYSTEM

The solar-energy system is computer controlled.
The computer control has five functions:
1. The determination of the heating output of the solar collector.
2. The control of the discharge system.
3. The safeguarding of the collector against freezing.
4. The gathering of relevant data.
5. The external communication (telephone) for supervision.

As to 1. When the collector temperature is 10 °C higher then the lowest temperature in the storage vessel, the solar system is filled with water and put into operation.
The vessel has two inflow openings. One practically on top and one midway. These openings serve to maintain the laminar temperature structure in the storage vessel.
If the difference between the collector temperature and the lowest temperature in the vessel has dropped to 2 °C, the system is (automatically) put out of order.

By day the control waits minimally 10 minutes before the system is pumped out. By the end of the afternoon this waiting-time decreases to five minutes.
Times, temperatures and adjustments are variable.

As to 2. There are three hydraulic functions:
1. Filling and venting of the system
2. Being normally in operation
3. Discharging of the solar collectors.
Since the water level in the storage vessel is approximately 1 meter above the highest collector level, an extensive system of valves is required to put into operation the correct hydraulic action dependent on the function involved (see also figs.3 and 4).

As to 3. The computer pays special attention to the safeguarding against risk of frost, which treatens the solar collectors. At all times the collector will be pumped out.

As to 4. Operational data, such as average temperatures and hours of sunlight, are gathered in statistics.

As to 5. Communication by means of the public telephone network offers the opportunity for remote monitoring and controlling of the installation and for remote reading out of data.

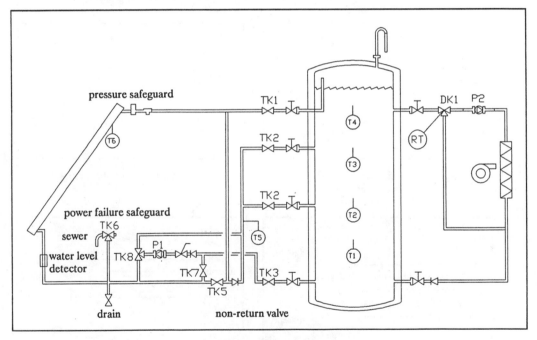

Fig. 4 . The solar collector and storage system

ENERGY BALANCE

By means of a dynamic computer simulation of the house and the solar system the energy balance for the house, tapwater, solar collector and storage vessel are calculated. This computer model DYWAG is based on a 1978 dissertation.(Ref.1. and 2)
In the mean time, with the help of some tests models, the model is validated.(Ref.3 and 4.)
The year 1964 is used as real outside climate for the calculations. This year is characterized by a severe winter and a hot summer in relation to the Dutch climate.
The quantity of hot tap water is based on a daily use of 160 litre water with a temperature of 45 °C.

Based on these calculations, the solar system will deliver 2,278 kWh out of the total of 2,514 kWh for tap water heating. The defecit, 236 kWh, is being obtained by electrical after-heating (fig.6).

The total yield of 70 m^2 solar collector area is equivalent to 24,921 kWh a year.
To determine the yield of the solar collector we used efficiency calculations, published by ISSO 14 (Ref.6.).
In this manner, calculated   average efficiency figures are reproduced in fig.5.
The solar system has been operating according to these calculations for the year 1964 during a 1,971 hour period.

The saving in energy with this system having 70 m^2 collector area and a 40 m^3 storage vessel of water (80 °C at the
end of the summer season, 30 °C at the end of the winter season) amounts to 11,874 kWh.
This is contributed to:

- 7,419 kWh   Active for heating the building.
- 2,177 kWh   Heatlosses of the storage vessel used for heating the building.
- 2,278 kWh   Heating warm domestic water.

The energy balance of the storage vessel of 40 m^3 is given in fig.7.

Also based on the standard figures for the climate glass walls of the dwelling, as publicated in ISSO 16 (ref.7.),the
total energy balance for heating the house is calculated (fig.8). In case of the severe winter of 1964 this would mean
an extra energy consumption of 531 m^3 natural gas (31,7 MJ/m^3).

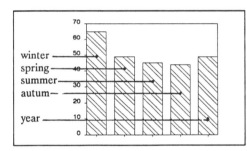

Fig.5 : Average efficiency of the solar water
collector

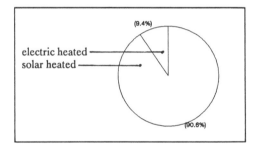

Fig.6 : Energy balance of hot tap water

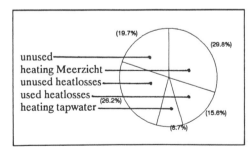

Fig.7 : Energy balance of the storage vessel

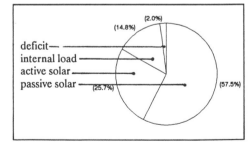

Fig.8 . Energy balance of "Meerzicht"

POINTS OF ATTENTION

We based the construction of the system on the laminar temperature structure of the water in the storage vessel.
The temperature of the top layer is the highest, which is . . .
W h y   the heat exchanger has been placed in that layer.
When we tap drinking water, the water in the storage vessel cools around the heat exchanger.
This results in undesirable turbulence, which in turn causes loss of energy.
Further research into the best place for heat exchangeris desirable.

It must be possible to (automatically) reverse the tap installation from summer to winter position, and vice versa, in a user-friendly way.

CONCLUSION

In the house "Meerzicht"it has been shown since 1988 that a system that heats tap water via a heat exchanger in a storage vessel containing hot water for space heating will function properly. When in the winter season the cold water for shower and tap is per-heated by means of the heat exchanger, optimal benefit can be gained from solar energy. With this system, moreover, an electrical boiler, having a small capacity, will suffice for additional heating of the shower water. This boiler need only be switched on in the period between 1st December and 1st March. The system could be optimized in several respects.

REFERENCES

Bruggen, R.J.A. (1978).   Energy Consumption for Heating and Cooling in relation to Building Deisgn.
Lammers, J.T.H. (1978).   Human factors, energy conservation and design practice.
Haaster, P.H. van (1989). Testset ten behoeve van de fysische kwaliteit van rekenmodellen voor energiegebruik.
Krijger, G.K. (1989).   Toetsing van DYWAG voor energiegebruik van woningen met behulp van testsets.
VNI ISSO 14. (1983).   Zonneboilers ontwerp en uitvoering.
VNI ISSO 16. (1989).   De jaarlijkse warmtebehoefte van woningen, energiegebruiksberekening per vertrek.

ASPECTS OF LOW TEMPERATURE LATENT HEAT STORAGE BY USING SODIUM SULPHATE DECAHYDRATE SALT MIXTURE

I.Memon and B.M.Gibbs

Department of Fuel and Energy, University of Leeds
Leeds LS2 9JT, UK

ABSTRACT

This paper presents the heat transfer characteristics of a cylindrical, latent heat store utilizing an inorganic salt mixture, (88 per cent of sodium sulphate decahydrate, 2.64 per cent of di-sodium tetraborate decahydrate and 9.36 per cent of bentonite clay melting point 32.4 °C). The cylindrical heat storage container consists of a vertical pipe with a heat transfer tube at the centre. Four circular fins are fixed in the bottom section of the vertical heat transfer tube in order to improve the convective heat transfer. An important feature of this design is the employment of an oil to protect the sodium sulphate decahydrate salt mixture from degradation by the atmosphere and to measure the change in salt level during heating or cooling. An upper cylinder indicating the oil level is connected to the lower cylindrical heat storage container by an inter connecting pipe. Water is used as a heat transfer medium; it is circulated in a vertical copper tube in order to transfer heat from the water to the storage medium or vice versa. Experiments have been conducted at three different temperatures and four different flow rates (1.0 to 2.5 lit/min). The heat transfer rates during melting of the heat storage material are measured. It is shown that heat transfer to the heat storage material is largely influenced by natural convection in the liquid layer.

KEYWORDS

Latent heat of fusion; thermal energy storage; inorganic salt; transformer oil; heat flux.

INTRODUCTION

Thermal energy storage is an important facility for the effective utilization of solar energy, heating or cooling applications, thermal electric power generation, and many other processes of current interest. The major problem in thermal energy storage is the selection of a material having suitable thermophysical characteristics in which solar energy can be stored in the form of heat. The second major problem is the design of an efficient heat storage unit.

A compact heat storage container used for air-conditioning as well as heating purposes has been studied by (Gawron, 1977; Herrick, 1979; Telkes and Reymond, 1949). Inorganic hydrous salts were used in their studies. However, these studies focused on the development of new heat storage materials. As a result, there are insufficient data on heat transfer essential for the design of heat exchangers in heat storage containers. Experimental results on heat transfer in latent storage material have been presented in few reports (Bathelt, 1979; Carlsson, 1980; Sparrow, 1978). Data were collected in our study by conducting heat transfer experiments using sodium sulphate decahydrate as a storage material, which possesses a melting point of 32.4 °C, convenient for air-conditioning as well as heating purposes. The experimental system consists of a simple unit with a heat transfer copper tube vertically inserted at the centre of a cylindrical heat storage container. This unit may be used for a wide range of practical applications.

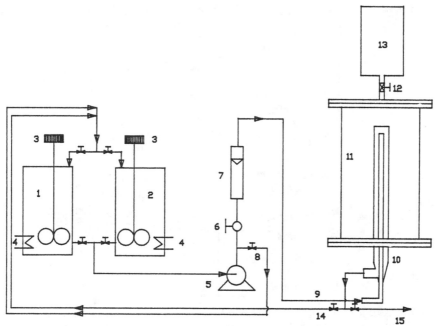

Fig. 1. Schematic Diagram of Experimental Set-up

1. Hot water tank; 2. Cold water tank; 3. Motor stirrer;
4. Electric immersion heater; 5. Water flow pump; 6. Flow control
valve; 7. Flow meter; 8. By-pass valve; 9. Water inlet line;
10. Heat transfer tube; 11. Heat storage cylinder;
12. Interconnecting valve; 13. Transformer oil tank; 14. Water
outlet line; 15. Load; 16. Thermocouple.

EXPERIMENTAL SETUP AND METHOD

A schematic diagram of the experimental system is shown in Fig. 1. A copper heat transfer tube (28.25mm I.D and 780mm long) is centrally inserted from the bottom into a cylindrical heat storage container made of stainless steel. The heat transfer tube basically consists of two tubes: an inner tube and an outer tube. Water flows in through the inner tube and out through the space between the inner tube and outer tube. The cylindrical heat storage container, which is 190.50mm I.D and 900mm long, contains the heat storage material, 33.86 kg of sodium sulphate decahydrate salt mixture. Because the rate of heat transfer is very poor at the bottom of the heat transfer tube, four circular copper fins were soldered to the bottom of the heat transfer tube, each fin being 150.8mm in diameter and 0.92mm thick. The distance between each fin is 28mm. Twelve chromium-aluminum thermocouples are fitted at three axial levels (150mm, 450mm and 750mm from the bottom of cylinder) and at four radial positions (19.88mm, 39.44mm, 59.15mm and 79.03mm from the surface of the heat transfer tube, respectively). One thermocouple at each axial level is located on the outer surface of the heat transfer tube in order to measure the wall temperature of the heat transfer tube.

Two thermocouples were used to measure the inlet and outlet temperature of water. Transformer oil was filled from the top of heat storage material in order to measure any volume change during melting or solidification and to prevent the evaporation of water vapour from the melted salt crystal. The oil container is installed at the top of the heat storage container and connected by an interconnecting pipe.

Table.1 Thermophysical properties of $Na_2SO_4.10H_2O$ and Water

	$Na_2SO_4.10H_2O$	Water
Melting point (^0C)	32.4	0
Latent heat of fusion (kJ/kg)	254	334
Specific gravity (kg/m^3)	1485 (s)	917 (s) 998 (l)
Thermal conductivity ($w/m.^0C$)	0.544 (s)	2.20 (0 ^0C) 0.626 (40 ^0C)
Specfic heat ($kJ/kg.^0C$)	1.93 (s) 3.26 (l)	2.00 (s) 4.18 (l)
Co-effecient of viscosity (Pa.s)	----	6.62×10^{-4}

The thermophysical properties of sodium sulphate decahydrate are shown in Table 1. Pure sodium sulphate decahydrate melts incongruently, so that it requires a nucleating agent Telkes (1949) such as borax and some thickening agent Farid (1989) such as bentonite clay to prevent settling of the higher density anhydrous ions at the bottom of container. However, this mixture reduces the storage density to some extent.

RESULTS AND DISCUSSION

Figure 2. shows the variation in the total heat flow rate Q_t to the heat storage material in relation to time. Symbol T_m in the Fig. 2 represents the melting point of the heat storage material and m represents the mass flow rate of hot water in kg/sec. T_i represents the inlet water temperature which flows into the heat transfer tube. The heat flow rate, Q_t is comparatively large at first, and decreases sharply within the time=30-150 min. The heat flow than tends to level off before decreasing again. This behaviour is particularly pronounced at the lower inlet temperature and is related to the latent heat absorption during the melting process.

Figures 3 and 4 show the radial temperature profile at the bottom and top sections of the heat storage cylinder. The temperature distribution was measured vertically at 150mm and 750mm from the bottom of the heat storage cylinder, respectively. T_i and m represent the inlet temperature and mass flow rate of hot water which flows through the heat transfer tube. From Fig. 3, the heat transfer rate decreases progressively through the material, towards the shell wall. The heat transfer rate in the radial direction depends on conduction and natural convection. From the copper tube to the copper fins it is conduction heat transfer, and from the copper fins and tube to the heat storage material it is convective heat transfer. This figure reveals that a large temperature gradient exists in the storage medium, especially between the first and second radial positions This is due to the very high resistance caused by increases in radial depth as well as the thickness of the solid layer. Nevertheless, the radial temperature gradient decreases from the second radial position onward. Fig. 4 shows similar temperature gradients exist at the top of the storage unit to that at the bottom (Fig. 1), but there is a higher temperature level, indicating that considerably more heat is transferred to the storage medium at the top of unit than at the bottom (Fig. 3) of the a unit. Clearly buoyancy/natural convection effects within the medium lead to this higher heat transfer at the top, and this is promoted by the finned tube at the lower section of the container. Basically the heat transfer rate at this bottom section is very low due to low temperature of water, so that fins lead to the creation of uniform heat transfer at bottom section of heat transfer tube.

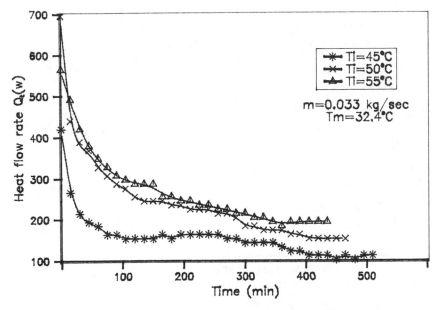

Fig. 2. Heat flow rate vs time.

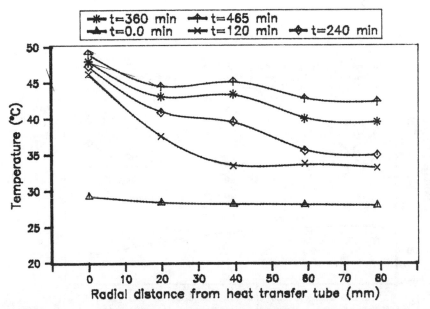

Fig. 3. Radial temperature profile from bottom of the heat transfer tube
T_i=55°C, m=0.025 kg/s, Z=150 mm.

Figure 5 shows the axial temperature profiles at the positions Z=150mm, Z=450mm and Z=750 within the salt. The temperature of the heat storage material decreases in the middle, Z=450mm, but increases towards the top, Z=750mm. This parabolic temperature behaviour is caused by the extended surface and very high inlet water temperature respectively. But the heat

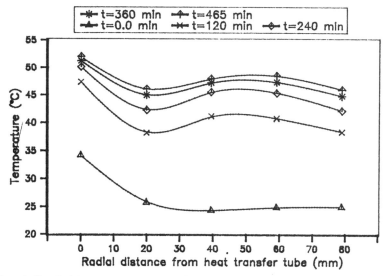

Fig. 4. Radial temperature profile at the top of the heat transfer tube
Ti=55°C, m=0.025 kg/s, Z=750 mm.

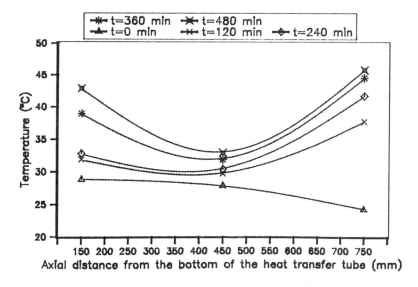

Fig. 5. Axial temperature difference from the bottom of heat transfer tube
Ti=55°C, m=0.017 kg/sec, r=39.44 mm.

transfer rate at the middle position is very poor due to low temperature of water and high thermal resistance of solid layer. In addition, the minimum temperature evident at the centre shows that this region melts more slowly compared to the outer vertical regions, which rapidly become fully melted.

CONCLUSION

1. Heat transfer from the central tube to the heat storage material is largely influenced by natural convection at the melting liquid layer section.
2. The heat transfer rate obtained at the top and bottom sections of the heat transfer tube are very high, due to influence of high inlet water temperature and extended surface, respectively.
3. The axial heat flux at the middle section of the heat transfer tube is very poor due to the low water temperature and resistance of the solid layer of heat storage material.

ACKNOWLEDGEMENT

One of the authors (I.Memon) wishes to thank the Ministry of Science and Technology, Government of Pakistan, for providing a postgraduate scholarship for this investigation.

REFERENCES

Bathelt, A. G., and R. Viskanta (1979). Latent heat of fusion energy storage: Experiment on the heat transfer form cylinders during melting. Heat transfer.J., 101, 453-458.

Carlsson, B., and G. Wettermark (1980). Heat transfer properties of a heat-of-fusion store based on $Cacl_2 .6H_2 O$. Solar Energy.J., 24, 239-247

Farid, M., and K. Yaccob (1989). Performance of direct contact latent heat storage unit. Solar Energy.J., 43, 237-251.

Gawron, K., and J. Schroder (1977). Properties of some salt hydrates for latent heat storage. Energy Research. Vol. 1, 351-363.

Herrick, C. S. (1979). A rolling cylinder latent heat storage device for solar heating/cooling. ASHRAE. Trans., 85. PH79-5.

Sparrow, E. M., R. R. Schmidt, and J. W. Ramsey (1978). Experiment on the role of natural convection in the melting of solids. Heat Transfer.J., 100, 11-16.

Telkes, M., and E. Reymond (1949). Storing solar heat in chemicals, heat and ventilating, 80-86.

THERMAL ANALYSIS OF A SOLAR COOKER

D.P. Rao, S. Karmakar and T.C. Thulasidas

Department of Chemical Engineering,
Indian Institute of Technology, Kanpur 208016, INDIA.

ABSTRACT

The "box-type" of solar cooker has gained acceptance in India. In brief, the cooker is an insulated rectangular box with two glass covers on the top. Up to four cylindrical vessels with flat ends, loaded with materials to be cooked, are kept in the cooker. Modelling and simulation of the cooker is presented. The dominant mode of heat transfer is found to be from the bottom plate to the vessel base by conduction. The cooking time is governed by the uncovered area of the bottom plate and its thickness. The modelling would be of help in the design and better utilization of the cooker.

KEY WORDS

Solar Cooker, Solar Energy, Thermal Analysis, Cooking

INTRODUCTION

Among the various types of solar cookers proposed in literature, the "box-type" cooker became popular in India. The design of the cooker appears to be based on heuristics rather than sound principles. The understanding of the heat transfer processes within the cooker can lead to a design that can overcome the deficiencies of the present-day cooker, like long cooking time, lack of homogenity of the cooked food, etc. We present the modelling and simulation of the cooker using the results of the previous work on flat collectors.

MODEL

The schematic diagram of the cooker along with heat exchange processes is shown in Fig. 1. The convective heat fluxes are shown by the firm lines, where as the radiative fluxes are represented by the broken lines. To model these processes, the following simplifying assumptions are made:
1. The air inside the cooker, the side plates, the glass covers, the vessel cover and the vessel and its contents are at different but uniform temperatures. 2. Heat conduction between the absorber and side plates and between the vessel cover and the vessel side is negligible. 3. There is no radiation exchange between the side plates and the vessel cover. 4. Solar radiation incident on the side plates and the vessel sides is negligible. 5. Shading of the absorber due to the vessel is negligible.

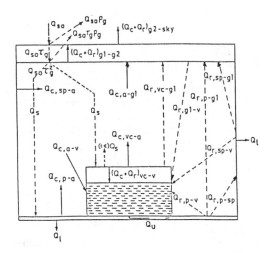

Fig. 1. Various heat exchange processes in a cooker.

Energy Balance Equations

The unsteady-state energy balance equations for the various parts of the cooker are as follows:

Vessel and its contents

$$(M\, C_p)_v\, \frac{dT_v}{dt} = Q_u + Q_{c,a-v} + Q_{r,p-v} + Q_{r,g1-v} + (Q_c + Q_r)_{vc-v} + Q_{r,sp-v} \qquad (1)$$

Vessel cover

$$(M\, C_p)_{vc}\, \frac{dT_{vc}}{dt} = Q_s - Q_{c,vc-a} - Q_{r,vc-g1} - (Q_r + Q_c)_{vc-v} - (1-\varepsilon)Q_s \qquad (2)$$

Air inside the cooker

$$(M\, C_p)_a\, \frac{dT_a}{dt} = Q_{c,p-a} + Q_{c,vc-a} + Q_{c,sp-a} - Q_{c,a-v} - Q_{c,a-g1} \qquad (3)$$

Side plate

$$(M\, C_p)_{sp}\, \frac{dT_{sp}}{dt} = Q_{r,p-sp} - Q_{r,sp-g1} - Q_{c,sp-a} - Q_{r,sp-v} - Q_l \qquad (4)$$

Glass cover 1

$$(M\, C_p)_{g1}\, \frac{dT_{g1}}{dt} = Q_{c,a-g1} + Q_{r,vc-g1} + Q_{r,sp-g1} + Q_{r,p-g1} + (1-\varepsilon)Q_s$$
$$- Q_{r,g1-v} - (Q_c + Q_r)_{g1-g2} \qquad (5)$$

Glass cover 2

$$(M\, C_p)_{g2}\, \frac{dT_{g2}}{dt} = (Q_c + Q_r)_{g1-g2} - (Q_c + Q_r)_{g2-sky} \qquad (6)$$

These coupled differential equations can be solved to obtain the temperatures at various parts of the cooker. However, this requires the average absorber plate temperatures underneath the vessel base and of the uncovered area.

Absorber Plate Temperature

The absorber serves as the collector of solar radiation as well as a fin to transfer heat to the vessel base by conduction. The cooker could be loaded with one or more vessels. Then the temperature profile of the bottom plate can be obtained by solving the unsteady-state differential energy balance equations with appropriate boundary conditions. But it poses a considerable mathematical complexity due to the geometry of the plate-vessels configuration. To simplify the problem, the following assumptions are made: 1. The temperature is uniform along the thickness of the absorber. 2. Insofar as the temperature profile over the absorber is concerned, the absorber is at a pseudo-steady state condition at any time. 3. When one or more vessels are kept on the absorber, its area can be apportioned to each vessel, taking the advantage of symmetry. Kuan and co-workers(1984) presented a method of approximating the rectangular fin and circular or complicated tube geometries as an equivalent circular fin and circular tube, and gave the associated error in the fin efficiency. Using their method, we can find the equivalent radius of the circular fin, R_2 for each vessel to be $R_2 = \sqrt{lm/n\pi}$, where n is the number of vessels and l and m are the length and breadth of the absorber plate, respectively. Now the differential energy balance equation for the fin can be obtained as(Karmakar, 1988)

$$\frac{d^2 T_p^*}{dr^2} + \frac{1}{r}\frac{dT_p^*}{dr} - A\, T_p^* = 0 \qquad \text{for } R_1 \le r \le R_2 \tag{7}$$

where

$$A = \frac{U_1}{Kb} + \frac{K_i}{Kb\delta_i}\ , \quad B = \frac{1}{Kb}\left(S + U_1 T_a + \frac{K_i T_{amb}}{\delta i}\right), \quad T_p^* = T_p - B/A$$

For the absorber underneath the vessel, it can be obtained as

$$\frac{d^2 T_p'}{dr^2} + \frac{1}{r}\frac{dT_p'}{dr} - C\, T_p' = 0 \qquad \text{for } 0 \le r \le R_1 \tag{8}$$

where

$$C = \frac{U}{Kb} + \frac{K_i}{Kb\delta_i}\ , \quad D = \frac{1}{Kb}\left(U\, T_v + \frac{K_i T_{amb}}{\delta_i}\right)$$

Equations (7) and (8) have been solved subjected to the following boundary conditions,

$$\frac{dT_p^*}{dr} = 0 \quad \text{at } r = R_2, \qquad \frac{dT_p'}{dr} = 0 \quad \text{at } r = 0,$$

$$T_p^* + B/A = T_p' + D/C \quad \text{at } r = R_1 \quad \text{and} \quad \frac{dT_p^*}{dr} = \frac{dT_p'}{dr} \quad \text{at } r = R_1$$

to obtain

$$T_p = B/A + D_1\, I_0(r\sqrt{A}) + D_2 K_0(r\sqrt{A}) \qquad \text{for } R_1 \le r \le R_2 \tag{9}$$

$$= D/C + D_3\, I_0(r\sqrt{C}) \qquad \text{for } 0 \le r \le R_1 \tag{10}$$

where I_0, I_1 are the modified Bessel functions of the first kind of order zero and one and K_0, K_1 are the modified Bessel function of the second kind of order zero and one respectively. The constants D_1, D_2 and D_3 are listed in the appendix. The area average plate temperature can be obtained as

$$T_{p1} = B/A + \frac{2}{\sqrt{A}(R_2^2 - R_1^2)}\bigg(D_1 R_2 I_1(R_2\sqrt{A}) - D_1 R_1 I_1(R_1\sqrt{A}) - D_2 R_2 K_1(R_2\sqrt{A})$$

$$+ D_2 R_1 K_1(R_1\sqrt{A})\bigg) \qquad \text{for } R_1 \le r \le R_2 \tag{11}$$

$$T_{p2} = D/C + \frac{2 D_3}{R_1 \sqrt{C}} I_1(R_1 \sqrt{C}) \qquad \text{for } 0 \le r \le R_2 \qquad (12)$$

SIMULATION

The model has been used to simulate the cooker performance. The simulation requires a number of parameters of the cooker, heat transfer coefficients, view factors, etc. These are given in Table 1. The cooker parameters are chosen based on the ones available on the market. Some of the transfer coefficients were based on the general heat transfer considerations, as these are not available in the literature. A few of these(marked by *) were based on the measurements(Karmakar,1988). In practice, the vessel temperature never exceeds $100^{\circ}C$ due to the evaporation of the water in the vessel. However, it was considered that no evaporation takes place even beyond $100^{\circ}C$ to asses the maximum temperature that can be attained.

TABLE 1 Different Parameters of the Cooker

a_v	$= 0.0327 \text{ m}^2$	K	$= 250.0 \text{ W/m }^{\circ}K$
a_{vc}	$= 0.0327 \text{ m}^2$	K_i	$= 0.034 \text{ W/m }^{\circ}K$
$a_{g1,g2}$	$= 0.4428 \text{ m}^2$	l	$= 0.82 \text{ m}$
a_p	$= 0.4428 \text{ m}^2$		
$a_{v \text{ side}}$	$= 0.04486 \text{ m}^2$	m	$= 0.54 \text{ m}$
b	$= 0.003 \text{ m}$	R_1	$= 0.102 \text{ m}$
*$h_{c,p-a}$	$= 5.0 \text{ W/m}^2 \text{ }^{\circ}K$	R_2	$= 0.3754 \text{ m}$
$h_{c,vc-a}$	$= 4.0 \text{ W/m}^2 \text{ }^{\circ}K$	*U	$= 90.0 \text{ W/m}^2 \text{ }^{\circ}K$
$h_{c,a-g1}$	$= 5.0 \text{ W/m}^2 \text{ }^{\circ}K$	U_1	$= 15.0 \text{ W/m}^2 \text{ }^{\circ}K$
$h_{c,sp-a}$	$= 5.0 \text{ W/m}^2 \text{ }^{\circ}K$		
$h_{c,a-v}$	$= 5.0 \text{ W/m}^2 \text{ }^{\circ}K$	ε_{steel}	$= 0.50$
$h_{c,g1-g2}$	$= 2.0 \text{ W/m}^2 \cdot \text{ }^{\circ}K$	*τ_{glass}	$= 0.86$
$h_{c,g2-sky}$	$= 10.0 \text{ W/m}^2 \text{ }^{\circ}K$	δ_i	$= 0.10 \text{ m}$
*$h_{c,vc-v}$	$= 4.0 \text{ W/m}^2 \text{ }^{\circ}K$		

RESULTS AND DISCUSSION

Considering the cooker is initially at $25^{\circ}C$, and it was set for operation from 9:00 a.m, the temperature profile with time has been found and presented in Fig. 2. It can be seen that there is considerable phase difference in temperature of the various parts of the cooker. An ideal design should ensure minimum phase difference among the temperature profiles for fast cooking. If we consider the material gets cooked if it is maintained at $95^{\circ}C$ or beyond for 40 minutes, the vessel profile indicates that it takes about 120 minutes if there is only one vessel in the cooker.

The different heat transfer rates to the vessel with time are presented in Fig. 3. The dominant mode of heat transfer appears to be from the vessel base by conduction. The black paint generally applied to the vessel appears to be unnecessary, as radiative heat transfer plays a minor role.

The heat transfer through the vessel base is governed by the uncovered plate area and the plate thickness. We have simulated the vessel temperature for different equivalent radii per vessel for 3 mm thick plate and are presented in Fig. 4. If short duration of cooking time is required, the large equivalent radius is required. For definitive guidelines for design and better utilization of the cooker, further simulation studies and experimental validation are warranted.

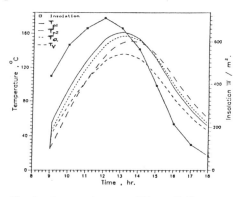

Fig.2: Temperature profiles of the parts of the cooker

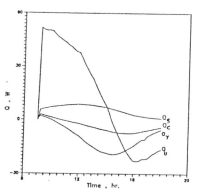

Fig.3: Heat transfer rates to the vessel with time

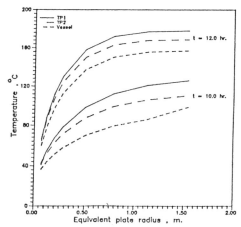

Fig.4: Variation of plate temperature with equivalent plate radius

CONCLUSIONS

Modelling and simulation of a box-type solar cooker are presented. The preliminary results indicate the dominant mode of heat transfer is throgh vessel base by conduction. The uncovered plate area and bottom plate thickness are the important parameters that govern the duration of the cooking time.

APPENDIX

$$D_1 = \sqrt{C} \ I_1(R_1\sqrt{C}) \ K_1(R_2\sqrt{A}) \ (D/C - B/A)/Z$$

$$D_2 = \sqrt{C}\ I_1(R_1\sqrt{C})\ I_1(R_2\sqrt{A})\ (D/C - B/A)/Z$$

$$D_3 = \left[\sqrt{A}\ I_1(R_1\sqrt{A})\ K_1(R_2\sqrt{A}) - \sqrt{A}\ I_1(R_2\sqrt{A})\ K_1(R_1\sqrt{A})\right](D/C - B/A)/Z$$

$$Z = \sqrt{C}\ I_o(R_1\sqrt{A})\ I_1(R_1\sqrt{C})K_1(R_2\sqrt{A}) + \sqrt{C}\ I_1(R_1\sqrt{C})I_1(R_2\sqrt{A})K_o(R_1\sqrt{A}) +$$
$$\sqrt{A}\ I_o(R_1\sqrt{C})\ I_1(R_2\sqrt{A})\ K_1(R_1\sqrt{A}) - \sqrt{A}\ I_1(R_1\sqrt{A})\ I_o(R_1\sqrt{C})\ K_1(R_2\sqrt{A})$$

NOMENCLATURE

a	area, m^2
b	thickness of the absorber plate, m
C_p	specific heat, J/kg ^{o}C
h_c	convective heat transfer coefficient, $W/m^2\ ^o$K
I_o	modified Bessel function of first kind of order zero
I_1	modified Bessel function of first kind of order one
K	thermal conductivity of absorber plate, $W/m\ ^o$K
K_o	modified Bessel function of second kind of order zero
K_1	modified Bessel function of second kind of order one
K_i	thermal conductivity of insulating material, $W/m\ ^o$K
l	length of absorber plate, m
m	breadth of absorber plate, m
M	mass, kg
Q_c	rate of heat transfer by convection, W
Q_l	rate of heat lost from the insulation, W
Q_r	rate of heat transfer by radiation, W
Q_s	rate of incidence of solar radiation on the absorber, W
Q_u	rate of heat transfer from plate to the vessel, W
R_1	radius of the vessel, m
R_2	equivalent radius of the absorber plate, m
S	insolation, W/m^2
T	temperature, ^{o}C
t	time, hr
U	overall heat transfer coefficient, $W/m^2\ ^o$K
U_1	overall heat transfer coefficient, plate to air, $W/m^2\ ^o$K

Greek Letters

ε	emissivity	,	τ	transmissivity
δ_i	thickness of insulation, m			

Suffixes

a	air		p	absorber plate
amb	ambient		sp	side plate
g1	glass cover 1		v	vessel
g2	glass cover 2		vc	vessel cover
sky	sky			
-	heat exchange between two surfaces is denoted by a dash			

REFERENCES

Karmakar, S. (1988). Studies on the heat transfer aspects of solar cookers. M.Tech thesis, IIT Kanpur.

Kuan, D.Y., R. Aris, and H.T. Davis (1984). Estimation of fin efficiencies of regular tubes arrayed in circumferential fins. Int.J.Heat Mass transfer, 27, 148-151

2.18 Concentrating Collectors

OPTIMIZED HEAT PIPE FOR APPLICATION IN INTEGRATED CPCs

M. Collares-Pereira ∗, F. Mendes ∗∗
O. Brost∗∗∗, M. Groll∗∗∗, S. Roesler ∗∗∗

∗ - CCE, Estrada de Alfragide, Praceta 1, 2700 AMADORA, Portugal.
∗∗ - DER/LNETI, Estr. Paço do Lumiar 22, 1699 LISBOA CODEX, Portugal.
∗∗∗ - IKE/Univ. Stuttgart, Pfaffenwaldring 31, 7000 STUTTGART 80, Germany.

ABSTRACT

Previous work on a test program for comparison and selection of a heat pipe, which is intended to be the absorber of an integrated CPC (1.3 X) type collector, has been published (Collares-Pereira and co-workers, 1987) and preliminary results presented. A gravity assisted longitudinally grooved heat pipe has been shown to be the best solution, and was selected for optimization. In the present paper we present the test results of copper tubes with different number and shape of the grooves and with different construction procedures, like insertion of an helicoidal wire against the internal wall. In addition, results of qualitative observations in a glass thermosyphon, simulating one of the copper tubes, are also presented. The comparison of thermal behaviour of those tubes confirms the preliminary results; a grooved tube with a small number of grooves has proved to be a good choice for the CPC application.

KEYWORDS

Two-phase closed thermosyphon; grooved heat pipe; glass thermosyphon; helicoidal wire inserts; fill ratio; inclination angle; thermal resistance; vacuum solar collector; CPC.

INTRODUCTION

The use of heat pipes in solar energy thermal applications, presents a lot of advantages over classical solutions (Akyurt, 1984). Its utilization as absorber in a CPC(1.3X) solar collector implies a reduction in the number of heat pipes per unit of aperture area and, in the present case, of CPC shaped evacuated glass tubes, Fig. 1, a solution with no bellows and only one metal-glass seal is possible.

The test program we are working on comprises the study of copper tubes as containers, combined with water as internal working fluid, and with different structures on the inner wall, in order to obtain a good behaviour according to the operating conditions imposed by the present application: i) slow and fast start-up; ii) 80 W as the maximum heat load to be transferred; iii) low ($\beta < 5°$) and high ($\beta = 38°$, Lisbon latitude) inclination angles, in case of E-W or N-S collector orientation and iv) low fill ratio (F<5%) determined by security considerations. Optimization of the CPC determines also the caracteristic lenghts of the tube as well as its outer diameter.

The maximum heat load of the heat pipes, results from the solar irradiation on a clear day, at noon. The low fill ratio is associated with the possibility of system failure : in case of collector stagnation the temperature would be too high and a great mass of evaporated water causes an exponential increase of the inside pressure; but with a small quantity of water, all of it would be easily evaporated, and after that the pressure would grow linearly to a value sustainable by commercial tubes.

The tube should provide an easy return of the liquid from the condenser to the evaporator and a good wetting of the internal wall, even with the small quantity of available fluid, in order to achieve a small thermal resistance. The construction of the heat pipe should also be as easy as possible, and it would be preferable that a sole solution for the two possible collector orientations is found.

Our previous work (Collares-Pereira and co-workers, 1987,1988) showed, with this condition. , that : i) the screen wick heat pipe is absolutely inadequate; ii) the smooth thermosyphon could be used with high inclination angles and iii) the longitudinally grooved tube proved to be the best solution, which needed, however, further work on its optimization. For that optimization we tested tubes with different number of grooves and groove shape, Fig.2, and with different construction procedures, like insertion of a helicoidal wire, of small diameter, against the wall. This last solution was compared with a tube provided with a helicoidal groove, and its operation was observed in a similar glass tube, which permitted qualitative analysis of the related phenomena.

Next we describe the experiments - test mounting, type of tests and dimensional characteristics of the tested tubes, and after that we discuss the results obtained for each tube. Lastly we present our conclusions.

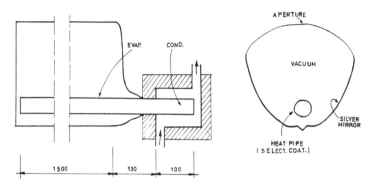

Fig. 1 - Integrated CPC vacuum tube.

DESCRIPTION OF THE EXPERIMENTS

Experiments were conducted on a test rig, Fig. 3, with variable inclination angle; heat input to the heat pipes was provided by an electrical heater tape wrapped around the evaporator, and power was adjusted by a transformer; heat extraction at the condenser was achieved by circulation of water at the desired temperature (T=10°C and T=85°C) and the whole was with foam insulation. In the case of the copper tubes, temperature was measured with 11 T type thermocouples: 8 of them were attached along the external wall of the tubes - 6 in the evaporator and 2 in the condenser, 2 in the external refrigeration circuit - in and out - of the condenser and, finally, 1 for the ambient temperature. With a data acquisition system, the dissipated and transferred energy, Q, the temperature difference between evaporator(Te) and condenser(Tc), $\Delta T(=Te-Tc)$, and the thermal resistance $(R=\Delta T/Q)$ of the heat pipe were recorded as well as plots of temperature evolution were obtained regularly.

Table 1   Characteristic dimensions of the tested tubes.

[m]	Long.grooves	Long.grooves	Long.grooves	Helic.groove	Helic. wire	Helic.wire
Container	copper	copper	copper	copper	copper	glass
Inner diam.	16X10E-3	16X10E-3	16X10E-3	16X10E-3	16X10E-3	16.5X10E-3
Evap. length	1.5	1.5	1.5	1.5	1.5	.8
Adiab. length	.18	.18	.045	.18	.18	.3
Cond. length	.115	.115	.15	.115	.115	.8
Groove shape	triangular	rectangular	rectangular	rectangular	———	———
Groove width	.5X10E-3	.5X10E-3	.5X10E-3	.5X10E-3	———	———
Groove height	.3X10E-3	.3X10E-3	.3X10E-3	.3X10E-3	———	———
Numb. of groov.	51	16	8	1	———	———
Spiral pitch	———	———	———	———	.15	.05/.15
Wire diameter	———	———	———	———	.5X10E-3	.5X10E-3

Each of the tubes (characteristic dimensions are listed in Table 1), was first tested with a very low fill ratio (F=2% or 3%) and for each temperature (T=10°C or T=85°C) of the external refrigeration water, we adopted a low ($\beta<5°$) or high ($\beta=38°$) inclination angle. 80W power were applied; the thermal behaviour of the tube was analysed in terms of the thermal resistance and time constant. When successful behaviour was observed, the power was increased and its influence on thermal resistance was analysed.

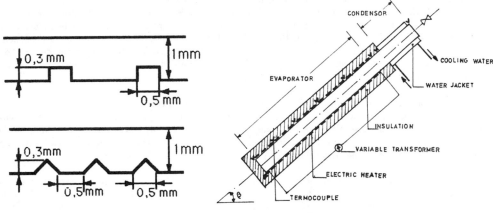

Fig. 2 - Groove shapes.   Fig. 3 - Test rig.

DISCUSSION OF RESULTS

Heat Pipes with Triangular Longitudinal Grooves

Results previously obtained with a heat pipe with a lot of small triangular longitudinal grooves, indicated that it could be used satisfactorily in our application, with a low value of thermal resistance for $\beta=38°$ even with a very small fill ratio, F=2%; however, for a low inclination angle the minimum value of F would be greater than 4%. Both results have been confirmed now, and they result from the fact that with a lot of grooves all the inside wall is wetted, reducing thermal resistance; as the grooves are longitudinal, the return of water from the condenser towards the evaporator is facilitated and aided by a high inclination angle. However, the surface of contact and consequently the friction resistance to the movement is relatively high, and while this has no importance for high β, it seems to be the reason for the dry-out at the bottom which occurs for those heat pipes with low fill ratio and low inclination angle.

Heat Pipes with Rectangular Longitudinal Grooves

Heat pipes with a lower number of longitudinal grooves with rectangular cross section were tested next in order to improve operation at low inclinations. We tried first a tube with 16 of those grooves, and we found very encouraging results. As before, when successful operation was achieved, this implies low thermal resistance denoting good wetting of the inside wall. For $\beta=2°$, the tube exibits good performance only with 5% fill ratio but for $\beta=3°$ the same result can be obtained with F=4% and almost with F=3% (for which occurred dry-out at the bottom). At high inclination angles the tube worked without problems for the different temperatures of the condenser refrigeration fluid. When operating, this tube showed an easy start-up at low or high inclination angles.

With these results, we tried a longitudinal rectangular grooved tube with 8 grooves, which could still operate without problems for $\beta=38°$ and operated also in a promising way, at low tilts : for $\beta=3°$ with F=3% and for $\beta=2°$ with F=4%; with this last tilt angle dry-out only appeared at the bottom of the tube filled with 3%, after 100W power were applied. However, the thermal resistance of this tube is greater than that of the tube with 16 rectangular longitudinal grooves. in similar situations ; this clearly results from the fact that the wetting of the walls is worse in the 8 grooves case.

Glass Tube with Helicoidal Wire

A smooth glass tube was first operated, for comparison with the smooth copper tube and visualization of the associated transfer phenomena, for fill ratios up to 10%. Dropwise condensation instead of film condensation is the dominant operation mode. When the drops grow, they slip along the wall, and they join at the inferior directrix on their way to the evaporator. There is also no film in this zone : there is a more or less stretched pool according to the inclination angle of the tube, and in the other part of the evaporator wall there are also some drops sprayed from the pool in nucleate boiling régime or remaining there after the return from the condenser. There is a discontinuous return of condensate, which reaches the bottom of the tube, or not, depending on the fill ratio (below F=10%, for β=2°, and below F=8%, for β=38°, the condensate no longer reaches the bottom). So the dropwise condensation model is more appropriate than the filmwise condensation model, for the smooth glass tube. This, however, does not hold for metal tubes.

The glass tube was tested next with a helicoidal wire insert, of inox steel, for two pitch values and for fill ratios less than 10%. At low inclination angles the insert with big pitch and small diameter helps operation because it induces small pools all along the evaporator avoiding accumulation of fluid at the bottom; this resistance to the movement, however, can be fatal with the very low fill ratios we are interested in or with a helicoidal wire with small pitch.

In case of high tilt angle, the action of the wire is positive because water doesn't return so fast to the evaporator, getting sprayed when it finds a wire step, therefore wetting all the evaporator wall. This effect is improved with a helix of small pitch, whose resistance to the liquid movement has no importance here. In conclusion, the helicoidal wire creates a lot of pools along the evaporator and acts like a nucleation inducer.

Heat Pipe with Helicoidal Wire

Bearing in mind the results of the work done in (Groll and co-workers, 1987) and those obtained with the glass tube, we tested a copper tube with a helicoidal inox wire insert. As expected, the tube worked at β=38°, for a fill ratio of 3%; however, the performance was better (lower thermal performance) for F=4%, showing regular operation without overheating along the evaporator.

When tilted at 2°, the tube performs poorly with this helicoidal wire for low fill ratios, as expected from the glass results. Only from 6% fill ratio on does it work but with a difficult start-up (almost one hour time constant), and a great thermal resistance : the condensate return flow is very irregular 0 which causes temporary dry-out in some locations of the evaporator above the pool. For low tilt angles case, the wire makes the condensate return difficult; only for fill ratios greater than 12% do we have a sufficient inventory of working fluid to obtain the pools mentioned above.

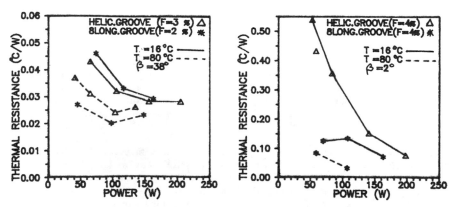

Fig. 4 - Performance comparison between the heat pipes with small number of grooves, operating at the smallest fill ratios.

Heat Pipe with Helicoidal Groove

Combining the results obtained with the longitudinally grooved tubes and the glass tube, we tested one tube with a helicoidal groove similar to the rectangular grooves of the other tubes. At $\beta=38°$ and F=3% it worked with low thermal resistance and low time constant (=5 min). We obtained good reproducibility; performance was maintained even when transferring around 200W at different temperatures.

At low inclination angle ($\beta=2°$) the tube worked without problems at 4% fill ratio, which is a similar result to the one obtained with the heat pipe employing 8 longitudinal rectangular grooves; like this one,it also operated at 3% fill ratio, but drying-out under a heat load of 100W.

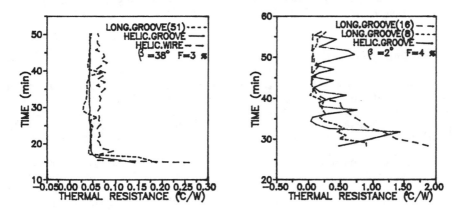

Fig.5 - Evolution of thermal resistance with time in the heat pipes
at low temperature.

CONCLUSIONS

As stated, good wetting of the inside wall and easy return of the fluid from the condenser to the evaporator is essential for the good operation of the heat pipe at low thermal resistance. Our application imposes also that this must be done at the lowest possible fill ratio (F < 4 %) and the lowest possible inclination angle (β < 5°), as well as at $\beta=38°$. For this last tilt angle, and if we think in separate solutions for the two collector orientations (E-W and N-S), we can achieve our goal with different internal structure of the tube wall: helicoidal or longitudinal rectangular or triangular grooves can be used, but the lowest thermal resistance is associated with the greatest number of grooves.

The tests with low values of the tilt angles showed that, with a low fill ratio, it is very difficult to obtain good operation below 3 ° tilt; the heat pipe must be assisted by gravity, but this action only works efficiently above this tilt value. The heat pipe with rectangular helicoidal groove,or with 8 longitudinal grooves,performed satisfactorily. In the future we intend to carry out more experiments using tubes with a small number of triangular and rectangular longitudinal grooves and with many triangular micro grooves.

REFERENCES

Akyurt, M. (1984). Development of heat pipes for solar water heaters. Solar Energy, Vol. 32, No. 5. .

Collares-Pereira, M., F. Mendes, O. Brost, M. Groll and S. Roesler (1987). A heat pipe comparison for a CPC solar collector application. 10th Biennial Congress of the ISES, Hamburg.

Collares-Pereira, M. and F. Mendes (1988). Tubo de calor para aplicação em colector solar do tipo CPC. 4º Congresso Ibérico de Energia Solar, Porto.

Groll, M., O. Brost, D. Heine and S. Roesler (1987). Development of advanced heat transfer components for heat recovery from hot waste gases" - IKE/Universitat Stuttgart, Report No. IKE - 5TF - 790 -87 March 1987.

EVALUATING IMPROVEMENTS TO A LOW-CONCENTRATION-RATIO, NON-EVACUATED, NON-IMAGING SOLAR COLLECTOR

K.G.T. Hollands   A.P. Brunger   I.D. Morrison

Solar Thermal Engineering Centre, Dept. of Mechanical Engineering,
University of Waterloo, Waterloo, Ontario, CANADA, N2L 3G1

ABSTRACT

In a non-imaging unit-concentration-ratio CPC type trough-like solar collector, in which the reflector is a semi-circle and the flat absorber is perpendicular to the aperture, free convection from the absorber plays the dominant role in fixing the collector's heat loss coefficient U_L. Methods to reduce this convective heat loss, as well as benefits to be obtained from increasing the reflector's reflectance, are evaluated in this paper. Three methods of decreasing the free convection are examined: (i) placing an FEP plastic film around the absorber; (ii) changing the position of the absorber to horizontal; and (iii) increasing the concentration ratio to 1.25. Collector efficiency calculations based on detailed free convection heat transfer measurements show that method (i) is most effective in increasing collector efficiency, and method (iii) is the least effective. Method (ii) is the most cost effective way of increasing the thermal efficiency of this type of collector.

KEYWORDS

Non-imaging optics; natural convection; large collector arrays; CPC; collector thermal performance.

INTRODUCTION

The development of a unit concentration ratio, non-imaging (CPC type), concentrating solar collector with a fully-illuminated flat receiver, of the optical class described by Gordon (1986) has been described by Hollands, Brunger, and Mikkelsen (1988a, 1988b). Called, for brevity, the '$M\epsilon$ collector', this very long trough-like collector (Fig. 1) is intended for large installations containing many hundreds of square meters, providing hot water at moderate temperatures $(40$–$80^{\circ}C)$ for commercial and industrial purposes, and possibly for annual storage district heating systems. In the past, collector arrays for such very large installations have been constructed by connecting together a large number of individual factory-built flat plate collector modules, each having an aperture of say $3m^2$. Wilson, Svensson, and Karlsson (1985), however, have described methods whereby very large modules (20 to $100m^2$) can be assembled on-site, using an absorber plate manufactured by the method described by Olsson, Thundal, and Wilson (1981). (This absorber plate arrives in long strip-like rolls, roughly $0.15\ m$ wide, the flattened integral fluid passage being inflated in the field.) While highly advantageous (for example it greatly reduces the number of plumbing connections), this "mega collector" concept has had some problems in its realization: the need for clear weather and skilled crews, and the difficulty of tilting the collector after assembly on the ground. In addition, it contains the same number of components as the standard flat plate collector.

The $M\epsilon$ collector draws on the Wilson, Svensson, and Karlsson concept, adapting it to the CPC approach described by Gordon. In its original concept, it has only three components: a semi-circular reflector, which doubles as the collector box, the Olsson, Thundal, Wilson absorber, "selective-surfaced" on both sides, and a plastic Hostaflon glazing (Hostaflon is a Hoechst AG registered trademark), held in tension by the stiffness of the aluminum reflector

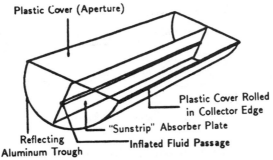

Fig. 1. Sketch of standard $M\epsilon$ collector.

and rolled into the aluminum at the edges. All three components are available in rolls, so the collector can be fabricated continuously by a specially-built machine. Thus the labor cost for manufacture is expected to be minimal, and since, for example, it uses only half the absorber material as a flat plate collector, material costs are also expected to be lower. Nor should tooling-up costs be excessive; a recent estimate to construct a machine to continuously roll-form the semi-circular reflector was $35,000 US. Manufacturing experts have demonstrated the feasibility of the entire machine, indicating that the major cost of the machine should lie in this roll-former. Very long lengths (30 m or more) are envisioned for these collectors, which, because of their lightness (roughly 5 kg/m^2) can be readily carted to the support rack, where they can be affixed by simple methods and plumbed with a minimum number of connections.

Several studies have been completed on the $M\epsilon$ collector. An algorithm, incorporating some ray-tracing, has been developed to calculate the collector's optical efficiency (Mikkelsen, Hollands, and Brunger (1988)). Basic studies (Hollands and Moore (1991)) have (experimentally) characterized the free convective heat loss from the absorber, by means of equations. Prototype collectors have been built and tested for thermal performance. Finally, a theoretical model (Moore, Hollands, and Brunger (1989)) has been assembled, and its first-principle predictions of the collector's efficiency curve have been shown to agree within about 4% with the measured efficiency of the prototypes.

These results have shown that the expected annual efficiency for the collector used for heating service hot water from mains temperature to domestic hot water temperatures is about 10% below that of a standard selective-surfaced flat plate collector. This relative efficiency decrement will increase as the collection temperature increases. Although, for low temperature applications, this decrement in efficiency should be more than compensated for by a reduction in installed cost, there is a clear incentive to raise the collector's efficiency. These figures are based on using a reflector with a specular reflectance of 77% (and a total reflectance of 85%). To raise the collector's efficiency it is necessary to raise this reflectance or to lower the collector's U_L, the dominant component of which has been shown to be the free convective heat transfer coefficient h from the absorber to the reflector/glazing. The present paper evaluates three methods (Fig. 2) of reducing this coefficient (which is referenced to the aperture area): (i) adding an internal glazing of 25 μm thick transparent FEP film, spaced 11 mm from the absorber; (ii) orienting the absorber plate horizontally, regardless of the orientation of the aperture (ordinarily the absorber is perpendicular to the aperture); (iii) increasing the ideal concentration ratio CR of the collector from unity to 1.25, by increasing the aperture area, leaving the absorber unaltered, and changing the shape of the reflector from semi-circular. In addition it examines the benefits to be obtained from raising the reflector's reflectivity.

METHOD

An important first step in the method consisted of repeating the experiments of Moore (1989) and Moore and Hollands (1991) with appropriate changes made to their physical model of the $M\epsilon$ collector. Shown in Fig. 3, this all-copper model has an isothermal semi-circular enclosure (simulating the reflector and the glazing as a single unit) and a flat isothermal plate (simu-

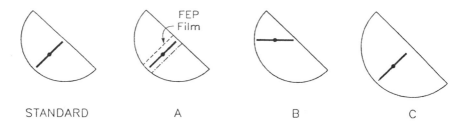

Fig. 2. Cross-sections of the standard (S) $M\epsilon$ collectors configuration as well as of the three altered configurations evaluated: (A) the FEP film configuration; (B) the horizontal absorber configuration; and (C) the $CR = 1.25$ configuration.

lating the absorber) located symmetrically along the enclosure's center line. The enclosure had tubes soldered to it, through which circulated water from a constant temperature bath, maintaining its temperature (of about $25°C$) constant and uniform to within $0.1\ K$. The plate consisted of three parts: two end plates and a heater plate. The end plates contained integral passages through which circulated water from another constant temperature bath, maintaining their temperature (of about $35°C$) also constant and uniform to within $0.1\ K$. The heater plate contained an imbedded electrical resistance whose (measured) wattage was adjusted until the heater plate temperature equalled that of the end plates. Under this condition, essentially all of the electrical power passes to the enclosure. Thus the heat flow q' per unit axial length could be determined, and after subtracting the radiant component of this flow, one obtains the convective heat flow for entry into the Nusselt number: $Nu = q'/k\Delta T$ where k is the thermal conductivity of air and ΔT is the temperature difference. The experiments were conducted with the model contained in a pressure/vacuum vessel. By repeating the heat transfer measurement at different air pressures, the radiant heat flux was readily separated out, and a large range of Rayleigh number $Ra = g\beta\Delta T H^3/\nu\alpha$ (where H is the plate width, g the acceleration of gravity, and β, ν, and α are the expansion coefficient, kinematic viscosity and thermal diffusivity of air) could be covered.

The work of Moore and Hollands (1991) covered a range of values for the spacing S between plate and semi-circle: in dimensionless terms, this range was from $S/H = 0.042$ to $S/H = 0.091$. Using their results, Morrison (1991) estimated the optimal S/H from the point of view of overall $M\epsilon$ collector efficiency of the standard design. Both optical efficiency and thermal losses decrease with increasing S/H, so there is an optimum, which Morrison showed was close to $S/H = 0.042$. Thus this latter value of S/H was used in all the experiments reported in the present study. The present study does, however, cover the same range in Ra, from $Ra = 10^2$ to $Ra = 10^8$, and also the same range in the angle θ (the angle between the normal to the aperture and the gravity vector), namely $0 \leq \theta \leq 90°$. The exception is the horizontal absorber configuration, where the results are limited to $\theta = 45°$.

This new natural convection data was correlated into dimensionless equations, which, in turn, were inputted into the theoretical model for the $M\epsilon$ collector, yielding efficiency plots illustrating the advantages of each configuration, as well as of the improved reflectivity.

RESULTS

The measured Nu vs. Ra curves, exemplified in Fig. 4, were fitted closely by the equation

$$Nu = \left\{ Nu_c^n + [(A_1\ Ra^p)^m + (A_2\ Ra^q)^m]^{n/m} \right\}^{1/n} \tag{1}$$

where A_1, A_2, m, n, p, and q are parameters depending upon θ and upon the configuration. The conduction Nusselt number, Nu_c, was obtained by numerically solving the steady heat diffusion equation and found to be as follows: $Nu_c = 10.95$ for the FEP film configuration, $Nu_c = 12.23$ for the horizontal absorber configuration, and $Nu_c = 8.47$ for the $CR = 1.25$

Table 1. Experimental Values of Parameters in Eqn. 1

Configuration	Angle	A_1	A_2	p	q	m	n
	0°	5.97	0.222	0.071	0.289	8.11	15.88
	30°	7.04	0.319	0.054	0.267	6.71	23.03
FEP Film	45°	8.17	0.228	0.037	0.285	3.97	25.00
	60°	6.76	0.301	0.055	0.267	7.60	23.64
	90°	5.79	0.383	0.067	0.248	16.21	21.71
Horizontal Absorber	45°	0.606	0.136	1/4	1/3	22.8	2.52
	0°	0.975	0.221	1/4	1/3	23.3	3.51
	30°	0.947	0.221	1/4	1/3	22.5	3.54
$CR = 1.25$	45°	0.900	0.216	1/4	1/3	16.0	3.35
	60°	0.857	0.200	1/4	1/3	22.2	3.78
	90°	0.679	0.154	1/4	1/3	21.2	2.89

configuration. Moore and Hollands (1991) used equation (1) with $p = 1/4$ and $q = 1/3$ to correlate their data, but in the present study, use of these values for p and q would not give a close fit. Table 1 gives the best values of the parameters for each configuration at each angle. The rms and maximum deviation between eqn. (1), and the experimental data were typically 1.5 percent and 3 percent respectively, which was well within experimental error. Figure 4 illustrates a typical fit of eqn. (1) to the data.

The main practical result of this study lies in how the convective loss in each configuration compares to that in the standard $M\epsilon$ configuration. This comparison should be made at the Rayleigh numbers actually applying in a $M\epsilon$ collector under typical operating conditions, which range from $Ra = 10^7$ to $Ra = 10^8$. To make the comparison fair, we compare the convective heat transfer coefficients based on unit of aperture area. Assume the absorber length remains constant, this is characterized by the quantity R, defined as

$$R = \frac{(Nu\,H/a)_S - (Nu\,H/a)_x}{(Nu\,H/a)_S} \times 100\% \tag{2}$$

where a is the aperture length and subscript S stands for the standard configuration and x stands for, one of the other configurations. The results are shown in Table 2. The reduction in free convection afforded by each configuration change ranges from about 50% for the FEP film configuration to about 15% for the $CR = 1.25$ configuration.

Figure 5 shows the projected efficiency curves for the three configurations, as well as for the standard configuration. The set of curves on the left is for the reflector that has been used in several prototypes, and which is available directly from an aluminum supplier, without any additional surface treatment. Its specular reflectivity is 0.77 and its total reflectivity is 0.88. If a specially prepared specular silver reflective film (ECP–305 solar reflective film, available from 3M) were laminated onto the base aluminum, the specular and total reflectance could be raised to 0.94. The set of curves on the right in Fig. 5 show the efficiency plots for each configuration, with this improved reflector property.

Simply orienting the absorber in the horizontal position is seen to be the best (least-cost per unit improvement) method of improving the thermal efficiency of the $M\epsilon$ collector, since this change should involve a negligible change in manufacturing costs. The use of a high-reflectance film applied to the inside of the reflector trough would (at a $(T_i - T_a)/G$ value of 0.06 Km^2/W) increase the energy collection efficiency of the horizontal-absorber $M\epsilon$ by another 12 percent.

Table 2. Reduction in Convective Heat Transfer Afforded by the Three Configurations Tested.

Configuration	Angle θ	Per Cent Reduction R at	
		$Ra = 10^7$	$Ra = 10^8$
FEP Film	0°	54	56
	30°	53	54
	45°	52	53
	60°	52	53
	90°	45	46
Horizontal Absorber	45°	28	32
$CR = 1.25$	0°	16.3	19.6
	30°	15.4	13.2
	45°	15.4	13.8
	60°	17.8	15.8
	90°	19.4	15.8

ACKNOWLEDGEMENT

The authors wish to express their gratitude to the Energy, Mines and Resources Canada and the Natural Sciences and Engineering Research Council Canada for their financial support of this work. We also thank Mr. Bert Habicher for his assistance in the laboratory.

REFERENCES

Gordon, J.M., (1986), "Non-Imaging Solar Energy Concentrators (CPC's) with Fully Illuminated Flat Receivers", *ASME J. of Solar Energy Engineering, 108*, pp. 252–256.

Hollands, K.G.T., A.P. Brunger, and J.V. Mikkelsen, (1988a), "Unit Concentration Ratio, Long-Ground Based Solar Collector: Recent Developments", *Proceedings, North Sun '88*, Swedish Council for Building Research, pp. 345–350.

Hollands, K.G.T., A.P. Brunger, and J.V. Mikkelsen, (1988b), "The Mega-Epsilon Collector: A Low-Concentration, Non-Imaging Solar Collector for CIPH in Canada", *Proceedings, 1988 Annual Conference, Solar Energy Society of Canada, (SESCI)*, pp. 150–154.

Hollands, K.G.T., and G.A. Moore, (1991), "Natural Convection Heat Transfer From a Plate in a Semi-Circular Enclosure", submitted, *ASME Journal of Heat Transfer*.

Mikkelsen, J.V., K.G.T. Hollands, and A.P. Brunger, (1988), "Concentration-Transmission-Absorption Product for Non-Imaging Solar Collectors Using Ray Tracing", *Proceedings, 1988 Annual Conference, SESCI*, pp. 155–160.

Moore, G.A., (1989), "Natural Convection from an Isothermal Plate Suspended in a Semi-Circular Enclosure", M.A.Sc. Thesis, Department of Mechanical Engineering, University of Waterloo, Waterloo, Canada.

Moore, G.A., K.G.T. Hollands, and A.P. Brunger, (1989), "Predicting the Thermal Performance of a Unit-Concentration Non-Imaging Solar Collector", *Proceedings 1989 Annual Conference of SESCI*, pp. 230–234.

Morrison, I.D., (1991), "Thermal Improvements of Low Concentration Ratio Non-Imaging Solar Collectors", M.A.Sc. Thesis, Department of Mechanical Engineering, University of Waterloo, Waterloo, Canada.

Olsson, G., B. Thundal, and G. Wilson, (1981), "Advanced Solar Absorber of Metallurgically Bonded Copper and Aluminum", *Proceedings of Solar World Forum*, 1981 ISES Congress, Vol. 1, pp. 163–167.

Wilson, G.L., L. Svensson, and B. Karlsson, (1985), "The Long-Ground-Based Flat Plate Collector", *Intersol '85, Proceedings of 1985 ISES Congress*, Vol. 2, pp. 1097–1101.

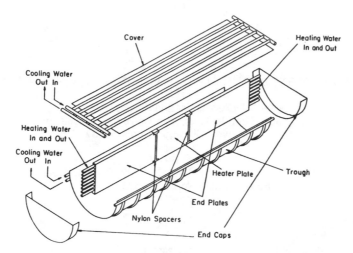

Fig. 3: Sketch of experimental apparatus for evaluating free convection inside the collector.

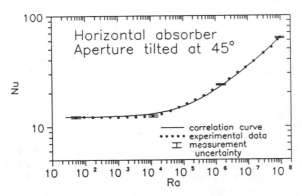

Fig. 4. Typical plot of Nusselt number Nu vs Rayleigh number Ra, as obtained experimentally and as correlated by eqn. (1).

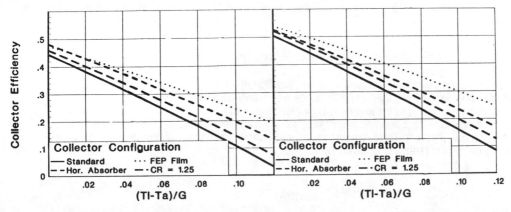

Fig. 5: Efficiency plot of standard $M\epsilon$ collector, and collectors modified according to configurations (i), (ii) and (iii) with low reflector reflectance (left) and high refelectance (right). Units of $(T_i - T_a)/G$ are Km^2/W.

THE INTEGRATED CPC: SOLAR THERMAL ENERGY FOR THE NINETIES

Joseph J. O'Gallagher and Roland Winston
The Enrico Fermi Institute, The University of Chicago
5640 S. Ellis Ave., Chicago, Illinois 60637 USA

William Duff
Solar Energy Applications Lab, Colorado State University
Fort Collins, Colorado 80523 USA

ABSTRACT

The advanced Integrated CPC (ICPC) is the only simple and effective method for delivering solar thermal energy efficiently in the temperature range from 100°C to about 300°C without tracking. It provides an efficient source of solar heat at these temperatures and makes practical and economical several cooling technologies which are otherwise not viable. When driven by these collectors, such advanced cooling technologies as double-effect or multi-stage regenerative absorption cycle chillers or high temperature desiccant systems will have significantly improved performance and can achieve an overall system co-efficient of performance high enough to be economical in a wide variety of applications. Even the best flat plate collectors cannot deliver the required high temperatures and thermal performance. Furthermore, tracking parabolic troughs are not likely to be practical in residential scale systems or effective in many environments where cooling is desired. The ICPC is the only high temperature nontracking option available. In addition to its potential for driving cooling systems, this technology also provides a highly versatile solar source for virtually all thermal end uses including general purpose space and domestic hot water heating as well as industrial process heat. In this paper we review the background and basic principles of these concentrating solar thermal collectors, compare their performance and operating advantages with other advanced collectors and present in preliminary form some recent quantitative results on their energy collection in non-optimum deployment configurations.

KEYWORDS

Compound Parabolic Concentrator, Cooling, Evacuated Tube, Thermal efficiency, Long-term performance

REVIEW OF CONCEPT AND BACKGROUND

More than fifteen years has passed since the invention of what has come to be called the Compound Parabolic Concentrator or CPC (1). CPCs are one family of a whole class of optical devices based on nonimaging design principles (2) which maximize the geometric concentration ratio (collecting aperture to absorber area ratio) for a given angular field of view. In particular, in a 2-dimensional (trough like) geometry, C_{max}, the maximum concentration permitted by physical conservation laws, is given by

$$C_{max} = \frac{1}{\sin(\theta_c)} \qquad (1)$$

where θ_c is the design half-angle of acceptance of the optical system. All previously known conventionally designed concentrators fall short of this limit by at least a factor of 2 to 3. The great advantage of CPC collectors for solar energy collection is that θ_c can be made large enough so that active tracking of the sun can be eliminated while still achieving enough concentration to be useful. A choice of θ_c equal to ± 35° yields a value for C_{max} of 1.7 X and permits collection of direct beam insolation for a minimum of 7 hours per day throughout the year with a completely fixed collector mount. It is this configuration that formed the baseline for the design evolution of the CPC. The first versions, developed at Argonne National Laboratory, used external reflectors coupled to evacuated dewar type absorbers and lead to commercial collectors manufactured by the Energy Design and Sunmaster Corporations. Shortly, later, it became clear that considerable improvement in performance and potential manufacturability could be had by integrating the reflector and absorber into an evacuated tubular module (3). The University of Chicago in collaboration with GTE Laboratories began an intensive program to develop an advanced evacuated solar collector integrating CPC optics into the design. An experimental pre-prototype panel of these "Integrated CPC" tubes achieved the highest operating efficiency at high temperatures ever measured with a non-tracking stationary solar collector (4). Details of the optimum materials and performance parameters are given in a great many previous reports (5). Arrays of such ICPC tubes have the potential to become the general purpose solar thermal collectors of the future.

COMPARISON WITH OTHER THERMAL COLLECTORS

A) Operational Considerations

For other than the highest temperature applications, solar thermal collectors, fall into one of three general categories, flat plates, evacuated tubular collectors and parabolic troughs. In the following discussion, we have emphasized the Integrated CPC since it represents the state-of-the-art evacuated tube. To compare the basic operational features, we show schematically, in Figures 1-3, the collection geometry and deployment configurations for each collector type. As Figure 1 shows, a typical flat plate sees all the direct beam insolation while the sun is in front of the collector plane (subject to cosine effects) and also sees all of the available diffuse insolation. The ICPC, shown in Figure 2, has a full angle of acceptance of 70° and is oriented with its long axis in the E-E direction and deployed so that the plane of its aperture is tilted by an angle equal to the local latitude. In this configuration it will accommodate the full range of solar declination (± 23.5°) as well as the additional 12° motion below and above the noontime values on winter and summer solstice respectively. In addition, if one assumes that the diffuse component is distributed isotropically, the CPC will collect a fraction equal to 1/C (where C is the geometric concentration ratio). Most stationary CPC designs under consideration for an ICPC are truncated to C = 1.5 X or so, so that the diffuse fraction collected is at least 2/3. There is considerable evidence, both from recent more sophisticated insolation models and from preliminary studies of CPC performance as a function of diffuse fraction, that this represents a lower limit on the collectibility of diffuse for a CPC. This is particularly important in the great many climates which have a high percentage of diffuse, such as the northeastern U.S. or many tropical locations. Figure 3 illustrates, the two main disadvantages for parabolic trough for use as a general purpose thermal collector. The effective acceptance angle for a moderate to high concentration (20X- 40X) linear trough is about 0.5° to 1° so that (a)the entire collector must be moved so that the optic axis points continually at the sun and (b) the collection of diffuse radiation is negligible.

B) Thermal Performance

The peak thermal performance calculated for representative versions of these three collector types is presented in Figure 4. The double-glazed Flat Plate has a selective coating to reduce reradiation losses, but its area is too large (it's effective concentration is ≤ unity) and it has high convection and conduction heat losses so its efficiency drops rapidly at temperatures much above that for boiling water. Vacuum greatly reduces the heat losses resulting in much better high temperature performance. Two curves for evacuated tubes are shown to illustrate the effect of even the moderate amount of concentration incorporated into the ICPC. The "evacuated flat plate" is a horizontal fin in an evacuated tube and, although surround by vacuum and selectively coated, its heat loss area (from both sides of the fin) is larger than the illuminated area so that

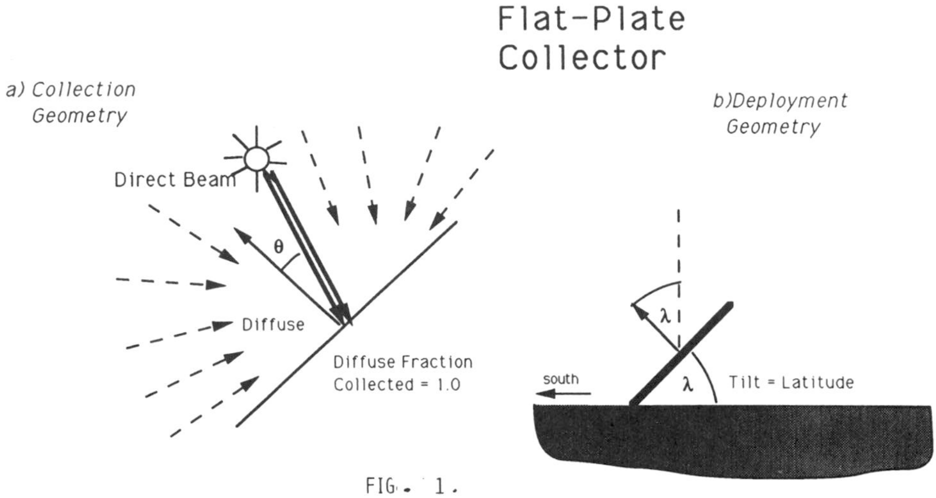

Flat-Plate Collector

a) Collection Geometry

b) Deployment Geometry

Direct Beam

θ

Diffuse

Diffuse Fraction Collected = 1.0

south

λ

λ

Tilt = Latitude

FIG. 1.

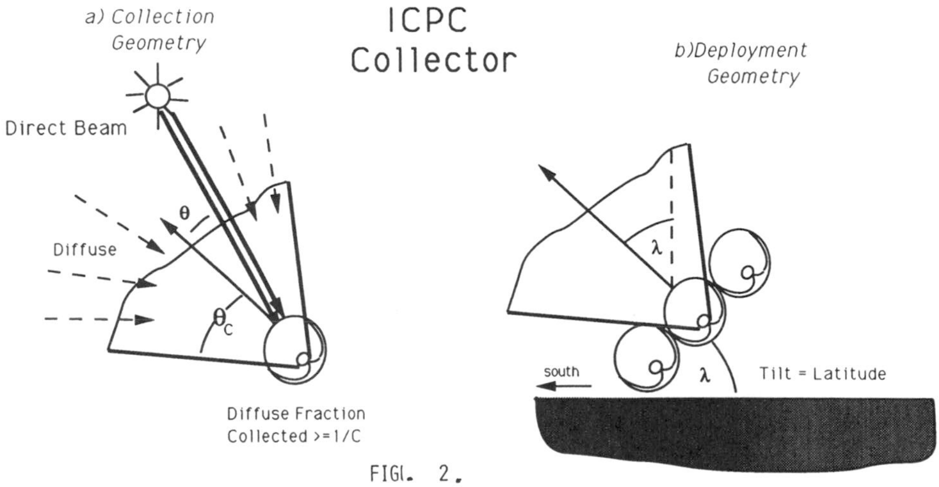

ICPC Collector

a) Collection Geometry

b) Deployment Geometry

Direct Beam

θ

Diffuse

θ_C

Diffuse Fraction Collected >=1/C

λ

south

λ

Tilt = Latitude

FIG. 2.

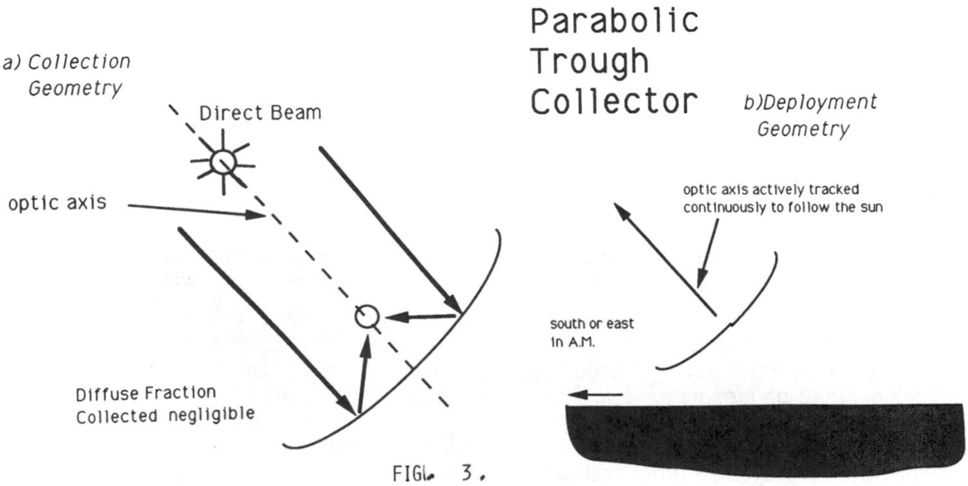

Parabolic Trough Collector

a) Collection Geometry

b) Deployment Geometry

Direct Beam

optic axis

Diffuse Fraction Collected negligible

optic axis actively tracked continuously to follow the sun

south or east in A.M.

FIG. 3.

its effective concentration is also < 1. In contrast, the ICPC has C= 1.5X and the collecting aperture is 1.5 times the <u>circumference</u> of the tube so that the heat losses are almost three times lower and the high temperature efficiency is excellent. Because of the high concentration, the heat losses for the parabolic trough are very low and the efficiency curve is almost flat with respect to increasing temperature. However because of the loss of diffuse, exposed metal reflector surfaces, and the losses in the glazing around the absorber tube the optical efficiency relative to global insolation is somewhat low (these curves are calculated assuming an instantaneous diffuse fraction of 15% of the total). When the effects of both optical efficiency and low heat losses are combined the ICPC emerges as the single collector having the highest peak operating efficiency over the broadest range of temperatures and is suitable not only for cooling and other high temperature applications but for space heating and domestic hot water as well as industrial process heat. Long term performance calculations show that the ICPC is superior under this basis of comparison as well.

CPC DEPLOYMENT FLEXIBILITY

Since the CPC does have a limited angular acceptance, it should be mounted as close as possible to the nominal design configuration shown in Figure 2. However considerable flexibility is possible. During the last five years, in cooperation with the group at Colorado State University much of the research effort has been directed towards performance analysis and systems studies. Computer modeling has shown quantitatively that the effects of non-standard deployment orientations on the long term projected collected energy of CPCs may be quite acceptable over a relatively large range of directions. For example, in Figure 5 we show the energy collected by a $\pm 35^{\circ}$ CPC in Phoenix, Arizona as calculated from hour by hour simulations based on Typical Meteorological Year data. The results are presented as a function of misalignment angle for three different orientations of the misalignment axis; i) perpendicular to the collector aperture, ii) parallel to the ground and pointing due south, and iii) perpendicular to the ground (i.e. vertical). Data for cases i) and ii) is shown for all angles out to $\pm 90^{\circ}$, while case iii) is plotted only for $\pm 20^{\circ}$ in the region where all three curves are indistinguishable. The important conclusions are that for deviations of $\pm 10^{\circ}$ the drop-off compared with ideal orientation is less than 3% , at $\pm 20^{\circ}$, it is about 5% and it remains less than 10% out to $\pm 30^{\circ}$. Similar behavior is found at other locations such as Boston and Miami so that in general it can be concluded that while they are more sensitive to alignment that flat plate, CPCs collectors retain a considerable flexibility in deployment.

PRESENT STATUS

More than ten years has elapsed since the University of Chicago, in collaboration with GTE Laboratories, began an intensive program to develop an advanced evacuated solar collector integrating the CPC optics into the design. An experimental pre-prototype panel of these "Integrated CPC" tubes achieved the highest operating efficiency at high temperatures ever measured with a non-tracking stationary solar collector (4). For example, an efficiency of 44% was measured at an average collector temperature of 270° C. However, since that time, market conditions and remaining technical barriers to high volume manufacturing have slowed progress and delayed the transfer of this very promising technology to industry and the commercial sector in the US. The long-term goal of all of our work has been the development of a high efficiency, low cost solar collector to supply solar thermal energy at temperatures well above 200°C. Our studies have indicated that a mass-producible collector, incorporating the same basic concepts as our prototype, can be developed which will deliver excellent performance across a broad range of temperatures, extending from about 50°C to above 250°C. This wide temperature interval makes it possible to use this one collector type for virtually all thermal energy needs for residential and commercial end uses. For example, domestic hot water, as well as space heating and cooling, industrial process steam and even power generation can be supplied from one basic collector design.

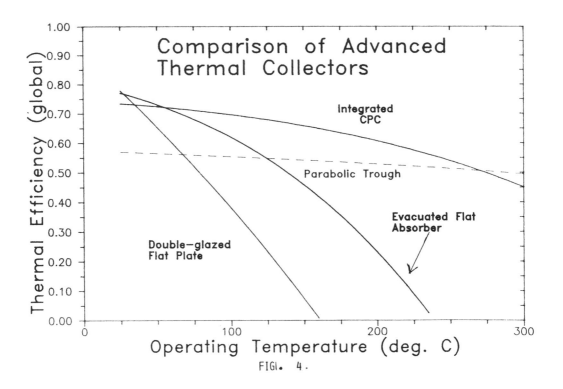

Comparison of Advanced Thermal Collectors

Integrated CPC

Parabolic Trough

Evacuated Flat Absorber

Double-glazed Flat Plate

Thermal Efficiency (global)

Operating Temperature (deg. C)

FIG. 4.

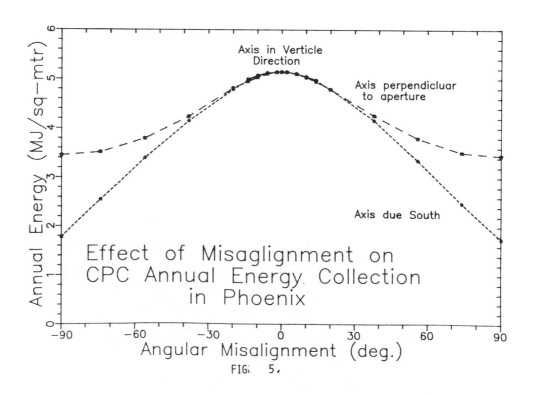

Axis in Verticle Direction

Axis perpendicluar to aperture

Axis due South

Effect of Misaglignment on CPC Annual Energy Collection in Phoenix

Annual Energy (MJ/sq-mtr)

Angular Misalignment (deg.)

FIG. 5.

Based on both performance and operational considerations it is clear that the ICPC is the optimum collector. Some remaining technical barriers to inexpensive high volume production as well as volatile market conditions over the past decade have prevented the commercial development and deployment of these collectors until now. However recent world events and changing expectations for the future of conventional energy sources make it imperative that the progress already made on this unique, highly adaptable solar thermal collector be continued. In the meantime, extensive optical and thermal design studies have provided the basis for a sound understanding of collector performance parameters, and recent research on varieties of Integrated CPC designs suggests strongly that these types of collectors can be fabricated and deployed inexpensively and in high volume at a cost equal to or less than that for current flat plate collectors. It should be noted that, in Japan, early experiments with this concept have been followed with larger scale field tests and studies of high volume production designs and techniques (6).

The current mid-east crises has refocused world attention on the fragility of conventional energy sources. Conventional energy supplies are expected to become limited again in the 1990's. There remains an urgent, long-term need for simple, general purpose solar-thermal energy source operating at moderate to high temperatures. The present time offers a unique "window of opportunity" to embark on a renewed effort to overcome the remaining barriers to production.

ACKNOWLEDGEMENT

Special thanks to Ken Mellendorf for carrying out the misalignment sensitivity analysis. This work was supported in part by the U.S. Department of Energy under grant DE FG03-85SF-15753.

REFERENCES

1) R. Winston, "Principles of solar concentrators of a novel design," Solar Energy, 16, 89, 1974.

2) W. T. Welford and R. Winston, The Optics of Nonimaging Concentrators, Academic Press, New York (1978).

3) J. D. Garrison, "Optimization of a fixed solar thermal collector," Solar Energy, 23, 93, (1979).

4) K.A. Snail, J. J. O'Gallagher and R. Winston, "A stationary evacuated collector with integrated concentrator," Solar Energy, 33, 441, (1984).

5) J. O'Gallagher, R. Winston, W. Schertz, and A. Bellows, "Systems and applications development for integrated evacuated CPC collectors," Proc. of the 1988 Annual Meeting of the American Solar Energy Society, Cambridge, Mass., 469, 1988 (and references therein).

6) Japan's Sunshine Project: 1987 Annual Summary of Solar Energy R&D Program: Solar Industrial Process Heat Projects of NEDO, (1988).

EVALUATION OF HEAT COLLECTING PERFORMANCE FOR
A STATIONARY MID-TEMPERATURE CPC SOLAR COLLECTOR

T. Kotajima, H. Konno, A. Suzuki, S. Yamashiro, and T. Fujii

Koto Electric Co.,Ltd., Aoki, Kawaguchi, Saitama, 332, Japan

ABSTRACT

Evacuated tubular collectors of CPC(Compound Parabolic Concentrator) type were designed and fabricated for mid-temperature industrial thermal applications. Collector elements of 100mm in diameter were fabricated in two different lengths, 3m and 5m. Heat-collecting tests were conducted to evaluate the solar thermal efficiency in the outdoor exposure in Kumamoto, Japan. In the experiments, the inside temperature of the thermal absorbing tube without a heat-transfer fluid readily rose to over 380°C. Thermal collecting efficiency of 0.55 has been achieved in water at 90°C on sunny days. Further, effective absorber surface emissivity and linear heat loss coefficient were estimated as e=0.04 and u$_p$=0.97, respectively.

KEYWORDS

Solar thermal, CPC, ISEC collector, Evacuated solar collectors, Mid-temperature collection,

INTRODUCTION

At the previous congress held in Kobe, Japan in 1989, we presented a paper on the design of our integrated solar collectors with their optical and mechanical properties(Yamashiro and others, 1989). We proposed two types of CPC shape in the paper; one had optically superior shape but mechanically weak shape and the other had mechanically strong shape and optically a good property. The latter type was assembled into collector panels for evaluating heat collecting proper-ty. Table 1 shows the cross-section of the latter type which is strong against outside pressure and ideal for evacuating inside of the glass tube.

At first 3m long collector elements were fabricated. Longer collectors de-crease heat loss and increase optical efficiency. After further improvement on the production technology, we were able to fabricate 5m long collector elements successfully. 5m long collector elements were also assembled into panels and heat collecting tests were carried out on both 3m and 5m long col-lectors with identical cross-section.

In this paper, we present the results of heat collecting tests on the above two types of integrated CPC collector panels. In conclusion, the 5m long collector exhibited very high optical performance and very small heat-loss coefficient.

Table 1 Cross-Sectional Geometry of Present Integrated CPC Collector

----
Half-acceptance angle	30°
Diameter of absorber pipe	20 mm
Aperture width*	108 mm
Height*	118 mm
Geometrical concentration	1.59
Gap interval**	2.5 mm
----

*   involving glass tube thickness
** distance between an absorber pipe and the
   bottom of CPC shaped glass tube

DESIGN OF COLLECTOR PANELS

Two types of our collector panels which are assembled with 3m long and 5m long
collector are shown in Figs.1 and 2, respectively. A panel consists of four
collector elements. A small gap between the elements is 5mm, so that the width
of transparent part between the elements is about 13mm. This small gap makes
vain ray losses penetrating through the panel low in comparison with the other
evacuated type collector panels.

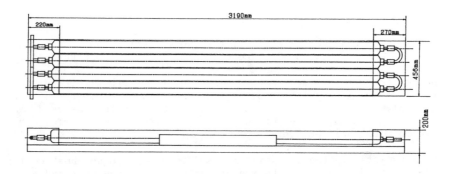

Fig. 1. Collector panel with four 3 m elements

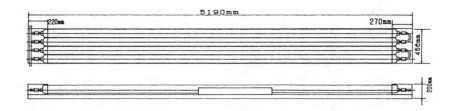

Fig. 2. Collector panel with four 5 m elements

TABLE 2 Heat Collecting Performance and Dimensions of Present
         Integrated CPC-Collectors

	Total length (m)	Area (m^2)	Effective area (m^2)	Optical efficiency (-)	Linear heat-loss coefficient (W/Km^2)
#1	3.19	1.45	1.08	0.61	1.33
#2	5.19	2.36	1.87	0.66	0.89

In Table 2, dimensions and parameters of heat-collecting performance on the two collector panels are shown. Values of the optical efficiency and the heat-loss coefficient are the average measured on some of the panels.

A long collector panel is expected to have two favorable features for heat-collection; one is the small heat loss per panel area and the other is better optical efficiency. Both effects are realized by the small fraction of the manifold area relative to the overall panel area.

As for the flow path of the heat transfer medium in the collector panel, two tubes are connected so that the medium flows in parallel.

EXPERIMENTS AND DISCUSSIONS

Heat Collecting Experiments

Heat collecting experiments of the two collector panels were carried out in Kumamoto, Japan(Latitude=33°, Longitude=131°) throughout the year, 1990.

The collector panels were set to face the true south direction with the inclination of 35° from the ground; at the noon on equinoxes, the direct solar rays vertically strike the collector panels. Since water was used as the heat-collecting medium, collector output temperature was controlled not to be higher than 95°C to avoid boiling.

The insolation intensity which is used to calculate the heat-collecting efficiency was derived from the measurements of a pyrgeometer and a pyrheliometer mounted on the same plane of the collector panel.

Figures 3 and 4 show typical efficiency curves on the collector panels with 3m and 5m long elements, respectively. The axis of ordinate represents heat-collecting efficiency, and the unit of abscissa is a parameter obtained by dividing temperature difference between the mean heat-collecting medium and the ambient temperatures by the solar insolation intensity.

The inclination of the efficiency lines represent the degree of heat loss from the absorber. In the figures, gradients of two lines are nearly the same though the value for 5m long collector is a little smaller than that 3m long one. This result leads us to the conclusion that both collectors have equivalent properties on thermal insulation.

On the other hand, there is an apparent difference in the optical efficiency which is the other important parameter for evaluating heat-collecting capability of a solar collector; the 5m long collector is optically superior to the 3m long one in performance.

This leads us to the conclusion that the advantage of a longer collector is in increasing the optical efficiency rather than decreasing the heat-loss per panel area.

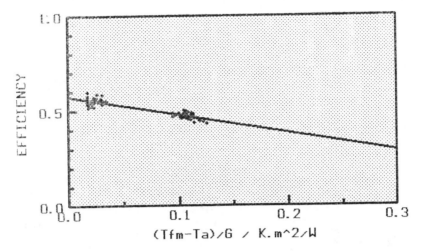

Fig. 3. Efficiency curve of a 3m long collector panel

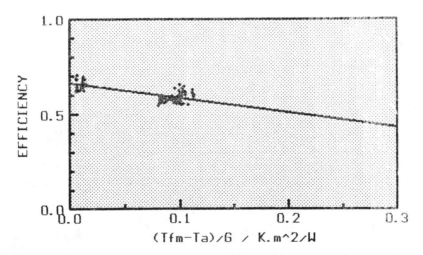

Fig. 4. Efficiency curve of a 5m long collector panel

Stagnant Experiments

Stagnation temperature measurements on collector panels were made to get the heat insulation properties. A C-A thermocouple was inserted in the center of an absorber tube and its output was monitored. Strictly speaking, the obtained values were not exact stagnation temperature of the absorber surface, but a good approximation was realized.

Figure 5 shows an output of the stagnant experiments. Since the measurement took place after noon, the insolation intensity decreased with time. At around 830W/m^2 of insolation, the stagnation temperature approached to 400°C so the experiment was stopped to avoid discharge gasses of the metal component of the collector.

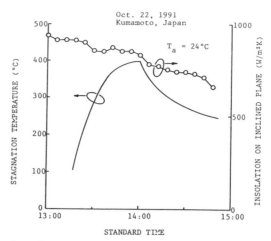

Fig. 5. An example of a stagnation temperature change with time. Experiment was stopped not to exceed the temperature above 400°C.

Figure 6 shows the stagnation temperature under different insolation conditions. Each point in the figure represents the temperature at steady condition.

Here, a heat balance model for expressing heat-collecting process is considered as follows(Winston and Hinterberger, 1975):

$$q = E_{opt}G - es(T_p^4 - T_a^4) - u_p(T_p - T_a) \tag{1}$$

where

e (-)	:	effective emissivity on absorber surface,
E_{opt} (-)	:	optical efficiency,
G (W/m^2)	:	insolation intensity on panel surface,
q (W/m^2)	:	heat gain per unit panel area,
s (W/K^{4}m^2)	:	Stefan-Boltzmann constant = 5.67×10^{-8},
T_a (K)	:	ambient temperature,
T_p (K)	:	temperature of absorber surface, and
u_p (K)	:	linear heat-loss coefficient per unit panel area.

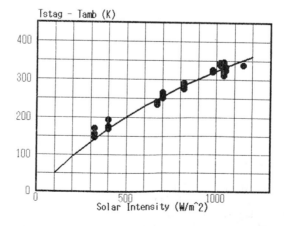

Fig. 6. Insolation dependence of stagnation temperatures

When q=0, equation (1) expresses stagnant condition, we can obtain e and u_p from Fig. 6 by use of the least squares method; results are e=0.04 and u_p=0.97. The solid line in the figure was drawn with these values. By considering the light concentration ratio of our CPC geometry, the emissivity of the absorbing surface is evaluated as 0.06. This value probes that our selective surface coating is at an excellent level. This result is also supported by a direct measurement value of 0.08 which was obtained by a certain organization in Japan.

CONCLUSION REMARKS

This project,launched with the aims of developing a high performance stationary solar collector in three years, has most programs completed so far. At the present stage, our 5m long panels have recorded above 55% of heat collecting efficiency at 90° operating temperature.

From the fact of very high stagnation temperature, our collector panels would be suitable for gaining higher temperatures than $90^{\circ}C$. According to Suzuki(1988), one of the authors of the paper, the optimum operating temperature for a solar collector with a stagnation temperature $T_{st}(K)$ can be approximated as follows:

$$T_c = (T_{st}T_a)^{1/2} \tag{2}$$

where T_c is the temperature of absorber surface. In Fig. 5, if the solar panel exhibited $400^{\circ}C$ of the stagnation temperature at $800W/m^2$, equation (2) results in T_c=$173^{\circ}C$ assuming T_a=$23^{\circ}C$.

From the view point of this discussion, our solar panel could be used for gaining process heat, heat source for absorption refrigerator, and further, as the heat source for electric generation.

Generally, solar thermal-electric generation plants have large load to operate fluid pumps. If 5m long collector elements were employed, the system may have small load for fluid pump as well as high heat insulation properties. Since collecting temperature is not so high, high generation efficiency may not be expected. However, the cost-effective system can be designed by using our long stationary collector if operating load within the system is small.

ACKNOWLEDGMENTS

This work is supported in part by New Energy and Industrial Technology Development Organization, Japan.

REFERENCES

Suzuki, A. (1988). A fundamental equation for exergy balance on solar collectors. Trans. ASME J. Solar Energy Eng., 110, 102-106.
Winston, R. and H. Hinterberger (1975). Princeples of Cylindrical Concentrators for Solar Energy. Solar Energy, 17, 255-258.
Yamashiro, S., H. Konno, T. Kotajima, and A. Suzuki (1989). Development of a stationary mid-temperature solar-collector with CPC-reflector. Proc. Int. Solar Energy Congress in Kobe, vol.2, Pergamon Press. pp.1443-1447.

A UNIFIED MODEL FOR OPTICS AND HEAT TRANSFER IN
LINE-AXIS CONCENTRATING SOLAR ENERGY COLLECTORS

P.C. Eames and B. Norton
PROBE
centre for Performance Research On the Built Environment
Department of Building and Environmental Engineering
University of Ulster, Newtownabbey BT37 0QB
Northern Ireland

ABSTRACT

A two-dimensional steady-state computer-based numerical model was developed to simulate both the optical and thermal behaviour of line-axis concentrating solar energy collectors. A Mach-Zehnder interferometric study was performed to validate the model. Realistic experimental boundary conditions for the solar energy input were obtained by the use of a novel solar simulator. Using the validated model, parametric analyses were performed.

KEY WORDS

Line-axis concentrating, Mach-Zehnder, Finite element, Free convection

INTRODUCTION

The line-axis solar energy collector may validly be considered two-dimensional. To model the convective fluid flow a primitive variable formulation, using 8-node elements, quadratic in velocity and temperature and linear in pressure (Gartling, 1977) was used.
The simplified Navier-Stokes equations for incompressible flow in two-dimensions are, for:
Continuity

$$\frac{\partial u}{\partial x} + \frac{\partial v}{\partial y} = 0 \tag{1}$$

Conservation of momentum

$$u\frac{\partial u}{\partial x} + v\frac{\partial u}{\partial y} = \frac{-1}{\rho_o}\frac{\partial p}{\partial x} + \gamma\frac{\partial^2 u}{\partial x^2} + \frac{\partial^2 u}{\partial y^2} \tag{2}$$

$$u\frac{\partial v}{\partial x} + v\frac{\partial v}{\partial y} = \frac{-1}{\rho_o}\frac{\partial p}{\partial y} + \gamma\frac{\partial^2 v}{\partial x^2} + \frac{\partial^2 v}{\partial y^2} + g\beta(T-T_o) \tag{3}$$

Conservation of energy

$$u\frac{\partial T}{\partial x} + v\frac{\partial T}{\partial y} = \alpha\frac{\partial^2 T}{\partial x^2} + \frac{\partial^2 T}{\partial y^2} \tag{4}$$

The Boussinesq approximation was assumed, thus variations in the density were only incorporated in bouyancy forces.

THE EMPLOYED BOUNDARY CONDITIONS

No-slip boundary conditions were applied for the fluid, i.e., u,v=0 at the surface of absorber, reflector and aperture cover. Energy inputs, losses and extractions from, and thermal conduction in, the reflector boundary material are shown in Fig. 1. The energy fluxes were incorporated in the boundary conditions for the energy equation. The net influx or outflux was calculated as the sum of the energy input and the convective loss per unit area i.e.

$$q_{net} = q_{ins} - h(T-T_{env})$$ (5)

The energy input was found by performing a ray trace and evaluating the incident energy distributions at the collector surfaces (Eames and Norton 1991). From this the energy flux at element boundaries was calculated.

The incorporation of conduction in the boundaries was important in that it, for the absorber and reflector, will reduce any propensity to localised over-heating. Conduction in the reflectors will result in a decrease in the convective flow within the system.

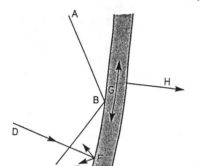

A – Incident Solar Radiation – Specular
B – Absorbed Solar Radiation
C – Reflected Solar Radiation – Specular
D – Incident Long Wave Radiation – Diffuse
E – Reflected Long Wave Radiation – Diffuse
F – Absorbed Long Wave Radiation
G – Conduction in Reflector
H – Back Loss Coefficient

Fig 1. The energy fluxes at the reflector

Conduction into the material of the reflector and absorber, may be neglected due to the high conductivities of the thin materials employed generally (i.e. Copper and aluminium, less than 1 millimetre and several microns thick respectively). One-dimensional behaviour was thus incorporated by the superimposition of one-dimensional quadratic elemental conduction matrices on to the global energy equations.

Basic Mesh Specification

In producing a finite-element idealisation of the compound parabolic concentrating collector's geometry, an example of which is shown in Fig. 2, the major considerations were that,
(i)   there were sufficient elements of the correct dimensions to model the behaviour realistically,
(ii)   the equation bandwidth was kept down to a reasonable size, and
(iii) excessive element distortion was avoided.

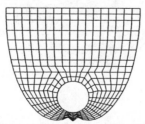

Fig. 2. An example of a finite element mesh employed

Since the momentum equations are non-linear, they cannot be solved directly by a method such as Gaussian elimination. However by linearising the system using the Newton-Raphson technique, the solution can be found. Rall(1969) presents the theory of the Newton-Raphson method, together with its derivation and requirements for convergence. A variation of the frontal solution method was then used to solve the linearised system of equations.

VALIDATION

The Mach-Zehnder interferometric technique was used to determine the temperature distributions, at a selection of inclinations, that existed within the cavities of several different CPC solar energy collectors when exposed to simulated insolation.

Parameters that were to be specified as inputs into the model comprised,
- (i)   insolation input - intensity and angular distribution,
- (ii) ambient environmental temperatures,
- (iii) heat transfer coefficients from _external_ surfaces of the aperture cover and the rear of the reflector,
- (iv) internal absorber heat transfer rate and the temperature of the heat sink to which energy is lost, and
- (v)   thermal conductivity and thickness of the boundary materials (i.e. aperture cover, reflector and absorber).

A Moll-Gorczynski pyranometer was used to measure the intensity of the simulated insolation, incident at the aperture cover of the CPC solar energy collector laboratory test section. The ambient environmental temperature was measured using thermocouples. Estimates of heat transfer coefficients were made based on the known thickness of insulation and the speed of the forced cooling (simulating wind), over the aperture cover.

A Sample Validation.

In the validation both the predicted isotherm and experimentally obtained interference fringe spacing denotes a temperature difference of 1.1K.
From Fig. 3 good agreement can be seen between the experimentally obtained interferogram of a 60° acceptance half-angle CPC inclined at 20° to the horizontal, and the corresponding simulation. This has been also shown for a representative range of CPC geometries and inclinations.

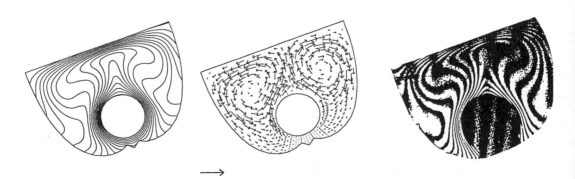

Fig. 3. Experimental and predicted isotherms for a 60° acceptance
half-angle CPC collector inclined at 20°

PARAMETRIC ANALYSES

Two sets of boundary conditions were considered, one being an "ideal" collector possessing near perfect reflectance at the mirror, no conduction in the mirror material and a thickness of backing insulation (so that the heat loss in this direction was effectively zero), the other having mirrors whose reflectance was 0.95, a thickness typical of vacuum deposited aluminium (i.e., 3 microns), and a back loss of 0.5W/m²/K.

A representative sample of isotherm plots and velocity vector diagrams demonstrate the way in which the thermal behaviour in a CPC cavity differs with angular inclination and CPC acceptance angle.

The two sets of boundary conditions for the scenarios investigated are given in table 1.

TABLE 1 The boundary conditions and boundary material
properties used in scenario 1 and scenario 2.

	Boundary condition or material description	SCENARIO 1	SCENARIO 2
ABSORBER	Heat extraction	1 W/m²/K	1 W/m²/K
	Material	Copper	Copper
	Thermal conductivity	385 W/m/K	385 W/m/K
	Thickness	0.5mm	0.5mm
	Absorptance	1	1
REFLECTOR	Heat loss	0.5 W/m²/K	0
	Material	Aluminised polyester film	Aluminised polyester film
	Thermal Conductivity	238 W/m/K	0
	Thickness	3 microns	3 microns
	Reflectance	0.95	0.95
	Absorptance	0.05	0.05
APERTURE COVER	Heat loss	5 W/m²/K	5 W/m²/K
	Material	Low iron glass	Low iron glass
	Thermal conductivity	1.05 W/m/K	1.05 W/m/K
	Thickness	3mm	3mm
	Extinction coefficient	0.03512/mm	0.03512/mm

The properties of air used in the simulations were those at 293 K and atmospheric pressure. The energy inputs at the cover and absorber become more important when the heat extraction from the absorber is large. However, the general thermal behaviour will be similar if the heat extraction at the absorber is not larger than the energy input.

A Sample of Thermal Plots and Velocity Vector Diagrams.

The contours on the thermal plots are at one degree intervals, and the velocities can be scaled against the arrow to the left of the collector profile which represents a velocity of 10cm/s, the absorber diameter is 15 mm.

The boundary conditions employed are those for scenario 1 in table 1. It can be seen in Fig. 4. that the flow is bicellular for the inclinations of 0° and 30°, and unicellular for an inclination of 60°.

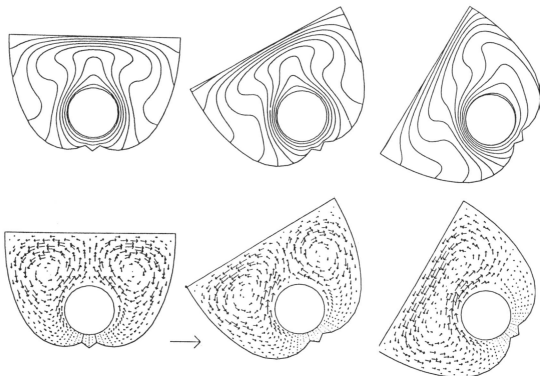

Fig. 4 The calculated isotherms and velocity vector diagram
for the 60° acceptance half-angle CPC inclined at 0°, 30° and 60°
with the boundary conditions for scenario 1

The calculated isotherms and velocity vector diagram for the 60° acceptance
half-angle CPC inclined at 30° with the boundary conditions for scenario 2, as
given in table 1, are shown in Fig 5. It can be seen that the flow is
unicellular.

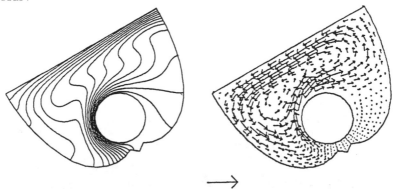

Fig. 5. The velocity vector diagram and isothermal
plot for a 60° acceptance angle collector inclined
at 30° with the boundary conditions of scenario 2.

Fig 6. shows the isothermal and velocity plots for 45° and 30° acceptance half-
angle collectors inclined at 30°, the flow can be seen to be unicellular.

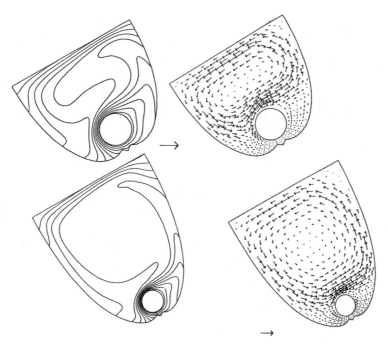

Fig 6. The isothermal and velocity plots for 45° and 30° acceptance half-angle collectors inclined at 30°, with boundary conditions of scenario 1.

From this small sample of velocity vector diagrams and isothermal plots it can be seen that the free convection that occurs in CPC solar energy collectors is dependent on cavity aspect ratio, angle of inclination and the boundary conditions imposed

ACKNOWLEDGEMENT

This work was supported by the Science and Engineering Research Council (Swindon, UK.).

REFERENCES

Eames P.C., and B. Norton (1991). The effect of sky conditions on the partition of incident solar energy between the components of a CPC solar energy collector, Proceedings of ISES World Solar Congress, Denver USA.
Gartling D.K. (1977). Convective Heat Transfer Analysis by the Finite Element Method. Computer Methods in Applied Mechanics and Engineering 12, 365-382.
Rall L.B., (1969). Computational Solution of Non-Linear Operator Equations, John Wiley and Sons Inc. New York.

NOMENCLATURE

h	Convective heat transfer coefficient	[W/m^2/K]
P	Pressure	[N/m^2]
q_{ins}	Insolation input	[W/m^2]
q_{net}	Net energy flux	[W/m^2]
T	Temperature	[K]
T_{env}	Environmental temperature	[K]
u,v	Components of velocity	[m/s]
x,y	Coordinate directions	
α	Thermal diffusivity	[m^2/s]
β	Coefficient of cubical expansivity	[K^{-1}]
γ	Kinematic viscosity	[m^2/s]
ρ_o	Reference density	[Kg/m^3]

THE EFFECT OF SKY CONDITIONS ON THE PARTITION OF INCIDENT SOLAR
ENERGY BETWEEN THE COMPONENTS OF A CPC SOLAR ENERGY COLLECTOR

P.C. Eames and B. Norton
PROBE
centre for Performance Research On the Built Environment
Department of Building and Environmental Engineering
University of Ulster, Newtownabbey BT37 0QB
Northern Ireland

ABSTRACT

Using ray trace techniques accurate predictions of local areas of high incident
flux, and overall optical performance can be determined with respect to the
nature of the insolation, the geometry of the system, and the properties of the
materials from which the collector is fabricated. The results are reported of
such an analysis performed for several CPC solar energy collectors with
(i)   different acceptance half-angles,
(ii)   reflector optical properties,
(iii)   direct insolation off angle effects,
(iv)   reflector truncation and
(v)   three different diffuse angular skyward insolation distributions.

KEYWORDS

Energy Flux; CPC; Truncation; angular diffuse distributions; Ray trace;

INTRODUCTION

The factors affecting the optical performance of solar energy collectors are the
incident insolation, the geometry of the system and its material properties.
The absolute and relative magnitudes of the diffuse and direct insolation
components, and the nature of the distribution of the diffuse (e.g. isotropic,
gaussian or a hybrid)(Prapas and others, 1987), are influenced by the clearness
indicies and the global location of the test site. Alternative skyward angular
distributions of the diffuse component of insolation are shown in Fig. 1 (Prapas
and others, 1987).

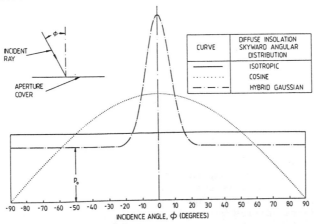

Fig. 1. Skyward angular distributions of
the diffuse component of insolation employed.

The reflector-absorber unit considered, an example of which is shown in Fig. 2, was a compound parabolic concentrating system with a tubular absorber. The optical and radiative characteristics of materials used for collector fabrication, i.e. transmittance, reflectance and absorptance, are assumed to be constant across the complete spectrum of interest.

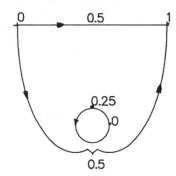

Fig. 2. The loci around the absorber, reflector and cover used for
the energy inputs to a CPC solar energy collector

RAY TRACING : THE BASIC PREMISE.

In ray tracing, all reflections are assumed to be specular, so that the laws of reflection apply, (i.e. the angle of incidence equals the angle of reflection, and that the normal to the plane, the incident ray and the reflected ray are co-planer). Example ray traces are shown in Fig. 3. In order to utilise a ray tracing technique for a line-axis solar energy collector, the entrance aperture was assumed to be of infinite length, end effects could then be ignored, additionally the energy incident could then be considered to be distributed solely in one plane.

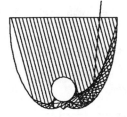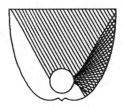

Fig. 3. Example ray trace diagrams for a 45° acceptance half-angle CPC
collector, rays incident at angles of 90°, 75° and 60° to the aperture cover.

THE ENERGY ABSORBED WITHIN THE SYSTEM

Absorption in the Glass Cover

The Fresnel equations govern the reflection of light at the glass cover (Rabl 1985). When a ray passes through the aperture cover, either on entry or exit, some energy is absorbed in the glass. Absorption can be a wavelength determinant characteristic. An approximation to the energy absorbed is given by,

$$E_{gc} = \sum_{i=1}^{n} W_i I_r \exp^{-\alpha_i x} \tag{1}$$

When a ray strikes the glass perpendicularly to its surface, the path length is the thickness of the glass. If the incident ray is not perpendicular, then the path length is given by L/sin θ, where θ is the angle between the horizontal and the path of the refracted ray in a clockwise direction. For simplicity n was taken to be unity implying uniform absorption across the spectrum. The total amount of energy absorbed in each glass segment was obtained by the summation of the energies absorbed from all the rays that entered each segment, (i.e., diffuse and direct). This was only undertaken for one segment, since the incident insolation was uniform. Though a ray path might go through several segments, this may be ignored as due to the uniform radiation field, the amount lost to one segment would be gained from another. The amount of energy absorbed from a ray passing through the glass into the collector was subtracted from the total energy of the ray.

For rays exiting from the collector, the energy absorbed in the glass could be calculated by equation 1, however, since the radiation field was not uniform, the point of inlet and outlet from the glass was calculated and the energy absorbed divided equally into each element of the cover that the ray passed through.

Absorption at the Reflector

In order to find the energy distribution around the surface of the reflector it was divided into a series of regions. Each intersection of a ray with the reflector was contained in a region. The value for the total energy incident in a region was the sum of the individual amounts absorbed from each ray that was reflected in the region. The amount absorbed from each ray was

$$E_{re} = E_{in}\ \alpha_r \qquad (2)$$

This value was subtracted from E_{in} to give the amount of energy remaining in the ray. This was undertaken for all rays that intercept the reflector, including those which undergo multiple reflections.

Absorption at the Absorber

The energy reaching the absorber was all assumed to be absorbed there. The absorber was divided into segments, since the point of intersection of the ray with the absorber was known, the relevant segment was determined and the energy level for this segment could then be increased.

Calculation of the Energy Fluxes

The energy flux due to the direct insolation was found by tracing the paths of 1000 rays spaced equally across the aperture cover, incident at the required angle through the collector, calculating the energy absorbed at each reflection. In order to model the diffuse insolation the two dimensional semi-circular sky is divided into 180 equal angular divisons, 1000 rays spaced equally across the aperture cover were then traced for each angular divison. The normalised angular distribution function was used to weight the energy values assigned to the rays.

THE EFFECT OF COLLECTOR ALIGNMENT TO THE SUN ON THE ENERGY INPUTS TO THE CPC COLLECTOR SYSTEM

The loci used to describe the energy inputs around the absorber and reflector and across the aperture are shown in Fig. 2. The energy distribution at the aperture cover along the reflector and around the absorber are shown in Fig. 4. when the direct insolation angle of incidence is 90° 75° and 60°. The most important feature is the increase in the magnitude of the peak energy density. When the collector is misaligned by 30°, the largest peak of energy density is greater than 27 times that of the incident energy (i.e., 2700Wm⁻² for a 100Wm⁻² input). The average concentration ratio is 1.414 for a 45° acceptance angle collector. The set of corresponding ray trace diagrams is shown in Fig. 3.

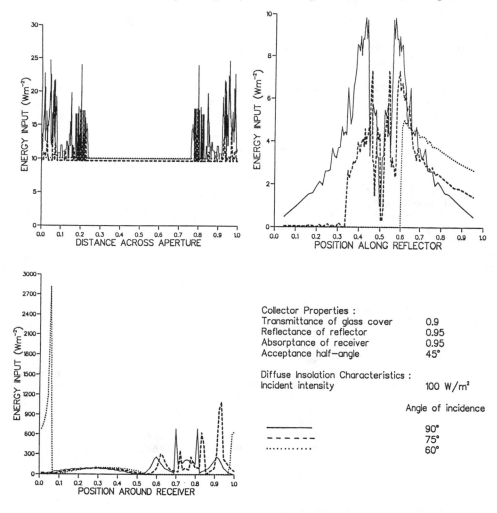

Fig. 4. The energy distributions across the aperture cover, along the reflector and around the absorber of a 45° acceptance half angle CPC, for incidence angles of 90°, 75° and 60°.

THE EFFECT OF THE SKYWARD ANGULAR DIFFUSE ENERGY DISTRIBUTION ON THE ENERGY INPUTS TO THE CPC SOLAR ENERGY COLLECTOR

Three skyward angular diffuse energy distributions; isotropic, cosine and hybrid gaussian (Prapas and others, 1987) were used as test cases to investigate the effects that different assumed distributions have on the energy flux at the compound parabolic concentrating collector's cover, reflector and absorber. The shape of these three distributions is shown in Fig.1. A 45° acceptance half-angle CPC collector with a 15mm absorber, with truncation of the reflector in the cusp region as shown in Fig. 2, was used initially to investigate these effects.
The total energy inputs into the aperture cover, reflector and absorber for these diffuse angular skyward distributions are given in Table 1.

TABLE 1   The variation in the partition of diffuse insolation energy inputs between components of a 45° acceptance half-angle CPC solar energy collector with assumed diffuse distribution.

Diffuse Distribution	ENERGY ABSORBED AT			Total incident energy	Average number of reflections
	Aperture W	Reflector W	Absorber W	W	
Isotropic	0.857	0.425	2.673	6.664	0.97
Cosine	0.829	0.394	3.750	6.664	0.97
Hybrid gaussian	0.830	0.428	3.063	6.664	0.97

THE EFFECT OF MIRROR REFLECTIONS ON THE ENERGY INPUT INTO THE COLLECTOR SYSTEM

For a perfect mirror (i.e., with $\rho = 1$), all incident energy is reflected and none is absorbed. With decreasing reflectance energy is increasingly absorbed by the reflector. For n multiple reflections, the energy input at the reflector is a function of $(1-\rho)^n$, thus the increase in energy absorbed as the reflectance decreases is non-linear. A summary of the total energy inputs to the aperture cover, reflector, and absorber for varying reflector reflectance are given in Table 2. The absorber diameter was 15mm and a hybrid gaussian diffuse distribution was used.

TABLE 2 The variation in the partition of diffuse insolation energy inputs between components of a 45° acceptance half-angle CPC solar energy collector with assumed reflectance of the reflector.

Reflectance of reflector	ENERGY ABSORBED AT			Total incident energy.	Average number of reflections
	Aperture W	Reflector W	Receiver W	W	
1.00	0.861	0.000	3.229	6.664	0.97
0.95	0.830	0.428	3.063	6.664	0.97
0.90	0.803	0.823	2.903	6.664	0.97
0.85	0.778	1.189	2.749	6.664	0.97
0.80	0.756	1.527	2.600	6.664	0.97

THE EFFECT OF REFLECTOR TRUNCATION

Table 3 shows that for a reflector truncated up to half full-height the total energy input and concentration ratio are not affected greatly for a 45° acceptance half-angle CPC solar energy collector with a 15mm diameter absorber. The diffuse distribution used was hybrid gaussian.

TABLE 3 The effect of reflector truncation on the partition of diffuse insolation energy inputs between components.

Reflector height. % of maximum	ENERGY ABSORBED AT			Total incident energy.	Average number of reflections
	Aperture W	Reflector W	Receiver W	W	
100	0.830	0.429	3.063	6.664	0.97
75	0.824	0.353	3.055	6.580	0.91
50	0.780	0.295	3.040	6.290	0.85

THE EFFECT OF ACCEPTANCE HALF-ANGLE ON THE ENERGY INPUT

As the acceptance half-angle decreases, energy input to the aperture cover increases at a greater rate than the total energy incident on the aperture; this is a logical result of a greater percentage of diffuse radiation being reflected back out of the collector and thus passing out through the aperture cover. The partition of energy is shown for 60°, 45° and 30° acceptance half-angle CPC solar energy collectors in Table 4. The incident insolation was at 100W/m², with hybrid gaussian diffuse distribution assumed.

TABLE 4 The partition of diffuse insolation energy inputs between components

Acceptance angle	ENERGY ABSORBED AT			Total incident energy.	Average number of reflections
	Aperture W	Reflector W	Receiver W	W	
60°	0.619	0.271	3.059	5.441	0.88
45°	0.830	0.428	3.063	6.664	0.97
30°	1.264	0.826	3.349	9.425	1.09

CONCLUSION

It has been shown that if a low concentration ratio CPC is misaligned with respect to the direct insolation component, localised high flux concentrations may occur on the absorber. The calculated amount of diffuse insolation, reaching the absorber of a 45° acceptance half-angle CPC, can vary by more than 25% dependent on the diffuse insolation distribution assumed. The amount of diffuse insolation absorbed at the absorber, as a percentage of that incident at the aperture cover, increases with increased truncation. This arises from the acceptance of insolation incident outside the collector's acceptance half-angle. As this half-angle decreases, the concentration of diffuse insolation at the absorber rises slightly, however, the amount of the diffuse energy reaching the absorber, as a percentage of that at the aperture cover, decreases rapidly.

ACKNOWLEDGEMENT

This work was supported by the Science and Engineering Research Council Swindon, UK.

REFERENCES

Prapas, D.E., B. Norton, and S.D., Probert (1987). Optics of Parabolic-Trough Solar-Energy Collectors Possessing Small Concentration Ratios. Solar Energy., 39, 541-550.
Rabl, A. (1985) Active Solar Collectors and Their Applications. Oxford University Press, New York.
Winston, R. (1980) Cavity Enhancement by Controlled Directional Scattering. Applied Optics, 19, 195-197.

NOMENCLATURE

E_{gc}	energy absorbed in glass cover from a ray	[W]
E_{in}	energy remaining in a ray	[W]
E_{re}	energy absorbed at reflector from a ray	[W]
I_r	energy per ray	[W]
L	thickness of glass cover	[m]
W_i	percentage of total energy present in wavelength band i	
x	path length	[m]
α	extinction coefficient	[m^{-1}]
α_r	absorptance of reflector	
ρ	reflectance	

2.19 High Flux

OPTIMIZATION OF TWO-STAGE CONCENTRATING SYSTEMS FOR HIGH TEMPERATURE AND HIGH PHOTON FLUX DENSITY APPLICATIONS

U. Schöffel and R. Sizmann

Sektion Physik, Ludwig-Maximilians-Universität München,
Amalienstr. 54, D 8000 München 40, FRG

ABSTRACT

A terminal concentrator (secondary concentrator) provides a second optical element in a solar radiation concentrating system. It allows the concentration ratio to better approach the thermodynamic upper limit of concentrating devices. The increased flux density permits use of smaller receiver apertures. This reduces convection losses and, in particular, thermal radiation losses, which become dominant at high receiver temperatures.

In CRS plants terminal concentrators appear to be mandatory to achieve receiver temperatures considerably higher than ca. 800 degrees centigrade. Such high temperatures are required for solar process heat applications, e.g. solar thermochemistry and high photon flux density photochemistry.

KEYWORDS

Terminal Concentrator, Secondary Concentrator, Genetic Algorithm, Ray Tracing.

INTRODUCTION

The maximum usable heat flux at receiver temperature T_R and ambient temperature T_a is
$$\dot{Q} = I_H - S_R \left[\epsilon_R \sigma (T_R{}^4 - T_a{}^4) + U_L(T_R - T_a) + ... \right]. \tag{1}$$
$I_H = \alpha_R \, \eta_{op} \, \gamma \, \rho_H \, \eta_{cos} \, \eta_{obs} \, S_H \, E_b$ is the product of

E_b direct insolation flux density from the sun,

S_H mirror area of the heliostat field or aperture area of dish concentrator,

η_{obs} obstruction factor: the fraction of the incoming light directed to a heliostat field mirror (or dish) and not obstructed by receiver, terminal concentrator (and tower),

η_{cos} averaged cosine factor (for parabolic dish: $\eta_{cos} = 1$),

ρ_H mean reflectivity of the field,

γ fraction (intercept factor) of the flux $\rho_H \, \eta_{cos} \, \eta_{obs} \, S_H E_b$ intercepted by the receiver,

η_{op} optical efficiency of terminal concentrator (= 1 without terminal concentrator),

α_R absorptivity of receiver,

ϵ_R emissivity of receiver, approaches unity for deep cavities,

S_R aperture area of receiver,

σ Stefan-Boltzmann constant (= $5.67 \cdot 10^{-8} \text{W/m}^2\text{K}^4$).

$S_R U_L (T_R - T_a)$ is the convective heat loss with U_L as specific heat loss coefficient. Other sources of losses are neglected. The maximum thermal efficiency is:

$$\eta = \frac{\dot{Q}}{E_b S_H} = \eta_0 - \frac{L}{C} \quad \text{with} \tag{2}$$

$$\eta_0 = \alpha_R \, \eta_{op} \, \gamma \, \rho_H \, \eta_{cos} \, \eta_{obs} \tag{3}$$

$$L = \frac{\epsilon_R \sigma (T_R{}^4 - T_a{}^4) + U_L (T_R - T_a)}{E_b} \tag{4}$$

$$C = S_H / S_R \tag{5}$$

The thermodynamic upper limit of concentration is for the case of biaxially focusing systems (Welford, 1989):

$$C_{\max} = 1/\sin^2 \theta, \tag{6}$$

where θ is the half opening angle the radiation subtends. With imaging optical systems this limit cannot be attained — with parabolic dishes approximately only one fourth of this limit. With an additional element, the *terminal concentrator*, in principle about 90% of the limit could be attained (i.e., for a parabolic dish of longer focal ratio combined with an ideal concentrator) (Welford, 1989, p. 197).

The optical efficiency η_{op} of a terminal concentrator is an intricate function of the geometry of the reflecting surfaces of the terminal concentrator, their reflectivities and surface errors. In particular, η_{op} is dependent on the spatial and angular distribution of the incident radiation. (An optical element not affecting, i.e., not focusing the incident radiation has $\eta_{op} = 1$.)

While L is given as input and C can be calculated directly from design parameters, calculating η_0 requires integration over "many" rays.

OPTIMIZATION

The physical problem is to design an optical system consisting of a primary and a terminal concentrator for maximum efficiency η. For this an optimization calculation is necessary. The function to be optimized is of the form:

$$\eta(\mathbf{p}) = \eta_0(\mathbf{p}) - \frac{1}{C(\mathbf{p})} L \tag{7}$$

where concentration $C(\mathbf{p})$ can be determined directly from the geometrical parameters $\mathbf{p}$ of receiver, primary and terminal concentrator. In general $\eta_0(\mathbf{p})$ is an integral over an at least four-dimensional domain (Schöffel, 1990). Within the concentrator, multiple reflections can occur; rays from some directions might leave the terminal concentrator through the entrance aperture after one or several reflections. Therefore a detailed ray tracing through primary and terminal concentrator is necessary.

Because of   practical reasons Monte-Carlo integration has been used. The error estimate of the numerical integral is in proportion to $1/\sqrt{N}$ and therefore controllable by the number of rays N. The integral values appear to be noised, unless the same random number sequence is always used. If there is an optimization algorithm which eliminates noise, systematic errors of a deterministic integral evaluation are also avoided. Precautions have to be made to ensure that the noise is widely independent of $\mathbf{p}$. This is a precondition for identifying optimal parameters of the noised function as optimal parameters of the (unknown) exact function.

The optimization algorithm used now is a Genetic Algorithm inspired by (Goldberg, 1989). (The previously employed stochastic gradient method proved to be too limited in its applicability to the present problems.) Optimizing with Genetic Algorithms requires mapping of the

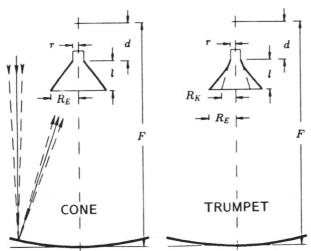

Fig. 1: Parabolic dish configuration for optimizing (b) and (c) with conical terminal concentrator (left). For optimizing (d) and (e) with hyperboloid of revolution (trumpet concentrator, right).

parameter vector to a bit string, which is considered as a (haploid) chromosome of an individual; its fitness is the (rescaled) objective function value. For the present case coding was performed by discretizing the parameter ranges of each dimension (using 6 or 7 bits); these bits are concatenated. The Genetic Algorithm repeatedly processes a population of bit strings, that are subjected to the mechanisms selection according to fitness, crossing-over between pairs of selected "parents", and occasionally mutation.

RESULTS

With computer code OPTEC (an Optimization Program for TErminal Concentrators (Schöffel, 1990)) configurations concerning parabolic dishes as well as CRS systems have been optimized for maximum thermal efficiency.

Parabolic Dish Configurations

Parameters for the parabolic dish configurations have been taken essentially from (O'Gallagher, 1987), Tab. 2-1, p. 8: Flux density distribution with Gaussian profile, standard deviation of the one-dimensional distribution: 0.6° (also considered here: 0.5°); direct insolation 800 W/m²; diameter of dish: 11 m; receiver temperature ((O'Gallagher, 1987), Fig. 3-4, differing from Tab. 2–1): 1200°C; ambient temperature: 20°C; convection loss coefficient: 16 W/m²K; reflectivity: 0.9 for parabolic dish, 0.95 for terminal concentrator; absorptivity and emissivity of receiver: 1.0.

Considered terminal concentrator shapes (Fig. 1): a cone as well as a hyperboloid of revolution (trumpet). Remarks: (i) the trumpet concentrator used here differs from the classical one described in (O'Gallagher, 1986): the throat of the trumpet is allowed to lead into the receiver entrance plane non-perpendicularly, i.e., the throat of the trumpet may be truncated; (ii) the receiver entrance plane may differ from the nominal focal plane.

In particular we have optimized five cases (Fig. 1) considering two opening angles for each

Table 1: Results of optimization calculations for one- (a) and two-stage parabolic dish configurations with conical (b,c) and trumpet-concentrator (d,e), respectively (see text). In brackets: given or fixed values. $\alpha_R = 1.0$, $\rho_H = 0.9$.

	Half opening angle 0.5°					Half opening angle 0.6°				
	(a)	(b)	(c)	(d)	(e)	(a)	(b)	(c)	(d)	(e)
	Optimal parameters / [m]									
F	4.4	[4.4]	5.21	[4.4]	5.11	4.6	[4.6]	6.0	[4.6]	5.8
d	0.0	[0.0]	0.03	[0.0]	0.03	0.0	[0.0]	0.06	[0.0]	0.08
r	0.125	[0.125]	0.119	[0.125]	0.117	0.135	[0.135]	0.127	[0.135]	0.127
R_E	—	0.47	0.96	0.70	0.81	—	0.93	1.37	0.84	1.14
l	—	0.16	0.53	0.27	0.43	—	0.43	1.07	0.38	0.83
R_K	—	—	—	0.47	0.53	—	—	—	0.66	1.0
	Efficiencies and efficiency factors									
η_{obs}	≈ 1.0	0.99	0.97	0.98	0.98	≈ 1.0	0.97	0.94	0.98	0.96
γ	0.87	0.99	≈1.0	0.99	0.99	0.82	0.99	0.99	0.99	0.97
η_{op}	[1.0]	0.91	0.92	0.91	0.92	[1.0]	0.88	0.89	0.88	0.90
η_0	0.78	0.80	0.80	0.80	0.80	0.74	0.76	0.74	0.77	0.75
L/C	0.18	0.18	0.16	0.18	0.16	0.21	0.21	0.19	0.21	0.19
η	0.60	0.62	0.64	0.62	0.64	0.53	0.55	0.55	0.56	0.56

case: (a) one-stage (without terminal concentrator): optimized parameters focal length F of the parabolic dish, radius r of the circular receiver entrance aperture, displacement d of the receiver location off the nominal focal plane; (b) one-stage (as in (a)), then, without changing the optimal parameters found under (a), we optimized parameters R_E, radius of the entrance aperture, and length l of a conical terminal concentrator; (c) two-stage: as in (a) and (b), but simultaneously; (d) one-stage (as in (a)), then parameters radius R_E of entrance aperture and length l of a trumpet as well as entrance radius R_K of the cone which is tangential to the trumpet at its throat; (e) as in (d), but all parameters simultaneously.

Table 1 shows results (parameters, efficiency factors, efficiencies). Thermal efficiencies have been calculated with typically 40 000 rays (equivalent to a standard deviation of ca. 0.0015 in efficiency) for the optimal parameters. In addition efficiency factors are listed. Fixed parameters and factors are in brackets.

CRS-Configuration

The SSPS-CRS, Plataforma Solar de Almeria (PSA), Spain, originally consists of 93 heliostats, distributed in a north field aiming at the center of a receiver at 43 m above ground; each heliostat has a reflective area of 39.3 m² with reflectivity 0.911.

From the construction report (Becker, 1983) of the SSPS-CRS the following parameters are taken for the design point at equinox noon: Gaussian flux density distribution with one-dimensional standard deviation (including primary shape deviations): 3.1 mrad; direct insolation: 920 W/m²; absorptivity and emissivity of receiver: 1.0.

For the present investigations the following additional conditions were chosen: receiver temperature: 1000°C; ambient temperature: 25°C; convection loss coefficient: 16 W/m²K (see Parabolic Dish Configurations); atmospheric attenuation: 0.99; reflectivity of terminal concentrator 0.9.

Table 2: Results of optimization calculations for one- (a) and two-stage CRS configurations with conical (b, two different parameter domains) and trumpet concentrator (c), respectively (see text). In brackets: given value. $\alpha_R = 1.0$, $\rho_H = 0.911$, atmospheric attenuation 0.99.

Case	Optimal parameters / [m]				Efficiencies and efficiency factors						
	r	R_E	l	R_K	η_{obs}	η_{cos}	γ	η_{op}	η_0	L/C	η
(a)	0.99	—	—	—	1.0	0.95	0.90	[1.0]	0.77	0.15	0.62
(b)	0.83	4.45	5.25	—	1.0	0.95	0.99	0.93	0.79	0.11	0.68
	0.96	2.2	2.05	—	1.0	0.95	0.99	0.97	0.82	0.14	0.68
(c)	0.83	4.15	4.2	2.8	1.0	0.95	0.99	0.95	0.80	0.10	0.70

The receiver is modeled with circular (instead of octagonal) entrance aperture of a cavity, inclined by 28° (manually calculated from the extent of the field). Heliostats are modeled as single-piece rectangles (6.3 m × 6.25 m) with spherical curvature that are individually tracked (This is similar to the on-axis alignment of computer code HELIOS (Biggs, 1979), used in (Becker, 1983) for flux density calculations). Cones and trumpets are considered as terminal concentrators. Table 2 shows results. All parameters of the terminal concentrators were optimized simultaneously (see Fig. 1). The receiver position has been held fixed.

CONCLUSIONS

General

- The optical efficiency η_{op} is a quantity not characterizing a terminal concentrator and can be considerably smaller than its reflectivity when it is part of an optimized two-stage system: the fraction of rays reflected backwards through the entrance aperture of a terminal concentrator after several reflections cannot be neglected. When optimizing the optical system the fixed parameters (like T_R) determine the influence of the individual efficiency factors.

- The optima appear to be not strongly pronounced — or there even exist several parameter combinations with approximately maximum thermal efficiency. This means more freedom in designing terminal concentrator / receiver systems.

- Before calculating optimal trumpet concentrators one should first calculate optimal conical concentrators. In our calculations with trumpet concentrators for certain parameter domains the optimization algorithm ran into a local optimum which yielded results worse than those with conical concentrators. Changing parameter domains then yielded (slightly) better results as is to be expected, since the cone can be regarded as a limit of a generalized trumpet.

  Optimization could be enhanced by selection of a set of free parameters that incorporates appropriate scaling relationships.

Dish Configuration

- The shading effect of a terminal concentrator cannot be neglected in a two-stage dish system (with on-axis focus, Fig. 1). However, in practice shading caused by the receiver is larger than that induced by its entrance aperture alone — as was assumed here, and may exceed that of the terminal concentrator.

- The receiver plane in an optimized parabolic dish configuration is almost exactly the nominal focal plane of the dish.

- Results given in (O'Gallagher, 1987), p. 16, Fig. 3–4 for nonimaging concentrators could not be confirmed using conical and trumpet concentrators. The reported maximum thermal efficiencies for 1200°C receiver temperature and Gaussian insolation distribution (one-dimensional standard deviation 0.6°) are (interpolated from the mentioned figure): 0.52 (without terminal concentrator) and 0.59 with terminal concentrator; here we found: 0.53 and 0.56 (with trumpet), respectively.

CRS Configuration

- The enhancements in efficiency may be regarded as being not very high. However, there can be a considerable decrease in receiver entrance aperture: about 30% off the area of a single receiver.

- Calculated flux density distributions absorbed by the terminal concentrator surfaces showed not only axial but also azimuthal variations. Preliminary results with a slightly more inclined conical terminal concentrator show azimuthally smoother flux densities at approximately the same efficiency. Therefore the inclination angle should be included in the set of free parameters.

- Maximum flux densities absorbed by the terminal concentrators range from 10 to 15 suns.

The terminal concentrators considered here are the (mathematically) most simple ones. Other shapes are to be investigated, especially CPCs.

ACKNOWLEDGEMENTS

We would like to thank the Bundesministerium für Forschung- und Technologie, Bonn, and the Deutsche Forschungsanstalt für Luft- und Raumfahrt (DLR) in Köln for financing and support under the SOTA R & D program.

REFERENCES

Becker, M., Ellgering, H., and Stahl, D., 1983. *Construction Experience Report for the Central Receiver System (CRS) of the International Energy Agency (IEA) Small Solar Power Systems (SSPS) Project.* IEA-SSPS Operating Agent DFVLR, Köln, 1983.

Biggs, F. and Vittitoe, C.N., 1979. *The Helios Model for the Optical Behavior of Reflecting Solar Concentrators.* SAND76-0347.

Goldberg, David E., 1989. *Genetic Algorithms in Search, Optimization, and Machine Learning.* Addison Wesley, Reading, Mass. (1989).

O'Gallagher, J. and Winston, R., 1986. Test of a "Trumpet" Secondary Concentrator with a Paraboloidal Dish Primary, *Solar Energy* **36**, 37 (1986).

O'Gallagher, J. and Winston, R., 1987. *Performance and Cost Benefits Associated with Nonimaging Secondary Concentrators Used in Point-Focus Dish Solar Thermal Applications. A Subcontract Report.* SERI/STR-253-3113 (1987).

Schöffel, U. and Sizmann, R., 1990. *Final Report Optimization of Terminal Concentrators (Project SOTA).* DLR Report, Köln.

Welford, W.T. and Winston, R., 1989. *High Collection Nonimaging Optics.* Academic Press, San Diego (1989).

Influence of Sunshape on the Flux Distribution in Concentrators

M. Schubnell and H. Ries

Paul Scherrer Institut, CH-5232 Villigen-PSI, Switzerland

Abstract

Solar limb darkening and scattering in the atmosphere determine the angular distribution of the direct incident solar radiation. This distribution, referred to as sunshape, affects the flux density distribution in concentrating systems such as e.g. a solar furnace. Here we investigate this influence on our two stage solar furnace as well as on parabolic concentrators. Comparing various sunshapes with the corresponding flux density distribution in our furnace we show that this influence is usually negligible as compared to distortion due to astigmatism. However, in more accurate optical systems such as highly concentrating parabolic dishes the flux density distribution in the focal spot is a more accurate image of the sun. Thus it is sunshape that basically determines the maximum concentration ratio as well as the spot size in a parabolic concentrator.

Keywords

Sunshape, Solar Furnace, Concentration Ratio, Flux Density Distribution, Circumsolar Ratio, Intercept Factor.

Introduction

It is well known that solar limb darkening and scattering in the atmosphere determine the angular distribution of the direct incident solar radiation. This distribution, referred to as sunshape, affects the flux density distribution in concentrating systems such as e.g. a solar furnace. In this paper we investigate this influence on the solar furnace at Paul Scherrer Institute (PSI) as well as on parabolic concentrators. The furnace at PSI is a two stage concentrator. The first stage is a 51.8 m^2 glass-heliostat with a focal length of 100 m. The second stage is a parabolic dish with an area of 5.7 m^2 and a focal length of 1.93 m. The distance between the heliostat and the parabolic dish amounts to 73 m. Oblique reflection at the heliostat leads to an astigmatism, distorting the image of the sun to an ellipse and reducing the concentration ratio. Depending on season and daytime the peak flux density in the focal region reaches $200 - 700 \ W/cm^2$ [1].

The sunshape is recorded with a CCD-camera mounted on a tracker with a long term tracking accuracy of $\pm 3 \ mrad$. To characterize the sunshape we use the circumsolar ratio, CSR, which we calculate according to

$$CSR = \frac{\int_{a_s}^{\rho_0} B(\rho) \cdot \rho \cdot d\rho}{\int_{0}^{\rho_0} B(\rho) \cdot \rho \cdot d\rho} \tag{1}$$

Here a_s denotes the angular width of the solar disk as seen from earth, $a_s = 4.65$ mrad, ρ_0 the viewing angle of the CCD-camera and $B(\rho)$ the brightness of the sun at angular distance ρ from the center of the sun. The viewing angle of our camera, ρ_0, amounts to 55 mrad. The A/D-converter in our image processing system has a resolution of 8 bit. Thus we neglect radiation originating more than 55 mrad off the center of the sun as well as radiation of irradiance less than 0.5 % of that of the sun.

The flux density distribution in our furnace is measured with a similar CCD-camera by imaging a white lambertian target placed in the focus. The camera in conjunction with the target was calibrated by simultaneously measuring the flux density with a high precision radiometer. The experimental setup is shown in Fig. 1.

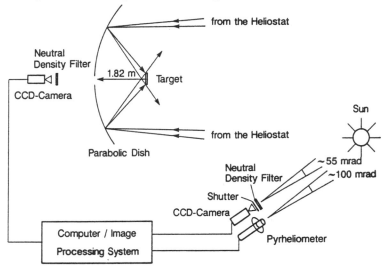

Fig. 1: Experimental setup for measuring sunshape and flux density distribution in the focal region of the two stage solar furnace at PSI.

Results and Discussion

During summer 1990 about 50 different sunshapes have been recorded and analysed. In Fig. 2 we show 3 typical sunshapes with CSR-values of 6.7 %, 12.8 % and 16.7 %. We also show the ideal 'pillbox-sunshape' as well as the theoretical extraterrestrial sunshape calculated according to the Kuiper distribution [2]. This distribution describes the limb darkening of the solar disk. It is due to the fact that in the center of the disk an observer perceives radiation from deeper (hotter) layers, near the limb; however, from more superficial (colder) layers. Therefore, the limb of the solar disk is reddened and darkened as compared to the center [3]. The sunshapes shown in Fig. 2 are normalized to the same direct irradiation.

It is very difficult to analyse the influence of sunshape on the flux distribution experimentally because one seldom finds various sunshapes with the sun in the same position. However, this would be necessary because the flux distribution in our furnace depends strongly on the position of the sun [1]. We therefore determine the influence of sunshape on the flux distribution using a ray tracing computer program [4]. To validate this method we compare the flux density distribution measured on October 10, 1320 UT, with the flux distribution as calculated by ray tracing assuming the same position of the sun and the observed sunshape.

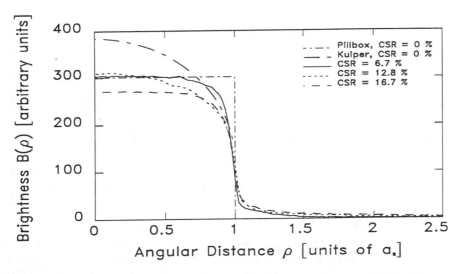

Fig. 2: Extraterrestrial sunshape, pillbox sunshape and three typical sunshapes measured at PSI. The angular distance, ρ, is given in units of the aperture of the solar disk as seen from the earth, $a_s = 4.65 \ mrad$.

Fig. 3 shows, that these two distributions are in good agreement. Due to 'blooming' of the CCD-camera the lower right part of the measured flux distribution is distorted. The net flux across the focal spot bounded by the $100 \ W/cm^2$ contour line amounts to $14.6 \ kW$ and $15.5 \ kW$ for the measured and the calculated distribution, respectively. The spot size is $62 \ cm^2$ (measured) and $60 \ cm^2$ (calculated).

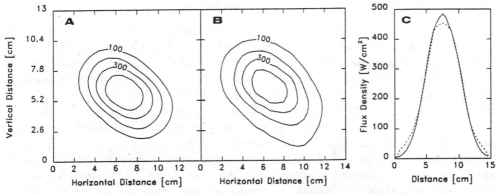

Fig. 3: Flux density distribution in our solar furnace on October 10, 1320 UT. Flux densities are given in $[W/cm^2]$. A calculated data, B measured data, C horizontal cross sections, solid line calculated, broken line measured data.

We now computed the flux distribution for various observed sunshapes and compared these results with the flux distribution assuming a pillbox shaped sun. Horizontal cross sections of the flux distributions on August 24, 1400 UT, September 28, 1200 UT and October 10, 1320 UT are depicted in Fig. 4. Solid and broken lines denote the results assuming a pillbox shaped sun and the observed sunshapes (c.f. Fig. 2), respectively.

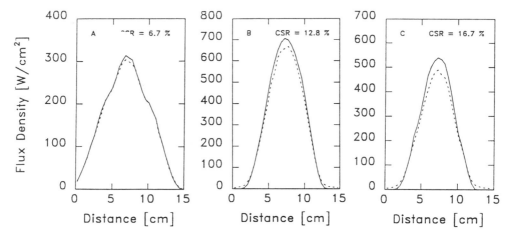

Fig. 4: Horizontal cross sections of the flux distribution in our solar furnace assuming various sunshapes (solid lines) as well as a pillbox shaped sun (broken lines). A, B and C represent data on August 24, 1400 UT, September 28, 1200 UT and October 10, 1320 UT, repectively.

With increasing CSR-values the image of the sun broadens as compared to the image of a pillbox shaped sun. Fig. 4 shows that this broadening is negligible for CSR-values up to 13 %, basically because the imaging errors due to astigmatism dominate. For CSR-values larger than 13 %, however, the flux density in the center of the focal spot is decreased by more than 6 %. As shown in Fig.4 the size of the focal spot bounded by the 100 W/cm^2 contour line is not affected by sunshape. Thus the net flux within the focal spot decreases with increasing circumsolar ratio. Assuming a CSR-value of 16.7 % the flux is lowered by 9 % as compared to the flux assuming a pillbox shaped sun.

In more accurate optical systems, such as highly concentrating dishes, the flux density distribution in the focal spot is a more accurate image of the sun. But even in a perfect parabolic dish the image is blurred particularly at its rim by spherical aberration [5]. Nevertheless, the flux distribution in optically ideal reflecting systems is basically determined by sunshape. In Fig. 5 we show the concentration ratio as calculated by ray-tracing of a parabolic dish with a numerical aperture of 0.64 assuming various observed sunshapes (c.f. Fig. 2) as well as a pillbox-shaped sun. The radial distance is given in units of the radius of a perfectly imaged pillbox-shaped sun, r_{img}. The influence of spherical aberration can be estimated by comparing the ideal image and the image assuming a pillbox sunshape. In each case the image of the sun is broadened as compared to the ideal image. To characterize the flux distribution in a parabolic dish we use the circumspot ratio, which we define analogously to (1) as

$$CSPR = \frac{\int_{r_{img}}^{\infty} C(r) \cdot r \cdot dr}{\int_0^{\infty} C(r) \cdot r \cdot dr} \tag{2}$$

Here r_{img} denotes again the radius of an ideal image of a pillbox shaped sun and C(r) is the concentration ratio at radial distance r from the center of the focal spot. The figure shows that the values for the circumspot ratio converge to the circumsolar ratio with increasing

CSR-values. This indicates that in this case the influence of spherical aberration on the flux distribution is negligible as compared to sunshape.

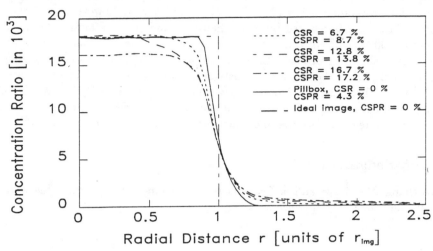

Fig. 5: Concentration ratio in the focal spot of a parabolic concentrator (numerical aperture 0.64) for various sunshapes.

The ratio between the radiation flux impinging on the aperture of the reactor and the radiation flux, referred to as intercept factor, is an important parameter for the design of a solar reactor. The intercept factor can easily be calculated for a parabolic dish. In Fig. 6 it is depicted as a function of the reactor aperture for various sunshapes (c.f. Fig. 2). Sunshape increases total power losses by up to 35 % if the reactor is operated in the highly concentrated regime of the flux density distribution (i.e. reactor aperture < 1).

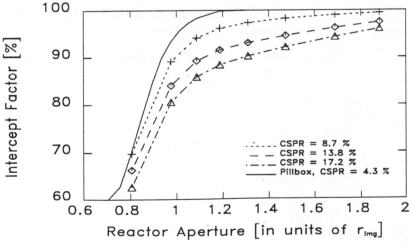

Fig. 6: Intercept factor of a reactor in the focal plane of a parabolic dish for various sunshapes (c.f. Fig. 2). The aperture of the reactor is given in units of the radius of the perfectly imaged sun, r_{img}.

Conclusions

In this paper we investigated the influence of sunshape on the radiation flux distribution on our two stage solar furnace and on parabolic concentrators. We showed that it is negligible in our furnace for CSR-values below 13 %. For larger CSR-values, however, the flux density in the center of the spot is reduced by more than 6 % and the flux across the focal spot bounded by the 100 W/cm^2 contour line is reduced by more than 9 % as compared to the values assuming a pillbox shaped sun. In more accurate optical systems such as highly concentrating parabolic dishes the flux density distribution in the focal spot is a more accurate image of the sun. Therefore it is basically determined by sunshape. Thus sunshape has to be considered in the design of experiments running in the focal region of a parabolic dish. Depending on reactor aperture and sunshape radiation losses of up to 35 % can be attributed to it.

Acknowledgements

We wish to thank P. Haueter and S. Leutenegger for technical assistance. This work was supported by the Swiss Federal Office for Energy.

References

[1] H. Ries and M. Schubnell. The optics of a two stage solar furnace. *Sol. Energy Mater.*, 21, 1990.

[2] F. Biggs and Ch. N. Vittitoe. *The Helios Model for the Optical Behaviour of Reflecting Solar Concentrators.* Technical Report SAND76-0347, Sandia National Laboratories, 1979.

[3] M. Minnaert. *The Photosphere*, in G. P. Kuiper (Ed.), *The Sun, the Solar System*, Vol. I, University of Chicago Press, 1971.

[4] M. Schubnell, J. Keller, and A. Imhof. Flux density distribution in the focal region of a solar concentrator system. *to be published in JSEE*, 1991.

[5] W. T. Welford and R. Winston. *High Collection Nonimaging Optics.* Academic Press, New York, 1990.

BRIGHTER THAN THE SUN

R. Winston, D. Cooke, P. Gleckman, H. Krebs, J. O'Gallagher, and D.Sagie
Department of Physics and The Enrico Fermi Institute
The University of Chicago, 5640 Ellis Ave.,
Chicago, Illinois   60637, USA

ABSTRACT

Sunlight at the surface of the sun has an intensity of ~63 W/mm^2 but by the time the light reaches the surface of the earth the intensity has fallen to about 0.8 to 1.0 mW/mm^2. In a recent set of experiments we have concentrated terrestrial sunlight by a factor of 84,000 + 3,000 to intensities of 72 W/mm^2. This establishes a new world record for solar flux concentration, exceeding our previous result by a factor of 1.5. Such high concentrations cannot be attained with a conventional imaging device alone; our approach relies on a device designed according to principles of the new field of nonimaging optics. Nonimaging concentrators are essentially "funnels" for light, and the enormous intensities of sunlight we have attained with them actually exceed the intensity at the surface of the sun itself by about 15 percent. In other words, we have produced the highest concentration of sunlight in the solar system. Our approach offers the promise of making efficient solar-pumped lasers.

KEYWORDS

Nonimaging, laser, high flux, concentration, sapphire.

INTRODUCTION

Phase space conservation[1] and thermodynamic arguments[2] place a limit on the concentration of sunlight any optical device can attain, namely

$$C_{max} = n^2/\sin^2\theta, \qquad (1)$$

where n is the index of refraction at the target surface and θ is the semiangle subtended by the sun. At the surface of the earth $\theta = 0.27^o$, and so with an ordinary refractive material (n $\simeq$ 1.5) one obtains an upper limit on concentration of about 100,000.

The concentrations attained with conventional imaging devices fall far short of this limit, because of abberations[3]. A parabolic mirror, for example, produces a perfect image on axis, but blurs and broadens the image off axis. To attain high concentrations we relax the requirement of forming images. Simply put, a laser crystal or a solar furnace does not care about receiving a picture-perfect image of the sun; all either care about is receiving the maximum power per area. By dispensing with image-forming requirements in applications where no image is required, we can attain concentrations that approach the theoretical limit.

NONIMAGING CONCENTRATION

The relatively new field of nonimaging optics provides a framework for designing such powerful concentrators.[3] We design our concentrator by an approach called the "edge ray" method[3], in which all light rays entering the device at the maximum collection angle are directed after one reflection at most to the rim of the exit aperture[4] (Fig. 1). In this way the other rays are reflected within the aperture itself.

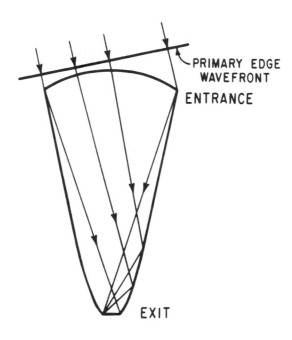

Fig. 1. In a nonimaging concentrator designed by the edge ray method all light rays entering the device at the maximum collection angle are directed after one reflection at most to the rim of the exit aperture.

In principle we could attain high concentrations of sunlight with a nonimaging concentrator alone, but such a device would be quite large and unwieldy. Our approach for concentrating sunlight utilizes a two-stage system incorporating a parabolic mirror and a nonimaging concentrator (Fig. 2). The overall concentration for a two-stage system is simply the product of the concentration of the parabolic mirror with that of the nonimaging concentrator:

$$C = n^2\cos^2\phi/\sin^2\theta, \qquad (2)$$

which falls short of the theoretical limit (equation (1)) only by the factor of $\cos^2\phi$, where ϕ is the rim angle of the mirror (the inverse tangent of the radius of the mirror divided by its focal length). The effects can be minimized by making the rim angle small; i.e., choosing a mirror with a relatively large f number.

Employing a technique based on this concept, our group previously achieved a record concentration of sunlight of 56,000 times the intensity at the surface of the earth.[5] We have now modified and improved that apparatus to reach still higher concentrations. We employ a 40.6-cm-diameter silver coated telescope mirror with rim angle $\phi = 11.5^{\circ}$ (f = 2.5). Sunlight is concentrated to a one centimeter spot one meter away from the mirror. The light is further compressed to a one millimeter spot by a nonimaging concentrator made out of sapphire. Sapphire was chosen for its low absorption and high index of refraction, n = 1.76. (The concentrator in our previous experiment was a silver-coated vessel filled with an immersion oil having an index of refraction of 1.53.) Substituting these values into equation (2) gives the theoretical limit to the concentation of our new two-stage system:

$$C = [1.76\cos(11.5^{\circ})/\sin(0.27^{\circ})]^2 = 137,000 \qquad (3)$$

This is a 31 percent increase over the theoretical upper limit of 104,000 in our earlier device.

Another modification we have made is that our concentrator now works by total internal reflection[6]. In the previous experiment the walls of the concentrator were silvered, and as such there were losses due to absorption. Total internal reflection, on the other hand, is nearly loss-free, and thereby increases the efficiency of the concentrator, allowing a greater fraction of the theoretical limit to be attained.

CALORIMETRIC MEASUREMENT

Measuring the actual concentration required some ingenuity. The intensity of the light exiting the tip of the sapphire concentrator would damage conventional power meters. Moreover, the light sprays out in a wide cone (filling nearly 2π sr), and it is contained in a dielectric. To surmount these difficulties as in our previous measurement[7], we have relied on a 19th century technique: calorimetry (Fig. 3), although we have redesigned and improved the calorimeter as well. Light exiting the tip of the concentrator passes through a sapphire window and into a thermos bottle filled with a Cargille index matching fluid (n = 1.76). The tip of the concentrator is optically coupled to the sapphire window with a drop of a second, different matching fluid, tin chloride in glycerol.[8]

To measure directly the sun's energy passing through the target area at the tip of the concentrator we allow sunlight to enter the calorimeter for a 30-second interval. Three thermocouples inside the calorimeter permit us to record the equilibrium temperature of the liquid before and after each 30-second interval, and thus determine the increase in temperature ΔT_s. The liquid is allowed to cool, and then to calibrate the temperature increase a heater inside the calorimeter of known power P_H is turned on for a 30-second interval, and the change in temperature ΔT_H is measured. The power delivered by the sun through the tip of the concentrator during a 30-second interval can therefore be determined by

$$P_s = (\Delta T_s/\Delta T_H)P_H, \qquad (4)$$

independent of the properties of the fluid.

A separate experiment was done to verify that all the sunlight entering the calorimeter does indeed pass through the tip. Employing a masked solar cell we directly measured a small amount of light that does not come from the tip, corresponding to 3.8 percent of the total power. This leakage probably comes from the tiny pool of tin chloride solution spilling over at the tip, which reduces the effectiveness of total internal reflection.

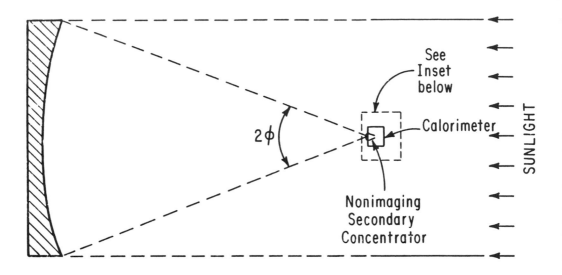

Fig. 2.   Two-stage approach to concentrating sunlight is illustrated.

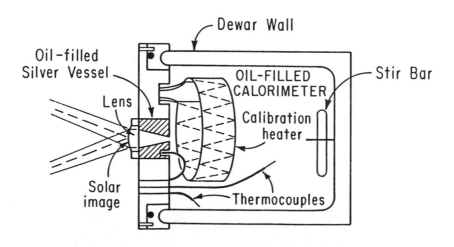

Fig. 3.   Calorimeter, or thermos bottle, traps sunlight passing through the tip of the concentrator. The temperature rise of the liquid inside is calibrated with the electric heater.

In a series of trials over several cold, crisp days in January and February, 1990 we measured an average power of 72 watts of sunlight passing into the calorimeter, after correcting for the 3.8 percent leakage. Dividing by the area of the tip, 1.0 mm^2, gives an intensity of 72 W/mm^2. This intensity is 84,000 times greater than the direct incident intensity, 0.86 mW/mm^2. Based on the reproducibility of our results and analysis of possible sources of error we estimate the uncertainty to be $\pm$ 3,000. This establishes a new world record for the concentration of sunlight, which is 1.5 times higher than our previous record of 56,000 $\pm$ 5,000. Furthermore, the light passing through the 1 mm tip of our concentrator is the highest intensity of sunlight anywhere in the solar system, including the value of 63 W/mm^2 at the surface of the sun itself! The measured value falls short of the theoretical limit primarily because of losses at the reflecting surface of the primary mirror and blockage from the calorimeter.

APPLICATIONS

Potential uses for such high levels of solar flux are just beginning to be explored. We have already employed nonimaging concentrators with lower indices of refraction (n = 1.5) to solar-pump Nd:YAG and Nd:Cr:GSGG laser crystals from one end. In the past it has proved difficult to make efficient solar pumped lasers because of the limited materials available and the lower levels of concentration of sunlight that could be achieved[9,10]. We believe our approach should increase the attained efficiencies[5]; the new sapphire concentrators should allow us to pump Nd:Cr:GSGG crystals with efficiencies of a few percent. Moreover, the high concentrations should give enough intensity to exceed the threshold of alexandrite, with the aim of making a tunable solar-pumped laser. Other applications include the destruction of hazardous waste[11] and the processing of specialized materials that require high temperatures.

ACKNOWLEDGEMENT

We thank the US Department of Energy, in particular the Solar Energy Research Institute and the Office of Basic Energy Sciences, for support of this work. We are grateful for all the help from the University of Chicago Central Shop; in particular, we wish to thank Bob Wentzel.

REFERENCES

1. Winston, R. *J. Opt. Soc. Am.* **60**, 245 (1970).
2. Rabl, A. *Sol. Energy* **18**, 93 (1976).
3. Welford, W. T. & Winston, R. *High Collection Nonimaging Optics* (Academic, New York, 1989).
4. The edge-ray principle is used in a slightly different form for the concentrator described here, to ensure that light is reflected by total internal reflection. Hence, the maximum exit angle of the light is reduced slightly, from 90° to 86°, which reduces concentration by one half percent.
5. Gleckman, P. *Appl. Opt.* **27**, 4385-4391 (1988).
6. Winston, R. *Appl. Opt.* **15**, 291-292 (1976).
7. Gleckman, P., O'Gallagher, J. & Winston, R. *Nature* **339**, 198-200 (1989).
8. Huff, L. "Sun Pumped Laser," Technical Report (1971)
9. Arashi, H. *et al Jap. J. Appl. Phys.* **23**, 1051-1053 (1984).
10. Weksler, M. & Shwartz, J. *IEEE J. Quantum Electron.* **24**, 1222-1228 (1988).
11. Couch, W. A. (ed.) *Proc. A. Solar Thermal Technology Research and Development Conf.* Washington, DC (US Department of Energy, 1989).

ULTRA HIGH FLUX CONCENTRATION USING A CPC SECONDARY AND THE LONG FOCAL
LENGTH SERI SOLAR FURNACE

J. O'Gallagher, R. Winston, C. Zmola, L. Benedict and D. Sagie
The University of Chicago, 5640 S. Ellis Ave., Chicago, IL 60637

A. Lewandowski
Solar Energy Research Institute, 1617 Cole Blvd., Golden, CO 80401

ABSTRACT

Since 1984, the University of Chicago and the Solar Energy Research Institute (SERI) have been pursuing an effort to develop techniques and applications for generating as high a flux of concentrated solar radiation as possible. In the past, furnaces based on imaging optical designs have produced fluxes in the range 5000 to 16000 "suns". The angular subtense of the sun, together with physical conservation principles, sets a limit on the maximum achievable geometric concentration of sunlight at a factor of about 46,000 (or 46,000 "suns"), if the index of refraction at the target surface is unity (i.e. in air). SERI recently designed and constructed a 10 kilowatt solar concentrating furnace facility, which uses a modified long focal length off-axis design, so that when combined with a nonimaging secondary element the resulting configuration can approach this allowed limit. A ray-trace analysis of a variety of nonimaging secondary designs was used to select a prototype secondary for a series of high flux experiments. The selected design is a Compound Parabolic Concentrator (CPC) which has an acceptance angle 14° and is truncated to 80% of its full height.   The entrance and exit aperture diameters are 6.0 cm. and 1.48 cm. respectively corresponding to a geometric concentration ratio C = 16.4 X. Two mechanically identical water-cooled prototypes were constructed using an alloy of tellurium and copper. In a series of experiments conducted at SERI during the summer of 1990, the performance of both secondary units was measured using a small cold-water calorimeter. The optical models for the full system predict that an average concentration in excess of 20,000 suns at a net power of about 3.5 kilowatts should be delivered at the CPC exit aperture. Net flux values corresponding to intensities of 17,000 to over 19,000 suns were obtained with the first unit exceeding the previous highest values achieved in air. Greater care was taken in preparing the reflecting surface of the second unit and on several occasions, flux concentrations at multikilowatt levels in excess of 22,000 suns were achieved, surpassing the previous values by more than 30%. Some potential applications include decomposition of toxic waste materials, pumping of high powered lasers and processing of a variety of materials.

KEYWORDS

Concentrator; high flux; solar furnace; nonimaging; ray trace.

BASIC PRINCIPLES

The maximum achievable geometric concentration of sunlight on earth is determined by the well known "thermodynamic limit", according to which, the geometric concentration limit that can be achieved (with 100% intercept) by an optical system having a conical field of view of half-angle $\pm\theta$ is

$$C_{max} = \frac{n^2}{\sin^2\theta} \qquad (1)$$

where n is the index of refraction at the target surface. Concentrations approaching this limit can be achieved only by applying the techniques of nonimaging optics (1). A conventional imaging primary, for instance a single stage paraboloid with a convergence angle (or rim angle) ϕ falls short of this limit by a factor of $\dfrac{n^2}{\cos^2\phi\sin^2\phi}$. This factor is a minimum for $\phi = 45^\circ$, meaning that the shortfall is at least 4 n^2. In practice, the best configuration for approaching the limit of Eqn. (1) is some form of two-stage optical system which employs a nonimaging secondary element in the focal plane. In this case the practical geometric limit is substantially increased above that for a single stage to

$$C_{2\text{-stage}} = \frac{n^2\cos^2\phi}{\sin^2\theta} \qquad (2).$$

which comes close to the limit of Equation (1) for small ϕ, i.e., for large focal length to diameter (F/D) ratios.

Since 1984, the University of Chicago and SERI have undertaken an effort to explore these concepts for delivering as high a flux of concentrated solar radiation as possible. The geometric limits can be approached in practice by reducing optical errors to values very much smaller than θ_s, the half-angular subtense of the sun, and designing a two-stage system to have as small a value of ϕ as practical, i.e. by using a long focal length primary. Putting $\theta = \theta_s$, (4.6 milliradians) in Eqn. (1) tells us that the maximum achievable geometric concentration of sunlight on earth is 46,000X for $n = 1$ and about 110,000 for $n = 1.5$. The net concentration levels attainable will, of course, be lower than the geometric values due to reflection, absorption, and other unavoidable losses in any real set of optical components.

Recently, at the University of Chicago, laboratory scale versions of two such systems, with $n = 1.53$ and $n = 1.77$, were used to establish new records for the concentration of sunlight of 56,000X and 84,000X respectively, (2,3). Following through with a long-standing desire to explore the development of these techniques for larger scale, higher power applications, the Solar Energy Research Institute (SERI) designed and constructed a scaled-up solar concentrating furnace facility which uses a modified long focal length design and is capable of delivering up to 10 Kilowatts to the focal zone and in particular to the entrance aperture of a secondary concentrator (4). Aside from the larger scale, the main optically significant difference between the design of the SERI primary and the smaller versions constructed at Chicago is that the incident and target directions are off-axis. The major advantage of this configuration is that the secondary concentrator and target, as well as any associated equipment (e.g., cooling, data acquisition etc.), do not shadow the primary.

The University of Chicago solar energy group designed and constructed an optimized CPC reflecting secondary concentrator for use with the SERI primary and, in collaboration with the SERI staff, tested two prototype units during the summer of 1990. On several occasions, flux concentrations at multikilowatt levels in excess of 22,000 suns were achieved, surpassing the previous values achieved in air ($n = 1.0$) by more than 30%. Further details concerning the design and construction of the SERI facility are given elsewhere (4,5). The remainder of this paper is concerned with the design and fabrication of the CPC secondaries and the measurement and analysis of the high flux performance.

SECONDARY OPTIMIZATION

For the first large scale experiments it was decided to use a simple reflecting CPC type secondary, since these are easy to cool and the concentrated flux can be used in a low index ($n \geqslant 1$) target. A ray trace model for analyzing the characteristics of nonimaging secondary elements to be used with the SERI facility has been developed at the University of Chicago. The inputs to this model are "ray files" generated by SERI. Each of these is a set of rays chosen by a Monte-Carlo procedure to characterize the focal plane spatial and directional distribution for a particular set of primary concentrator parameters. An analysis of the optical throughput and net concentration ratio of a large number of different CPCs covering a wide range of design parameters was carried out by tracing 5000 or 10000 rays through each. The peak and average flux concentration and total power delivered were investigated as a function of CPC design acceptance angle, entrance aperture intercept factor, degree of truncation and secondary surface reflectivity. In addition, information on the position and directional distribution of concentrated solar flux in the target plane at the CPC exit aperture was generated. It should be recognized that the much larger scale of the heliostat and primary mirrors results in accumulated optical error distributions which broaden the effective value of θ_s, and that this, together with the choice of a reflecting (n =1) CPC limits, for now, the achievable flux levels to values considerably lower than the record values achieved earlier.

The optimization procedure can be understood by reference to Figure 1. Each point represents the predicted average flux concentration ratio at the exit aperture of a particular CPC plotted versus the corresponding total power delivered. The shape of each CPC is specified completely by three parameters, the aperture entrance radius, R_{ent}, the design acceptance angle ϕ_c, and the truncation fraction, f_t. The effects of all reflection losses are included in the ray trace calculations. For purposes of this calculation, a conservative value of $\rho_2 = 0.91$ was used for the secondary reflectivity. Points for several families of CPCs are shown. The SERI primary consists of 23 facets, each with a spherical contour and hexagonal perimeter, arranged in a roughly rectangular array and each is individually aligned to point towards the same off-axis design aiming point. The maximum convergence angle ϕ at this aiming point for the outer edges of the corner facets is 17.3°, while 19 of the 23 facets are contained within a cone of half angle 15°. The resulting flux distribution at the secondary entrance plane is approximately rotationally symmetric with 90% of the energy lying within a circle 10 cm. in diameter. The maximum secondary geometric concentration ratio for the reflecting (n = 1.0) CPC decreases with increasing ϕ_c according to

$$C_2 = \frac{1}{\sin^2\phi_c} \tag{3}.$$

For fixed acceptance angle, the average flux concentration in the entrance aperture decreases with increasing radius (and therefore decreases in the exit aperture) while the total intercepted power increases as can be seen from the dashed curves in Figure 1. The points represented by the (solid squares) are for untruncated (full height) CPCs of the indicated values of ϕ_c and r_{ent}. At fixed small entrance aperture(e.g. r_{ent} = 2 cm.) the average concentration achieved increases rapidly with decreasing ϕ_c at relatively small sacrifice in power, while for large apertures, the net concentration is not strongly dependent on acceptance angle but the total power is, as ϕ_c increases to "see" the entire primary.

Since one goal of these early experiments was to demonstrate flux levels higher that ever achieved previously in air and at the same time to deliver these levels at a multi-kilowatt level over an appreciable area, the design selected represented a compromise between the two extremes of high flux/low power and low flux/high power. We chose a CPC with an entrance aperture radius of 3 cm. and a design acceptance angle of ϕ_c = 14°. The profile for the untruncated version is shown in Figure 2a.

Finally, the effect of truncation is illustrated in Figure 1 for 6 cases (ϕ_c = 14° and 15°, and R_{ent} = 2, 3, and 4 cm) by the points labeled with an asterisk. As each CPC is truncated to a fraction f_t of its full height, the profile was rescaled to keep the entrance aperture fixed. Thus the exit aperture increases accordingly and the achieved flux eventually decreases. For each case values of f_t from 1.0 down to 0.5 are plotted. For small truncations the effect is negligible and in some cases is more than offset by decrease in the net reflection loss with respect to the untruncated cases.

Based on this analysis, the final secondary design selected has an acceptance angle ϕ_c = 14° and was truncated to 80% of its full height(see Figure 2b), corresponding to a geometric concentration ratio C = 16.4. The nominal design entrance and exit aperture diameters are 6.0 cm. and 1.48 cm. respectively. As can be seen from Figure 1, the combined optical models for

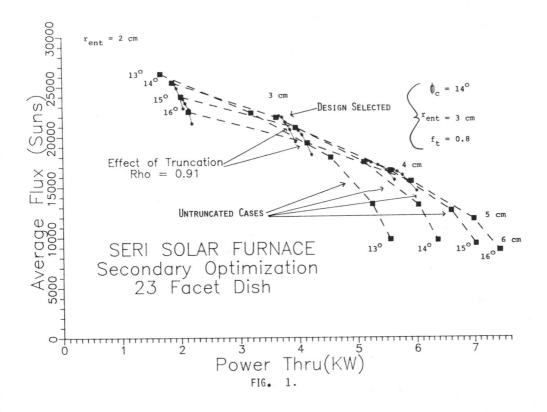

Average Flux (Suns)

r_{ent} = 2 cm

13°
14°
15°
16°

3 cm

→ DESIGN SELECTED

ϕ_c = 14°

r_{ent} = 3 cm

f_t = 0.8

Effect of Truncation
Rho = 0.91

UNTRUNCATED CASES

4 cm

5 cm

13°   14°   15°   6 cm
                    16°

SERI SOLAR FURNACE
Secondary Optimization
23 Facet Dish

Power Thru(KW)

FIG. 1.

the furnace and secondary predict that, when deployed in the focal plane of the SERI high flux solar furnace, an average concentration in excess of 20,000 suns at a net power of about 3.5 kilowatts will be delivered at the exit aperture.

FABRICATION OF PROTOTYPE CPC SECONDARY CONCENTRATORS AND COLD WATER CALORIMETER

Two mechanically identical water cooled CPCs were constructed. The material selected for the concentrator body was an alloy of tellurium and copper which has the thermal properties close to those of pure copper (high conductivity and heat capacity) but is much more readily machined. A cross-section drawing is shown in Figure 3a. The concentrator profile was machined out of the inside of a solid cylinder. Spiral grooves were cut around the outside surface to channel the cooling water.

To measure directly the radiant energy emerging from the CPC exit aperture, a small cold water calorimeter was designed and built. A cross section (not to scale) is shown in Figure 3b. The basic design consists of a cavity absorber whose walls are made of the same tellurium-copper material as the CPC. Grooved fins were machined in its outer surface to distribute cooling water evenly around the outsides and bottom of the cavity walls and the whole piece was inserted into an insulating cylindrical container made of Teflon. The components were held together by an O-ring seal and insulating cover, machined so that the CPC exit aperture was flush with and precisely matched to the cavity opening. The inside of the cavity was painted black with Pyromark™ paint ($\alpha \geq 0.95$) and from the aspect ratio of the cavity the effective absorptivity of the cavity aperture is calculated to be 0.998. Additional details concerning the measurement techniques and analysis are given in (6).

PROFILES OF CPC SECONDARY DESIGNS

$\Phi_c = 14^\circ$, $r_{ent} = 3.0$ cm

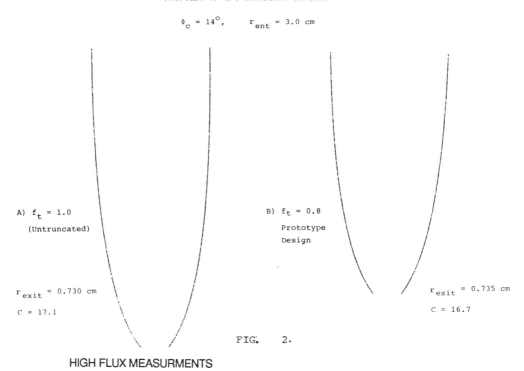

A) $f_t = 1.0$
  (Untruncated)

B) $f_t = 0.8$
  Prototype
  Design

$r_{exit} = 0.730$ cm

C = 17.1

$r_{exit} = 0.735$ cm

C = 16.7

FIG. 2.

HIGH FLUX MEASURMENTS

In a series of experiments conducted at SERI during the summer of 1990, the performance of both secondary units was measured. Detailed quantitative analysis of the data is still underway so that at this time we report only preliminary results. On June 27 and 28, 1990 values of net concentration from 17000 to over 18000 suns were measured for several time intervals of several minutes or more exceeding previous highest values of 16000 suns. Although these measurements with the first unit fell somewhat short of predictions, they did represent a new record for the concentration of sunlight in air ($n = 1$). The major reason for the lower than predicted values with this first prototype unit is thought to be non-specularity of the CPC reflecting surface due to insufficient polishing of the reflector substrate and some characteristics of the nickel undercoat. Other possible factors contributing are source broadening due to circumsolar radiation and short time scale tracking instabilities caused by wind. The second CPC , whose reflector surface was machined and polished with exceptional care, was tested on August 16 and 17th and produced values over 22,000 suns. Although this is very slightly higher than ray trace predictions, based on nominal heliostat, primary and secondary reflectivities, the result is considered to be in excellent agreement with expectations.

FUTURE PLANS AND POTENTIAL APPLICATIONS

These most recent values are more than 25% above previously attained values and represent a new domain of high intensity solar fluxes which can be made available through nonimaging optical designs for delivery to a variety of targets and absorbers with relatively little technical difficulty. In the near future, experiments will be continued with the two prototype CPC secondaries in order to gain a full understanding of the characteristics of the beam and the operation of the secondary. The ray trace code is being used to examine methods of coupling the high flux directly to experiments for studies of material behavior irradiated under extremely high intensities. Various modifications of the secondary shape to limit the emergence angle at the exit are being considered. Planned experiments include investigation of methods for decomposition of toxic waste materials, pumping of high powered lasers and processing of a wide variety of materials.

A. WATER COOLED CPC   B. WATER COOLED CALORIMETER

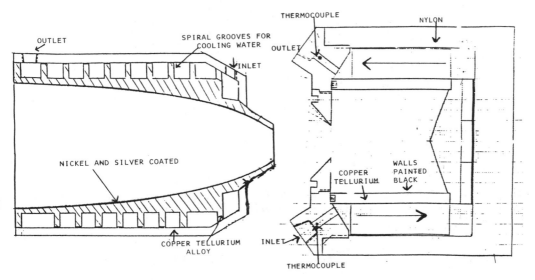

FIG. 3.

VI. ACKNOWLEDGMENT

This work was supported in part by the U.S. DOE under SERI subcontract Nos. XK-4-04070-03 and XX-6-06019-02.

VII. REFERENCES

Welford, W.T. and R. Winston. High Collection Nonimaging Optics. Academic Press, New York (1989).

Gleckman, P., J. O'Gallagher, and R. Winston. "Concentration of Sunlight to Solar-Surface Levels Using Nonimaging Optics." Nature 339, 198 (1989).

Cooke, D., P. Gleckman, H. Krebs, J. O'Gallagher, D. Sagie, and R. Winston. "Sunlight Brighter than the Sun." Nature 346, 802, (1990).

Lewandowski, A.. "The Design of an Ultra High Flux Solar Test Capability." Presented at the 1989 Intersociety Energy Conversion Engineering Conference, August 1989, Davos, Switzerland.

Lewandowski, A., C. Bingham, J. O'Gallagher, R. Winston, D. Sagie, "Performance Characterization of the Solar Energy Research Institute High Flux Solar Furnace." Proceedings Int'l. Energy Agency, 5th Symposium on Solar High Temperature Technologies, August 27-31, 1990, Davos, Switzerland.

O'Gallagher, J. , R. Winston, C. Zmola, L. Benedict, D. Sagie and A. Lewandowski, "Attainment of High Flux-High Power Concentration Using a CPC Secondary and the Long Focal Length SERI Solar Furnace," 1991 American Society of Mechanical Engineers, Solar Energy Division Conference, Reno, Nevada, March 17-22, 1991.

PERFORMANCE AND CHARACTERIZATION OF THE PSA SOLAR FURNACE

Del Arco, J.A.; Rodriguez, J
Plataforma Solar de Almeria
P.O.Box: 22
04200 TABERNAS (Almeria)
SPAIN

ABSTRACT:

A Solar Furnace with a high concentration to obtain very high temperatures is currently under development at the Plataforma Solar de Almeria (P.S.A), a solar energy research center located in Southern Spain.

Our main motivation is the large variety of new possibilities to use the sun's high quality energy. The project goal consists of offering the scientific and engineering community a powerful device to do physical, chemical and material science research experiments as well as developing new Furnace Techniques. The Solar Furnace project is part of the European Community "Access to Large Scale Scientific Installations Program". The primary objective is to melt Zirconium-dioxide and hence pull it into fibers. The concentrating mirror consists of more than 80 facets mounted on a truss structure. Therefore precise alignment of the facets is important. The laser ray tracing method was found to be suitable for the planned arrangement of the PSA solar furnace and a proposal has been made on how to put this system into practice. A computer program for calculation of the reference points was developed as a tool to design the target plane. Furthermore, starting with the basic optical laws of an astigmatic system the development of an aim point strategy for a facet mirror was carried out.

Further progress is necessary to establish reliable methods to measure the temperature and flux density.

KEYWORDS

Solar Furnace, High temperatures, Thermal properties of materials, Zirconia, Optical Alignment.

INTRODUCTION:

The idea of using concentrated solar radiation for high temperature material testing and chemistry is very old but has become much more important in the last few years. Steam reformation, coal gasification and hydrogen production are possible applications for the temperature range achievable in a central tower system as installed at the P.S.A. The research program is directed to the fundamental understanding of the interaction between photons and matter; treatment of materials which are classified as strategic or advanced and with final extreme and hostile applications. For developing materials with the best mechanical properties by quenching high temperature, a special arrangement is needed. The PSA is planning to install a Solar furnace, which is the best option to achieve quick heating in a clean environment. The need for a high concentration ratio close to 10.000 suns makes it necessary to use a dish with exceptionally high optical quality and large rim angle. To verify the performance of the PSA solar furnace for these experiments a study on the temperature distribution of a Z_rO_2 probe in the focus of the parabolic dish concentrator has been carried out. A special sample chamber for testing high temperature ceramic materials has been investigated.

THE PSA SOLAR FURNACE DESIGN:

The main components of the system are: Tracking mirrors (heliostats), concentrator, target table and shutter.

A four plane heliostats field is tracking the sun's movement and reflects the sun light continuously towards the parabolic dish concentrator located 80 m. from the heliostats inside a house, which concentrates the collimated incoming insolation into its focus. A three axis movable testbed is placed at the focal point on a rigid pedestal. The experiments are mounted onto the testbed. An attenuator in the north face close to the furnace housing between the testbed and the heliostat field adjusts the incoming flux. The attenuator also acts as a front door of the furnace building. The control room and work shops are located on the east side of the furnace housing with direct access to the testbed. Figure 1 shows a sketch of the Furnace design.

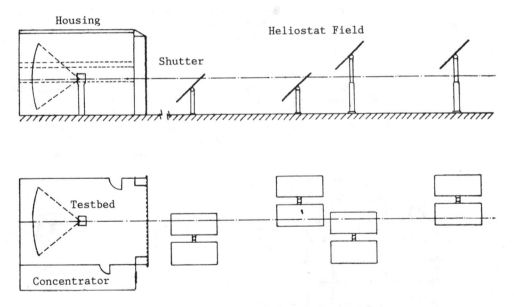

Fig. 1. Solar Furnace Final Global Distribution

THE SOLAR FURNACE COMPONENTS

HELIOSTATS

The heliostats chosen were the MBB-System. They are normally used for central receiver systems. They have been selected from the MBB heliostat field of the PSA. Communication with the heliostats was reestablished after five years repose. The surface of the mirror facets are spherical and the arrangement of the truss structure is not co-planar. The possibility to achieve plane heliostat facets were studied. The structural deformation of the heliostat caused by the influence of gravity was calculated in order to compensate them when canting the heliostat. The heliostats were disassembled, and to adapt them for the solar furnace purpose, planar facets were manufactured and mounted in a co-planar configuration. The size of the reflecting area is 7,44 m. high and 8,21 m. wide, not enough to illuminate the whole dish with one heliostat. Therefore it was necessary to chose four MBB heliostats and to distribute them in the way shown in figure 1. Because of the height of the dish (11 m), two of the four heliostats have to be mounted on a special pedestal. The heliostat surface, their distribution and their height have been calculated to ensure at any working time (10.00 - 14.00 solar time) that the concentrator area is always covered with insolation. The position of the sun is computed every second and a command is communicated from the control to the motor to move accordingly.

Heliostat Field: - 4 MBB heliostats, each consisting of
                    - 16x3.35 m^2 plane, non-concentrating sandwich facets.
                    - Total heliostat reflecting area 53.61 m^2
                    - 90% surface reflectivity

CONCENTRATOR

The parabolic dish concentrator is an altered McDonald Douglas Stirling Dish Concentrator. The dish consists of curved glass mirror facets, a mirror support or truss structure and a pedestal. There are a total of 89 curved facets giving a total reflection area of 98,51 m^2.

The original purpose of the dish is to provide an even flux over the receiver surface of a Stirling engine, rotating an induction generator which converts this mechanical energy into electric energy at 480 V, 3 phase and 60 Hz.

For the PSA furnace only a rigid structure in 9 vertical columns and 12 horizontal rows, so that the whole module is 11,2 m wide and 11,89 m high. There are 5 groups of facets each one with different radius of curvature as shown in fig. 2.

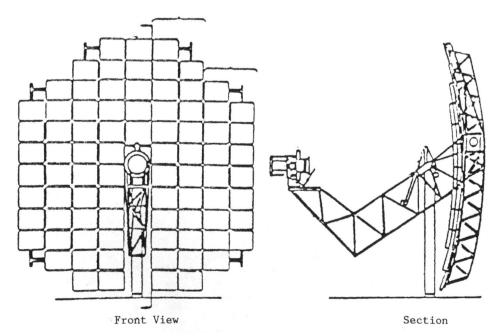

Front View   Section

Fig. 2. McDonnell Douglas Glass/Metal Collector.

The ideal arrangement of each facet group would be in concentric circles around the dish center. The radius of curvature increases with increasing distance from the center, so that an ideal focus can be achieved. Because of the rectangular shape of the facets, the center points are not located on circles, so that the real system is only an approximation. Too many different radius of curvature would be needed to adapt them ideally to each location.

Each facet is mounted at the truss structure by three bol studs, to allow an easy adjustment and alignment.

Concentrator:
- Dimensions: 11.01 m x 10.41 m
- 89 sandwich facets (0,91 m x 1.21 m) with five different curvature radii affixed to a partial surface of a segmented cylinder with individual segment displaced perpendicularly with respect to the cylinder axis.
- Total reflecting surface area: 98.51 m^2
- Reflectivity ρ = 92%
- Focal length f = 7.45 m
- Focal height H_f = 6.09 m
- Estimated peak concentration of about C_{peak} = 8000
- Peak power: 88 kW
- Estimated focus size, about FWHM = 10 cm
- Black body temperature attainable: T = 3,350 K

SHUTTER

The attenuator design is the original design of SANDIA Shutter used in their new solar furnace. The only change will be made in the motor, gear mechanism and the computer control. This is located between the heliostats and concentrator, is an array of horizontal metal slats. It is remotely-controlled and can vary the power from zero to full. Necessary closing speed, flux adjustment precision, back lash and movements are calculated. A communication test between the shutter and the control system is under development. During the course of the experiments the shutter costs minimum shadow on the concentrator. In case of an emergency the shutter is also activated and it blocks the concentrator beam.

Shutter:
- Dimensions: 11.44 m x 11.20 m
- Consists of 30 slats 5.60 m x 0.93 m
- 15896 positions between 0° (open) and 55° (closed), yielding angles accurate to 0.00346° and flux regulation accuracy of $8.01 \cdot 10^{-5}$ and $4.94 \cdot 10^{-5}$ respectively.

TEST TABLE

The sample of material to be tested has to be located at the focus of the dish 7.45 m in front of the vertex and 6,5 m above the ground. To handle the sample and position it at the focal area, two different systems are required, one for the long distance movement (to lower down the whole experiment and to reach it from the ground) and another, very precise one, to put it into the point of the maximum flux-intensity. Different systems for this purpose, hydraulic and electrical were considered. For the intensity-measurement in the focus, the sample is replaced by a radiometer. The set-up must allow for a quick exchange of sample and measuring instrument. All these requirements must be satisfied without too massive construction not to produce a big shadow area. All obstacles between the dish and the heliostats low down the effective aperture area.

Test Table: - Dimensions: 0.7 m x 0.6 m
                     - Movable on 3 axes
                     - Displacement paths x= 0.92 m, y= 0.66 m (along the optical
                       axis), z= 0.50m.

OPTICAL ALIGNMENT:

For alignment and testing, various dish optical analyzing methods have been
developed in the last years. Till now, there is no method favourized as a general
allpurpose system. The prealignment of heliostat mirror facets is normally done
by inclinometer, a very handsome and precise method. The optical method used to
align the facets of the Solar Dish is *"On-Axis Alignment Method"*.

The so called on axis alignment is the most evident way of adjusting the mirror
facets of the dish, to gain a high flux peak in the focal area. The procedure is,
to cover all facets except the one to be aligned. The parallel incident sunlight
from the heliostats will be reflected onto a scatter plate, located at a distance
equal to the focal length (f=7.45m) in front of the dish vertex.

An image will be visible near or at the focal point. To align the facet, the nuts
of the three bolt studs have to be turned· until the maximum of the intensity
distribution at the scatter plate will coincide with the focal point. With the
naked eye it is impossible to recognize the maximum of the intensity of the image.
"A high-intensity flux mapper has been developed at the Solar Energy Research
Institute (SERI Golden, Colorado). This instrument provides an accurate
description of the two dimensional distribution of flux at the focal plane of
concentrating solar collectors where the flux density may be as high as 1000
suns". The flux mapper consists of a ceramic scatter plate, video camera with
appropriate filters and data acquisition system. It was determined to have a
standard deviation from perfect cosine (Lambertian) scattering. The
proportionality of the intensity of the scattered light to the flux intensity at
this point is applied to map the one-dimensional flux intensity distribution.
To know the absolute value of the radiation, an exact calibration of the system
is necessary. In this case only the maximum of the distribution has to be figured
out so that the relative values can be used and no calibration is necessary. A
similar system has been developed by the DLR, called HERMES (*Heliostat and
Receiver Measurement System*) and could be used for this purpose.

The disadvantage of this method is that the accuracy depends on the parellelity
of the sunlight reflected from the heliostats. The optical influences of the
atmosphere (scattering) and the tracking errors of the heliostats are increasing
the uncertainty of the measurement. An alternative to the sunlight would be an
artificial light source, for example the headlight of a car or a ship, located far
away from the dish. The problem would be the shadows caused by the heliostats and
the shutter. For testing of the alignment, some other methods based on the
principle of the inverted ray path have been developed.

HIGH-TEMPERATURE MATERIAL EXPERIMENTS:

Advanced ceramics are currently of great interest to the materials community.
Several sensors for this interest in ceramics are their high strength-to-weight
ratio at high temperature and at tractive tribological properties.

The first tests are planned to be conducted by visiting French and Spanish
organizations. The Departamento de Física Universidad de Sevilla (Spain)
researches on advanced ceramics in close cooperation with European and USA
research groups. The proposed research shall analyze Z_rO_2 fibre production and

characterization. The ultimate use of such Z_rO_2 fibers would be in the production of a shock and fatigue resistant ceramic composite for high temperature applications. It is generally accepted that the best mechanical properties are founded on multiphase materials, of which Z_rO_2 is a clear example. Appropriate alloying with other oxides (MgO, CaO, Y_2O_3) can produce a material in which two or three of the Z_rO_2 allotropic forms (monoclinic, tetragonal and cubic) coexist. Since the mechanical properties are heavily dependant not only on the existence of the phases themselves, but on their relative volume and also the size of the tetragonal particle during the cubic phase, control of particle growth and the "nucleation" process is essential. Over 2000°C thermal treatment during the cubic phase, followed by quick tempering is intrinsic to the achievement of this goal.

This need for speed and also for a clear atmosphere, make the solar furnace the best option for developing this material. Also, as a parallel process, Z_rO_2's melting point (2600°C) must be reached. The fiber produced will be introduced into a ceramic matrix, obtaining a composite with better physical properties that the present C or SiC fiber whisker matrix.

Another organization interested in solar furnace experiments is the Laboratoire de Physique de Materiaux - C.N.R.S Meudon Cedex (France). They are interested by a system allowing them to heat, treat or melt Zirconia permitting to quench high temperature phase retaining random distribution of stabilizing atoms (Ca, Y, Mg...)

REFERENCES:

Biggs, F. C.N. Vittitoe " The HELIOS Model for the Optical Behavior of Reflecting Solar Concentrators", Sandia National Lab., Albuquerque, Report Nº SANDIA 76-0347 page 51-53 1976.

Craig, E. Tyner "Proceedings of the Solar Thermal Technology Conference" SAND 87-1258.UC-62 August 1987.

Gaul, H.W. "A High-Intensity Flux Mapper for Concentrating Solar Collector", available from: National Technical Information Service U.S. Department of Commerce 5285 Port Royal Road Springfield, VA 22161.

Henker, H. "A Solar Cooker for Nepal", II. Seminario de Energía Solar Universidad del Norte, Antofogasta, Chile; Marzo de 1.981.

Jeter, S.M. "The Distribution of Concentrated Solar Radiation in Paraboloidal Collectors", Jornal of Solar Energy Engineering Vol. 108/219-225 August 1986.

Kamadu, O. "Theoretical Concentration and Attainable Temperature in Solar Furnaces" Solar Energy Vol. 9 Nº 1/p.39-40 1965.

Masterson, K. Gaul "Optical Characterization Method for Concentrating Solar Collectors Using Reverse Illumination" Golden, Colorado SERI/TP-641-1179 July 1981.

Walton, J.D. "Energy Technology Handbook", Mc.Graw-Hill Book Company Copyright 1977.

William, B. Ph.D. Stine "Progress in Parabolic Dish Technology", SERI /SP-220-3237.

THE SERI HIGH FLUX SOLAR FURNACE

Allan Lewandowski and Carl Bingham

Solar Energy Research Institute
1617 Cole Boulevard
Golden, Colorado 80401 USA

ABSTRACT

This paper describes a unique new solar furnace at the Solar Energy Research Institute (SERI) that can generate a wide range of flux concentrations to support research in areas including materials processing, high-temperature detoxification and high flux optics. The furnace is unique in that it uses a flat, tracking heliostat along with a long focal length-to-diameter (f/D) primary concentrator in an off-axis configuration. The experiments are located inside a building completely outside the beam between the heliostat and primary concentrator. The long f/D ratio of the primary concentrator was designed to take advantage of a nonimaging secondary concentrator to significantly increase the flux concentration capabilities of the system. A turning mirror allows the beam to be redirected onto a horizontal target. Flux characterization results are reported for various configurations of the system.

KEYWORDS

Solar furnaces, solar concentrators, high flux optical systems, primary concentrators, secondary concentrators, high flux measurements, solar materials processing, high temperature solar detoxification.

INTRODUCTION

Since December 1989, SERI has been operating a unique new solar furnace capable of generating extremely high flux concentrations. Aerial photographs of the facility are shown in Fig. 1. The High Flux Solar Furnace (HFSF) has supported a number of research experiments since it became operational. One of the significant results was the demonstration of flux concentrations of 21,000 suns using a nonimaging secondary concentrator (O'Gallagher and co-workers, 1991) designed and built at the University of Chicago. Even higher concentrations are anticipated with new nonimaging devices under development (Gleckman, Winston and O'Gallagher, 1988).

Specific experimental work has already been conducted in materials processing and high-temperature, gas-phase detoxification of hazardous wastes. The objective of the materials-processing effort is to utilize the high heating rates and photon properties made available by concentrated solar flux to modify surfaces of selected materials in order to obtain special desirable properties. Of particular interest are corrosion/erosion-resistant films, tribological coatings, and electronic and superconducting materials. Successful experiments have been conducted in a number of these areas (Pitts, 1990). Recently, researchers were able to demonstrate high-temperature, gas-phase detoxification of hazardous waste, in this case trichloroethylene (TCE). A destruction efficiency of 99.99% was measured.

Fig. 1. Photographs of SERI's High Flux Solar Furnace. View (a) is from the east, showing flat, tracking heliostat (right), control building (center), and primary concentrator building (left) in an off-axis configuration. Experiments are located inside the control building. View (b) is from the north, showing the faceted primary concentrator (background).

In addition to the detoxification and materials-processing work, a number of tests have been conducted to define the flux capabilities of the HFSF. The remainder of this paper will address the design of the HFSF, its major components, and selected results of flux characterization experiments.

FURNACE DESIGN

The performance objectives set for the HFSF resulted in a unique design (Lewandowski, 1989). To enable support of varied research objectives, designers made the HFSF capable of achieving both extremely high flux concentrations and a wide range of flux concentrations. The ability to maintain a stationary focal point was necessary because of the nature of many anticipated experiments. It was also desirable to move the focal point off axis. An off-axis system allows for considerable flexibility in size and bulk of experiments and eliminates beam blockage and consequent reduction in power.

In particular, achieving high flux concentration in a two-stage configuration (an imaging primary in conjunction with a nonimaging secondary concentrator) dictates a longer f/D for the primary than for typical single-stage furnaces. Typical dish concentrators used in almost all existing solar furnaces are about $f/D = 0.6$. To effectively achieve high flux concentration, a two-stage system must have an $f/D \sim 2$. The final design of the HFSF has an effective f/D of 1.85.

At this f/D, it was also possible to move the focal point considerably off axis (30°) with very little degradation in system performance. This is because of the longer f/D and partly because of the multifaceted design of the primary concentrator. This off-axis angle allows the focal point and a large area around it to be completely removed from the beam between the heliostat and the primary concentrator.

The major HFSF components include the heliostat, faceted primary concentrator, attenuator, turning mirror, shutter, XYZ positioning table and a PC-based data acquisition and control system (Bingham and Lewandowski, 1991).

Both the heliostat (31.8 m^2) and primary concentrator (11.5 m^2) are front-surface, UV-enhanced, aluminum reflectors on a glass substrate. An aluminum reflector was chosen because of its relatively high reflectance in the UV portion of the solar spectrum. A thin-film coating both enhances the UV reflectance and protects the surface from environmental effects. The heliostat surface is flat, with 10 reflector panels, each with four facets. The primary concentrator has 23 hexagonal facets mounted in a plane, each with an identical radius of curvature (14 m). The pointing of each primary facet is adjustable so that various flux patterns can be obtained at the target plane. Maximum system power is 9.4 kW at a typical direct normal irradiance value of 1000 W/m^2.

The attenuator provides control of the flux reaching the focal plane. The attenuator consists of two 1.0-m^2 (140-cm x 73-cm) vertically opposing plates located 1.8 m in front of the focal plane and 5.2 m from the center facet of the primary concentrator. At this location, the attenuator does not require active cooling. In automatic mode, the attenuator can be controlled as a function of direct normal irradiance (measured by an Eppley Normal Incidence Pyrheliometer) to maintain a constant flux, or as a function of sample temperature to maintain a relatively constant temperature.

Compared to the attenuator, faster response is given by a separate actively cooled shutter. The shutter consists of two air-actuated 37-cm x 97-cm plates located 83 cm in front of the focal plane. The shutter can provide either an emergency cutoff of the flux from the focal point in a fraction of a second, or precise time control of flux exposure for certain experiments. The shutter can be actuated manually with a switch in the control room or automatically from the data acquisition system, either as a function of time or sensor input such as temperature.

Recently, a new turning mirror system was installed and tested. The turning mirror intercepts the beam approximately 0.5 m in front of the focal plane and redirects it downward. This allows targets to be placed in a horizontal orientation. The mirror (46 cm x 76 cm) is held to a flat, actively cooled platen by vacuum. Several different front surface reflectors on glass substrates are available for use. When not needed for nominal furnace operations, the turning mirror can be rotated out of the way.

A three-axis positioner (the XYZ table) is used to move the experiment in and around the focal point. The table is capable of a preprogrammed motion sequence with adjustable speed and acceleration with a precision of 0.25 mm. Capacity of the table is 880 kg.

The data acquisition and control system uses three PCs. One 386 PC is dedicated to data acquisition, which collects data but also transmits commands for control of the attenuator

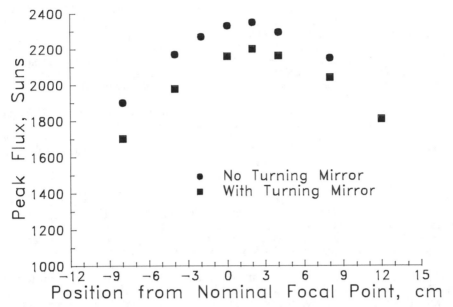

Fig. 2. Maximum instantaneous peak flux measurements from circular-foil calorimeters. Data are from single-stage tests both with and without a turning mirror.

and shutter. Another 286 PC controls the heliostat, while a third 386 can multi-task the XYZ table control and the flux mapping system. The latter two PCs are switched to a single monitor and keyboard. The central element of the data acquisition system is a Hewlett Packard 3497A Data Acquisition/Control Unit. It is currently wired for more than 100 sensor inputs and control outputs, including digital and analog signals.

The video monitoring capability at the HFSF includes three black-and-white surveillance cameras monitoring the heliostat, the primary concentrator, and the target, and two color cameras aimed at the target (one with remote zoom and focus). The color video output can be time-stamped and recorded on a VHS-format video recorder.

The flux mapping system uses a black and white video camera and frame grabber in conjunction with the commercial image analysis software package for flux profile analysis. Both absolute flux measurements and flux maps are taken from a single, actively cooled, aluminum oxide-coated flux plate.

PERFORMANCE RESULTS

We have characterized the flux at the HFSF in single- and two-stage configurations (Lewandowski and co-workers, 1991). The results presented in this paper are selected samples of the data that have been acquired and analyzed. All data so far have been taken with all primary facets aimed at a single target point, defined as the nominal focal point.

The objective of characterizing the flux in a single-stage configuration is to map the concentration profile over a volume of space around the nominal focal point. We have conducted a series of experiments to define the peak flux and the flux profiles as a function of position along the optical axis. These tests have been conducted in several series since the HFSF became operational.

Representative results of these tests along the optical axis are shown in Fig. 2. Peak flux data were acquired at positions as much as 8 cm from the focal point (+ direction is away from the primary).   The figure shows the concentration in suns (incident calorimeter intensity/direct normal irradiance) for tests over several successive days.   These results indicate a broad peak in maximum concentration. These data indicate that the location of maximum peak flux exists somewhere in the range of 0-4 cm beyond the nominal focal point.   This is probably due to the difference between actual and modeled design parameters (e.g., slight primary facet pointing errors, real systematic optical errors vs. modeled random errors). Peak values of almost 2400 suns were achieved.

Performance of the HFSF with the turning mirror has also been characterized. The results of tests are also shown in Fig 2. In this case, a front-surface, UV-enhanced, aluminum reflector was installed on the turning mirror system. The peak flux obtained in these tests is reduced by the reflectance of the mirror. The turning mirror, as expected by analytical modeling, does not contribute to a spread of the beam.

The objective of the two-stage tests was to demonstrate the extremely high flux capability of the HFSF.   The first step toward this goal was to design and fabricate a reflective-type secondary concentrator.   This task was accomplished at the University of Chicago (O'Gallagher and co-workers, 1991). The secondary concentrator was a compound parabolic design with a 14° acceptance angle, truncated to 80% of its full height. It has an entrance aperture of 6 cm and an exit aperture of 1.4 cm. The reflective surface is protected silver.

A cold-water calorimeter was designed to mate with the secondary concentrator to measure the power at the exit of the device. Tests of this calorimeter using a known heat source indicated that the necessary flow and temperature measurements were accurate to within 2%. Data were acquired with the cold-water calorimeter until a sufficient length of steady-state performance was achieved. The concentration at the exit of the secondary concentrator is the power measured by the calorimeter divided by the area of the exit divided by the direct normal irradiance. This value is the average concentration at the exit.

To assess the performance of the secondary concentrator as a function of position along the optical axis, we conducted a series of tests from +6 to -4 cm from the nominal focal point. This range covers the positions where maximum single-stage flux occurred in the previously described tests. The secondary concentration as a function of position along the optical axis is shown in Fig. 3. At the optimum locations, concentrations of more than 21,000 suns were demonstrated.   The data shown are maximums of time-averaged (over a 10-s period) instantaneous concentrations.   The concentrations reported for the secondary tests are average concentrations. Based on analytical predictions, the peak flux at the exit of the secondary should be considerably more than 25,000 suns.   Both measured average concentration and peak flux concentration are higher than demonstrated anywhere else for reflective secondaries.

CONCLUSIONS

The HFSF has demonstrated the capability to achieve a wide range of solar flux concentrations.   A number of experiments have been conducted, including potentially attractive solar technologies (e.g., materials processing and detoxification) and flux characterization. The HFSF continues to be a valuable research tool which can assist in the implementation of a variety of attractive solar technologies and processes. Efforts are under way to allow use of this facility by outside groups that are interested in its special capabilities.

ACKNOWLEDGMENT

This research was supported by the U.S. Department of Energy under Contract No. DE-AC02-83CH10093. In addition, we would like to acknowledge the significant contributions of Joe O'Gallagher and Roland Winston and their colleagues at the University of Chicago. The critical participation of several SERI researchers, in particular Greg Glatzmaier, Gary Jorgensen, Roland Pitts, and George Yeagle, is appreciated.

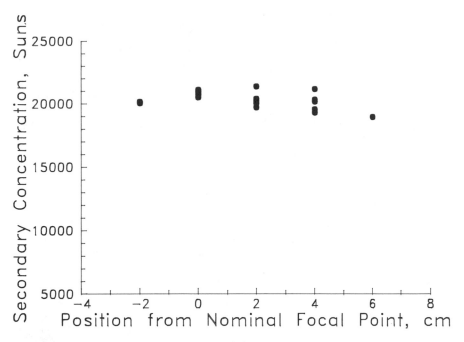

Fig. 3. Average concentration at exit of secondary concentrator plotted as 10-s averages of instantaneous data. Scatter in the data at a given position is due primarily to gusty wind conditions, which caused the beam location at the target to vary slightly.

REFERENCES

Bingham, C. and A. Lewandowski (1991), Capabilities of SERI's High Flux Solar Furnace, Proceedings of the 1991 ASME/JSME/JSES International Solar Energy Conference, Reno, Nevada, March 18-20.

Gleckman, P., R. Winston and J. O'Gallagher (1988), Approaching the Irradiance of the Surface of the Sun, Proceedings of the Annual Meeting of the American Solar Energy Society, 374.

Lewandowski, A. (1989), The Design of an Ultra-High Flux Solar Test Capability, 24th Intersociety Energy Conversion Engineering Conference, IECEC-89, 4, 1979. Institute of Electrical and Electronics Engineers

Lewandowski, A. C. Bingham, J. O'Gallagher, R. Winston, and D. Sagie (1991), One- and Two- Stage Flux Measurements at the SERI High Flux Solar Furnace, Proceedings of 1991 ASME/JSME/JSES International Solar Energy Conference, Reno, Nevada, March 18-20.

O'Gallagher, J., R. Winston, C. Zmola, L. Benedict, and A. Lewandowski (1991), Attainment of High Flux-High Power Concentration using a CPC Secondary Concentrator and the Long Focal Length SERI Solar Furnace, Proceedings of 1991 ASME/JSME/JSES International Solar Energy Conference, Reno, Nevada, March 18-20.

Pitts, R. (1990), Materials Processing Using Highly Concentrated Solar Radiation, 25th Intersociety Energy Conversion Engineering Conference, IECEC-90, 6, 262. Institute of Electrical and Electronics Engineers

2.20 Solar Heat Storage

ON THE IMPORTANCE OF TEMPERATURE
IN PERFORMANCE EVALUATIONS
FOR SENSIBLE THERMAL ENERGY STORAGE SYSTEMS

M. A. Rosen

Department of Mechanical Engineering, Ryerson Polytechnical Institute
Toronto, Ontario, Canada, M5B 2K3

ABSTRACT

The energy and exergy efficiencies for a simple sensible thermal energy storage (TES) system are derived and compared, and the differences between the two efficiencies demonstrated. The results indicate that exergy analyses should be used when analyzing the performance of TES systems because they weight the usefulnesses of heat at different temperatures on equivalent bases (i.e., exergy analyses account for differences in temperature levels). Energy analyses tend to present overly optimistic views of performance, since they neglect the temperature levels associated with a heat flow. The results are illustrated by examining several TES systems.

KEYWORDS

Thermal energy storage; efficiency; energy; exergy; temperature.

INTRODUCTION

Thermal energy storage (TES) systems have been utilized in energy systems for a number of years, and are continually being investigated (Bejan, 1982; Hahne, 1986; Beckman and Gilli, 1984). At present, there exist no generally accepted evaluation criteria for TES systems. In addition, most of the existing performance measures disregard the temperature associated with the heat injected into and recovered from a TES.

The author feels that exergy efficiencies must be examined when determining how advantageous is one TES relative to another. Examining energy efficiencies alone can result in misleading conclusions because such efficiencies weight all thermal energy equally. Exergy efficiencies acknowledge that the usefulness of thermal energy depends on its quality and quantity, where quality is related to the temperature level associated with the heat.

Exergy analysis is based on the first and second laws of thermodynamics. Unlike energy, exergy is not a conserved quantity; it is consumed due to process irreversibilities. Flows of exergy are associated with flows of matter, heat and work. Details on the theoretical and applied aspects of exergy analysis are presented in many references (Moran, 1982). Other sources discuss the application of exergy analysis to TES systems (Bejan, 1982; Hahne, 1986; Beckman and Gilli, 1984). The present author has applied exergy analysis to many facets of TES performance (Rosen and Hooper, 1987, 1990, 1991a, 1991b; Rosen and co-workers, 1988, 1989), including, as reported in another paper in this proceedings, stratification (Rosen and Hooper, 1991b).

This paper discusses and demonstrates the importance of temperature levels in performance evaluations for TES systems. The energy and exergy efficiencies for TES systems are examined and compared. An illustrative example is provided.

THEORY

Balances for Energy and Entropy

Consider a process involving only heat interactions, and occuring in a closed system for which the state is the same at the beginning and end of the process. Balances of energy and entropy, respectively, can be written for the system as follows:

$$\sum_r Q_r = 0 \tag{1}$$

and

$$\sum_r (Q_r/T_r) + \Pi = 0 \tag{2}$$

where Q_r denotes the heat transferred into the system across region r on the system boundary at a temperature T_r, and Π denotes the entropy created within the system. For irreversible processes $\Pi > 0$, and for reversible processes $\Pi = 0$. The left sides of Eq. 1 represents the net amount of energy transferred into the system, and the first and second terms on the left side of Eq. 2 represent respectively the net amount of entropy transferred into the system with heat and the entropy created within the system. The right sides of Eqs. 1 and 2 represent the amounts of energy and entropy, respectively, accumulated within the system (zero in the present case). Clearly, energy is conserved, while entropy is not.

Balance for Exergy

The exergy associated with a heat flow Q at a constant temperature T is denoted E^Q, and can be expressed as

$$E^Q = \tau Q \tag{3}$$

where τ denotes the "exergetic temperature factor," defined as

$$\tau \equiv 1 - T_o/T \tag{4}$$

Here, T_o is the temperature of the environment.

It is useful at this point to investigate further the relation between several temperature parameters. The exergetic temperature factor τ is illustrated as a function of the temperature ratio T/T_o in Fig. 1. In addition, the parameters T/T_o and τ are compared with the temperature T in Table 1 for above-environmental temperatures (i.e., for $T \geq T_o$). This temperature range includes the values of interest for most TES systems.

A balance can be written for exergy by multiplying Eq. 2 by T_o and subtracting the result from Eq. 1:

$$\sum_r E_r^Q - I = 0 \tag{5}$$

where I denotes the exergy consumption, which can be expressed as

$$I = T_o \Pi \tag{6}$$

The first term on the left side of Eq. 5 represents the net input of exergy associated with heat, and the second term the exergy consumption. The balance demonstrates that exergy is subject to a nonconservation law.

The Reference Environment

The reference environment is postulated to be in stable equilibrium. It acts as an infinite system, and is a sink and source for heat and materials. It experiences only internally reversible processes in which its intensive state (e.g., its temperature T_o) remains constant. In the present undertaking, it is only necessary to consider the value of T_o.

Efficiencies

Energy efficiency (η) and exergy efficiency (ψ) definitions for processes are often written according to the following general statements:

$$\eta = \frac{Energy\ in\ products}{Total\ energy\ input} = 1 - \frac{Energy\ loss}{Total\ energy\ input} \tag{7}$$

and

$$\psi = \frac{Exergy\ in\ products}{Total\ exergy\ input} = 1 - \frac{Exergy\ loss\ and\ consumption}{Total\ exergy\ input} \tag{8}$$

Exergy efficiencies often give more illuminating insights into process performance than energy efficiencies because they weight energy flows according to their exergy contents, and they separate inefficiencies into those associated with effluent losses and those due to irreversibilities (3). In general exergy efficiencies provide a measure of potential for improvement relative to the ideal.

ANALYSIS OF A GENERAL TES SYSTEM

Model

Consider the overall heat storage process for the general TES system shown in Fig. 2. In this simplified model, a quantity of heat Q_c is injected into the system at a constant temperature T_c during a charging period. After a storing period, a quantity of heat Q_d is recovered from the system at a constant temperature T_d during a discharging period. During all periods, a quantity of heat Q_l leaks from the system at a constant temperature T_l, and is lost to the surroundings.

For normal TES applications, the temperatures T_c, T_d and T_l range from as low as the environment temperature T_o to as high (in theory) as infinity. In addition, the discharging temperature can not exceed the charging temperature. Hence, the exergetic temperature factors for the charged and discharged heat are subject to the constraint

$$0 \leq \tau_d \leq \tau_c \leq 1 \tag{9}$$

Analysis

The energy and exergy balances in Eqs. 1 and 5, respectively, can be written for the modelled system as

$$Q_c = Q_d + Q_l \tag{10}$$

and

$$E_c^Q = E_d^Q + E_l^Q + I \tag{11}$$

With Eq. 3, the exergy balance can be expressed as

$$Q_c \tau_c = Q_d \tau_d + Q_l \tau_l + I \tag{12}$$

Following the general energy and exergy efficiency statements in Eqs. 7 and 8, the energy efficiency can be written for the modelled system as

$$\eta = \frac{Q_d}{Q_c} \tag{13}$$

and the exergy efficiency as

$$\psi = \frac{E_d^Q}{E_c^Q} = \frac{Q_d \tau_d}{Q_c \tau_c} = \frac{\tau_d}{\tau_c} \eta \tag{14}$$

Equations 3 and 13 were used in obtaining the alternate forms for ψ in Eq. 14.

Comparison of Energy and Exergy Efficiencies

An illuminating parameter for comparing the efficiencies is the following ratio: ψ/η. For the general TES system above, Eq. 14 implies that the energy-efficiency-to-exergy-efficiency ratio can be expressed as

$$\frac{\psi}{\eta} = \frac{\tau_d}{\tau_c} \tag{15}$$

With Eq. 4, Eq. 15 can be alternatively expressed as

$$\frac{\psi}{\eta} = \frac{(T_d - T_o)T_c}{(T_c - T_o)T_d} \tag{16}$$

The ratio ψ/η is plotted against τ_d for several values of τ_c in Fig. 3. It is seen that ψ/η varies linearly with τ_d, for a given value of τ_c. Figure 3 also shows that if the product heat is delivered at the charging temperature (i.e., $\tau_d = \tau_c$), $\psi = \eta$, while the product heat is delivered at the temperature of the environment (i.e., $\tau_d = 0$), $\psi = 0$ regardless of the charging temperature. In the first case, there is no loss of temperature during the entire storage process, while in the second there is a complete loss of temperature. The largest deviation between values of ψ and η occurs in the second case.

The deviation between ψ and η is significant for most present day TES systems. This can be seen by noting that (1) most TES systems operate between charging temperatures as high as $T_c = 130°C$ and discharging temperatures as low as $T_d = 40°C$, and (2) a difference of about 30°C between charging and discharging temperatures is utilized in most TES systems (i.e., $T_c - T_d = 30°C$). With Eq. 4 and $T_o = 10°C$, the first condition can be shown to imply

$$0.1 \leq \tau_d \leq \tau_c \leq 0.3 \qquad \text{(for most present-day systems)} \tag{17}$$

Since it can be shown with Eq. 4 that

$$\tau_c - \tau_d = \frac{(T_c - T_d)T_o}{T_c T_d} \tag{18}$$

the difference in exergetic temperature factor varies approximately between 0.06 and 0.08. Then the value of the exergy efficiency is approximately 50% to 80% of that of the energy efficiency.

ILLUSTRATIVE EXAMPLE

In this example, the ratio ψ/η is determined for a simple TES having charging and discharging temperatures ranging between 40°C and 130°C. The reference-environment temperature is set at $T_o = 10$°C.

The results are listed in Table 2. The energy and exergy efficiencies are equal only when the charging and discharging temperatures are equal (i.e., $T_d = T_c$). No values of the ratio ψ/η are reported for the cases when $T_c < T_d$ since such situations are not physically possible. The energy and exergy efficiencies differ (with the exergy efficiency always being the lesser of the two) when $T_d < T_c$, and the difference becomes more significant as the difference between T_c and T_d increases.

The energy efficiencies tend to appear overly optimistic, in that they only account for losses attributable to heat leakages, but ignore temperature degradation. In accounting for both of these factors, the exergy efficiencies are more realistic and meaningful.

CLOSURE

The results demonstrate the importance of temperature in performance evaluations for TES systems, and, consequently, the need for exergy-analysis-based TES evaluation methodologies. For TES systems, exergy efficiencies are more illuminating than energy efficiencies because they weight heat flows on an equivalent basis. Exergy efficiencies are sensitive to values of (1) the fraction of the heat injected into a TES that is recovered, and (2) the temperature at which heat is recovered from a TES, relative to the temperature at which it is injected. Energy efficiencies are only sensitive to the first of the above factors.

Energy efficiencies are good approximations to exergy efficiencies when there is little temperature degradation during the entire storage process. This result is due to the fact that in that case, equal quantities of heat have similar "qualities." Energy efficiencies are poor approximations to exergy efficiencies in most practical situations because for most practical situations heat is injected and recovered at moderate and low temperatures. Thus the use of energy efficiencies in practical situations can lead to erroneous interpretations and conclusions.

ACKNOWLEDGEMENTS

Support for this research was provided by Energy, Mines and Resources Canada, and the Natural Sciences and Engineering Research Council of Canada.

NOMENCLATURE

I exergy consumption
Q heat
T temperature
E^Q thermal exergy
η energy efficiency
Π entropy production
τ exergetic temperature factor
ψ exergy efficiency

Subscripts

c charging
d discharging
l loss
o environmental state
r region of heat interaction

REFERENCES

Beckman, G. and P.V. Gilli (1984). *Thermal Energy Storage.* Springer-Verlag, New York.

Bejan, A. (1982). Thermal energy storage." Chapter 8 of *Entropy Generation through Heat and Fluid Flow.* Wiley, Toronto, pp. 158-172.

Hahne, E. (1986). Thermal energy storage: some views on some problems." *Proc. Int. Heat Transfer Conf.,* San Francisco, pp. 279-292.

Moran, M.J., (1982). *Availability Analysis: A Guide to Efficient Energy Use.* Prentice-Hall, Englewood Cliffs, N.J.

Rosen, M.A. and F.C. Hooper (1987). Evaluation and comparison of thermal Energy storage systems." *Seasonal Thermal Energy Storage Newsletter,* IX(3), 5-8.

Rosen, M.A. and F.C. Hooper (1990). Non-deceptive measures for the evaluation of the performance of thermal energy storage systems." *Sustainable Energy Choices for the 90's: Proc. 16th Annual Conf. of Solar Energy Soc. of Can.,* 18-20 June, Halifax, N.S., pp. 84-89.

Rosen, M.A. and F.C. Hooper (1991a). Meaningful performance evaluation criteria for thermal energy storage systems." *Proc. 11th Can. Congr. of Applied Mechanics,* 2-6 June, Winnipeg, Man.

Rosen, M.A. and F.C. Hooper (1991b). Evaluating the energy and exergy contents of stratified thermal energy storages for selected storage-fluid temperature distributions. *Proc. Int. Solar Energy Soc. 1991 Solar World Congress,* 17-24 Aug., Denver.

Rosen, M.A., F.C. Hooper and L.N. Barbaris (1988). Exergy analysis for the evaluation of the performance of closed thermal energy storage systems." *Trans. ASME J. Solar Energy Engineering,* 110, 255-261.

Rosen, M.A., F.C. Hooper, L.N. Barbaris and D.S. Scott (1989). Thermodynamic considerations in the evaluation of thermal energy storage systems." In Cai Ruixian and M.J. Moran (Eds.), *Thermodynamic Analysis and Improvement of Energy Systems.* Pergamon, Toronto. pp. 590-595.

TABLE 1. Relation between Several Temperature Parameters for Above-Environmental Temperatures[*]

T/T_o	T (K)	τ
1.00	283	0.00
1.25	354	0.20
1.50	425	0.33
2.00	566	0.50
3.00	849	0.67
5.00	1415	0.80
10.00	2830	0.90
100.00	28,300	0.99
∞	∞	1.00

[*] The reference environment temperature is set at $T_o = 10°C = 283$ K.

TABLE 2. Values of the Ratio ψ/η for a Range of Practical Values for T_d and T_c[*]

Discharging temperature, T_d (°C)	Charging temperature, T_c (°C)			
	40	70	100	130
40	1.00	0.55	0.40	0.32
70	–	1.00	0.72	0.59
100	–	–	1.00	0.81
130	–	–	–	1.00

[*] The reference environment temperature is set at $T_o = 10°C = 283$ K.

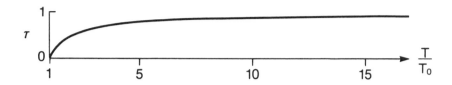

Fig. 1. The relation between the exergetic temperature factor, τ, and the absolute temperature ratio, T/T_o.

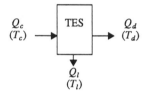

Fig. 2. The overall heat storage process for a general TES system. Shown are heat flows, and associated temperatures at the TES boundary (terms in parentheses).

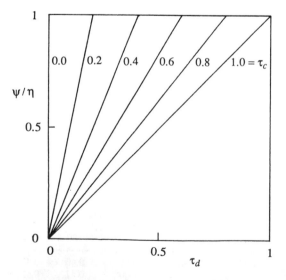

Fig. 3. Energy-efficiency-to-exergy-efficiency ratio, ψ/η, as a function of the discharging exergetic temperature factor τ_d, for several values of the charging exergetic temperature factor τ_c.

SALT CERAMIC THERMAL ENERGY STORAGE
FOR SOLAR THERMAL CENTRAL RECEIVER PLANTS

E. Hahne, U. Taut, U. Groß

ITW, Institut für Thermodynamik und Wärmetechnik
University of Stuttgart, Pfaffenwaldring 6, 7000 Stuttgart 80,
FRG

ABSTRACT

A new salt-ceramic composite storage material for high temperature application has been developed. The heat is stored both as latent and sensible heat. The new material is inserted into the storage unit only in certain regions at the top- and the bottom-end, while most of the storage material consists of conventional ceramic material as known from heat blast furnaces. With proper arrangement of the new material a buffer effect is achieved and a large increase in the storage capacity can be obtained.

KEYWORDS

Solar energy storage; latent heat storage; phase change; regenerator; salt-ceramic; high temperature storage.

INTRODUCTION

A new composite salt-ceramic storage material has been developed recently for application in gas cooled solar thermal central receiver plants. Temperature distribution and storage capacity in this and conventional ceramic material is investigated numerically. The salt-ceramic material consists of salt as the Phase Change Material (PCM) contained in the pores of a ceramic material. This new material is formed in checker bricks and arranged as shown in Fig. 1. A direct contact heat exchange between gas and surface of the storage material is achieved. The diameter of the flow ducts is 10 mm. Heat is stored as latent heat in the PCM and as sensible heat in both the ceramic material and the PCM (Tamme, 1986).

The technical concept of such new heat storage units, the geometrical shapes of the checker bricks and their arrangement inside the unit is the same as that used in regenerators for high temperature application (for example in a heat blast furnace). Numerical codes for the calculation of regenerators and their behaviour are available, and much experience exists. No experience however is available for hybrid material regenerator units.

In order to evaluate such hybrid units the simulation code HYBRID has been developed. This treats forced convection in the gas flow ducts and unsteady two dimensional (radial-axial) heat conduction with phase change in the storage material. The thermophysical properties of the composite material are temperature dependent.

Fig. 1. Salt-ceramic composite matrix.

The program uses an implicit finite difference scheme to solve the set of governing equations. It is tested on a conventional regenerator which stores sensible heat only. The results agree very well with measured data. A verification of the hybrid-material calculations will be performed on a test unit.

TECHNICAL APPLICATION

As an example the heat storage, designed for the 30 MW solar power plant PHOEBUS, was calculated. In a first case, it is assumed to be a conventional unit for sensible heat with a storage capacity of 250 MWh thermal. The operating data are as follows: Charging temperature 700°C, discharging temperature 185°C, maximum gas outlet temperature during the charging period 240°C, minimum gas outlet temperature during the discharging period 650°C (required for the steam turbine), charging time 8h, discharging time 3.5 h. The storage vessel is proposed to be 20 m high and 16 m in diameter (PHOEBUS, 1990). At charging the hot gas enters at the top, at discharging the heated gas leaves there.

In Fig. 2 the temperature distribution along the height of the storage vessel is shown for 9 different times during a charging process. Only a small temperature difference is obtained between gas temperature and mean temperature of the storage material. It should be noted that the stored heat is proportional to the area between the two boundary temperature curves (a,b).

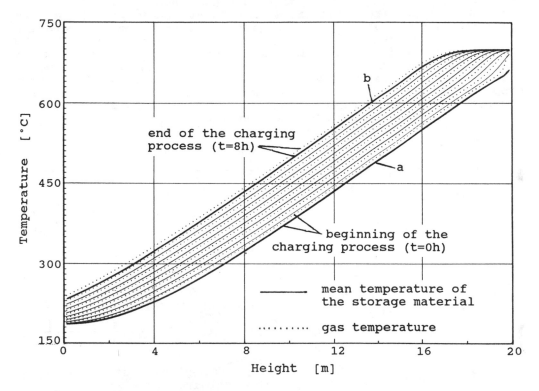

Fig. 2. Time dependent temperature profiles in the storage material and in the gas along the height of the storage vessel at charging for PHOEBUS project conditions.

Apparently it is not possible to increase the storage capacity under the given conditions of storage operation.
An improvement can be achieved by increasing the diffusivity $b = \sqrt{\rho c_p k}$. Thus, an increase in the specific heat of the material or the use of a PCM is advantageous. Temperature changes in the storage material will occur more slowly and the limiting gas outlet temperatures will proceed with a delay. This could also be achieved by metallic materials (e.g. cuttings) embedded in the ceramic matrix or by optimizing geometric configurations.

In a second case study, the same storage unit is assumed to be filled with the new hybrid material. The temperature outlet conditions and mass flows are the same; capacity and charging time are free parameters (even if it doesn't make much sense to charge a solar storage longer than 8h, it is the best way to show the effect). Now we get the following result: With salt-ceramic of, e.g., a melting temperature of 500°C filling the entire storage, only a very small amount of the material actually takes part in the phase change. This is shown in Fig. 3 where the horizontal part of the temperature curves at 500°C represents the phase change area. Only in the region between 10 and 13m a phase change does occur; above and below, the salt either remains liquid or solid all the time and there is no real improvement of the storage capacity. There might be the idea of introducing a large number of latent materials in a large number of layers with different melting points. In the given case more than 80 material layers would be necessary for all the PCMs to take part in the phase change which would increase the storage capacity by a factor of 3.5 (Tamme, 1990). This is, however, technically infeasible, for up to now there exist not enough salt-ceramic composites with appropriate melting points, which are thermally, chemically and mechanically stable. Hence this solution remains theoretical and a better idea must be practised.

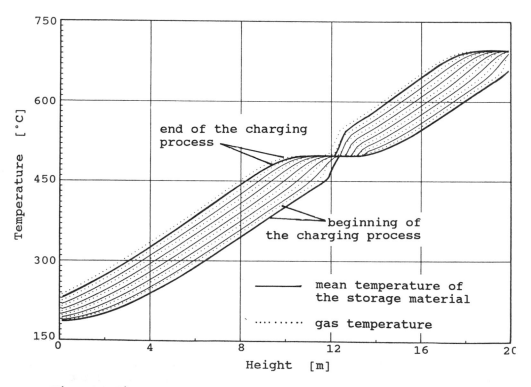

Fig. 3. Time dependent temperature profiles at charging for PHOEBUS project conditions with PCM of a melting temperature of 500°C.

ADVANCED STORAGE CONFIGURATION

A so-called advanced storage configuration is suggested with only a 2 meter layer of salt-ceramic at the top- (at 670°C melting temperature) and the bottom-end (at 220°C melting temperature) of the storage unit. Each layer would amount to 10% of the storage volume. The large middle part is packed with conventional ceramic material. It can be shown (in Fig. 4) that the salt-ceramic acts as a buffer which keeps the gas outlet temperatures at a constant level during a long time. This is exactly the desired behaviour. In this special case where the charging period lasts now 20h the sensible heat stored in the middle part can be increased very much and the overall heat capacity (including the latent heat of the PCM) exceeds to 620 MWh which is a factor of 2.5 compared to the first case. A possibility to keep the PHOEBUS design data (250 MWh capacity and 8h charging time) is to reduce the storage size to only 60% of the former volume. By this way also the pressure drop in the vessel can be reduced to 60%. Another way of operation with the original vessel could be with an increase of the minimum gas outlet temperature during the discharging process (e.g. from 650°C to 680°C). The capacity would be again 250 MWh in 8h charging time yet the efficency of the steam turbine becomes higher.

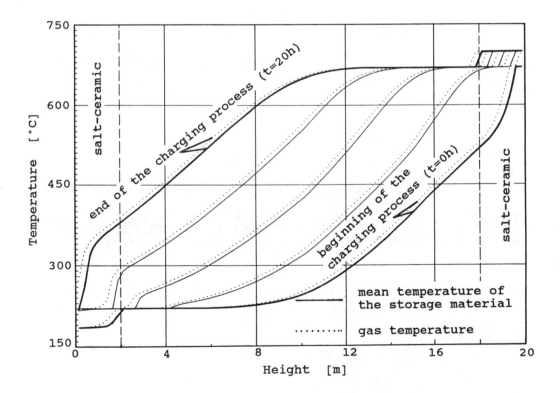

Fig. 4. Time dependent temperature profiles for the advanced storage configuration with 2m of PCM at the top and the bottom end of the storage vessel.

These results can cast a new light on storage technology. The aim would no longer be to achieve a maximum increase in the storage capacity by applying a large amount of PCM but to achieve the highest possible increase in sensible heat by using the least possible amount of PCM.

This effect can also be obtained, though to a smaller extent, by inserting PCM only at one end of the storage vessel. For example an already existing storage could be modified at the upper end without exchanging the entire storage material. If there occur problems with the mechanical stability of the salt-ceramic material a stress-free layer could be applied.

CONCLUSION

Even if the new material is more expensive (maximum factor 5 according to the manufacturer DIDIER Company), there exists a large economical advantage. An input of 20% salt-ceramic (as shown above) leads to 80% more cost but the capacity increases by a factor of 2.5. Only two different salt-ceramic materials are necessary for every application. A higher phase change enthalpy of the salt leads to even smaller layers (down to 1m). An optimum has to be found for every application.

ACKNOWLEDGEMENT

The authors acknowledge the financial support by the German Ministry for Research and Technology (BMFT) and also like to express their thanks to DIDIER Company for providing necessary data.

REFERENCES

PHOEBUS (1990). Phase Ib - Feasbility Study.

Tamme, R., P. Allenspacher and M. Geyer (1986). High temperature thermal storage using salt-ceramic phase change material. Proceedings 21st IECEC Meeting.

Tamme, R., U. Taut, and Ch. Streuber (1990). Energy storage development for solar thermal processes. Proceedings 5th IEA Symposium on Solar High Temp. Tech., Davos, Switzerland.

DEVELOPMENT AND TESTING OF ADVANCED TES MATERIALS FOR SOLAR THERMAL CENTRAL RECEIVER PLANTS

A.Glück[1),2)], R.Tamme[1)], H.Kalfa[2)], C.Streuber[2)], T.Weichert[2)]

[1)]DLR, Deutsche Forschungsanstalt für Luft- und Raumfahrt, Institut für Technische
Thermodynamik, Pfaffenwaldring 38-40, 7000 Stuttgart 80, FRG
[2)]DIDIER WERKE AG, 6200 Wiesbaden 1, FRG

ABSTRACT

For high temperature solar thermal applications there is a need for efficient thermal energy storage. The adaptation of the cowper concept, well known in the steel and glass industry, to the solar specific conditions seems to be the straight forward solution. As longer cycles are required, the development of new storage media with higher energy density than the existing refractory materials is necessary. To meet these requirements, the 'composite salt/ceramic thermal energy storage concept' has been proposed. It offers the potential of using phase change materials via direct contact heat exchange at temperatures well above 600°C. The paper presents a survey of the current status of material development, describes a pilot scale test facility, that has been built to investigate, among others, new developed hybrid materials in technical scale and it summarizes the planned tests and main targets of the program.

KEYWORDS

High temperature thermal energy storage; composite salt/ceramic thermal energy storage concept; salt/ceramic hybrid materials; sensible heat storage; latent heat storage; high temperature heat storage test facility; solar thermal central receiver plant; solar thermal electricity production.

INTRODUCTION

Like conventional power plants, solar plants must cover the energy demand of the consumer. That means solar plants have to guarantee a specific output. Since this requirement usually does not correspond with the energy input, limited by diurnal, seasonal and weather related insolation changes, solar plants must be constructed with thermal energy storage (TES) and/or fossil fired backup systems. Both have the functions to prolong operation times and to guarantee a defined output, necessary to shift the energy output from low price off-peak periods to peak periods and to get capacity payments [1].

As a result of the progress in the receiver development, the next generation of solar thermal central receiver (STCR) plants will be operated with air as the primary heat transfer fluid at temperatures between 600 and 1000°C. Based on this concept the preliminary design for the 30 MWel central receiver plant PHOEBUS was developed [2].

For this type of STCR plants the utilization of refractory materials as storage media is the most obvious solution. These high temperature TES materials are oxide ceramics used as checker bricks in an ordered stacked arrangement, called checkerwork, proven in regenerators or cowpers of the glass and steel industry. In contrast to industrial applications (continuous operation and complete cycle times of about 1-2 hours), only one complete charging and discharging cycle per day is feasible for solar applications. This significantly affects the specific costs of the solar TES system. Therefore improved storage materials with higher storage densities than the existing refractory materials, but with comparable material and manufacturing costs are required. This goal can be achieved with the advanced 'composite salt /ceramic TES media concept' [3].

STATUS OF MATERIAL DEVELOPMENT

Composite salt/ceramic TES media may be considered as hybrid materials consisting of a solid, porous ceramic matrix and of a salt as phase change material (PCM) embedded within the micro structure of the ceramic matrix. Charging and discharging in the range of the melting temperature of the PCM leads to a significantly improved storage capacity compared to pure ceramic material [4]. In order to use this improvement efficiently for STCR plant applications, where temperature differences between storage in- and outlet of several hundred degrees must be considered, a cascaded arrangement of different hybrid materials was proposed [3]. Their melting temperatures have to be adapted to the storage temperature profile of the respective application.

Salt ceramic development at DLR was started in 1986 with lab scale material research. At 1988 DIDIER WERKE AG joined the project, responsible for manufacturing the hybrid material in technical scale and investigating low cost processing procedures.

The presently used preparation method of lab scale materials is cold pressing of the raw mixture. Essentially, it consists of the basic ceramic, the salt and further additives, mainly chemical binders and water. Usually, we prepare cylindrical pellets with diameters varying between few milimeters for thermoanalytical investigations and 26 or 40 millimeters for testing thermophysical properties and thermal, chemical and mechanical stability. Main influence on the stability of the composite materials is caused by particle size distribution of the ceramic, type and quantity of the binders, and by temper- and burning processes [3].

Detailed investigations have been conducted with the salt/ceramic systems Na-BaCO$_3$/MgO and Na$_2$SO$_4$/SiO$_2$. They included essentially chemical, thermal, mechanical and X-ray investigations and single pellet tests in laboratory furnaces under ambient atmosphere. For both systems, the results have demonstrated thermal and chemical stability of the composite materials. Their properties are summarized in Table 1.

TABLE 1 Thermophysical properties of 2 selected lab scale hybrid materials

	Na$_2$SO$_4$/SiO$_2$	Na-BaCO$_3$/MgO
Weight % PCM	45	45
Density, g/cm^3	2.0	2.7
Fusion Temperature T$_F$, °C	880	700
Specific Heat (600°C), J/gK	1.1$_5$	1.1$_9$
(800°C), J/gK	1.1$_7$	1.3$_5$
(990°C), J/gK	1.2$_3$	-
Latent Heat, J/g	80	82
Storage Capacity, J/g (ΔT=100K)	200	210
Hot Crushing Strength, N/mm^2 (100K > T$_F$)	>0.25	0.11

With the sodium sulfate silica composite material, DIDIER has started manufacturing in technical scale. Fig. 1 shows a survey of several test materials, at the left side technical scale bricks of about 2000 g, at the right side lab scale pellets of about 20 g each. Preliminary results concerning technical scale Na$_2$SO$_4$/SiO$_2$ indicate excellent agreement with lab scale results. The compressive strength of the porous ceramic matrix, prepared by dissolving the salt phase, was determined to be about 6-8 N/mm^2. From these results, we conclude that the hot crushing strength of the hybrid material will be significantly higher than we need for investigations with the storage test facility (0.04-0.06 N/mm^2) and for future large scale solar thermal applications (PHOEBUS plant design: 1 N/mm^2).

Fig. 1 Different test bricks of Na_2SO_4/SiO_2 composite material

HIGH TEMPERATURE STORAGE TEST FACILITY

The following chapter describes the test facility and the possible operation conditions [5]. The plant has been built at DLR in Stuttgart, Fig. 2. It is operated automatically by a central control station, which also serves as data acquisition system. The desired data can be transferred to a PC. Data reduction and evaluation can be done on the PC or on a linked host computer.

The whole plant can be divided into an open, pressureless charging cycle and a closed discharging cycle, which can be put under the maximum pressure of 21 bar. Connecting part of both cycles is the storage. Its maximum dimensions can be described as a pillar with 0.5m in diameter and a height of 2m (usable storage volume 0.4m³). The axial and radial temperature profiles can be measured with 88 thermocouples distributed individually within the checkerwork during its erection. In order to enable quick and easy checkerwork exchange a lifting device is placed below the storage. The storage containment can be opened at the bottom and the checkerwork can be lowered.

Charging of the storage is done with the flue gas of a 200kW-gas burner, Fig. 3. The adjustable temperature range is between 100 and 1300°C. The gas flow can be chosen between 100 and 1000m³i.N./h. The independent choice of both parameters is guaranteed by

Fig. 2 High temperature storage test facility

separate fresh air, added to the flue gas within the combustion chamber. Mass flows in the three mains leading to the combustion chamber (natural gas, combustion air and fresh air) are measured with orifice plates. Addition of the three values yields the total charging gas mass flow. The gas enters the storage at the top (max. temperature: 1300°C) and leaves it at the bottom (max. 700°C). There are two ways how the charging mode is stopped: first, the reaching of a certain outlet temperature, or second, the lapse of the chosen charging time.

Discharging is done in a closed cycle with uncoupled burner, Fig. 4. It is designed for pressures up to 21bar. Before starting the discharge mode the cycle is set under the chosen pressure. The air is now circulated by a hot gas blower. It enters the storage at the bottom with the given inlet temperature and leaves it at the top (max. 1300°C). Since the air circulates in a closed cycle the absorbed heat must be transferred to a 'consumer' before it reenters the storage. This is achieved through a so called buffer which is in fact a second storage filled with conventional ceramics. Such a regenerative heat exchanger is not able to guarantee the required constant and eligible outlet temperature together with a given flow. Therefore a bypass has been built through which part of the air can pass the buffer. The temperature of the two reunited flows corresponds to the chosen storage inlet temperature. The buffer is cooled with ambient air during the following charging cycle in order to have similar conditions at the beginning of the next discharging cycle. Measuring and controlling of the total circulated air flow is done by means of two venturi tubes of different size. Similar to the charging mode there are two limiting conditions for discharging. The first is once again the discharge time, the second occurs when the outlet temperature at the storage top falls beyond a given margin, that means when the 'consumer' cannot use the heat anymore. The pressure is now released and the air leaves the system through two sound absorbers.

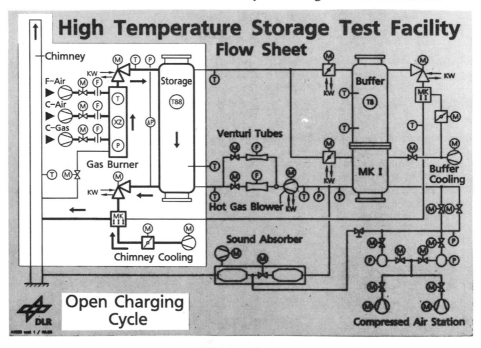

Fig. 3 Flow Sheet of Test Facility (Charging Mode)

PLANNED TESTS AND MAIN INTENTIONS WITH THE TEST FACILITY

In the first test period starting in march, a checkerwork with the charateristic data of the proposed PHOEBUS storage system will be investigated. The used bricks consist of an almost conventional refractory material (66% Al_2O_3, 33% SiO_2) storing only sensible heat. Their special geometry results from the compromise between good heat transfer and low pressure losses. It meets the requirements set up for the PHOEBUS plant design (8h charging, 3h discharging).

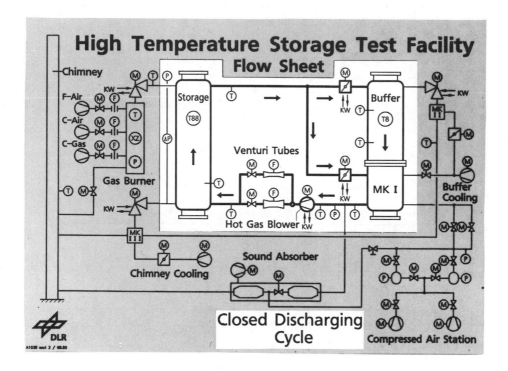

Fig. 4 Flow Sheet of Test Facility (Discharging Mode)

For the first tests with this checkerwork parameters corresponding to the PHOEBUS plant will be used. The main intention with these tests is the validation of the simulation code used for optimizing the checkerwork design. This code has already been successfully used describing the behaviour of cowpers. Characteristic for these cowpers is a big air temperature difference at the hot end between charging and the beginning of discharging and an almost linear decrease of the outlet temperature during discharging, Fig. 5. Calculations with the program have shown, that for the new bricks the results are different. The discharging outlet temperature starts with almost no difference to the charging temperature and its decrease looks very much like an exponential function, Fig. 6. This was achieved because of the substantially higher specific heating surface of the new bricks and because of a laminar flow in the storage which is caused by the smaller channel diameter (same gas velocity provided). The experimental proof of these theoretically determined storage behaviour should be done in the test facility. Because these discharge characteristics are not only desirable in the solar case, additional experiments concerning different industrial applications will follow. Moreover, investigations of heat transfer within such storage systems are included. They may lead to empiric, dimensionless equations describing the charge/discharge behaviour. With regard to the upcoming hybrid material tests, so called reference tests will be conducted. These tests serve to work out the specific influence of phase transition on material behaviour.

In the next period the checkerwork will be built up of new-developed hybrid materials. Primary goal of these experiments is to proof the thermal, chemical and mechanical stability of the technically manufactured bricks. These lifetime tests serve to demonstrate the realization of the salt/ceramic concept and the transferability of the lab scale results to technically manufactured products. Another important point is the validation of a computer code simulating storage systems with hybrid material. The use of the experimental data is also essential for this validation, because no other possibilty for comparing simulation with experimentally determined results is yet available.

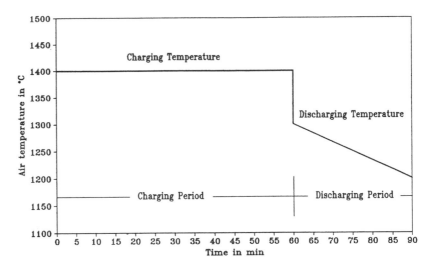

Fig. 5 Charging inlet and discharging outlet temperature of a typical cowper cycle vs. time

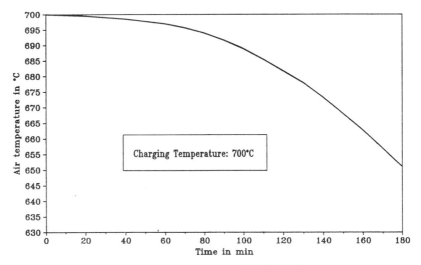

Fig. 6 Discharging outlet temperature of the proposed PHOEBUS-1 storage system vs. time

REFERENCES

[1]   M.Geyer, R.Tamme, H.Klaiß, 'High Temperature Thermal Storage in Solar Plants', Proc. ISES Solar World Congress, Hamburg, 1987

[2]   PHOEBUS, Phase 1B - Feasibility Study, March 1990

[3]   R.Tamme, 'Energy storage development for solar thermal processes', Proc. 5th Symp. on Solar High Temperature Technologies, Davos, 1990

[4]   R.Tamme, 'Advanced regenerator media for industrial and solar thermal applications', Proc. 25th IECEC, Reno, 1990

[5]   A.Glück, R.Tamme, H.Kalfa, C.Streuber. 'Investigation of high temperature storage materials in a technical scale test facility', Proc. 5th Symp. on Solar High Temperature Technologies (Davos, 1990)

HEAT TRANSFER ENHANCEMENT OF A GAS-SOLID COUNTER-CURRENT THERMAL STORAGE SYSTEM

A. Bashir and B.M. Gibbs

Department of Fuel and Energy
The University of Leeds
Leeds LS2 9JT England.

ABSTRACT

Sensible heat storage in fluidizing solids is a cheaper way of storing solar energy in a gas-solid contact system, compared to a latent heat storage unit utilizing analytical grade chemicals. In order to obtain an enhancement in such a gas-solid system, the concept of introducing packing materials, such as Raschig rings, Lessing rings or Pall rings, is relatively new. Operational limits for the various parameters, such as gas flow rate, solids flow rate, packing size and packing height depend strongly on the hydrodynamic stability of the system. In a previous study, the hydrodynamics of our system have been discussed; on the basis of these results, the thermal performance of a direct contact heat transfer between gas (air) and solids (sand) has been experimentally investigated utilizing 15 mm diameter pall rings. A comparison is presented for gas-solid heat transfer between free falling solids system (Raining bed) and the system with packing of pall rings (Raining packed bed). Heat transfer enhancement due to the presence of packing is reported. Results show that, for a free falling solids system, solids follow a straight vertical path; therefore, most of the heat is transferred in the central periphery. Whereas, presence of packing counteracts this tendency and disperses the solids and gas throughout the interstices causing a greater disturbance in the gas film surrounding the particle surface. Furthermore, presence of packing considerably increases the gas-solid contact because of an increase in the number of collisions per unit time for a given solids flow rate.

KEYWORDS

Pall rings;packing material;Raining bed;particulate solids

INTRODUCTION

The efficiency of a gas-solid heat storage system depends on the hydrodynamic stability of the operation. In a previous study (Bashir and Gibbs, 1991), it is shown that pall rings packing improves the gas-solid contacting in a Raining bed system without a considerable increase in pressure drop. On the basis of hydrodynamics findings, heat transfer experiments were carried out in a direct contact gas-solid system.

Heat transfer between gas and particulate solids has been extensively studied by various researchers in different types of contactors. Bergougnou et al. (1981) have reported that solids, such as sand, constitute an attractive heat storage medium because of its high heat capacity and low vapour pressure. Harker and Martyne (1985) have studied the granular solids of 15 mm diameter as the heat storage medium in the temperature range of 303-320 °k. Kamimato and Tani (1976) have studied the effect of conduction promoters, such as stainless chips and fins for heat transfer performance of a heat storage unit. It is generally believed that the dominating mode of heat transfer is convection, though conduction may occur in situations where particles come into contact with the walls of the contactor, and radiation may take place at higher temperatures. According to Howard (1989), in gas-solids systems, the heat transfer coefficient depends on the flow regime in the voids surrounding the particles. As the gas velocity through the voids increases, the heat transfer coefficient increases.

Because of the very large surface area of particulate solids, heat exchange from gas to solid is very rapid. This heat transfer primarily occurs by conduction through a gas film surrounding the individual particle. It is believed that major resistance to heat transfer from gas to solids is due to the presence of this fluid film. The presence of adjacent particles, and hence their collisions with each other, affect the thickness of the fluid film, thereby increasing the heat transfer. Also, the number of collisions per unit time are proportional to the solids concentration. An increase in the solids concentration decreases the film thickness resulting in higher heat transfer rates. On the other hand, gas by passing zones of the bed will adversely affect the rates of heat transfer between the gas and surrounding solids. The degree of by-passing, however, depends on the bed material, gas flow rate, solids flow pattern and design of the apparatus.

EXPERIMENTAL SETUP AND PROCEDURE

The experimental set-up consisted of a 158 mm diameter, 1 m high mild-steel column. Air was introduced from the bottom via a stand pipe type distributor. Air was preheated prior to entering into the Raining bed. The experiments were carried out at one bar pressure. The inlet and outlet temperatures of both the gas and the solids were measured with the help of Chromel-Alumel thermocouples. The ambient air temperature, together with axial temperature profiles at various heights, were also measured.

A digital temperature indicator equipped with a multichannel switch system was used to monitor the temperatures of both the solids and air at different positions. To minimize heat losses from the system, kaowool, 25.4 mm thickness, was uniformly applied all around the Raining bed. Sand of 262 μm was used as a particulate solids medium; solids flow rate was controlled with the help of a slotted valve assembly, which was connected to the outlet of the top hopper. Pall rings of 15 mm diameter were used as a packing material for the various experimental conditions.

In making a test run, the flow of air was adjusted and stabilized before switching on the electric supply for the preheater . At a particular air inlet temperature, the system was allowed to run till a steady state was achieved. Sand of 262 μm was then dropped down from the hopper through three downcomers of a fixed distributor, some readings were taken with a rotating distributor, and a comparative analysis was done for the effectiveness of the mode of solids distribution. Once a steady state was achieved with the flowing solids, the temperatures at that condition for the air and solids were logged. On the basis of these temperatures, gas-solid heat transfer coefficient was evaluated using the following formula:

$$h = \frac{solids\ mass \times c_p \times solids\ temperature\ rise}{surface\ area \times residence\ time \times LMTD}$$

where LMTD is the log mean temperature difference and c_p is the solids specific heat.

RESULTS AND DISCUSSION

Heat transfer coefficient increases with increasing air mass flux at a given air inlet temperature for 15 mm diameter pall rings (Fig. 1). The relative position of the curves show that at low air mass flux, the heat transfer coefficient is low compared to high air mass flux. At higher air mass flux, inter-particle mixing increases, and the frequency with which the particles move also increases. The turbulence set-up, at higher air flow rate, results in breaking the laminar film surrounding the particle, therefore a higher heat transfer coefficient is obtained.

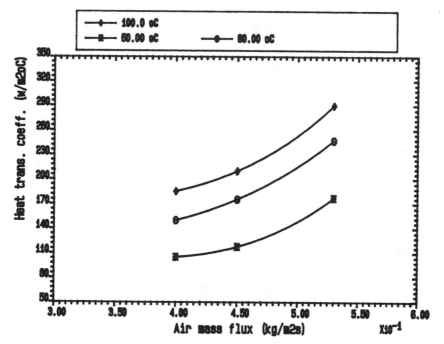

Fig. 1 Effect of air mass flux on heat transfer coefficient for various air inlet temperatures, S=0.76 kg/m2s.

It has been observed (Bashir and Gibbs, 1991) that the presence of packing improves the gas-solid interactions by increasing the contact area of the particles. There are more inter-particle contacts, and air-particle mixing and the rate of particles replacement in the vicinity of the surface rises. The position of the curves (Fig. 2) shows that there is a considerable enhancement in heat transfer coefficient in the presence of pall rings packing. In the presence of packing, however, air passes through the sheet of solids in a cross-current manner, contacting all the particles surface area, thereby reducing the resistance to heat transfer by thinning the laminar film thickness.

It is believed that there exist two resistances for heat transfer in a gas-solid contact system, i.e. the surface resistance (due to the presence of fluid film) and the particle internal resistance. The criterion for deciding the effectiveness of resistance is the measurement of Biot number for a particular set of experimental conditions. Biot number is a dimensionless group defined as:

Bi = (Interior resistance/Surface resistance)

Bi = (hL/k)

where;

h = heat transfer coefficient ($w/m^{2°}K$)
L = characteristic length (1/3 of particle radius) (m)
Ks= thermal conductivity of particles ($w/m°K$)

It is generally accepted (Levenspiel, 1984) that particle internal resistance can be neglected if Bi<0.1. For the present experimental conditions, however, the Bi number is well below the above limit, hence it is reasonable to assume that particle internal resistance is ineffective.

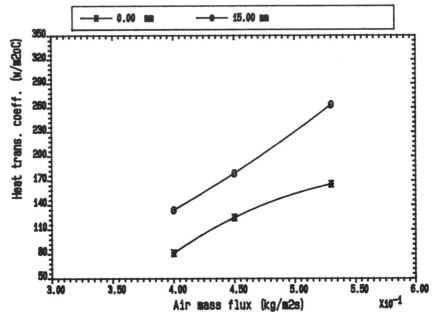

Fig. 2   Effect of air mass flux on heat transfer coefficient
for different packing sizes, S=1.06 kg/m2s, T=80oC.

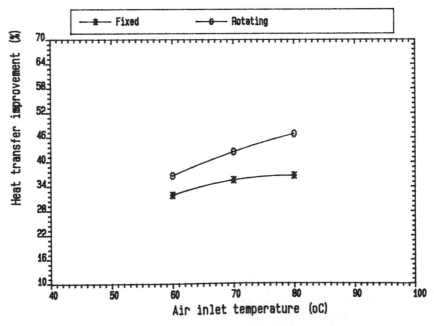

Fig. 3   Effect of temperature on heat transfer for different solids distribution, G=0.53 kg/m2s, S=1.06 kg/m2s.

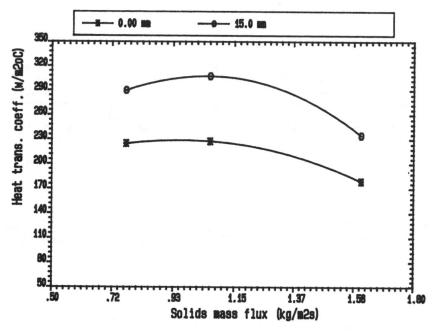

Fig.4. Effect of solids mass flux on heat transfer coefficient
for various packing sizes of pall rings, G=0.53 kg/m2s, T=100 oC.

In Figure 3, the lower curve represents the results with a Fixed distributor, whereas, the upper curve shows the results with a Rotating distributor. It is clear from the two curves that there is a considerable improvement in the heat transfer between gas and solids with a Rotating distributor. This is because the rotation of the distributor creates zones of dense solids phase, spiralling inside the Raining bed, thereby forcing the gas to contact the solids over a greater area. On the other hand, with a Fixed distributor, the particles fall in straight paths covering less cross sectional area compared to the Rotating distributor, therefore the chances of gas-solid contacting are less because of an increase in gas by-passing zones.

From the results presented in Figure 4, it is clear that heat transfer coefficient increases as the solids mass flux is increased from 0.76 $kgm^{-2}s^{-1}$ to 1.06 $kgm^{-2}s^{-1}$, then afterwards there is a gradual decline in the heat transfer coefficient as the solids mass flux is increased from 1.06 $kgm^{-2}s^{-1}$ to 1.61 $kgm^{-2}s^{-1}$. It has been mentioned by Verver (1986) that for a better gas-solid contact in counter-current operations, the solids loading ratio (Ws/Wg) should be between 0.5 to 2.0. In the present studies on the gas-solid heat transfer, best results are obtained when the solids loading ratio is between 1.7 to 2.4. The data presented in Figure 4 show that heat transfer coefficient increases as the solids loading ratio increases from 1.7 to 2.3, but with a further increase in solids loading ratio to 3.2, there is a gradual decrease in heat transfer coefficient (h). The increase in `h' with solids mass flux can be attributed to the increased solids surfaces available in the system. Furthermore, when the particle concentration (solids mass flux) is increased, there is more turbulence as a result of enhanced inter-particle mixing and collisions.

CONCLUDING REMARKS

In a Raining bed system without packing, heat transfer coefficient increases with an increase in air mass velocity. Particle internal resistance can be neglected as Bi<0.1. There was a considerable improvement in gas-solid heat transfer in the presence of 15 mm diameter pall rings. A Rotating distributor appeared to be more effective compared to the Fixed distributor.

The best results for gas-solid heat transfer coefficient were obtained with a solids loading ratio between 1.7 to 2.4.

ACKNOWLEDGEMENT

One of the authors (A. Bashir) would like to thank, Ministry of Science and Technology, Pakistan, and State Cement Corporation of Pakistan for providing an opportunity for this research.

REFERENCES.

Bergougnou, M.A., J.S.M. Botterill, J.R. Howard, D.C. Newey, A.G. Salway, Y. Teoman (1981). High Temperature Heat Storage. 3rd Int. Conf. on Energy Cons., London.
Harker, J.H. and E.J. Martyne (1985). Energy storage in gravel beds. J. of Inst. of Energy.
Howard, J.R. (1989). Fluidised bed Technology, Adam Higler, New York.
Kamimoto, M. and Tani, T (1980). Effect of conduction promotors and Fins on Heat Transfer in Latent Heat Storage unit. Bul.Elec.Tech.lab., 44(4).
Levenspiel, O. (1984). Engineering Flow and Heat Exchange, Plenum press, New-York.
Verver, A.B. and W.P.M vanSwaaij (1986). The Heat Transfer Performance of Gas-Solid trickle flow over a regularly stacked packing. 45, Powder Technology.

A GAS-SOLID COUNTER-CURRENT THERMAL STORAGE SYSTEM - HYDRODYNAMIC CHARACTERISTICS

A. Bashir and B.M. Gibbs

Department of Fuel and Energy
The University of Leeds
Leeds LS2 9JT England.

ABSTRACT

The efficiency of a particulate solids thermal storage unit depends strongly on the system hydrodyn amics. Therefore, the hydrodynamic characteristics of particulate solids flowing counter-ci rently to a gas flow in a Raining packed bed system have been examined. The flow properties f the solid phase were observed, using different sizes of pall rings. Results on the pressure d 'p and the hold-up are reported together with the experimentally determined operating limits. The experiments were carried out in a 0.026 cubic metre glass column. Particulate solids were observed to collide regularly with the packing as well as with the column walls, thereby resulting in a zig-zag type of flow. At low air flow rates, the flow of particles, under the action of gravity, is fairly constant, however, at high air flow rates, the entrainment of particles from the column becomes unavoidable. Though at high air flow rate the hold-up is increased, no real flooding or loading phenomena have been observed. Results reveal that hydrodynamic characteristics of a Raining packed bed system depend not only on the physical properties of the solid phase but also on the voidage of the packing material.

KEYWORDS

Hydrodynamics;packing material;trickle flow;jet pump;Raining bed

INTRODUCTION

In a gas-solid counter-current Raining bed system without packing (RB), particulate solids fall down like rain droplets under the action of gravity in a vertical duct through a moving gas stream (Figure 1a). The gas velocity is below the terminal velocity of the falling particles,or otherwise the solids will be blown out of the system with the air stream. Distribution of these solids,together with the gas-solid contact time in the Raining bed,are the key design parameters. A proper distribution for both, solid particles and gas, are difficult to achieve in an RB, moreover, the RB is prohibitively large due to poor contacting between gas and solid particles. The difficulties related to a RB can be overcome in a Raining bed system with packing (RBP). In a RBP, solid particles fall down at their terminal velocity in a vertical column through an uprising gas stream (Figure 1b). A section of the vertical column is filled with some packing material similar to gas-liquid packed bed systems. The packing could be of any size and shape but depends on the size range of the solids in use; packing of Pall rings and Raschig rings is suitable for small solid particles. The operation of a RBP depends strongly on the system hydrodynamics, which entails the visual observations together with the measurements of pressure drop and hold-up of the system. For the present studies, pall rings of three different sizes are selected in order to carry out an extensive comparison and to evaluate the extent to which packing size affects the system hydrodynamics.

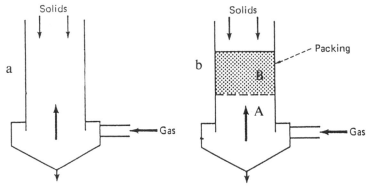

Fig. 1.   Raining bed Heat Exhchanger.
(a) A Raining bed without Packing.
(b) A Raining bed with Packing.

EXPERIMENTAL SETUP AND PROCEDURE

The experimental set-up consisted of a 8 mm thick glass column, which was 158 mm in diameter and 1500 mm in length. The bottom part of the glass column was flanged to a conical mild-steel section to facilitate easy removal of the particulate solids. After passing through the glass column, particulate solids were collected in a fluidized bed, in which a metered quantity of air was provided to create a state of fluidization. Particulate solids from the fluidized bed were pneumatically conveyed into the top hopper with the help of a jet pump.

Air was supplied from a main compressor and,after passing through oil and water filters, was split into three streams. One stream was passed to the fluidized bed, whereas, the other two streams were connected to the inlet of the RBP and the jet pump,respectively. For a uniform air distribution in the RBP, a conical shaped stand pipe type distributor was used. Sand of 262 μm was used as a particulate solids medium, solids flow rate was controlled with the help of a slotted valve assembly, which was connected to the outlet of the top hopper. Solids from the hopper , through the slotted valve, were distributed in the Raining bed with the help of a rotating distributor. The speed of the motor, connected with the rotating distributor, was adjusted in a manner to provide minimum channelling and wall flow of solids.

Three different sizes of pall rings, i.e. 15 mm, 25.4 mm, 38.1 mm were used for the hydrodynamic studies, which are divided into three parts, covering visual observations together with the pressure drop and hold-up measurements. The variables tested during this investigation were air flow rate, solid flow rate, packing size and packing height.

RESULTS AND DISCUSSION

The solids flow behaviour for different sizes of pall rings was visually observed at various operating conditions. In the presence of pall rings packing, when the particles were allowed to fall down, two clear regions were observed (Fig. 1). One below the packing (A) and the other in the packed section (B), in region 'A', solids fall in straight vertical streams. But in region 'B', solids were observed to move down slowly through the packing in the shape of rivuletes. Immediately after coming out from the distributor, solids strike with the top packing surface. This provides trickles of solids flow of varying size and voidage. The number of these trickles mainly depends on the geometry and the voidage of the packing. For 15 mm diameter pall rings, the size of trickles was small and the number of contacts with the packing were more. Furthermore, inter particle (particle-particle) collisions, in the vicinity of the packing, were quite frequent. At some locations, solids were seen falling in a zig-zag pattern, however, the length of this zig-zag path appeared to be less in case of 15 mm diameter pall rings compared to 25.4 mm or 38.1 mm diameter pall rings, which is because of the compactness of the former.

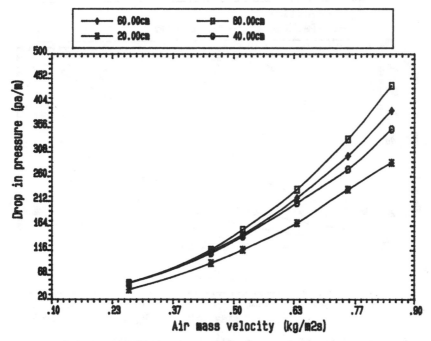

Fig. 2, Effect of air mass velocity on pressure drop for various packing heights of pall rings, S=0

At low air mass velocity (0.45 $kgm^{-2}s^{-1}$), solids flow was unhindered and continuous, but when the air mass velocity was increased to 0.64 $kgm^{-2}s^{-1}$ the trajectory of the particles changed from near vertical to some what horizontal. With a further increase in air mass velocity to 0.72 $kgm^{-2}s^{-1}$, some entrainment of particles was observed but even at this high air mass velocity, there appeared no flooding or loading points.

For the present studies, the total pressure drop consisted of the pressure drop caused by the friction between gas and packing plus column wall and the pressure drop due to friction between gas and solids. Figure 2 shows the drop in pressure per unit column length for various air flow rates. The sharp rise in the pressure drop, after 0.64 $kgm^{-2}s^{-1}$, is mainly attributed to the start of a turbulent region. From the results, it is revealed that the trend in variations in pressure drop, as a function of air mass velocity, across the packing, is similar to that found in gas-liquid system (Eckert, 1974), Treybal (1968), has also mentioned that pressure drop of a gas flow through a packed column is proportional to the air flow rate. In Figure 3, pressure drop due to the presence of packing is plotted against air mass velocity for three different sizes of pall rings. It is clear from the results that 15 mm diameter pall rings offer more pressure drop

compared to 25.4 mm and 38.1 mm diameter pall rings. This higher value is due to the compactness of 15 mm size pall rings, which have a packing to column diameter ratio of 0.09. In Fig. 4, pressure drop due to the flowing solids is plotted against the air mass flux. Pressure drop due to flowing solids is more complex because of the path that solids follow while falling down through the packing. It is clear from Fig 4 that with an increase in air mass flux pressure drop rises at a fixed solids mass flux, but there is only a slight increase in pressure drop value when the packing height is increased, from 20 cm to 80 cm, at a fixed air mass flux. It is envisaged that presence of packing reduces the pressure drop caused by the solids to quite an extent; this is because a part of the solids is supported by the packing, and the other part being supported by the air.

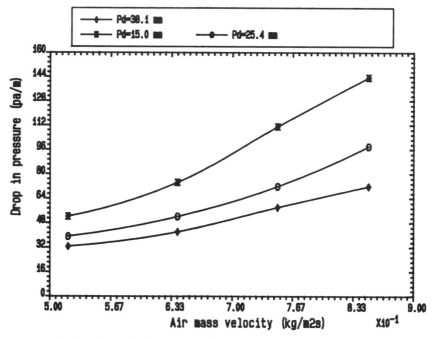

Fig. 3. Effect of air mass velocity on pressure drop for
various packing sizes of pall rings, S=0, PH=80 cm.

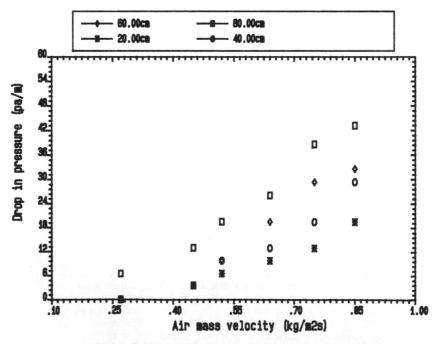

Fig. 4. Effect of air mass velocity on pressure drop due to
the flowing solids, S=1.064 kg/m2s.

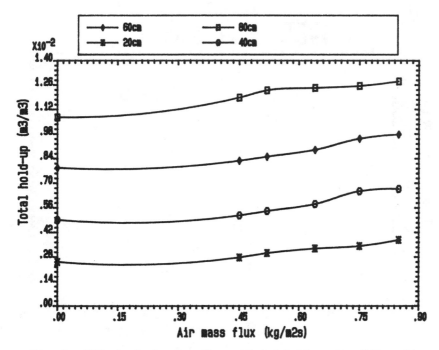

Fig. 5. Effect of air mass flux on total hold-up for different heights of pall rings.

In Fig. 5, total solids hold-up (Ht), which accounts for both the static and the operational hold-ups, is plotted against air mass flux for different packing heights at a solid mass flux of 1.61 $kgm^{-2}s^{-1}$. At low air mass flux, the increase is fairly constant, but at high air mass flux, a slight upward bend in 'Ht' indicates just the begining of turbulence in solids flow, Clause et al.(1976) have also reported the same behaviour. No clear loading point is observed, and the shape of the curve is similar to the gas-liquid system for the range before loading (Perry, 1973). For comparing the effect of different packing sizes of pall rings on 'Ht', packing to column column diameter ratio is plotted against 'Ht' (Fig. 6). There is a sharp decrease in hold-up values as the ratio is increased from 0.09 m to 0.15, but after 0.15, there is almost a constant value showing the relative ineffectiveness of packing for hold-up values.

CONCLUDING REMARKS

In a counter-current gas-solid, Raining bed system, solids were observed to trickle down uniformly through the packing. Trickles were of varying size and voidage; the number of these trickles was dependent on the packing diameter and geometry. Pressure drop was found to be proportional to the air flow rate, solids flow resulted in a slight increase in pressure drop, but the major drop in pressure was observed to be due to the presence of packing. Solids hold-up varied directly with air flow rate; no sharp increase was observed which indicated an absence of loading or flooding points.

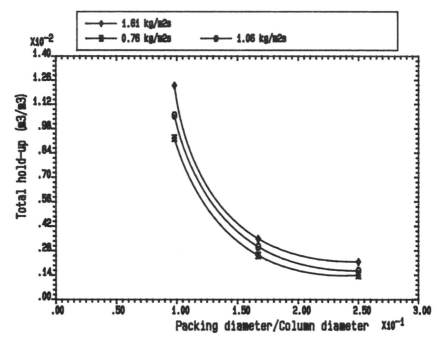

Fig. 6. Effect of packing/column diameter ratio on total hold-up
at various solids flow rates, G=0.52 kg/m2s.

ACKNOWLEDGEMENT

One of the authors (A. Bashir) would like to thank, Ministry of Science and Technology, Pakistan, and State Cement Corporation of Pakistan for providing an opportunity for this research.

REFERENCES.

Clause, G., F.Vergnes and P.L.Goff (1976). Hydrodynamics study of gas and
    solids flow through a screen packing. Can. J. Chem. Eng., vol.54.
Eckert, J.S. (1974). Chem. Eng. 82.
Perry, R.H. and C.H. Chilton (1973). Perry's Chemical Engineers'
    Handbook McGraw-Hill, New York.
Treybal A. (1968). Mass Transfer Operations, second edition,
    McGraw-Hill, New-York.

EVALUATING THE ENERGY AND EXERGY CONTENTS OF STRATIFIED THERMAL ENERGY STORAGES FOR SELECTED STORAGE-FLUID TEMPERATURE DISTRIBUTIONS

M. A. Rosen

Department of Mechanical Engineering, Ryerson Polytechnical Institute
Toronto, Ontario, Canada M5B 2K3

F. C. Hooper

Department of Mechanical Engineering, University of Toronto
Toronto, Ontario, Canada M5S 1A4

ABSTRACT

Analytical expressions for evaluating the energy and exergy contents of a stratified thermal energy storage (TES) are developed and compared. Linear, stepped and continuous-linear storage-fluid temperature distributions are considered. The authors feel that, since exergy is a measure of usefulness or quality of energy, exergy quantities are more meaningful than energy quantities, and that exergy analysis should be considered in the evaluation and comparison of stratified and non-stratified thermal storages. The application of the results to an example indicates how (1) for a single temperature distribution, the quantities of energy and exergy contained in a stratified TES differ; (2) the exergy content of a TES increases as the degree of stratification increases, even if the energy content remains fixed; and (3) the energy and exergy contents of a TES differ when no stratification is present.

KEYWORDS

Thermal energy storage; efficiency; energy; exergy; thermodynamics; temperature.

INTRODUCTION

Thermal energy storage (TES) systems are incorporated in a wide range of energy conversion systems (Bejan, 1982; Beckman and Gilli, 1984; Hahne, 1986). In many thermal storages, a stratified storage-fluid temperature distribution is desired, since maintaining stratification avoids the thermodynamic losses associated with the mixing of fluids of different temperatures. Many researchers have attempted to develop standards for the assessment of the performance of thermal storages, with stratified or mixed storage fluids (Bejan, 1982; Beckman and Gilli, 1984; Hahne, 1986). At present, however, no valid and generally accepted standards have been established for the comparison of the performance of stratified thermal storages. Furthermore, it is clear that in many instances conventional energy-based performance measures can be misleading.

The present work is directed towards developing simple methods for evaluating and comparing the energy and exergy contents of stratified thermal storages, and is part of a long-term program by the authors to develop and apply exergy-based performance measures for TES systems. This paper is an extension of a previous report (Rosen, 1991). Other results from this program have been reported elsewhere (Rosen and co-workers, 1988, 1991). In this paper, general expressions are developed for the energy and exergy contents of a stratified TES, for any one-dimensional temperature distribution in the vertical direction. Specific expressions are then developed for three particular storage-fluid temperature distributions which are realistic yet mathematically simple. Finally, the results are discussed and compared, and an illustrative example is presented.

GENERAL STRATIFIED TES ENERGY AND EXERGY EXPRESSIONS

The energy E and exergy Ξ in a TES can be found by integrating over the entire storage-fluid mass m within the TES as follows:

$$E = \int_m e \, dm \tag{1}$$

$$\Xi = \int_m \xi \, dm \tag{2}$$

where e denotes specific energy, and ξ specific exergy. For an ideal liquid, the e and ξ are functions only of temperature T, and can be expressed as follows:

$$e(T) = c(T - T_o) \tag{3}$$

$$\xi(T) = c[(T - T_o) - T_o \ln(T/T_o)] = e(T) - cT_o \ln(T/T_o) \tag{4}$$

where c denotes the specific heat of the TES fluid, and T_o the temperature of the reference environment. Both c and T_o are assumed constant here.

Consider now a TES of height H which has only one-dimensional stratification. Specifically, temperature varies only with height h (i.e., in the vertical direction). The horizontal cross-sectional area of the TES is assumed constant. A horizontal element of mass dm can then be adequately approximated as a function of dh as

$$dm = \frac{m}{H} dh \tag{5}$$

Since temperature is a function only of height (i.e., $T = T(h)$) for the one-dimensional stratification considered here, the expressions for e and ξ in Eqs. 3 and 4, respectively, can be written as

$$e(h) = c(T(h) - T_o) \tag{6}$$

$$\xi(h) = e(h) - cT_o \ln(T(h)/T_o) \tag{7}$$

With Eq. 5, the expressions for E and Ξ in Eqs. 1 and 2, respectively, can be written as

$$E = \frac{m}{H} \int_0^H e(h) \, dh \tag{8}$$

$$\Xi = \frac{m}{H} \int_0^H \xi(h) \, dh \tag{9}$$

With Eq. 6, the expression for E in Eq. 8 can be written as

$$E = mc(T_m - T_o) \tag{10}$$

where

$$T_m \equiv \frac{1}{H} \int_0^H T(h) \, dh \tag{11}$$

Physically, T_m represents the temperature of the TES fluid when it is fully mixed (i.e., uniform, or non-stratified). This can be seen by noting two points. First, the energy of a fully mixed tank E_m at a uniform temperature T_m can be expressed, using Eq. 3 with constant temperature and Eq. 1, as

$$E_m = mc(T_m - T_o) \tag{12}$$

Second, by the principle of conservation of energy, the energy of a fully mixed tank E_m is the same as the energy of the stratified tank E. That is,

$$E = E_m \tag{13}$$

Comparing Eqs. 10, 12 and 13 confirms that T_m represents the temperature of the mixed TES fluid.

With Eq. 7, the expression for Ξ in Eq. 9 can be written as

$$\Xi = E - mcT_o \ln(T_e/T_o) \tag{14}$$

where

$$T_e \equiv \exp\left[\frac{1}{H} \int_0^H \ln T(h) \, dh\right] \tag{15}$$

Physically, T_e represents the equivalent temperature of a mixed TES that has the same exergy as the stratified TES. In general, $T_e \neq T_m$, since T_e is dependent on the degree of stratification present in the TES, while T_m is independent of degree of stratification. In fact, $T_e = T_m$ is the limit condition reached when the TES has no stratification, that is, when it is fully mixed. This can be seen by noting (with Eqs. 2, 14, 12 and 13) that the exergy in the fully mixed TES, Ξ_m, is

$$\Xi_m = E_m - mcT_o \ln(T_m/T_o) \tag{16}$$

The difference in TES exergy between the stratified and fully mixed (i.e., at a constant temperature T_m) cases can be expressed with Eqs. 14 and 16 as

$$\Xi - \Xi_m = mcT_o \ln(T_m/T_e) \tag{17}$$

EXPRESSIONS FOR SELECTED TEMPERATURE DISTRIBUTIONS

Three stratified temperature distributions are considered: linear (denoted by a superscript L), stepped (denoted by a superscript S), and continuous-linear (denoted by a superscript C). For each case, the temperature distribution as a function of height is given, and expressions for T_m and T_e are derived. The distributions considered are simple enough in form to permit energy and exergy values to be obtained analytically, but complex enough to be relatively realistic. Although other temperature distributions are, of course, possible, these were chosen because closed form analytical solutions could readily be obtained for the integrals for T_m in Eq. 11, and T_e in Eq. 15.

The Linear Temperature Distribution

The linear temperature distribution varies linearly with height h from T_b, the temperature at the bottom of the TES (i.e., at $h = 0$), to T_t, the temperature at the top (i.e., at $h = H$), and can be expressed as

$$T^L(h) = \frac{T_t - T_b}{H} h + T_b \tag{18}$$

By substituting Eqs. 18 into Eq. 11 and 15, it can be shown that

$$T_m^L = \frac{T_t + T_b}{2} \tag{19}$$

which is the mean of the temperatures at the top and bottom of the TES, and that

$$T_e^L = \exp\left[\frac{T_t(\ln T_t - 1) - T_b(\ln T_b - 1)}{T_t - T_b}\right] \tag{20}$$

The Stepped Temperature Distribution

The stepped temperature distribution consists of k horizontal zones, each of which is at a constant temperature, and can be expressed as

$$T^S(h) = \begin{cases} T_1, & h_0 \leq h \leq h_1 \\ T_2, & h_1 < h \leq h_2 \\ \ldots \\ T_k, & h_{k-1} < h \leq h_k \end{cases} \tag{21}$$

where the heights are constrained as follows:

$$0 = h_0 \leq h_1 \leq h_2 \cdots \leq h_k = H \tag{22}$$

It is convenient to introduce here x_j, the mass fraction for zone j:

$$x_j \equiv \frac{m_j}{m} \tag{23}$$

Since the TES-fluid density ρ and the horizontal TES cross-sectional area A are assumed constant here, but the vertical thickness of zone j, $h_j - h_{j-1}$, can vary from zone to zone,

$$m_j = \rho V_j = \rho A (h_j - h_{j-1}) \tag{24}$$

and

$$m = \rho V = \rho A H \tag{25}$$

where V_j and V denote the volumes of zone j and of the entire TES, respectively. Substitution of Eqs. 24 and 25 into Eq. 23 yields

$$x_j = \frac{h_j - h_{j-1}}{H} \tag{26}$$

With Eqs. 11, 15, 21 and 26, it can be shown that

$$T_m^S = \sum_{j=1}^{k} x_j T_j \tag{27}$$

which is the weighted mean of the zone temperatures, where the weighting factor is the mass fraction of the zone, and that

$$T_e^S = \exp\left[\sum_{j=1}^{k} x_j \ln T_j\right] = \prod_{j=1}^{k} T_j^{x_j} \tag{28}$$

The Continuous-Linear Temperature Distribution

The continuous-linear temperature distribution consists of k horizontal zones, in each of which the temperature varies linearly from the bottom to the top, and can be expressed as

$$T^C(h) = \begin{cases} \phi_1(h), & h_0 \leq h \leq h_1 \\ \phi_2(h), & h_1 < h \leq h_2 \\ \dots \\ \phi_k(h), & h_{k-1} < h \leq h_k \end{cases} \tag{29}$$

where $\phi_j(h)$ represents the linear temperature distribution in zone j:

$$\phi_j(h) = \frac{T_j - T_{j-1}}{h_j - h_{j-1}} h + \frac{h_j T_{j-1} - h_{j-1} T_j}{h_j - h_{j-1}} \tag{30}$$

The zone height constraints in Eq. 22 apply here. The temperature varies continuously between zones.

With Eqs. 11, 15, 26, 29 and 30, it can be shown that

$$T_m^C = \sum_{j=1}^{k} x_j (T_m)_j \tag{31}$$

where $(T_m)_j$ is the mean temperature in zone j, i.e.,

$$(T_m)_j = \frac{T_j + T_{j-1}}{2} \tag{32}$$

and that

$$T_e^C = \exp\left[\sum_{j=1}^{k} x_j \ln (T_e)_j\right] = \prod_{j=1}^{k} (T_e)_j^{x_j} \tag{33}$$

where $(T_e)_j$ is the equivalent temperature in zone j, i.e.,

$$(T_e)_j = \begin{cases} \exp\left[\dfrac{T_j(\ln T_j - 1) - T_{j-1}(\ln T_{j-1} - 1)}{T_j - T_{j-1}}\right], & \text{if } T_j \neq T_{j-1} \\ T_j, & \text{if } T_j = T_{j-1} \end{cases} \tag{34}$$

It is noted that, while the above results for the stepped and continuous-linear cases are valid for any temperature distributions, a monotonic temperature increase with height is necessary for physical stability.

ILLUSTRATIVE EXAMPLE

Consider a TES having water as the storage fluid, and a fixed energy content. The temperatures at the top and bottom of the TES are held fixed, but the temperature distribution between these locations is allowed to vary. Three types of temperature distributions are considered: linear, stepped and continuous-linear. Specified data on the TES and on the temperature distributions are listed in Table 1. For each distribution, values of the temperatures T_m and T_e, and of several energy and exergy parameters, are evaluated and compared. The results are listed in Table 2. Several important points about the results in Table 2 are worth noting:

- There is a significant difference in the energy and exergy contents of the TES for all temperature distributions considered, the exergy values being more than an order of magnitude less than the energy values. While the Second Law of Thermodynamics requires the exergy to be less than the energy for heat storages, the large difference in these cases is attributable to the fact that the stored heat, although great in quantity, is at near-environmental temperatures, and therefore is low in quality or usefulness.

- As expected from Eq. 13, the energy values show that $E = E_m$ for all temperature distributions. Furthermore, because of the symmetry exhibited by all three temperature distributions considered (in the sense that $T(h) + T(H-h) = 2T(H/2)$), in all cases the temperature T_m is the same (i.e., $T_m^L = T_m^S = T_m^C$), the mixed-TES energy content is the same (i.e., $E_m^L = E_m^S = E_m^C$), and the mixed-TES exergy content is the same (i.e., $\Xi_m^L = \Xi_m^S = \Xi_m^C$).

- For the different temperature distributions, the exergy contents vary (i.e., $\Xi^L \neq \Xi^S \neq \Xi^C$), and consequently the exergy differences $\Xi - \Xi_m$ also vary. The TES exergy content when the temperature distribution is stepped is greater than that when the distribution is continuous-linear distribution, which in turn is greater than that when the distribution is linear (i.e., $\Xi^S > \Xi^C > \Xi^L$). This observation is attributable the the fact that the linear distribution exhibits a less pronounced degree of stratification than the continuous-linear distribution, which in turn exhibits a less pronounced degree of stratification than the stepped distribution.

CLOSING REMARKS

Analytical expressions for evaluating the energy and exergy contents are developed and compared for stratified thermal storages having linear, stepped and continuous-linear storage-fluid temperature distributions. The application of the results to an example has shown that, while energy evaluations are not changed by stratification, exergy values do change with the presence of stratification, giving a quantitative measure of the advantage provided by stratification. Specifically, the example application indicates how (1) for a single temperature distribution, the quantities of energy and exergy contained in a stratified TES differ; (2) the exergy content of a TES increases as the degree of stratification increases, even if the energy content remains fixed; and (3) the energy and exergy contents of a TES differ even when no stratification is present. The authors feel that, since exergy is a measure of usefulness or quality of energy, exergy quantities provide more significant comparisons of performance than do energy quantities, and that exergy analysis should be considered in the evaluation and comparison of stratified and non-stratified thermal storages.

ACKNOWLEDGEMENTS

Financial support for the research reported herein was provided by Energy, Mines and Resources, Canada, and by the Natural Sciences and Engineering Research Council of Canada, and is greatly appreciated.

REFERENCES

Beckman, G. and P.V. Gilli (1984). *Thermal Energy Storage*. Springer-Verlag, New York.
Bejan, A. (1982). Thermal energy storage. Chapter 8 of *Entropy Generation through Heat and Fluid Flow*. Wiley, Toronto, pp. 158-172.

Hahne, E. (1986). Thermal energy storage: some views on some problems. *Proc. Int. Heat Transfer Conf.*, San Francisco, pp. 279-292.

Rosen, M.A. (1991). On the importance of temperature in performance evaluations for sensible thermal energy storage systems. *Proc. Int. Solar Energy Soc. 1991 Solar World Congress*, 17-24 Aug., Denver.

Rosen, M.A., F.C. Hooper and L.N. Barbaris (1988). Exergy analysis for the evaluation of the performance of closed thermal energy storage systems. *Trans. ASME J. Solar Energy Engineering*, 110, 255-261.

Rosen, M.A., S. Nguyen and F.C. Hooper (1991). Evaluating the energy and exergy contents of vertically stratified thermal energy storages. *Proc. 5th Int. Conf. on Thermal Energy Storage (Thermastock '91)*, 13-16 May, Scheveningen, The Netherlands.

TABLE 1. Specified Data for the Example

Data for TES and for all temperature distributions	
Temperature at TES top, $T(h=H)$ (K)	353
Temperature at TES bottom, $T(h=0)$ (K)	313
Temperature of reference environment, T_o (K)	283
Height of TES fluid, H (m)	4
Mass of TES fluid, m (kg)	10,000
Specific heat of TES fluid, c (kJ/kg K)	4.18
Additional data for stepped distribution	
Number of zones, k	2
Height of zone-1 top, h_1 (m)	2
Additional data for continuous-linear distribution	
Number of zones, k	3
Height of zone-1 top, h_1 (m)	1.8
Height of zone-2 top, h_2 (m)	2.2
Temperature at zone-1 top, T_1 (K)	318
Temperature at zone-2 top, T_2 (K)	348

TABLE 2. Results for the Example

	Temperature distribution		
	Linear	Continuous-linear	Stepped
Temperatures (K)			
T_m	333.00	333.00	333.00
T_e	332.80	332.57	332.40
Energy values (MJ)			
E	2090	2090	2090
E_m	2090	2090	2090
$E - E_m$	0	0	0
Exergy values (MJ)			
Ξ	172.5	180.7	186.7
Ξ_m	165.4	165.4	165.4
$\Xi - \Xi_m$	7.1	15.3	21.3

2.21 Central Receivers

ANALYSIS OF A TWO-STAGE, LINEAR FRESNEL REFLECTOR SOLAR CONCENTRATOR

D. Feuermann and J.M. Gordon[*]

Center for Energy and Environmental Physics, Blaustein Institute for Desert Research,
Ben-Gurion University of the Negev, Sede Boqer Campus 84993, ISRAEL
[*]and the Pearlstone Center for Aeronautical Engineering Studies, Department of Mechanical
Engineering, Ben-Gurion University of the Negev, Beersheva, ISRAEL

ABSTRACT

The two-stage linear Fresnel reflector solar concentrator is analyzed via a study of an installed, 220 KW_t system. The concentrator includes: (1) a primary linear Fresnel reflector comprised of curved mirrors; and (2) a secondary CPC with tubular receiver. The principal practical design options for the secondary concentrator are evaluated. Via a computer simulation which includes ray-tracing of the primary reflector, we evaluate the sensitivity of energy output to: concentrator optical errors, system geometry, tracking mode, and the use of flat vs curved primary mirrors. Two-stage Fresnel concentrators can be considerably less expensive than parabolic trough collectors, but are found to deliver about 1/4 less yearly energy. However, much of this difference could be eliminated through the use of higher-quality CPC reflectors.

KEYWORDS

Fresnel reflector, solar thermal, secondary concentrator, CPC, flux map, simulation, ray-trace.

INTRODUCTION

This study evolved from the commissioned investigation of an installed, nominally 220 KW_t solar system for the production of 150 C steam in Sede Boqer, Israel (latitude 30.9° N) (Feuermann and Gordon, 1990). The two-stage optical design includes: (1) a linear tracking Fresnel reflector primary; and (2) a stationary nonimaging secondary (see Fig. 1). This type of solar concentrator was developed commercially in the 1970's and is reviewed by Rabl (1985).

Relative to parabolic trough collectors, Fresnel concentrators have potentially lower costs due to a stationary receiver assembly. This avoids flexible piping and reduces reliability problems. However, the Fresnel reflector incurs larger geometric losses. These losses will be quantified here. We will also address the proper optical design of the secondary concentrator, and the effect of principal design parameters on yearly energy delivery.

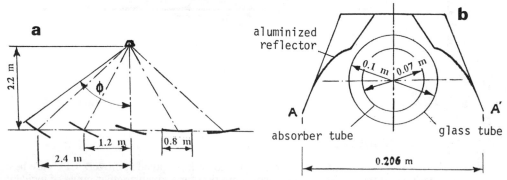

Fig. 1. Schematic of installed collector system. (a) View from the south, including primary mirror dimensions. ϕ denotes concentrator rim angle. (b) Second-stage receiver assembly.

The primary (Fig. 1a) is comprised of 5 rows of 0.8 m wide mirrors with circular curvature to enhance concentration. All 5 rows track solar motion about a horizontal north-south (N-S) axis, via an inexpensive, single mechanical linkage. The stationary receiver is 2.2 m above the mirror plane with a 0.206 m entrance aperture (Fig. 1b). A nonimaging reflector made of anodized aluminum further concentrates radiation onto a glazed, non-evacuated, selective-coated tubular receiver. The overall geometric concentration ratio (C = ratio of aperture area to actual absorber area) is 18.2; for the secondary only, the geometric concentration ratio C_2 is slightly below unity.

OPTICAL ANALYSIS

Primary Concentrator and Ray-Trace Simulation

Unlike *parabolic* concentrators, *Fresnel* reflectors are never in focus, and hence generate flux maps (normalized distributions of solar radiation intensity) that change with time. Selection of virtual absorber width, and of its height above the primary mirrors, must account for this. A computer ray-trace simulation was developed (Feuermann and Gordon, 1990) to predict solar flux as a function of position along a flat virtual absorber for arbitrary solar geometry. The effects of inaccurate tracking, misalignment, and mirror contour errors are lumped into one effective optical error, σ_{tot}. Ray-tracing for the secondary CPC was unnecessary in view of existing analytic solutions for estimating optical losses.

Fig. 2 shows a sample flux map, at solar noon, for the configuration and geometry of the installed collector. Three curves are shown: $\sigma_{tot} = 0.27°$ (width of solar disc), $1°$, and $2°$. The receiver aperture of 0.206 m is shown. These results illustrate how optical spillover increases with σ_{tot}.

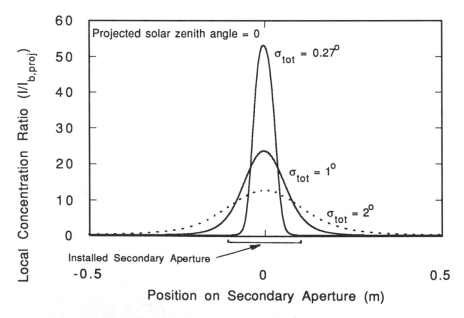

Fig. 2. Flux maps for projected solar zenith angle = 0, for 3 values of optical error σ_{tot}.

Secondary CPC Concentrator Design and Analysis

Background. The secondary concentrator is intended to collect optical spillover. CPCs are best suited to this task, since they offer maximum concentration for a given angular range of incident radiation on the secondary. Although the installed secondary appears to emulate a CPC-type concentrator, it can be improved upon, and was targeted for re-design.

The principles for the design of secondary CPCs, and for the specific case of tubular receivers, have been developed (Rabl, 1976; Winston, 1978; Winston and Welford, 1980; Kritchman, 1982; Baum and Gordon, 1984). The effective source is the primary mirrors. Their rim angle ϕ of around 55° is the design acceptance half angle for the secondary CPC.

Full CPC's are deep devices. The drawbacks include the expense of fabrication and mechanical loading, and high reflective losses. The upper reflector sections can be truncated, yielding slightly lower concentration at significant savings in reflector area and concentrator depth. Each of the designs illustrated below shows the full CPC, as well as a recommended truncated version.

Principal design options. The design here is complicated by the tubular glazing since the reflector involute cannot be brought to the absorber tube, as the ideal design requires. Three principal design options are (Rabl, 1985; Rabl, 1979; Winston, 1978, 1980; McIntire, 1980):
(i) Truncated involute; (ii) Design for a virtual (ice-cream-cone shape) receiver; (iii) Loss-less gap design. The scale drawings in Fig. 3 for these 3 principal designs show the full CPC with aperture FF', along with a recommended truncation to aperture TT' which yields 90% of full concentration. This small reduction in concentration reduces CPC depth by roughly a factor of 2.

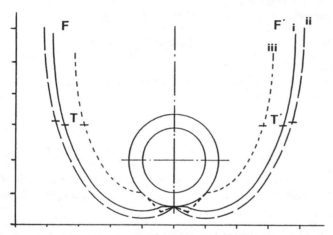

Fig. 3. Scale drawings of the 3 principal secondary CPC designs, both with full entrance aperture FF' and recommended truncated aperture TT' = 0.9 FF'. One unit = absorber radius. (i) Design with truncated involute. (ii) Design for ice-cream-cone virtual absorber (involute starts from bottom of glass tube). (iii) Design with loss-less (W-shaped) gap.

Although none of the 3 designs is patently superior to the other two, it can be shown (Feuermann and Gordon 1990) that, for anticipated collector operating conditions, designs (i) and (ii) turn out to be energetically superior to option (iii), as well as easier to fabricate. Due to a small edge in concentration, there is a modest energetic advantage for design (ii).

Final concentrator dimensions. For illustrative purposes, consider option (ii) (Fig. 3), with $C_2 = 1.17$ (truncated aperture TT'). Retaining the current receiver tubes, we then have TT' = 0.257 m. Simulation results show that for $\sigma_{tot} = 1^{\circ}$ this is wide enough to intercept about 60% of the yearly normal beam radiation. This figure exhibits a weak dependence on climate and latitude.

Estimate of secondary optical efficiency. For design (ii) (Fig. 3), $C_2 = 1.17$, and a CPC specular reflectivity of 85%, we estimate optical losses as: (1) CPC reflective losses = 11% (Carvalho and others, 1985); (2) Involute gap losses = 6.7% (Rabl, Goodman and Winston, 1979); (3) Absorptance and transmittance losses at normal incidence = 12%; (4) Incidence angle modifier losses = 8% (Carvalho and others, 1985). Hence secondary CPC optical efficiency = 83%, and the estimated overall receiver optical efficiency (including the secondary) is 67%. For the original secondary (Fig. 1b), the gap between the reflector involute and the glass tube, together with the incomplete V-groove design, would incur *additional* optical losses of about 10%.

SIMULATION RESULTS

Base-Case Simulation

The hourly numerical simulation also includes optical losses associated with the primary and secondary concentrators; optical spillover at the receiver; heat losses from the receiver; and threshold (turn-on/off) losses. Account is also taken of: (1) shading of primary mirrors by the secondary; (2) shading between mirror rows of the *same* (5-mirror-row) collector; and (3) shading between mirrors of different, adjacent collectors. For sensitivity studies, we define a base-case scenario, Table 1, and modify one system parameter at a time.

TABLE 1 Base-Case Scenario System Parameters

One-axis north-south horizontal tracking
Total primary mirror area = 730 m^2
Ground Cover Ratio = 0.606 (ratio of primary mirror area to gross land area)
Primary mirror reflectivity = 0.90
Total concentrator optical error σ_{tot} = 1^0
Climatic station: Bakersfield, CA, latitude = 35.4^0 N
Primary mirror dimensions and locations as installed (Fig. 1)
Secondary concentrator: CPC improved design option (i) (Fig. 3)
Secondary aperture width = 0.257 m
Optical efficiency of receiver assembly = 0.67
Absorber heat loss coefficient = 6 W/(K-m^2 of absorber area) (Rabl, 1985)
Constant load of 220 KW$_t$ (energy delivery in excess of 220 KW$_t$ not considered useful)
Backup in *parallel* with solar

Fig. 4 shows monthly energy delivery. The significant seasonal dependence arises from: (1) solar geometry effects; and (2) clearer weather during summer. Yearly energy delivery is 2.38 GJ$_t$/(m^2 of primary mirror). Not included are cool-down losses, which are estimated at 4% of this figure (Baer, Gordon and Zarmi, 1985). Yearly pumping energy is estimated at 0.055 GJ$_e$/m^2.

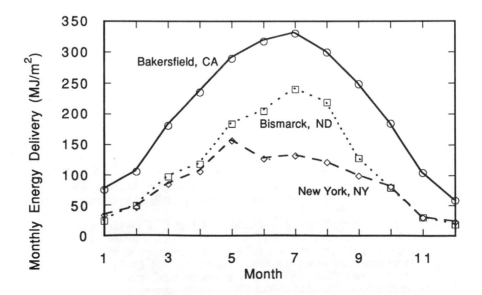

Fig. 4: Monthly energy delivery for base-case (Bakersfield, CA) and for two cloudier climates.

Principal Simulation Results

Receiver aperture width and receiver height. The sample flux map (Fig. 2) illustrated the magnitude of optical spillover for an individual moment. The dependence of yearly energy delivery on the width of the secondary entrance aperture is weak, changing a few percent for a range of width from 0.18 m to 0.28 m. Total shading losses turn out to be about 9% of yearly energy delivery. Larger secondary entrance apertures intercept more reflected radiation from the primary. However, they also result in larger shading losses, as well as larger reflective losses in the consequently deeper CPC. A broad optimum was also found for the receiver height. Within a range of 2.0 to 2.6 m, the energy delivery changes by less than 2% (the optimum being at 2.3 m).

Optical errors. Optical errors have not been measured. Based on field experience, a reasonable range for σ_{tot} is up to 2^O. Fig. 5 illustrates the sensitivity of yearly energy delivery to σ_{tot}.

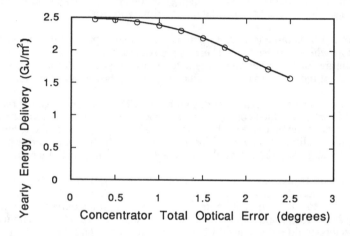

Fig. 5. Yearly energy delivery vs concentrator total optical error σ_{tot}.

Orientation of tracking axis. The tracking axis' N-S orientation yields a fairly smooth time-of-day energy delivery, but, as shown in Fig. 4, also yields a large seasonal dependence. The yearly energy delivery for a system with an *east-west* tracking axis is about 5% lower, with a more pronounced time-of-day dependence, but a more moderate seasonal dependence.

Climate. Energy delivery will scale roughly with the normal beam irradiance, since the solar concentrator accepts a negligible fraction of diffuse and ground-reflected radiation. Fig. 4 shows results for an intermediate cloudy and a cloudy climate besides the base-case figures.

Flat vs. curved mirrors. Relative to flat primary mirrors, curved mirrors increase concentration and reduce installation costs, but at the expense of increased capital investment. Retaining the same Ground Cover Ratio, collector rim angle, and concentration ratio, we find that the Fresnel concentrator with flat primary mirrors should have 19 rows of mirrors (Feuermann and Gordon, 1990). This permits use of the same secondary concentrator and receiver tube, and insures full utilization of the primary mirrors. The yearly energy delivery is 92% of the base case.

Comparison with Parabolic Trough

The parabolic trough collector intercepts about 7% more of the normal beam radiation than the Fresnel collector, depending on σ_{tot}. The difference increases to about 15% when σ_{tot} decreases to 0.27^O (width of the solar disc). At very large σ_{tot}, the Fresnel collector actually intercepts more radiation than the parabolic trough, because the parabolic trough has an acceptance angle function that decreases more rapidly with incidence angle than the Fresnel collector.

The Fresnel collector suffers from 17% additional reflective and gap losses in the secondary. Overall its optical gains are 23% lower than those of the parabolic trough. The difference in yearly delivered energy is 26% (accounting for threshold effects). However, since around half of this difference stems from secondary reflective losses, significant improvements are possible if higher-quality CPC reflectors would be used.

CONCLUSIONS

The linear Fresnel reflector concentrator - including a properly designed secondary CPC - can be a practical alternative to linear parabolic concentrators. Capital and maintenance costs can be significantly lower than for parabolic trough collectors, but at the price of reduced energy delivery. We have analyzed the performance potential and improved optical design of the linear Fresnel concentrator. Calculations of yearly energy delivery are based on an hourly numerical simulation which includes detailed ray-tracing of the primary mirrors.

The Fresnel concentrator is more tolerant to optical errors than a parabolic concentrator. This is an important feature when increased optical errors are accepted for the purpose of reducing costs. Also, broad optima are found for yearly energy delivery as a function of both receiver width (due to competing effects of increased shading vs reduced optical spillover) and receiver height.

System performance can be improved by re-designing the CPC secondary. Using flat, rather than curved, primary mirrors can reduce costs, but at a sacrifice in yearly energy delivery of about 8%. The Fresnel concentrator can be considerably less expensive, mainly due to the receiver assembly being stationary, which eliminates the need for flexible tubes. In addition, mirror profiles can be much lower, which can greatly reduce wind loading and can simplify mirror support structures.

Compared to parabolic trough collectors of the same concentration ratio, ground cover ratio, optical error and receiver, the two-stage Fresnel concentrator delivers 26% less yearly energy. The difference is due to: (1) inherent geometric losses (which account for around 1/3 of the difference); and (2) reflective and gap losses in the secondary concentrator. The use of higher-quality CPC reflectors could reduce this difference significantly. Thus the energetic losses must be weighed against cost reductions which may render the two-stage Fresnel collector an economically viable option for the production of intermediate temperature steam.

ACKNOWLEDGMENT

This work was funded by the Israel Ministry of Energy & Infrastructure

REFERENCES

Baer, D., J.M. Gordon, and Y. Zarmi (1985). Solar Energy, 35, 137-151.
Baum, H. P. and J. M. Gordon (1984). Solar Energy, 33, 455-458.
Carvalho, M.J., M. Collares-Pereira, J.M. Gordon & A.Rabl (1985). Solar Energy 35, 393-399.
Feuermann, D. and J.M. Gordon (1990). Analysis and evaluation of the solar thermal system at the Ben-Gurion Sede Boqer Test Center for Solar Electricity Generating Technologies. Final Report. Contract #88169101. Israel Ministry of Energy and Infrastructure.
Gerwin, H. J. (1979). A summary report: Suntec Solar Linear Array Thermal System (SLATS) test results. Sandia National Laboratories Report SAND79-0658.
Kritchman, E. M. (1982). J. Opt.Soc. Am. 72, 399-401.
McIntire, W. R. (1980). Solar Energy, 25, 215-220.
Rabl, A. (1976). Applied Optics, 15, 1871-1873.
Rabl, A., N.B. Goodman, and R. Winston (1979). Solar Energy, 22, 373-381.
Rabl, A. (1985). Active Solar Collectors and Their Applications. Oxford University Press, NY.
Winston, R. (1978). Applied Optics,17, 688-689.
Winston, R. (1978). Applied Optics, 17, 1668-1669.
Winston, R (1980). Applied Optics, 19, 195-196.
Winston, R. and W.T. Welford (1980). Applied Optics, 19, 347-351.

SOLAR CENTRAL RECEIVER PROJECT RESULTS

M. Castro, J. Carpio, F. Yeves, J. Bernárdez (*) and J. Peire (**)

Dep. Electrical, Electronics and Control Engineering
E.T.S.I.I./U.N.E.D. - Ciudad Universitaria, s/n - 28040 Madrid. Spain.
(*) Eléctrica de Moscoso - Pontevedra - Madrid. Spain.
(**) Dep. Electronics - E.T.S.I.I./U.C. - Santander. Spain.

ABSTRACT

This paper describes the results obtained in the "U.S.-SPAIN SOLAR CENTRAL RECEIVER JOINT PROGRAM", granted by U.S.-SPAIN JOINT COMMITTEE FOR THE SCIENCE AND TECHNOLOGY. This program studies the economic and technical aspects of the central receiver solar plants. The main objective is to define the main features of these plants and to study how they can be adapted to the energy demand requirements and the economic feasibility.

KEYWORDS

Solar power plants; test and operation analysis; flux measurement systems; control system automation; simulation analysis; operation strategies; grid solar integration; economic feasibility.

INTRODUCTION

The main goal of this project is focused in the comparative study of both the Solar One and the CESA-I solar plant. Thus, the main work developed has been a detailed study about the design conception, the final install equipment and the operation results, as well as the maintenance requirements during the routine plant operation of the two more relevant plant systems: the receiver and the energy storage. These two are the more critical from the technical point of view; the two more specific solar systems in all the solar power plants within the heliostat field. However, it has been also considered the rest of the plant systems, components and performances, including the plant operation strategies and the technical improvement required to fully automate the plant operation and thus reducing the produced energy cost.

Another joint research activity included in this program has been the development of mathematical models suitable to simulate any solar power plant with a reduced computer time and able to run in personal computers. With the developed computer models, a detailed study about plant operation strategies, has been performed.

The topics covered and the main related publications obtained in the Project are the following, covering this paper a description of them:
- Receiver and Storage Systems Study, (Baker, 1989; Castro, 1991).
- Solar Central Receiver Flux Measurement, (Castro, 1989d).
- Solar Central Plant Automation, (Castro, 1989b).
- Economic Feasibility and Operation Strategies of Solar Central Receiver Plant, (Castro, 1988a; Castro, 1988b; Castro, 1989a; Castro, 1989c).

RECEIVER AND ENERGY STORAGE SYSTEMS

This study is based on operating data from Solar One, the 10 MW_e experimental
solar central receiver plant located near Barstow, California, U.S.A., and CESA-
I, the 1.2 MW_e experimental solar central receiver plant located near Almeria,
Spain. Solar One, a joint U.S. Department of Energy and Industry project, began
operation in 1982. CESA-I, sponsored by the Spanish "Ministerio de Industría y
Energía", began operation in 1984, (Kuntz, 1986; Baker, 1989; Radosevich, 1985).

In the Solar One and CESA-I plants, a field of computer-controlled mirrors
(heliostats) reflect and concentrate solar energy on a receiver (boiler) on top
of a tall tower. Here the energy heats water and generates steam which is piped
to ground level where it is used to operate a conventional steam turbine, or the
energy is stored as sensible heat (in a mixture of oil, sand and rock, or in
molten salt) for use during periods of low insolation.

A comparative study about the receiver and energy storage systems has been
performed. The work carried out includes not only the design conditions but also
the performances and the efficiencies of both systems during the years of plant
operation. Major problems has been detailed pointed out specially those in direct
connection with the plant shutdown. Significant differences exist in the size,
design approach, and operating characteristics of the receiver and storage systems for the experimental plants. A comparison of the performance has increased
our understanding of the plant design variables and provides useful information
to improve the design of future C.R.S. plants, (Baker, 1989; Castro, 1991).

Comparison of receiver and thermal storage systems results.

The Solar One and CESA-I receiver and thermal storage systems offer several
contrasts in plant designs. Major differences include: external vs. cavity
receiver; thermocline vs. two-tank storage; and 10 MW_e vs. 1.2 MW_e plant size.
With these differences in mind we developed the following conclusions on a
comparison of the two systems.

Meteorological conditions. At the two plant sites were different. Both plants
experience insolation levels that were somewhat less than their design values.

Receiver systems: configuration. Measurements of the receiver efficiency confirmed that the CESA-I cavity receiver has a higher design point efficiency than the
Solar One external receiver. The results also showed that the CESA-I part-load
efficiency decreased faster than the Solar One receiver efficiency as the absorbed power decreased. The efficiency of both receivers changed with time as a
result of changes in the receiver panel surface absorptance. Periodic absorptance
measurements and repainting when necessary are desirable for both receivers.
The selection of an external or cavity design will depend on annual efficiency
values. The Solar One receiver average efficiency for over 500 days of operation
was only about 8 % points less than the peak efficiency. The Solar One receiver
also delivered power for over 70 % of the daylight hours these days. We did not
have sufficient data to assess the annual efficiency of the CESA-I.

Receiver systems: operations. The start-up times for the CESA-I receiver a forced
recirculation boiler with a steam drum, where longer than those of the Solar One
receiver, a once-through design. The steam drum acted as a storage unit which
must be charged at every start-up. During the course of testing, the CESA-I
receiver was sealed at night time by closing doors over its aperture and isola-

ting the steam drum from the receiver. This technique improved the CESA-I start-up time by keeping the receiver warm at night. However, even with this improvement the CESA-I start-up times were longer than the Solar One start-up times.

The transient response of the receivers to a cloud passage was similar. The only difference was the length of time the receiver could sustain steam production after the insolation had fallen to zero. The CESA-I receiver was able to maintain production for a longer time because of the energy stored in the steam drum. The Solar One receiver sustained production only at the expense of the thermal inertia in the receiver.

Receiver systems: life. Both receivers experienced leaks in their absorbing panels as a result of high thermal stresses. Reducing the amount of welding and developing improved techniques for attaching the panels to the support structure are desirable goals for future receiver designs.

Thermal storage systems: performance. The Solar One thermal storage system met or exceeded all its goal with respect to extractable capacity, heat loss rate, charging rate, and discharging rate. The CESA-I storage system met its heat loss goal but not its extractable capacity, charging rate and discharging rate goals. The CESA-I capacity was limited by three factors which were less than the plant's design values: salt mass, salt heat capacity and salt temperature swing. The reduced temperature swing was not an inherent storage system design problem but resulted from the less-than-design receiver steam temperature and pressure produced during the charging tests. The reduced temperature swing also limited the storage charging and discharging rates.

Thermal storage systems: use. The storage systems were effective buffers between the receiver and turbine generator during cloud operation, having storage charging and discharging systems in operation. They provided a steady flow of thermal energy to the turbine-generator for electrical power production. Nevertheless, after the functional operation of the two storage systems was demonstrated, they were used infrequently for electric power production.

Improving plant revenues, that is, generating electric power at night when it might be more valuable, could have incremented the use of thermal storage for power production. However, this operating strategy was not implemented until now.

SOLAR FLUX MEASUREMENT SYSTEMS

The performances of the main solar flux measurements systems installed in Solar One and CESA-I has also been analyzed in order to evaluate the technical viability for the measurement of the receiver efficiency. These systems are also compared with those developed for the other solar plants. The importance of this systems is due to the necessity of the knowledge of the energy which reaches the receiver, in order to calculate and evaluate its efficiency, (Becker, 1987; Castro, 1989d). The systems considered in this state of the art analysis are:

- Based in radiometers: HFD, (Heat Flux Distribution)
- Based in solar image analyzer: BCS, (Beam Characterization System)
                                   PAIS, (Solar Image Analyzer)
- Based in moving target: FAS, (Flux Analyzer System)
                                   HERMES, (Heliostat and Receiver Mea. System)
                                   OSIRIS, (System developed to the GAST Proj.)

Main characteristics and results

	HFD	BCS	PAIS	FAS	HERMES	OSIRIS
Resolution (points)	11 x 11	256 x 256	32 x 32	100 x 100	256 x 256	32 x 56
Radiometers	11	4	3	1	1	12
Video camera	No	CCD	CCD	CCD	CCD	VIDICOM
Time between measures (s)	80	25	-	60	5	16
Reflectance surface	No	Target	Target	Mobile bar	Mobile bar	Mobile bar
Receiver	Cavity	External	Cavity	Cavity	Cavity	Cavity
Plant used	Sandia	Solar One	CESA-I	SSPS	SSPS	GAST

SOLAR POWER PLANT AUTOMATION

The operation cost of the actual solar power plant is highly dependant of the labour costs. Thus, the control system design conception has been study, and new alternatives, basis on the advanced control systems theories, has been proposed, (Castro, 1989b; Kuntz, 1986; Peire, 1988). The technical evolution of control systems in solar central receiver plants has been developed as far as progress in digital communications and control has taken place. For that reason, the control system in the actual solar plants was installed with a very high advanced level of modern digital communication system. And due to the close contact among the project design leaders of the late build plants, the control philosophy is really similar among them.

The control system in a modern solar power plant is a digital distributed control system, covering the functions of:
- man-machine interface
- maintenance
- monitoring and diagnostics
- control strategies and system architecture

The Master Control Systems provides an overall command, control and data acquisition capability. It also performs the management of the daily startup and shutdown, as examples of complex operations. However, most of the times these are operations are made manually, due to the difficulties to optimize the different functions involved. As an example, the Master Control System at Solar One uses five computers to supervise operation and data acquisition: an operational control system, (OCS), two heliostat array controllers (HAC), one data acquisition system, (DAS), and a beam characterization system, (BCS).

Comparison of recommended and actual plant control capabilities

- Only one computer language should be used for the control software.
- A full automatic control of the plant is recommended, extending the actual automation degree, but maintaining the manual control over the systems.

- A peripheral control system should be used, integrating all of these function from the other systems.
- New man-machine interfaces should be introduced, as light pen input, mouse driven menus, and data storage for off-line evaluation and reporting.

ECONOMIC VIABILITY OF SOLAR CENTRAL RECEIVER PLANTS. OPERATION STRATEGIES

There has been developed a set of computer models for the components and systems of any solar central receiver power plant. With this, the transient behaviour of CESA-I and Solar One plants have been simulated, evaluating receiver start-up and shutdown and determining the losses of energy during them, (Castro, 1988a).

The yearly production for the CESA-I, Solar One and other new design plants have been calculated in order to evaluate the expected energy produced by these two power plants. Transients and cloudy conditions are considered in order to evaluate the yearly expected production. Several studies have been carried out with these computer models. Such as: plant operation strategies, cost of the energy produced, technical and economical viability of each plant system, up-scale of larger solar power plant considering operation strategies energy produced and economical viability, (Castro, 1989a; Castro, 1989c; Crespo, 1978).

A complete study of the weekly planning of a large electric grid with solar power plants integration was carried into this study, having as scenery the Spanish market. The solar electric generation values was obtained from the simulation programs developed. The planning model used makes the generation planning hour by hour in periods of a week, or multiples. The weekly scheduling is done using all the units present in the spanish power system, including hydro and pump-storage ones, and analyzing the introduction of solar thermal and photovoltaic ones. The planning results allows to define the maximum solar central operation cost that allows to replace thermal units (mainly fuel, gas and coal ones), by solar ones, (Castro, 1988b; Carpio, 1987).

The study was developed in one week of high load and low solar energy production, (December), and one week of low load and high solar energy production, (June). In these two weeks, the planning was done with three cases: the first one is the reference one without solar units, used to account the average operating cost of the units committed. In the second case the solar energy input value is the clear sky day, (without clouds), allowing to know the higher solar energy production. Finally, in the third one the solar energy input is considered with clouds, having a more realistic measure of the solar energy available.

The result obtained is the **substitution equivalent cost,** defining this value as the generation cost ($/kWh) of the energy replaced when the solar energy units are introduced. The average substitution equivalent cost ranges from 0.041 to 0.043 $/kWh in December (high electric load and low solar insolation), and from 0.033 to 0.037 $/kWh in June (low electric load and high solar insolation). The weekly generation considered in Spain in these weeks are 2,425,418 MWh in December, and 2,117,299 MWh in June, with a solar energy generation of 22,050 MWh (without clouds) and 16,990 MWh (with clouds) in December, and of 29,680 MWh (without clouds) and 17,710 MWh (with clouds) in June.

CONCLUSIONS

The main goal of this Project was to improve the relationship between to different research groups, one private company in USA, and the other one University Department in Spain. The great level of relationship and of collaboration between the both teams was excellent.

In the technical aspects of the Project, the main conclusions are:

- Complete comparative study of the receiver and storage systems of the solar central receiver plants, emphasizing:
  - Tests and operational procedures developed.
  - Data analysis evaluation.
  - Different configuration and size of the systems compared.
- State of the art analysis in receiver flux measurement.
- Mathematic model and simulation programs developed for solar central plants.
- Operation strategies study of the solar central plants:
  - Real operation constrains.
  - Operation strategy optimization: energy, efficiency or economic revenue.
- Central receiver grid integration analysis.

ACKNOWLEDGEMENTS

The authors would like to express their gratitude to the financial support of the U.S.-SPAIN JOINT COMMITTEE FOR THE SCIENCE AND TECHNOLOGY, to the help of the personnel from the Solar One and CESA-I plants, from the "Instituto de Energías Renovables" from the CIEMAT, and finally, to the personnel whom collaborate in the Project, (J. L. Presa, J. Díaz and F. Martín).

REFERENCES

Baker, A. and others (1989). U.S.-Spain evaluation of the Solar One and CESA-I receiver and storage systems. Sandia National Labs., SAND 88-8262.

Becker, M. and Bohemer, M. (1987). Achievements of high and low radiation flux receiver developments. Ed. DFVLR.

Carpio, J., López, M. and Valcárcel, M. (1987). A model to solve the weekly planning problem in a hydrothermal system with pump storage and safety constrains. Proc. of the IEEE MELECON Conference, Rome, Italy.

Crespo, L., Stahl, D. and Kostrzewa, S. (1978). Evaluation and optimization of possible operational strategies for solar thermal power plants. Ed. INTERATOM.

Castro, M. (1988). Simulación de centrales de producción de energía eléctrica a partir de la energía solar - Aplicación a la gestión energética. PhD Thesis, UPM - Madrid, Spain.

Castro, M. and others (1988). Large grid solar energy integration analysis. Proc. of the IEEE IECON Conference, Singapore.

Castro, M., Carpio, J. and Peire, J. (1989). Análisis de la viabilidad económica de las plantas solares de torre central. UPM - Madrid, Spain.

Castro, M., Martín, F. and Peire, J. (1989). Automatización del funcionamiento de las plantas solares de torre central. UPM - Madrid, Spain.

Castro, M., Martín, F. and Peire, J. (1989). Establecimiento de las estrategias de operación óptimas en plantas solares de torre central. UPM - Madrid, Spain.

Castro, M., Martín, F. and Peire, J. (1989). Sistema de medida de flujo de energía solar incidente en plantas solares de torre central. UPM - Madrid, Spain.

Castro, M. and others (1991). C.R.S. receiver and storage systems evaluation. Solar Energy. Accepted publication.

Kuntz, P. (1986). A handbook for solar central receiver design. Sandia National Labs., SAND 86-8009.

Peire, J. and Castro, M. (1988). Large high temperature solar thermal power plants - State of the art and future trends. Proc. of the IEEE International Workshop on Control Systems and New Energy Applications, Madrid, Spain.

Radosevich, L. (1985). Final report on the experimental test and evaluation phase of the 10 MWe solar thermal central receiver pilot plant. Sandia National Labs, SAND 85-8015.

Solar Thermal Power Plants:
No Need for Energy Raw Materials -
Only Conversion Technologies Pose Environmental Questions

C.-J. Winter[*], W. Meinecke[**], A. Neumann[**]

[*]DLR - Energetics Research Department, Stuttgart and ZSW - Solar and Hydrogen Energy Research Center, Stuttgart/Ulm, Pfaffenwaldring 38-40, 7000 Stuttgart 80, Germany
[**]Dipl.-Ing. Wolfgang Meinecke and Dr.rer.nat. Andreas Neumann, both DLR, Management Services Division, Energy Technology Department, Postfach 906058, 5000 Cologne 90, Germany

ABSTRACT

Energy conversion chains have three classic sections: (a) Energy raw material recovery and processing; (b) Primary/secondary/useful energy conversion, interim transport and storage; (c) Residual/pollutant processing and safe and secure final disposal. For solar thermal power plants -- indeed for all solar energy conversion systems -- sections (a) and (c) do **not** occur. All efforts concentrate on section (b). No energy raw materials are required for these purposes, only energy technologies and financial capital. And only these can be the source of any ecological consequences and hazard potentials. Albedo changes are given for places where solar thermal power plants are installed. Material intensity, surface area availability and energy intensity gradients and the ecology of selected commodities (sodium, salts, thermal oils) are presented. Predictions about the carbon dioxide implications of solar technologies, a discussion of their inherent low risks (no toxic substances, no radioactivity) and of the non-technical parameters (internalizing of external and long-term follow-on costs, short construction times, independence of the energy raw material markets) are given. Finally, trends are predicted concerning the relationship between the proportion of "solar-specific" (heliostats, receivers) to "non-solar-specific" investments in power plants, and reference is made to the decidedly recyclable character of power plant components.

KEYWORDS

Solar thermal power, energy conversion without energy raw materials, albedo changes, materials intensity, energy intensity

ENERGY CONVERSION CHAINS

Traditionally, energy conversion chains have three sections: An energy raw materials section (a); a technological section comprised of primary-to-secondary energy conversion; storage, transmission and distribution of the secondary energy; as well as secondary-to-final energy conversion (b); and finally the pollutant/waste section (c).

Solar thermal power plants -- as well as other solar[1] energy conversion systems -- have energy conversion chains which lack sections (a) and (c). Solar thermal power plants do **not** require energy raw materials: Solar irradiance is in economic terms a free commodity. It comes from space, and heat of the same energy content -- assuming a

---

[1]   In the context of this paper, solar energy conversion systems are understood to be solar thermal and photovoltaic power plants, hydropower and wind energy converters, oceanic thermal and wave energy power plants, plus heat pump systems, all of which are free of energy raw materials. The same is also true for such nonsolar renewable energy power plants as tidal or geothermal plants. Only biomass converters are unique among the solar energy converters: They utilize an intermediate solar "energy storage system," biomass, as an energy raw material.

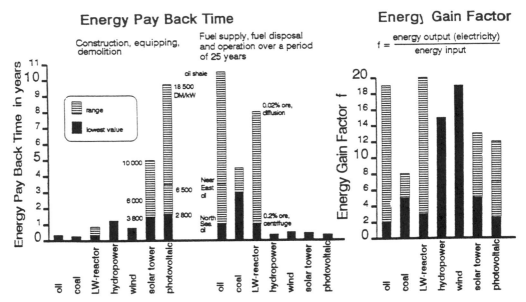

Sources: Rotty (1975), Moraw (1977), Meyers (1978), Wagner (1978, 1986), Enger (1979), Sandia (1981), Heinloth (1983), Voigt (1984), Aulich (1986), Hagedorn (1989)

Fig. 1. Net Energy Analysis of Power Plants

constant temperature in the geosphere -- returns to space, whether mankind makes use of it or not. Solar energy is closed loop energy; mankind has no influence over the course of the solar cycle. No waste or pollutants originating in an energy raw material can occur either. All the efforts required to make available the energy raw material in a form suited to the particular power plant, and all the potential ecological consequences of waste and pollutants arising from the energy raw material, have to add up to zero. Only the energy conversion, distribution and utilization **technologies** have to be taken into account, in order to estimate land availability, material and energy intensities and the efforts connected with maintenance, operation and final shutdown. In the end, a solar thermal energy conversion "system" reduces down to the power plant itself! Solar thermal power plants without energy raw material requirements are driven by technology and capital; they make possible an energy supply without energy raw materials, a "no-energy" energy supply.

No human intervention in nature can be entirely without ecological consequences, but it is a question of extent, of what is tolerable and what is responsible. Solar energy has a low energy density. The solar constant of 1,350 W/m^2 at the upper edge of the atmosphere drops to an annual average at ground level, depending on the latitude, of ≤ 300 W/m^2, the associated energy density being $\leq 2,500$ kWh/m^2a. The environmental challenge consists in carrying out solar energy conversion as efficiently as possible at each energy conversion step, using as little material as possible with as long a service life as possible, on as small an area as possible, and achieving, over the lifetime of the power plant, the highest possible energy gain (the factor reflecting how much more energy is produced by the power plant than was required for its construction, maintenance, operation and dismantling at the end of its service life). The energy gain factor[2] should be maximized; any material used should be reusable or recyclable, not toxic or radioactive.

Solar energy is not available everywhere at all times. Its discontinuous supply needs to be matched to human energy requirements, which are usually quasi-continuous, without creating any ecological problems. The current solution is solar/fossil hybrid operation, in which natural gas is used when there is no solar supply. This approach has both

---

[2]   Occasionally, energy gain factors for solar and fossil/nuclear energy converters are compared: Per se, fossil/nuclear energy converters **never** have an energy gain factor > 0; they are net depleters of limited energy raw materials.

TABLE 1 Land Area Intensities of Energy Conversion Chains with Power Plants

Energy Source	Capacity (MW$_e$)	Operation (h/a)	Service Life (a)	Land Area Intensity (m^2/GWh$_e$)
Lignite coal	600	6000	30	834
Bituminous coal	700	5000	30	80
Nuclear energy	1300	6000	25	81
Hydropower	0.1	5000	50	20
Wind				
- small plants	0.025	3000	20	67
- large plants	3	4190	20	40
Photovoltaic [1]				
- Central Europe	10	1050	30	700
- Southern Spain	10	1800	30	300
Solar Thermal [2] Power Plants				
- Trough	80	3600	30	425
- Tower	100	3600	30	575

[1] Technology of 2005   [2] Technology of 2005, location in Southern Spain

Source: U. Fritsche et al.: "Umweltwirkungsanalyse von Energiesystemen", GEMIS Gesamt-Emissions-Modell Integrierter Systeme, Öko-Institut Darmstadt, Kassel, Aug. 1989. ISBN 3-89205-0724-4 and J. Nitsch et al.: "Aufbaustrategien für eine solare Wasserstoffwirtschaft". TA-Gutachten für den Dt. Bundestag, Stuttgart, Juni 1990

advantages and disadvantages, also ecologically. It is of course a disadvantage that the greenhouse gases carbon dioxide and nitrous oxide are, in this case, being emitted also from "solar" power plants. The advantage is that the annual availability of the power plant is greater. Thus the financing and pay-back time for the initial investment can be spread over a longer annual operating period[3]. After a transition period, a future solution will be to use thermal storage for the role now filled by supplemental fossil firing. This will mean emission-free operation, since it would then be a "solar only" power plant. However, the proportion of the investment required for the concentrator field will then rise (solar multiple SM > 1) because it has to deliver energy both for the turbine and the storage system simultaneously. But with a combination of storage and hybrid operation there is the possibility to feed the grid at those peak load times which do not occur during the sunshine hours, taking advantage of the higher tariffs offered for peak load electricity. The environmental acceptability of the various storage media -- thermal oil, steam, rock, concrete, cast iron -- will have to be evaluated.

QUANTITATIVE RESULTS

Energy amortization times and energy gain factors for solar, fossil and nuclear power plant are shown in Fig. 1. The energy pay-back time for construction, initial inventory and final dismantling of solar power plants, measured as a few years, has to be compared with the energy pay-back times for fossil and nuclear powr plants, including their lifetime requirements for a fuel supply and their final disposal, which also have to be measured in years, so

---

[3]   Herein lies a cardinal difference between solar thermal and photovoltaic power plants: Solar thermal power plants can extend the annual operating period beyond the sunshine hours and into the medium load range (≤ 4000 h/a) by means of supplemental fossil firing or thermal storage. This cannot be done for photovoltaic plants because there are no electricity storage systems available of a size appropriate to a power plant. Thus their operating time is limited to the sunlight hours.

1984

that the following can be said about energy gain factors: (1) wind and hydropower plants are in the lead; (2) coal-fired plants are no longer in the running, and they are not yet even being confronted with the necessity of containing the CO_2 they produce; (3) the yields from nuclear and solar power plants are similar, and considering that safety restrictions will increase for the nuclear plants and costs will drop for solar power plants due to economies of mass production, the highest energy gain factors can ultimately be expected for the solar power plants.

Table 1 compares the land area intensities of the energy conversion chains of solar, fossil and nuclear power plants, as a factor of their output, capacity factor and lifetime. The land area intensity (m^2/GWh_e) of bituminous coal-fired and nuclear power plants, as well as of wind and hydropower plants is at 20-80 m^2/GWh_e lower by a factor of up to ten than that of lignite coal or photovoltaic and solar thermal power plants (300-834 m^2/GWh_e); small hydropower plants (20 m^2/GWh_e) have the lowest land area intensity, lignite coal-fired plants (834 m^2/GWh_e) the highest. The corresponding land area intensities for photovoltaic and solar thermal power plants are shown in Fig. 2 and 3. In Table 2 a comparison is made of energy carriers (kWh/m^2a) from solar energy converters (solar power plants and biomass). Technical solar energy converters have an energy yield higher than natural ones by one to two orders of magnitude.

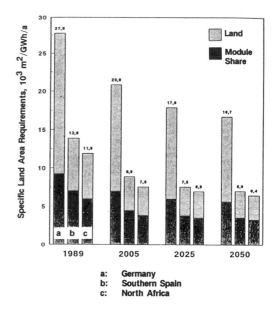

a: Germany
b: Southern Spain
c: North Africa

Source: Enquete-Kommission des Dt. Bundestages "Technikfolgen" zu: "Bedingungen und Folgen von Aufbaustrategien für eine solare Wasserstoffwirtschaft." DLR et al., Stuttgart, September 1989.

Fig. 2. Land Area Intensity of Large Photovoltaic Plants

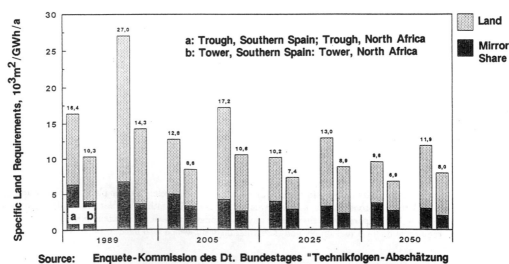

Source: Enquete-Kommission des Dt. Bundestages "Technikfolgen-Abschätzung und -Bewertung" zu: Bedingungen und Folgen von Aufbaustrategien für eine solare Wasserstoffwirtschaft" DLR u.a., Stuttgart, September 1989.

Fig. 3. Land Area Intensities of Solar Thermal Power Plants

TABLE 2 Energy Gain Factors from Technical (Solar Installations) and Natural (Biomass) Solar Converters

	Yield (kWh/m²a)	Type, Quality of Energy
Central Europe		
Flat collectors	220	heat < 100 °C
Photovoltaic (PV)	66	fluctuating electricity
PV + hydropower	50	burning/motor fuels
Wind	10[1]	fluctuating electricity
chopped wood	5	burning fuel
Ethanol from wheat	1.4(3.5)[2]	burning/motor fuels, (biogas)
Rape oil	1.5(2.9)[2]	motor (burning) fuels
Import from Southern Spain		
PV + electricity transport	95	fluctuating electricity
Solar thermal power plant + electricity import	70	diurnally fluctuating electricity
PV + hydropower + transport	76	burning/motor fuels

Land use factor for collectors and PV cells: 0.5, for solar thermal power plants: 0.25

[1] Surface area of wind parks allocated (related to rotor surface area 1800 kWh/m²a)

[2] Includes use of waste

Sources: L. Nitsch, J. Luther, Energieversorgung der Zukunft, Springer Heidelberg, 1990; L. Bölkow, M. Meliß, J. Ziesing, "Erneuerbare Energiequellen," Bericht für die Klima-Enquete-Kommission des Dt. Bundestages, 3/90.

TABLE 3 Estimated Albedo Changes from Solar Thermal Power Plant Surface Coverage (Intermediate-load plant on a desert site)

Without Solar Thermal Power Plant

Reflection	30	shortwave radiation	30
Soil Absorption	70	longwave radiation	35
		heat convection and conduction	35

With Solar Thermal Power Plant

Reflection	22	shortwave radiation	22
Soil Absorption	42	longwave radiation	21
Waste Heat Generation	28	heat convection and conduction	49
Electricity Generated	8	off-site thermal	8

Source: C.M. Bhumralkar, "Possible Impacts of Large Solar Energy Systems on Local and Mesoscale Weather," in "Climate and Solar Energy Conversion," Laxenburg, Austria 1977

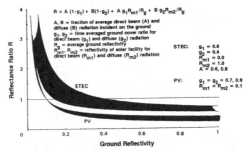

Source: J. Weingart, "Local Energy Balance of Solar Thermal Electric and Photovoltaic Power Plants," in "Climate and Solar Energy Conversion", Laxenburg, Austria 1977.

Wherever technical installations are placed on the earth's surface, the albedo is changed. This is also the case for solar power plants. Table 3 shows the changes in desert albedo caused by solar thermal intermediate load power plants and how the albedo changes when solar thermal or photovoltaic power plants are installed on ground having different levels of reflectivity. One sees that large albedo changes can only occur on ground with very high (> 0.8) or very low (< 0.1) reflectivity. Albedo changes are relatively minor in the cases of photovoltaic installations on ground with reflectivity values around 0.1 and of solar thermal installations on ground with values around 0.2-0.4. The inventories of synthetic oils or salts used as heat transfer media for solar thermal power plants have inherent ecological relevance. Table 4 shows the typical masses which, depending on the speed of degradation, have to be exchanged, regenerated or recycled.

The investment costs of solar thermal power plants can be subdivided into direct costs, indirect costs and credit costs during construction time. The solar-specific direct costs include the collector field, the receiver and storage systems. Table 5 compares the solar-specific costs for parabolic trough and tower power plants and shows how important it is to reduce further the solar-specific investment costs, especially for high output plants, which have a share of 60-85% of total direct investment cost today. Lightweight construction design and the use of materials and components with high specific strength and stiffness are of the utmost importance.

CONCLUSION

Solar energy conversion does not have anywhere near the ecological consequences of fossil or nuclear energy conversion. Nevertheless, as shown here using the example of solar power plants, they are not entirely environmentally neutral, but this is the case for any human intervention in nature.

TABLE 4 Design Data Comparison: Parabolic Trough and Central Receiver Systems

Parabolic Trough, Solar Only, With/Without Thermal Energy Storage

Name		SEGS VII 1)	SEGS VIII 1)	SEGS VII A	SEGS VIII A
Design Net Power	MW_e	30	80	30	100
Receiver Fluid		synthetic oil 2)	synthetic oil	synthetic oil	synthetic oil
Solar Multiple	---	1.2	1.2	1.3	1.2
Storage Full Load Hours	h	0	0	3	3
Reflective Area	m^2	201,432	510,774	220,660	638,000
Plant Land Area	km^2	0.66	1.55	0.71	1.93
Plant Land Use Factor	---	0.31	0.33	0.31	0.33
Erection Time	yrs	≤ 1	≤ 1	≤ 1	≤ 1
Sevice Time	yrs	≥ 30	≥ 30	≥ 30	≥ 30
Total Mass of Receiver Fluid	10^3kg	370	1,370	945	3,595
- Receiver Heat Transfer	10^3kg	370	1,370	370	1,710
- Storage	10^3kg	0	0	575	1,885
Specific Mass of Receiver Fluid	kg/MW_e	12,333	17,125	31,500	35,950
	m^3/MW_e	11.6	16.1	13.8	15.7
Mass of Energy Storage Material	10^3kg	⎤ 0		20,900 concrete + steel tubes	68,650 concrete + steel tubes
	kg/MWh	⎦ 0		79,180 "	79,180 "
	m^3/MWh			34.6 "	34.6 "

Central Receiver, Solar Only, With Thermal Energy Storage

Name		PHOEBUS 2nd Generation	PHOEBUS 2nd Generation	SANDIA SIT 3) 2nd Generation	SANDIA DIR 4) 2nd Generation
Design Net Power	MW_e	30	100	30	100
Receiver Fluid		air	air	nitrate salt 5)	nitrate salt 5)
Solar Multiple	---	1.2	1.8	1.4	1.5
Storage Full Load Hours	h	3	8	4.5	6
Reflective Area	m^2	190,608	856,937	202,649	721,950
Plant Land Area	km^2	2	5	1.6	4.5
Plant Land Use Factor	---	0.10	0.17	0.13	0.16
Erection Time	yrs	2.25	3	2.25	3
Sevice Time	yrs	≥ 30	≥ 30	≥ 30	≥ 30
Total Mass of Receiver Fluid	10^3kg	⎤ air	⎤ air	4,010	16,280
- Receiver Heat Transfer	10^3kg			500	1,100
- Storage	10^3kg			(4.5h) 3,510	(6h) 15,180
Specific Mass of Receiver Fluid	kg/MW_e	⎦ air	⎦ air	133,667	162,800
	m^3/MW_e	⎤ air	⎤ air	64.4	78.4
Mass of Energy Storage Material	10^3kg	6,660 ceramic	52,000 ceramic	3,510 salt	15,180 salt
	kg/MWh	26,429 ceramic	23,853 ceramic	9,486 salt	9,486 salt
	m^3/MWh	11.25 ceramic	10.15 ceramic	4.6 salt	4.6 salt

1) SEGS = LUZ Solar Electric Generation System; 2) Monsanto VP1; 3) SIT = Salt-in-Tubes; 4) DAR = Direct Absorption Receiver; 5) Density 2077 kg/m³ (20°C)

Quelle: W. Meinecke, DLR, 8/90

TABLE 5 Parabolic Trough and Central Receiver Power Plants: Direct Capital Cost Breakdown

Capital Cost Breakdown — Parabolic Trough Systems

Name	without thermal energy storage		with thermal energy storage	
	SEGS VII 1)	SEGS VIII	SEGS VII A	SEGS VIII A
Design Net Power, MW_e	30	80	30	100
Receiver Fluid	synthetic oil	synthetic oil	synthetic oil	synthetic oil
Solar Multiple	1.2	1.2	1.3	1.2
Storage Full Load Hours, h	0	0	3	3
Capital Cost Breakdown:				
% Solar-Specific Part	60-70	60-70	70-80	70-80
% Nonsolar-Specific Part	40-30	40-30	30-20	30-20

Capital Cost Breakdown — Central Receiver Systems

with thermal energy storage

Name	PHOEBUS 2nd Generation	PHOEBUS 2nd Generation	SANDIA SIT 2) 2nd Generation	SANDIA DAR 3) 2nd Generation
Design Net Power, MW_e	30	100	30	100
Receiver Fluid	air	air	nitrate salt	nitrate salt
Solar Multiple	1.2	1.8	1.4	1.5
Storage Full Load Hours, h	3	8	4.5	6
Capital Cost Breakdown:				
% Solar-Specific Part	65	75	60	66
% Nonsolar-Specific Part	35	25	40	34

1) SEGS = LUZ Solar Electric Generation System 2) SIT = Salt-In-Tubes 3) DAR = Direct Absorption Receiver

Quelle: W. Meinecke, DLR, 8/90

MINI POWER TOWERS FOR SOLAR THERMAL APPLICATIONS

Porter Arbogast

North American Sun, Inc.
P.O. Box 5015
Page, AZ 86040

ABSTRACT

A prototype of a "Mini Power Tower" concentrating solar collector has been constructed. Its focusing quality and cost have been found to be on par with more conventional concentrating solar collectors. Its fixed focal point technology offers enhanced versatility. A new servicing and refurbishing regime offers additional advantages.

KEYWORDS

Mini Power Tower; automatic servicing and refurbishing; fixed focal point concentrating collector; MODSYS; tracking error; solar thermal; solar power tower; concentrating solar collector.

INTRODUCTION

The Mini Power solar collector is basically a solar power tower in miniature. Since this system is a "fixed focal point" type system, as are the large solar power tower type systems, greater versatility is recognized.

The prototype was designed with several functions in mind. First, it serves as a test module from which data is gathered relative to performance characteristics such as focusing quality, trouble free operation, and durability. Second, the system was constructed in quasi-production manner to yield useful data relative to system cost. Third, it serves as a benchmark from which optimization techniques may be calibrated to provide a cost effective system in full scale production. Offered, herein, is a description of the prototype and test procedures, data characterizing the performance characteristics of the system and data characterizing the cost of the system. Also offered are conclusions which may be drawn from the data, a discussion of system configuration for domestic and industrial applications and finally a summary of topics for future study.

SYSTEM PRINCIPLE

The design of this solar collector is based on the following principle. Mirrors are attached perpendicularly to a rod. To function correctly, the rod should bisect the angle between the incident ray from the sun and the reflective ray to the receiver. It had been postulated that if mirrors were laid out in plane and that if focus was forced arbitrarily at any time of day, that the bisecting rods would lie row by row and column by column in planes projected north-south and orthoganally east-west. Thus, the mirrors could be supported, and the bisecting rods could be captured by trusses or frames running north-south and east-west. Further, these trusses could be linked in such a way as to be driven from one clock.

Fig 1. Concentrator Component

Inspection of the photograph of the concentrator component in Fig. 1 shows how the system is laid out. It can be observed that the bisecting rods pivot near the mirrors and slide within the trusses during the day. The system is counterbalanced for easier operation.

EXPERIMENTAL CONFIGURATION

The prototype observed in Fig. 1 has a place for thirty-two one foot square mirrors. For testing purposes, inch and half square mirrors were utilized. The focal length of 25 feet leads to a focal image approximately 4 inches in diameter for each test mirror. Figure 2 shows the test configuration. It includes three components; the solar concentrator, an invertor which flips the focal point back to the ground, and the target where various measurements characterizing the system are made.

EXPERIMENTAL MEASUREMENTS

The measurements made relate to several classes of tracking errors. First is the repeatability error, representing the random slop and slack in the system. The second is manufacturing and design error. The third is built-in error caused by the effects of imperfection in the principle of operation. The built-in error was identified by a computer simulation, MODSYS, designed for this kind of solar collector (Arbogast 1983).

With recent modifications to the simulation, good correlation has been established between the simulator and the data obtained from testing of the prototype. This has made the simulator a powerful tool in the design and analysis of these kinds of solar collectors. Some of the data provided here was created by the simulator which has been calibrated by the test results of the prototype.

Each mirror of the prototype can be separately tuned. Solar noon was selected as "Tune Time". In this way, repeatability error was measured. An image, on the target, six inches in diameter could be maintained and repeated from day to day at Tune Time (see fig. 3). Repeatability error is within a range of 1 to 5 milliradians.

Fig. 2. Test Configuration: Concentrator, Invertor,
and Target (barely visible in background)

Manufacturing error was measured directly on the prototype. The instruments used were accurate to approximately half of a degree. Since the prototype was the first off the production line, higher manufacturing errors were expected. Subsequent systems can be expected to have a greater degree of production repeatability. As a result of the lack of production repeatability and the relative inaccuracy of the measuring instruments, the set of manufacturing error data collected in this test is expected to be "heavier" than normal. Data in the graphs of Fig. 4 reflect tracking error with and without manufacturing error. The curves are also identified as representing error either in the vertical axis of the receiver or the horizontal axis of the receiver. The curves displayed represent an error corresponding to one standard deviation; sixty-eight percent of the sample lie within the value of error identified by the curve. Repeatability error is not represented on the graphs. Data was not perfectly symmetric about solar noon; however, the purpose here is to indicate trends in the tracking error. For brevity the afternoon data was omitted from the graph.

Graphs one and two in Fig. 4 correspond to a solar collector with a focal
height of 25 feet and a field of 480 mirrors. The field size is 32 ft. by 32
ft. The prototype, built to scale, represents one section of this concentrator with one fifteenth the mirrors. The data in the graphs was generated
by the simulator and calibrated with the prototype. The third graph in Fig. 4
corresponds to a concentrator with a focal point 250 feet high and field 100
times as big. This field has 100 individual concentrators, 32 ft. by 32 ft.,
all focused at one tower.

Fig. 3. Six inch diameter focal image.

All the curves in the graphs of Fig. 4 reflect effects of built-in error. The
curves identified as zero manufacturing error represent built-in error exclusively. In graphs one and two of Fig. 4, it can be observed that built-in
error is the dominate error. Comparing graph one and graph two, it can be
observed that the effects of manufacturing error are small compared to built-
in error.

In the third graph, the tracking error is not more than 3.25 milliradians.
This curve represents the effects of built-in error exclusively for a concentrator with a 250 ft. focal point. Here, the built-in error is more than an
order of magnitude less than the concentrators with a 25 foot focal length.
This indicates an important trend. When the collectors are broken up into
small modules relative to focal point length, built-in error is minimized and
other errors will begin to dominate.

It is concluded that in the case of   these concentrators with longer focal
lengths that the manufacturing error will be the dominate tracking error. It
is worth noting that the manufacturing errors can be controlled and therefore
minimized. It is expected that, in this configuration, generation of super-
heated steam in sufficiently high temperatures for efficient power generation
will not be a problem.

The less accurate shorter focal length systems can still be useful in many
applications such as the heating of buildings, domestic hot water, power
boosting of photo electric cells. There are also new techniques, under
consideration, which may increase the accuracy of the shorter focal length
systems by eliminating some of the built-in error.

COST ANALYSIS

This light weight concentrator has a correspondingly low cost. The materials
cost for the prototype was $454.84 and the labor cost was $638.03. (Costs do

not include cost of mirrors, control system, invertor, and receiver). In full production, these costs would drop significantly; an order of magnitude or more does not seem unreasonable. In the short term (production rates of thirty 480 square foot collectors per month), a cost of under $10,000 is anticipated for a complete hot water heating system (cost does not include storage). The system would be capable of providing most of the heating needs of an average sun belt home.

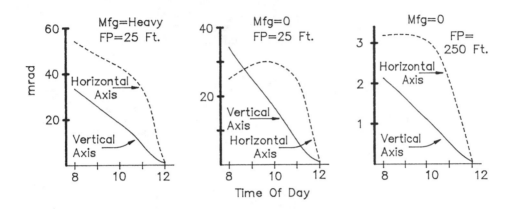

Fig. 4 One Standard Deviation Tracking Error

The system could be later augmented to provide some domestic electric power by splitting off some of the concentrated light with an invertor to power boost photo electric cells. The invertor can also be used to direct the focal point to a barbecue.

DURABILITY, AUTOMATIC SERVICING AND REFURBISHING

The concentrator was designed with durability in mind. Its expected life time is a minumum of thirty years. The mirror is not expected to be as durable as the concentrator. A special rack has been designed for the mirrors. During times of high wind, this rack feathers the mirrors. The mirrors go flat to provide an aerodynamic surface. The mirrors are held to the collector by a magnetic hinge which allows the mirrors to rotate from the operating position to the feathered position. The magnetic hinges also allow the mirrors to be simply removed intact in their racks. This facilitates storage of the mirror subsystem since the racks are stackable. In large systems, a simple vehicle can gather many racks and take them to a refurbishing center. Under severe weather conditions these racks may easily be removed, quickly and economically, preventing damage to the mirror subsystem. Because of the automated servicing system, the concentrator doesn't need to be designed for high load situations due to heavy snow and wind. The cost savings incurred, as a result of reduction in materials because of reduced loads, would much more than offset the cost of the servicing vehicle. The cost of the vehicle would be averaged into many collectors and could conceivably only add two or three percent to the cost of the concentrator. Also eliminated with automatic servicing, are the labor costs normally associated with installing a large amount of mirrors.

FUTURE STUDY

Work continues. Many aspects still need study. Control systems, mirrors, towers and receivers must be designed or selected. This concentrator has a slight tendency for lock up. Selection of alternate sliding surfaces should alleviate this problem. The focal characteristics especially relative to the manufacturing tracking error need to be better understood. There are other problems which need a great deal of attention; implementation of automated production, and other mass production techniques will lead to considerable cost savings. However, all that needs to be done is justified and necessary. The promise of the Mini-Power Tower offers much: heating, cooling, electric, and cooking for homes and business; large scale thermal applications, such as electric power production, as well as an endless list of other uses in industry.

ACKNOWLEDGEMENT

The author would like to express his gratitude to his wife, Karen Lawsing, who has patiently and unselfishly given her full support for many years of seemingly endless research.

REFERENCES

Arbogast, Porter (1983). Problem Report: An Analysis of Modularized Concentrating Solar Concentrating Collector Systems (MODSYS). West Virginia University, Morgantown, West Virginia.

SOLAR THERMAL TECHNOLOGY - IS IT READY FOR THE MARKET?

Bimleshwar P. Gupta
Operations Manager
Mechanical and Industrial Technology Division
Solar Energy Research Institute
1617 Cole Blvd.
Golden, Colorado 80401

ABSTRACT

In the short span of 15 years of research and development and industrial design of mass producible systems, the parabolic trough concentrating solar thermal system is now poised for substantial market penetration in the electric and process heat markets. Several of these systems have been installed in the United States, and are now operating in a commercial setting. In California, Luz International, Inc. is using these systems to generate electricity. Other companies, United Solar Technology and Industrial Solar Technology, Inc., have installed these systems to provide hot water economically to institutional users. Successful deployment of systems now producing about 350 MW of electricity and providing up to 100% of the hot water loads in institutions housing approximately 2,000 people, has proved the readiness of this technology for larger scale implementation in U.S. and world markets. The emerging new application for water detoxification offers another opportunity for commercial installations of parabolic-trough-type or similar concentrating systems.

Companion technologies, such as central receiver systems, and parabolic dish systems have also made considerable progress. A large number of concepts for heliostats and parabolic dishes and other components evolved from concepts to prototype hardware. These technologies form the next generation of solar thermal electric systems likely to become available in the late 1990s for commercial distributed and central electric generation.

INTRODUCTION

The uncertainty of today's energy environment makes predicting the future of all energy supply options tremendously difficult and imprecise. Fueled by the recent developments in the Middle East, new sources of energy are receiving increasing attention in shaping the policies for future supply options. Getting energy technologies from the concept stage to the point where they become major players in the energy market is a long-term effort. An energy technology usually takes decades to reach commercial entry into the market and a few decades to reach full potential. Such was the case with diesel power, hydropower, nuclear power, and others. Such will be the case with renewable energy technologies. Solar thermal technology is no exception. Over the past fifteen years, governments, in partnership with the universities and private industry around the globe, have supported research and development in solar thermal systems. These efforts have resulted in proving that solar thermal systems are technically viable for generating electricity and for producing heat for a multitude of applications. Heat applications range from the low temperatures (hot water for residential, commercial, institutional, and industrial needs) to the higher temperatures needed to produce process steam or concentrated radiant energy for industrial manufacturing processes.

TECHNOLOGY DESCRIPTION

The basic concept of using concentrated sunlight to generate heat and electricity by thermodynamic processes is well documented. In a current typical system, the concentrated energy is absorbed in a receiver, where it heats a fluid. Three main types of concentrating collectors have evolved for use in solar thermal systems -- parabolic troughs, parabolic dishes, and heliostats for central receivers. The modular parabolic dishes with an engine mounted at their focal point are classified as distributed systems, whereas the parabolic trough and the central receiver systems are classified as central generation systems. In this latter case, the energy collected either thermally or by redirection of the solar rays to a central receiver is brought to a central point for conversion from thermal energy to electrical energy.

All concentrating systems have their best annual output in regions where direct insolation is highest. Typically, this is the southwestern United States and other semiarid regions of the world. However, in regions with somewhat lower levels of direct insolation, these systems can be used for energy supply at somewhat higher costs.

SOLAR THERMAL ELECTRIC TECHNOLOGY STATUS

All types of solar thermal electric systems have been demonstrated in the past few years. Solar Thermal systems, operating either with storage in a hybrid mode with an auxiliary fuel or as part of a combined cycle system with gas turbines, offer significant potential to meet the needs of utility peaking or intermediate electric power generation.

Several systems using parabolic trough technology have been installed to generate electricity in southern California during the past six years. Building on the experience of smaller systems and developments of the past, Luz International, Ltd. further developed this most mature of the solar thermal technologies. They installed progressively larger systems to produce electricity for the Southern California Edison (SCE) utility grid. Figure 1 shows one of these systems. Today nine systems totalling 354 MWe are installed,

Fig. 1: Solar electric generating system (SEGS) plant installed by Luz International, Inc. in California

reliably operating in a hybrid mode and providing electric power in a utility setting. Figure 2 shows the annual installed capacity and cumulative capacity of solar thermal electric systems in the United States. The cost of electricity generation from these systems has fallen steadily from $0.24/kWh for the first 14 MWe system to an estimated $0.08/kWh for the 80 MWe plant installed in 1989 (Bazor 1989). Further system cost reductions can be achieved, which will make this technology cost-effective in an even wider range of U.S. and worldwide electricity markets.

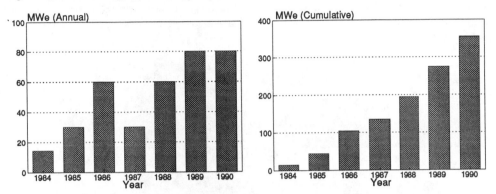

Fig. 2: Installed capacity of solar thermal electric systems in the United States

A 10-MWe experimental central receiver power plant was deployed by a joint government industry team and operated successfully by SCE on its grid for six years, achieving all the major objectives of the experiment. Extensive operating experience in a utility setting was obtained with this system. Further work showed that the change of working fluid in the receiver from steam to molten salt will reduce the cost of electricity generation from these systems (SNL; SERI 1989). Also, advances in the heliostat technology, in particular the development of membrane technology, are likely to reduce further the capital cost of these systems. Recent studies conducted by utilities project costs of $0.08 - $0.12 cents per kWh for plants of the 100-MWe size, reaching even lower costs of $0.05/kWh in the longer term after some years of experience, and for increased plant size of about 200 MWe.

Prototype parabolic-dish electric systems, totaling about 5 MWe, have also been operated in a utility setting in Georgia and in southern California. More recent development of a dish-Stirling generator concept has led to significant increases in system performance for modular electric systems (Stine 1989). These units, designed for producing electric power in the range of 5 to 25 kWe, are now being developed for field testing. A dish-Stirling engine generator module holds a record of 29% of overall system conversion from sunlight to electricity in a module size of 25 kWe.

SOLAR PROCESS HEAT TECHNOLOGY STATUS

A large fraction of the thermal energy used in the United States does not require generating very high temperatures. In the late 1970s and early 1980s, a number of solar thermal systems were installed at different sites to evaluate the possibility of providing the thermal needs of industry. These systems included low temperature flat-plate collectors and concentrating-type systems. Availability of low cost oil and natural gas prevented the widespread adoption of solar thermal systems for this application, however, systems for generating electricity found acceptance in the market. The recent events in the Middle East have reaffirmed the vulnerability of oil supply for providing our thermal energy needs. Hence, interest in solar thermal systems for process heat is reviving, not only to satisfy the needs of industrial users but also of institutional users with a constant need for hot water and steam.

Along with the energy efficiency measures adopted by industry in the past decade, solar thermal systems have also made impressive strides in efficiency improvement and cost reduction. Solar thermal systems recently installed in Brighton, Colorado, and Tehachapi, California, by Colorado's Industrial Solar Technology, Inc. have demonstrated the reliable performance of today's process heat systems. Figure 3 shows the California system in operation. These systems are also based on the more mature parabolic trough technology and make use of the advances in highly reflective polymer film to achieve higher efficiency and lower cost compared to older systems. The most recent system in California is achieving energy delivery cost low enough to attract investors, even with today's economic reality of competing with natural gas at $3.50/MBtu.

Fig. 3: Process heat system installed by Industrial Solar Technology, Inc. in California

NEW APPLICATIONS OF SOLAR THERMAL TECHNOLOGY

Beyond the use of solar thermal technology for generating electricity and process heat, there are several applications where the special attributes of sunlight are particularly useful (Gupta 1990). High-energy photons available in the near ultraviolet portion of concentrated sunlight make possible special reactions that destroy chemicals in water at near ambient conditions. When these high-energy photons are combined with the very high temperatures and high heating rates that solar thermal systems produce, the reactions can destroy toxic wastes in the gas phase or can produce space-age alloys.

Detoxification of chemicals in water is the first to receive increased attention in the United States, largely because of its immediate application potential and the possibility of using the most mature of the concentrating solar thermal systems. Ultraviolet (UV) radiation in the concentrated beam activates a catalyst (typically titanium dioxide, TiO_2) in the waste stream. Activated sites on the catalyst surface cause the breakdown of organic molecules. The organics are completely mineralized, forming carbon dioxide, water, and dilute concentrations of simple mineral acids (e.g., hydrochloric acid). The process is aimed at cleaning up large volumes of groundwater or industrial wastewater contaminated with low levels of toxic organic chemicals. Organic compounds such as trichloroethylene (TCE), and dyes, are actually destroyed using this process and not merely transferred to the air or other media.

Destruction of hazardous materials in the gas phase can be accomplished either by direct photolytic process, a photocatalytic process, or by means of the well-known catalytic steam reforming process. In the first, the destruction of the target molecules is enhanced when they absorb the near UV solar radiation.

In the second, a photocatalyst is used to enhance the destruction of compounds that do not directly absorb solar radiation. In the third, hazardous chemicals react with steam over a rhodium catalyst.

Highly concentrated solar energy, at intensities greater than 2000 times that of normal sunlight provides a controllable means of delivering radiant energy to solid surfaces. Modification of the surface region of these solids can provide many benefits: increased heat and wear resistance, improved corrosion resistance, surface alloying, surface transformation hardening, increased oxidation resistance, or superior optical or electronic properties of a material. Also at the same time, reduction in cost, consumed energy, and/or consumption of strategic materials in producing the desired materials are possible. Metal, ceramic, or composite materials can be treated with exposure to solar flux for seconds or less to obtain higher value materials with desired properties.

Recently, for the first time, the research team of Dr. Roland Winston and his colleagues at the University of Chicago achieved solar concentrations approaching 83,000 suns with a unique, high-index, nonimaging secondary concentrator placed at the focal point of a primary concentrator (Gleckman, O'Gallagher, and Winston 1989). This concentration level also reaches the energy threshold above which it is possible to efficiently use a solar beam to power a laser, which would normally require extremely high electric energy pulse to operate. The Solar Energy Research Institute (SERI) has developed and constructed a concentrator system that scales up the University of Chicago concept by a factor of 100. The concept provides a tool that will make it possible to learn more about the effect of the sun's energy on materials and chemicals.

MARKET OPPORTUNITIES

Many studies have been conducted to estimate market growth in the United States based on electrical energy demands in the various parts of the country. Using these general demand characteristics, a recent study by SERI estimates the potential contribution of each of the renewable energy technologies (SERI 1990). Solar thermal technology is expected to be one of the largest contributors in the western and southern regions of the country. Depending on the assumptions made, whether technological progress occurs at the current rate or at an accelerated pace, solar thermal technology could contribute up to 180 GWe (1 GWe is equal to 1000 MWe) of electric power by the year 2030, displacing about 9 Quads of an equivalent primary energy source.

This study does not include the potential for a significantly larger contribution possible in the international market. Arguments for this larger contribution can be supported by the fact that more than 4 billion of the world's population lives in areas without sufficient electric power, and a large fraction of those live in countries that are on the threshold of rapid industrial and economic development in the next century.

Some states in the United States are recognizing the tremendous potential of renewable energy technologies. Pushed by the need for generating additional electricity, while maintaining quality of life and maintaining a cleaner environment, they are introducing policy options that encourage the use of renewable energy systems where possible. New policies recently put in place in Nevada, and others under active consideration in California and other states, will pull solar technologies into these markets where they are marginally competitive today. Further, these early system deployments in the United States will allow companies to export the technology and systems to the emerging market throughout the world.

A recent projection for penetration in the process heat application market suggests that about 5 Quads of energy can be supplied just in the manufacturing sector of the U.S. economy by 2030, assuming an accelerated pace of development and increased market demand (Mueller 1990). Further, as more countries feel hostage to the supply interruption of oil, because they lack indigenous sources of natural gas, and face the reality of environmental implications of burning coal, they are likely to choose solar thermal technology as an alternative for providing their thermal energy needs.

CONCLUSION

Successfully operating parabolic trough systems are a testimony to the fact that this technology is reaching the level of maturity and field experience necessary to become a reliable source of electric energy and process heat in the near future. Continuing improvements in the design, operation, and maintenance techniques will increase the likelihood of lower capital and operating costs for future systems. A favorable environment created by progressive policies in some states, and decisions by forward looking investors and users willing to take some risks by deploying this new energy technology, are ingredients necessary to support capital investments by industry.

U.S. industry has proposed an aggressive program for commercialization of the other two concentrating type solar thermal systems. A number of companies are pursuing the development and future field testing of electric generating modules. These dish modules can be grouped into larger size electrical generating systems. Commercial deployment of 5 kWe dish-Stirling modules is expected in the mid-1990s after completing successful tests to verify reliable operation of the major components, in particular the Stirling engine. Larger modules of up to 25 kWe are also under development.

For the central receiver technology, the industry sees a 10 to 30 MWe size molten-salt based system as the next step towards commercial development of this technology. The commercial size for cost-effective energy delivery is estimated to be in the 100 to 200 MWe range. Reaching that stage will require orderly progression through one or more smaller size plants to gain the confidence of the decision-makers. Nonetheless, the promise of lower energy costs and larger size systems holds the key to make the vision of generating about 180 GWe power from solar thermal systems a reality by the year 2030.

REFERENCES:

Bazor, J., (1989). Testimony to the Tulsa, Oklahoma, National Energy Strategy Hearing, August 8, 1989.

Central Receiver Technology: Status and Assessment, (1989) Solar Energy Research Institute, Golden, CO., SERI/SP- 220-3314.

Estimating the Potential for Solar Thermal Applications in the Industrial Process Heat Markets, (1990), Contract final report, E. A. Mueller, Alexandria, Virginia.

Falcone, P. K., (1986) A Handbook For Solar Central Receiver Design, Sandia National Laboratories, Albuquerque, NM., SAND 86-8009.

Gleckman, P., J. O'Gallagher, and R. Winston (1989), "Concentration of Sunlight to Solar-Surface Levels using Nonimaging Optics," Nature, 339, May 18, 1989.

Gupta, B. P., and W. Traugott, eds. (1990) Solar Thermal Technology - Research, Development, and Applications, Proceedings of the Fourth International Symposium, Hemisphere Publishing Corporation.

Stine, W. B., (1989) Progress in Parabolic Dish Technology, Solar Energy Research Institute, Golden, CO., SERI/SP-220-3237.

The Potential of Renewable Energy: An Interlaboratory White Paper (1990) Solar Energy Research Institute, Golden, CO., SERI/TP-260-3674.

OPTIMIZATION OF THE SURFACE GEOMETRY OF A VOLUMETRIC FOIL RECEIVER

R. Pitz-Paal, J. Morhenne, M. Fiebig

Inst. f. Thermo- u. Fluiddynamik, Ruhr-Universität Bochum
Postfach 10 26 48, D-4630 Bochum 1, FRG

ABSTRACT

Optimal performance of volumetric foil receivers used in central receiver systems can be achieved by minimal channel cross sections and negligible channel wall strength in order to guarantee maximum convective heat transfer. However, these geometrical quantities are limited to a finite order of magnitude due to manufacturing constraints, creating highly loaded front surfaces. The resulting high front temperatures and solar reflective losses diminish energy conversion efficiency. Improvements can be achieved by a profiling of the front surfaces which influence the convective and radiative heat transfer.

Simulation using Navier-Stokes equations is performed in order to investigate the convective heat transfer in the profiled entrance region of the absorber. The effective receiver absorptivity is determined by the Monte-Carlo ray tracing method, and energy conversion efficiencies are evaluated by a one-channel receiver model depending on the ridge profile angle. The potential in efficiency improvement and decrease in front temperature is discussed.

KEYWORDS

volumetric receiver, convective heat transfer, profiled channel, thermal modelling, numerical simulation

INTRODUCTION

Volumetric receivers of different designs have been developed in the recent years for central-receiver-systems in order to deliver process heat or hot air for electricity generators by steam generation or Brayton cycle. One of the preferred design strategies is the so called 'foil' or 'multi-cavity' receiver. This receiver consists of parallel square channels (see fig. 1) in which concentrated sun light is entering. Since the heat absorbing surface is highly increased compared to the aperture plane, air can be used as heat transfer medium in spite of its poor heat exchanging properties. Air outlet temperatures of more than $800°C$ are reached in receiver experiments at the plataforma solar de Almeria (Spain) at mean solar fluxes of about 250 kW/m^2 (Böhmer,90) energy conversion efficiencies in the range of 60% were evaluated in those experiments. The heat loss mechanism in this conversions process can be split into two main parts:

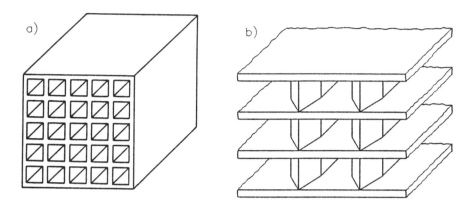

Figure 1: *Multi cavity volumetric foil receiver a) perpendicular and b) profiled front end surfaces*

1. Solar reflection losses occur as the high temperature resistant ceramic absorber material is not totally black, and high temperature resistant coatings can not be spread at low cross sections due to capillary effects with the danger of channel blockage.

2. Emission of thermal radiation occurs as the absorber is heated to at least the gas outlet temperature and even higher if the convective cooling is not sufficient.

Although best convective cooling is achieved when the channel cross section is minimal, this value is restricted by manufacturing constraints. Since we have to deal with finite extended channel walls in such receiver types, maximum solar load and reflection losses will occur on the front surfaces of the channels. This effect can be reduced by profiling these surfaces as it is shown in fig.1.

Improvements are achieved by two effects:

- Solar reflective losses are reduced as the reflected radiation is partly directed into the receiver.

- The front absorber temperature is decreased as the solar load on the surface is reduced compared to perpendicular front end surfaces.

However, the geometrical modification does not only influence the radiative heat transfer (solar + thermal) but also the convective cooling in the entrance region of the channels as the inlet flow is changed by the ridged profiles. Since there are no existing correlations describing the local heat transfer of such a complex geometrical shape in literature, a three dimensional Navier-Stokes simulation of the inlet flow is performed with the PHOENICS code (Cham,Lodon).

The results of these calculations for different ridge angles of the profiled surface are incorporated in a computer code which calculates the thermal efficiency for a one channel receiver model (Pitz-Paal,90). This model includes solar and thermal radiation, convection (by means of the given heat transfer coefficients) and conduction. The dependence of thermal efficiency of the receiver on the ridge angle of the profiled structure and on the solar absorptivity of the absorber material is discussed.

BOUNDARY CONDITIONS

It is obvious that the best potential for volumetric foil receivers can be achieved by combining minimum hydraulic diameters and negligible absorber wall strength. This increases the convective cooling and reduces the front surfaces, which are responsible to solar reflection losses making it possible for high solar flux densities to be attained. In order to improve the efficiency at any given channel diameter and given wall strength (which are in fact the real boundary conditions due to the manufacturing constraints), two further possibilities are open:

1. Using absorber material with high heat conductivity in order to conduct the heat to the rear parts of the channel which are not so highly irradiated (fin-effect) (Pirrot,87).

2. Profiling the front surfaces.

Investigations on the second point were made with the following design (see fig. 1b, Tab. 1) which is proposed for a receiver test at Plataforma Solar de Almeria in autumn 1991.

TABLE 1: Investigated Receiver Design

Material	SiSiC ceramic
channel cross section (d_h)	3 mm
channel length	100 mm
wall strength	0.8 / 3 mm
solar flux	1 MW/m^2
mass flow rate	0.7 $kg/m^2/s$
air inlet temperature	20°C
heliostate field opening half angle	45°

Siliconized Silicon-Carbid (SiSiC) ceramic has proved to be suitable as absorber material due to its high heat conductivity (≈ 150 W/mK at 20°C), temperature resistance (up to 1400°C) and an excellent thermoshock behavior. Wall strength and channel cross section are limited to the mentioned sizes in order to guarantee stability and prevent the channel from being blocked by the siliconization process. The solar flux specification is an upper limit in central receiver systems up to now, and the mass flow rate is determined by an desired outlet temperature of at least 800°C.

EFFECTS ON CONVECTIVE HEAT TRANSFER

The convective heat transfer coefficient was determined by a Navier-Stokes Simulation performed with the PHOENICS code (Cham,London) under the assumptions of constant wall temperature of 1000°C. A three dimensional boundary fitted coordinate grid which is refined near the channel walls is applied to the smallest symmetric element of the matrix structure (1/4 of one channel). As the Reynolds number ($Re = \frac{V \, d_h}{\nu}$) is in the order of 100, laminar flow can be expected. Inlet flow is assumed to be homogeneous with velocity of 0.58 m/s corresponding to assumed mass flow and air density. Temperature dependence of air

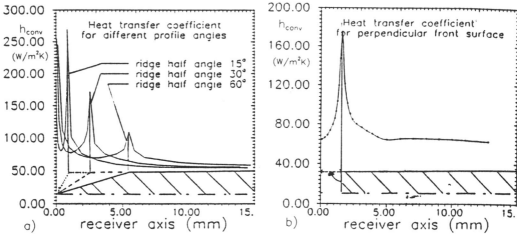

Figure 2: *heat transfer coefficient versus axial location for a) profiled front surfaces and b) perpendicular front surfaces evaluated for boundary conditions as given in Tab. 1*

density, kinematic viscosity and heat conductivity is incorporated. The local heat transfer coefficient is derived from temperature fields evaluated by PHOENICS as:

$$h_{konv} = \frac{\lambda(T_{wall} - T_{1.cell})}{\Delta x(T_{wall} - T_{bulk})} \qquad (1)$$

were λ represents the heat conductivity of air and Δx the distance between wall and center of the cell nearest to the wall.

Calculation were performed for three different profile angles ($30°, 60°, 120°$) and for the case of perpendicular front surface (here called $180°$). The local heat transfer coefficient is averaged over the channel height of the profiled channel and plotted against the axial location in the receiver. In fig. 2 a) the dependency is shown for different profile angles and in fig. 2 b) for the case of a perpendicular front surface. For all profiled geometries the heat transfer has a maximum at the stagnation point, decreases along the profile region and has a second maximum at the edge where the nominal channel cross section is reached. That corresponds qualitatively to experimental results of Kottke (1977) who investigated the flow at a flat profiled plate with finite wall strength at zero incidence.

In the case of perpendicular front surfaces a maximum value of h_{konv} is reached at the front edge. No separation of flow occurred, possibly because of the low flow Reynolds number.

EFFECTS ON THERMAL RADIATIVE LOSSES

In order to analyze the effect of the profiling on thermal radiative losses of the receiver, convective heat transfer data and boundary conditions as described above were transferred to one channel receiver model. In this model the channel is discretised into finite isothermal channel wall elements and finite volume elements, where the heat transfer medium is flowing. Steady state energy conservation implies that for each wall element the sum of convective, conductive and radiative heat transfer is zero:

$$\dot{q}_{conv} A_f + \dot{q}_{IR} A_f + \dot{q}_{Sol} A_f + \dot{q}_{hc} A_{contact} = 0 \qquad (2)$$

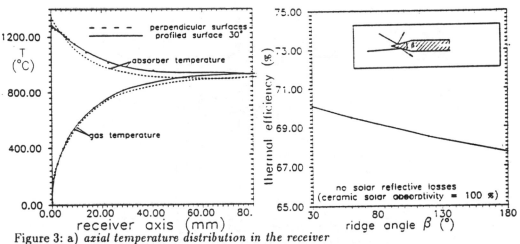

Figure 3: a) *axial temperature distribution in the receiver*
b) *Receiver energy conversion efficiency versus profile angle*

and for each volume element that convective heat transferred from all adjoining wall elements change gas enthalpy:

$$\dot{M}\, c_p\, \Delta T \;=\; \sum_i \dot{q}^i_{conv}\, A^i_f \tag{3}$$

Solar radiation distribution is evaluated by Monte-Carlo ray tracing, thermal radiation by the enclosure method applied to diffuse reflecting gray surfaces (e.g. Siegel 81/82), convective heat transfer by heat transfer coefficients as explained above and heat conduction by discretisation of Fourier's law. This set of non-linear equations is solved iteratively in the temperature. A more detailed description is given by Pitz-Paal (90).

In order to see the pure effect of geometry on convective and thermal radiative heat transfer, solar absorptivity is assumed to be 1.0. In figure 4 a) the effect on temperature distribution and 4 b) on receiver efficiency are shown. Front temperature of the receiver is reduced by $77°C$ for profile angle of $30°$, whereas receiver efficiency is increased by 2.4 %.

EFFECTS ON SOLAR REFLECTIVE LOSSES

Solar reflecting losses depend purely on the geometry and on the solar absorptivity of the material. Therefore they can be investigated totally independent of the heat transfer problem in the receiver. The effective solar absorptivity of the proposed design (fig. 1b, Tab. 1) was determined by a Monte-Carlo ray tracing algorithm as a function of the material absorptivity. Pure diffuse reflection characteristic was assumed. Specular reflecting properties of the material would increase the effective absorptivity beyond that for diffuse reflection, since no reflection losses out of the channel can occur. Incident radiation is assumed to have a uniformly distributed cone profile with a half angle of $45°$. Effective solar absorptivity of three different profiles ($30°, 60°$ and perpendicular) are shown in fig. 3 in dependence of the material absorptivity. High improvements of absorptivity can be shown for materials with poor solar absorptivity.

The authors would like to express their gratitude to the Deutsche Versuchsanstalt für Luft- und Raumfahrt e.V.(DLR), Cologne and the Federal Ministry of Research and Development (BMFT) for sponsoring this work.

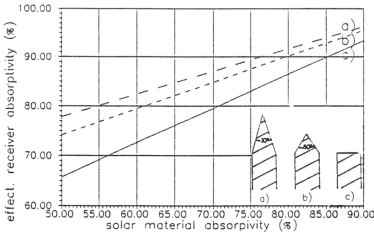

Figure 4: *Effective solar absorptivity of the receiver versus the solar material absorptivity*

CONCLUSIONS

Detailed numerical investigations proved that profiling of absorber front surfaces of volumetric receivers will improve energy conversion efficiency by two effects.

1. The effective solar absorptivity is increased.

2. Front temperature and thermal emissive losses are reduced.

The first point can be understood easily, as solar radiation is partly directed into the receiver. This effect is of high importance at low material absorptivities and will be increased further on if the reflectance characteristic of the absorber is of a more specular type.

The second one is determined by the interaction of different heat transfer mechanisms. Solar flux density is decreased on the tilted surfaces in comparison to perpendicular ones, the heat transfer coefficient is changed, and the thermal radiative losses of the profiled surfaces are different from either perpendicular surfaces or channel side wall elements, such that their cumulative effects can hardly be predicted without calculations. However, thermal modelling showed that energy conversion efficiency improvement of up to 2.5 % and a decrease in front temperature of $80°C$ % for the case considered can be achieved by this second effect.

The reduction in the maximum front temperature is an especially important result since it is usually a restrictive condition determining maximum solar flux density and maximum outlet temperature.

REFERENCES

Böhmer, M., C. Chaza (1990). The ceramic foil volumetric receiver. *Proceedings of 5th Symposium of Solar High Temperature Technologies, August 27-31,1990, Davos*

Kottke, V. (1977). *Wärme- u. Stoffübertragung, 10*, 159-174

Pierot, A. (1987). Contribution a l'etude des transferts de chaleur a haute temperature dans les milieux alveolaires. *These, universite de Perpingnan*

Pitz-Paal, R. J. Morhenne, M. Fiebig in M. Becker (Ed.) (1991), The Construction of a Volumetric Receiver with a Staggered Structure, *Solar Thermal Energy Utilization*, Vol 5, Springer Berlin

Siegel,R., J.R. Howell (1981/82). *Radiation Heat Transfer*, Macgraw Hill Book Company, Washington.

2.22 Dish Collectors

SOLAR FLUX DISTRIBUTION ON AXISYMMETRIC RECEIVERS AT THE FOCUS OF MULTI-FACETTED DISH CONCENTRATORS

I. Mayer

Energy Research Centre, Research School of Physical Sciences and Engineering, Institute of Advanced Studies, Australian National University, Canberra, A.C.T., Australia

ABSTRACT

This paper outlines the main considerations involved in the development of a computer program, FCAR, used to predict the distribution of concentrated solar radiation on the absorbing surfaces of an axisymmetric receiver of otherwise arbitrary shape, mounted at the focus of a dish concentrator made from multiple mirror facets. The facets could be plane, spherical, or pseudo-spherical. Errors that could arise in manufacture or operation of the collector can be taken into account. The aim is to help the designer select an optimum dish/receiver system, and subsequently 'fine tune' it during successive stages of the design process.

KEYWORDS

Point-focus distributed-receiver solar collectors; design analysis; ray-tracing; computer program; optimization.

INTRODUCTION

Point focus distributed receiver solar collectors can supply high temperature heat for applications such as high efficiency solar thermal electric power systems, production of fuels and chemicals, high temperature industrial process heat, and combined heat and power systems. A paraboloidal dish reflector with a plane receiver at its focus is perhaps the simplest embodiment of such a collector, but practical systems tend to be more complex than this idealisation. Rather than being a perfect paraboloid, the concentrator is frequently made from facets which may be plane, spherical, or pseudo-spherical, mounted, in various possible arrangements, on a surface which may be paraboloidal, plane or of some other figure. Each such configuration produces its unique three-dimensional distribution of concentrated solar radiation in the focal region.

Receivers likewise come in many forms, although they are usually axisymmetric. Cavity receivers offer high efficiency since significant radiant heat loss occurs only from the small area of the cavity aperture, whereas the actual heated area within the cavity can be much larger, resulting in lower, more manageable heat fluxes. A small cavity aperture should also minimise convective losses from the internal high temperature surfaces. If part of the receiver is to operate at a relatively low temperature, as a result of the absorption process occurring within its absorber wall,' then the cavity aperture can be made even smaller than the minimum cross sectional area of the concentrated solar radiation distribution, resulting in a 'semi-cavity' design (Kaneff, 1989). For example in a steam generating receiver, the low temperature feed water heating (or 'economiser') part can be located external to the cavity aperture.

The solar flux distribution over the absorber area of a receiver will depend on the shape of that surface, the receiver position in the focal region of the concentrator, and the three-dimensional flux distribution generated by the concentrator. The optimum flux distribution over the absorber surface will depend on several factors. In a steam generating receiver for example, it is important that the flux be low in the dry-out region to obtain an adequate service life from the receiver tubing (Mayer, 1988), whereas in a thermochemical receiver, it is desirable to match the flux to the reaction rate in different parts of the receiver to achieve a uniform temperature (Diver, 1987). The primary input to the analysis of receiver performance is the distribution of concentrated solar radiation incident on its absorbing surfaces.

To enable the collector designer to choose an optimum concentrator/receiver combination, it is necessary to calculate the flux distribution for each of the configurations being considered at the preliminary design stage, and then to 'fine tune' the chosen configuration during the subsequent stages of design and development. It is also necessary to be able to investigate the sensitivity of the design to off-design conditions that might result from manufacturing defects, tracking or alignment errors, or sunshape variations. The following sections outline the principal considerations which were involved in the development of a computer program, FCAR (or Facetted Concentrator Axisymmetric Receiver) designed to address this problem (Mayer, 1991). The work was done as part of a project at the Australian National University to develop large high temperature steam generating solar thermal systems, and was necessary because all of the previously published analyses which were examined either did not treat facetted collectors, or dealt only with plane receivers.

RECEIVER DESCRIPTION

The axisymmetric receivers considered here are usually of a close-packed helical construction, made by winding a metal tube around a former, resulting in a receiver surface which is corrugated in the axial direction. The heat transport or chemical reaction fluid is passed through the tube, which has the concentrated solar flux from the paraboloidal mirror impinging on its outer surface.

In the computer program FCAR described here the corrugations are ignored, and the receiver surface is treated as a smooth surface of revolution which touches the crests of the corrugations, or receiver tubing. An earlier version of the FCAR program treated this surface detail explicitly, but this degree of complexity was found to be unwarranted, since the increase in flux into the part of the tubing surface normal to the concentrated solar radiation tends to be balanced by the reduction in average flux which results from the larger surface area represented by the corrugations.

The receiver shape is specified by describing the surface of revolution which contains the helical centre line of the receiver tubing. Each part of this surface is generated by rotating a curve about the axis of the receiver. The generating curve can be of arbitrary shape. For computational purposes, the curve is represented by a series of straight line segments, so that the receiver surface is represented by a series of conical frustra. (In fact, receivers are often manufactured this way also, for ease of construction). Each of these frustra is, in turn, divided into one or more short sections, usually corresponding to individual coils of the receiver tubing, and computations are carried out at the mid-point of each of these sections.

GEOMETRIC IDEALISATIONS

To provide a 'bench mark' against which progressively more detailed computational schema could be checked, an idealised situation was treated first. The idealisations which were incorporated in this 'base case', in addition to the receiver geometry assumptions mentioned above, can be described as: point sun, perfect paraboloidal dish mirror, perfect pointing of the dish towards the sun, perfectly formed receiver, and perfect alignment of the receiver with the dish axis. In successive stages of program development, these idealisations, or departures from reality, were removed, one at a time, to deal with progressively more realistic situations.

The departures from complete realism can be regarded as 'errors', 'random errors' being those resulting from manufacturing imperfections or failures to achieve exactly the intended (as distinct from ideal) geometry, while 'systematic errors' are those resulting from the differences between the ideal geometry assumed in the calculations and the actual geometry. For example, assuming a point sun rather than one with finite apparent angular diameter is a systematic error; so is the assumption of a perfect paraboloidal mirror when the actual dish may be made from a collection of flat mirror facets attached to a paraboloidal former. In contrast, if the attachment procedure fails to achieve exact tangency of the flat mirror facets with the paraboloidal former, then each facet will have random slope errors associated with it.

As mentioned, this idealisation procedure provided a check as the program was developed through successive stages. For example, the results for a facetted dish should approach the perfect paraboloid case as the number of facets is increased. The procedure also helped to reveal the relative importance of the various steps from the ideal to the real, and showed that it is often not necessary to use all program features for a given task.

Point Sun

The point sun idealisation can be visualised by imagining the size of the sun to shrink to a point while its energy output remains the same. Although a physical impossibility, this concept simplifies the geometry since at any point in the collector, only a single ray, or sun direction, has to be considered. Because of the large sun-earth distance, all such rays incident on the collector are effectively parallel. With the further zero pointing error idealisation, they are also parallel to the dish axis. The apparent angular diameter of the sun, viewed from the earth, is about nine milliradians, and varies slightly throughout the year because of the eccentricity of the earth's orbit around the sun. Under hazy atmospheric conditions, a small proportion of the energy that would normally be in the direct beam from the sun's disc, is forward scattered so that it comes instead from a bright area of sky immediately around the sun - the solar aureole; furthermore, the apparent brightness of the sun is not necessarily uniform across its disc. The term 'sunshape' is frequently used to refer to both effects.

In the FCAR program, sunshape effects are dealt with by specifying, in the input data, a value for the apparent sun diameter, as well as the number of circular zones into which the apparent sun disc is to be divided, and the weights to be assigned to each zone to describe the sunshape distribution. The program then divides each circular zone into angular segments so that all segments in all zones are of equal area, and calculates the total irradiance on each receiver point from all sun segments.

Perfect Paraboloidal Dish Mirror

Large dish mirrors must be made from facets. Paraboloidal facets would be ideal, but then each zone of the dish would require facets of a different shape. To minimize cost, plane mirror facets can be used, mounted so that at their centre points they are tangent to an imaginary inscribed paraboloid. Spherical facets offer a smaller focal region, but need to be mounted in a stepped fashion which introduces another systematic error (or departure from the ideal) in that some of the incident or reflected (or both) rays associated with one facet will be blocked or shaded by another facet. The same is true for plane facets mounted in a plane (e.g. for constructional convenience). The result is that part of the aperture area of the dish is not effective. This same result occurs when (again for constructional reasons) the facets do not cover the whole area of the dish, as in the LaJet concentrator design.

A further variant that can be considered is to manufacture 'pseudo-spherical' facets by dividing a mirror facet of hexagonal planform into six equilateral triangles, and then bending each such sub-facet in one direction, thereby avoiding the difficulty of producing double curvature.

A separate analysis has to be incorporated into the program FCAR to deal with each different method of dish construction. After describing the point sun perfect parabola case, the analyses for flat facets and spherical facets are outlined below.

Receiver Errors

The main errors associated with the receiver are possible angular and lateral mis-alignments of its axis from that of the dish. Longitudinal mis-alignment (along the dish axis) is dealt with explicitly in the FCAR program, even for the ideal point sun perfect paraboloid case, since this is a variable which can readily be altered during design and construction of a collector.

POINT SUN, PERFECT PARABOLOID CASE

With a perfect paraboloidal mirror, perfectly aligned in the direction of a point sun, all reflected rays pass through the focus of the paraboloid. The central region of the dish is shaded by the receiver, so all reflected rays lie within a pair of co-axial cones having the focus as their common vertex.   The semi-vertex angle of the outer cone is commonly referred to as the rim angle, ϕ, of the dish, and ψ is the corresponding angle of the inner cone.

To determine the solar flux at a point on the receiver surface, it is first necessary to determine whether or not the point lies between the two reflected ray cones, either before or after the focus. In a practical dish/receiver configuration, only points on the external surface of the receiver can be irradiated by rays which have not yet passed the focus, and only internal receiver points can be irradiated by rays past the focus; this might be violated for receivers positioned impracticably close to or far from the dish along its axis.   If the line joining the receiver point to the focus makes an angle θ with the dish axis, then the point lies within the ray cones if

$$\psi < \theta < \phi . \tag{1}$$

If a point on the receiver surface satisfies this criterion, it will still not be irradiated if it is facing away from, rather than towards the dish.   If $\mathbf{s}$ is the unit vector in the direction of the outward normal to the receiver surface at the point, and $\mathbf{r}$ is the unit vector in the direction of the reflected ray through the point, then the surface is not irradiated if their inner product is positive, that is, if $\qquad \mathbf{s} \cdot \mathbf{r} > 0 . \tag{2}$

In terms of a cartesian reference frame having its origin in the entrance plane of the receiver, its z-direction pointing along the dish axis towards the sun, and its y-direction in the plane containing the receiver point, equation (2) implies that

$$m_r .m + n_r .n > 0, \tag{3}$$

where m and n are the direction cosines of the reflected ray in the y- and z-directions, and m_r and n_r are those of the outward surface normal, both cosines in the x-direction being zero in this symmetrical idealisation.

Finally, if the receiver point is internal, it will still not be irradiated if it is occulted by any part of the receiver closer to the dish.   To be able to test for this condition, the radius of each end of each internal receiver frustrum must be stored as the input data are read by the program.   Then for each internal frustrum closer to the dish than the receiver point, the radii of its ends can be compared with the distance from the axis of the ray through the receiver point as it passes through the receiver end planes.   That is, the receiver point will be occulted if either

$$y_1 < |y - r(z - z_1)|, \tag{4}$$

or

$$y_2 < |y - r(z - z_2)|, \tag{5}$$

where (y, z) are the coordinates of the receiver point, and (y_1, z_1) and (y_2, z_2) are the coordinates of the ends of the frustrum, r is the ratio m/n, and vertical bars denote an absolute value.

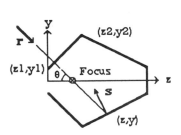

Fig. 1. Receiver geometry

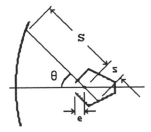

Fig. 2. Dish geometry

If all these tests reveal that the receiver point is, in fact, irradiated, then the solar flux on the receiver surface at that point is the scalar product of the unit vector representing the inward receiver surface normal with the concentrated solar radiation flux vector at the receiver point. This vector has the direction $\mathbf{r}$ of the ray from the mirror through the focal point to the receiver point. The magnitude of the concentrated solar radiation flux vector is equal to the magnitude of the incident solar flux (taken to be unity throughout this paper) times S^2/s^2, where S is the distance from the dish to the focus along the ray, and s is the distance from the focus to the receiver along the ray.   Now from the properties of a parabola and some algebra,

$$S = a \{2+2/\tan^2\theta - \sqrt{(1+2/\tan^2\theta)^2 -1} \} \tag{6}$$

where a is the focal length of the parabola.   Also,

$$s^2 = y^2 + (z + e)^2, \tag{7}$$

where e is the distance from the dish focus to the receiver entrance plane.   So finally, the contribution, F, to the flux on the receiver surface is given by

$$F = -(S^2/s^2)\mathbf{s} \cdot \mathbf{r} \tag{8}$$

PARABOLOID MADE FROM FLAT MIRROR FACETS

Point Sun, Zero Random Errors

Before incorporating random errors, all facets are assumed to be tangential, at their centre point, to a perfectly aimed perfect parabola. Now, however, only the ray from the centre point of a facet passes through the dish focus. All other rays reflected from a facet are parallel to this central ray, so that the concept is of a collection of beams of parallel rays irradiating the receiver, with each facet corresponding to a unique beam direction.

To determine whether or not a receiver point is irradiated by a particular facet, it is useful to introduce the idea of a 'backray'. This is an imaginary straight line emanating from the receiver point in the direction of the beam from the facet, but in the opposite sense, so that it is pointing from the receiver point towards the dish. If the backray is intercepted by any part of the receiver before it reaches the dish, or if it reaches the dish but meets it outside the facet being considered, then that facet makes no contribution to the irradiance at the receiver point. Again there is no contribution if the backray meets the facet, but the beam direction is out of, rather than into, the receiver surface, as determined by equation (2). The program must resolve all of these questions, taking into account input data on the planform of a facet, and the layout of facets over the dish.

If a facet does contribute to the flux at a receiver point, then its contribution is determined solely by the intercept angle between the beam and the receiver, since a flat mirror facet produces no concentration. That is, $F = -s \cdot r$ (9)
The total flux at a receiver point is then the sum of the contributions from all mirror facets.

In the perfect paraboloid case the complete axial symmetry means that it is only necessary to examine the geometry in one plane through the axis. The same is essentially true in the plane facets case, since rotation of the plane of study about the axis produces only small changes in the incidence angle between the beam from a facet and the receiver surface. Rotation of the plane can also cause the beam from a facet to be lost from the receiver point, but, on average, for every facet beam lost this way, another would be acquired, since each receiver point is irradiated by a large number (greater than the local concentration ratio) of facets. Consequently, examining the geometry of this case in one plane is again adequate.

Finite Sun Plus Random Errors

Because this case involves only reflection from plane surfaces, it is a relatively straightforward matter to remove the point sun, zero random error restrictions. A dish pointing error simply represents an angular correction to be added (vectorially) to the direction of all mirror facets. Facet attachment errors, chosen randomly from some assumed distribution, are added to each facet direction. The assumed sunshape is dealt with by calculating the flux contribution from each sun segment and adding the results. The angle subtended between the centre of a sun segment and the centre of the sun's disc is treated in the same way as a dish pointing error.

DISH MADE FROM SPHERICAL MIRROR FACETS

Point Sun, Zero Random Errors

Spherical facets, all of the same curvature, should be intermediate in cost between paraboloidal facets with a different curvature in each zone, and flat facets; at the same time they should give a higher overall concentration ratio than flat facets of the same size, or a similar concentration with fewer, larger facets. However, their use requires modification of the dish construction. All rays parallel to the axis of a spherical mirror are reflected through the axis, but not through a single focal point on the axis; rays from zones further from the axis are reflected closer to the dish. Consequently, instead of being attached directly to a paraboloidal former, zones of facets progressively further from the axis of the dish have to be brought progressively further ahead of the former, so that the central reflected ray from each dish still passes through the focus of the parabola. Then all rays from a facet will intersect the axis either at the focus, or close to it on either side.

If α is the angle subtended at the centre of curvature by the centre of a facet, then the central reflected ray will meet the axis a distance $a / \cos(\alpha)$ from the centre, where, as before, a is the focal length of the dish. Consequently, this zone of facets has to be brought forward of the paraboloid by a correction distance c, given by

$$c = a(1 - \sec(\alpha)) \qquad (10)$$

A receiver point will be irradiated by a given ring of spherical facets only if it lies within the fan of rays spread around the ray reflected from the facet centre. The centre ray passes through the dish focus and is inclined to the axis by the angle 2α. The ray fan subtends an angle $d\alpha$ on either side of this central ray, where $d\alpha$ is the angle subtended at the sphere centre by the ring of facets.

As before, even if the receiver point lies within the fan of rays, it will be irradiated by the facet only if the reflected rays are directed into, rather than out of the receiver surface. To find the direction of the reflected ray corresponding to a receiver point, the program first evaluates the polar coordinates (d, ρ) of the receiver point relative to the centre of the sphere, and then finds the angle γ, corresponding to the reflected ray by iteratively solving the equation

$$\sin \gamma = d \sin(\rho - 2\gamma) / R \qquad (11)$$

where R is the radius of curvature of the spherical facets.

If the reflected ray is directed into the receiver surface, the program then tests for occulting, as before, by calculating the distance of the reflected ray from the axis as it passes through each end plane of each receiver frustrum that is closer to the dish than the receiver point being examined.

If there is no occulting and the ray is directed into the surface, then, as before, the contribution to the flux on the receiver surface is given by equation (8), except that now, instead of being found from equation (6), the value of S must be derived from

$$S = R \sin \gamma / \sin(2\gamma) \qquad (12)$$

Finite Sun Plus Random Errors

In this case, the procedure to remove the point sun zero random error restrictions depends on the size of the mirror facets. If very small facets are used, it is sufficient to add the necessary corrections as in the plane mirror facet case. With large facets, however, the optical properties of the spherical surfaces must be taken into account, so that modifications must be incorporated in the equations given in this section.

CONCLUSION

The main considerations involved in the development of the computer program FCAR are outlined in the preceding sections. The program is being used in the design analysis of a family of large, high temperature, steam generating, point focus, distributed receiver, solar energy plants.

ACKNOWLEDGEMENTS

My thanks to Prof. S. Kaneff for guidance and support, and to others in the Energy Research Centre for helpful discussions.

REFERENCES

Diver, R. B. (1987). Receiver/reactor concepts for thermochemical transport of solar energy. *J. Solar Energy Engineering*, 109, 199-204.
Kaneff, S. (1989). On the characteristics and role of semi-cavity receivers for paraboloidal dishes. *Proceedings of the Annual Conference of the Australian and New Zealand Solar Energy Society*, Brisbane..
Mayer, I. (1988). Increasing the service life of high-temperature steam generating solar thermal receivers. *Proceedings of the Annual Conference of the Australian and New Zealand Solar Energy Society*, Melbourne.
Mayer, I. (1991). *FCAR, A Pascal Program to Calculate the Primary Solar Flux Distribution on Axisymmetric Receivers at the Focus of Multi-facetted Dish Concentrators*. Internal report EP-RR-54, in preparation.

VIABLE ARRAYS OF LARGE PARABOLOIDAL DISH SOLAR THERMAL COLLECTORS

S. Kaneff

Energy Research Centre, Research School of Physical Sciences and Engineering,
Institute of Advanced Studies, Australian National University, Canberra ACT Australia

ABSTRACT

Successful development of technology to produce large cost-effective solar thermal concentrating arrays is paving the way for major advances in many areas of solar energy application. A vast array of processes, products and systems can be driven by high temperature solar energy, including those which are dependent on heat alone, on photon energy, or on a combination of these. Useful applications include: high temperature process heat and electricity production; storage and transport of solar heat by thermochemical means; driving various photoreactions — including photochemical, photoelectrochemical, photo-enhanced catalytic and hybrid reactions which combine photon and thermal reactions in a synergistic manner, promising a major contribution to chemistry; and production of fuels and chemicals by solar chemical processing — offering the prospect of long term storage and transport and the production of energy-rich products and materials; hydrogen production; oil shale retorting; coal gasification; destruction of hazardous wastes; pyrolysis of biomass; and many others. The paper addresses costs and applications.

KEYWORDS

Solar thermal systems; paraboloidal dishes; high temperature solar collection; thermochemical systems; photon chemistry; viable solar power.

INTRODUCTION

Paraboloidal dish design must be considered in relation to system requirements and applications and to economic considerations. The past 10 years have seen major developments in this area.

However, although many attempts have been made, the goal of cost-effective paraboloidal collectors has proved elusive until very recently. Reasons for this situtation are complex but, in our opinion, the search has concentrated on the wrong aspects. Especially has the economy of size been neglected. Many elegant projects have been pursued leading to designs dependent on extensive computer and/or theoretical studies, but relatively little experimentation and even less real world field trials have been conducted.

One of the major inhibiting factors has occurred because certain design concepts have been assumed to be self-evident. For example, focal area mounting of engines has often been pursued almost to the exclusion of employing collector arrays to feed heat energy to a central plant, reducing the opportunity to take advantage of the enhanced efficiency of central plant as size increases and limiting efficiency to that of the focal region engines. It has also denied the opportunity to gain process heat or to utilize 'waste' heat, and to enter the applications area of thermochemical and photon chemical systems for which the characteristics and performance of dishes are well suited. Imitating concepts from space technology and astronomical telescope practice has occurred, leading to collectors which are mounted high above the ground, thereby being exposed to strong buffetting winds, requiring appropriate strength and resulting in unaccceptable cost.

These aspects have inhibited development of large units; even 'lightweight' collectors have been exposed to this problem and have accordingly not taken full advantage of the possible economy of lightness. Rigidity and vibration problems have led to costly dishes.

Yet there are fruitful approaches to the realization of cost-effective dishes. Bilodeau and co-workers (1987), for example, report on the Power Kinetics 'Square Dish' which, with its ANU Steam System, formed the basis for a 50 kWe module successfully tested in 1988. Another approach (Kaneff, 1989), based on learning from much past experience in the field, employs very rigid space frames which appear suitable for dishes with apertures up to 1500 m², lend themselves to factory mass production and straightforward field installation, and can be assembled into viable solar thermal arrays. Studies by Carden and Bansal (1991) have revealed that heat network transport losses and costs for such optimally-designed dishes can allow systems to be produced successfully in sizes of 100 MWe and larger.

These developments and related research are now paving the way for major advances in many areas of solar energy application, allowing a great array of processes, products and systems to be driven economically by high temperature solar energy, including processes which are dependent on heat alone, on photon energy, or on a (synergistic) combination of these.

COST OF SYSTEMS FOR PROCESS HEAT AND ELECTRICITY GENERATION

The following cost analysis is based on the ANU 334 m² aperture paraboloidal dishes reported in Kaneff 1989(b). Different size systems are considered, from 50 kWe dish to 100 MWe, providing process heat or generating electricity.

Tables 1 and 2 give details of process heat and electricity generating plant, using a real interest rate (difference between actual interest and inflation rates) of 8% per annum.

Life cycle cost for each candidate system is assessed using the parameters:

System Life =	25 years
Interest on Capital =	$i\%$ per annum
Inflation Rate =	$p\%$ per annum
Real Interest Rate =	$i^* = \frac{1+i}{1+p} - 1$
Real Interest Rate =	8%

- System output based on White Cliffs data: 2360 kWh/m²/annum insolation. The solar systems are assumed to be operational whenever 'available' solar energy is present.

- O&M costs per annum for the overall solar heat or electric power systems are taken as 2% of total equipment cost (White Cliffs experience has shown this to be pessimistic).

- The interest rate chosen is indicative of the real rates of return employed by electric utilities.

- Life cycle costs are calculated using the Long Run Marginal Cost Method (LRMC). This represents the cost of energy produced by the system. It includes in the analysis all economic costs and energy contributions in the future (long run). It is the present value of all financial outlays divided by the present value of energy produced. Present values are calculated in accordance with AS3595-1990, Energy Management Programs — Guidelines for Financial Evaluation of a Project (Australian Standard).

TABLE 1   Installed and Life Cycle Costs for Paraboloidal Dish Industrial Process Heat Installation (with no storage or backup); Process Steam Costs (**$US 1991**)

Dish System Details (Dishes each 334m² aperture)	Dish(es)	Auxiliaries (1)	Total System Costs	Heat Output at Plant MWh$_{th}$	Life Cycle Cost of Heat ($\rightarrow$ 500°C 7 MPa)
					i* =8%
	$	$	$		¢/kWh$_{th}$
(50kWe) 1 x 334m² (Demonstration Plant)	127 200	3 040	131 000	660	2.2
(50kWe) 1 x 334m² (Commercial Unit)	86 700	2 770	90 000	660	1.5
(200kWe) 4 x 334m²	343 000	3 950	347 000	2 640	1.4
(1MWe) 17 x 334m²	1.444m	15 800	1.460m	11 220	1.4
(10MWe) 127 x 334m²	9.16m	107 400	9.3m	83 840	1.1
(100MWe) 948 x 334m²	59.85m	671 500	60.5m	619 400	1.0

Auxiliaries: Feedwater pump, feedwater tank, controls, field piping.

TABLE 2 Installed and Life Cycle Costs for Paraboloidal Dish Electricity Generation (**$US 1991**); No Storage or Backup.

System Details	Total System Cost $	Output per Annum MWh$_e$	Cost/kW Installed	Average Annual Efficiency (Solar to Electricity) %	Life Cycle Costs* i*=8% ¢/kWhe
50 kWe Demonstration Unit (1 dish)	186 600	124	3 530	15.7	18.2
50 kWe Commercial Unit (1 dish)	140 200	124	2 650	15.7	13.0
200 kWe Commercial Unit (4 dishes)	517 900	558	2 150	17.7	11.0
1 MWe Commercial Unit (17 dishes)	1.89m	2 360	1 860	17.7	9.2
10 MWe Commercial Unit (127 dishes)	15.9m	29 140	1 510	24.0	7.5
100 MWe Commercial Unit (948 dishes)	107.4m	234 800	1 070	31.0	5.2

The costs in Table 2 may be compared with those reported for LUZ parabolic trough systems in Table 3 (from LUZ 1989). This comparison confirms the well-known advantages of using paraboloidal dishes (see for example Williams and co-workers 1987) and says much for LUZ technological success in establishing 384 MWe in California, with further hundreds of MWe to be installed. The LUZ systems employ natural gas augmentation to a degree not defined in Table 3. SEGS VIII output per year of 254 000 MWhe includes a natural gas component. Taking this into account, the life cycle solar only generation costs would probably exceed 10¢/kWhe. The addition of natural gas augmentation to the dish systems of Table 2 would reduce their generation costs well below those indicated.

Paraboloidal dish systems can take advantage of further technological advances and will then become even more attractive in producing medium and high temperature heat.

TABLE 3   LUZ Solar Electricity Parabolic Trough
Generating Systems $US 1989 (LUZ 1989)

Item	SEGS I 1984	SEGS II 1985	SEGS V 1987	SEGS VI 1988	SEGS VIII 1989	SEGS IX 1993
MW Capacity MWe	13.8	30	30	30	80	80–160
Peak Solar to Electricity Efficiency	22% (with Gas Superheater)	19%	22.2%	22.2%	24%	26–28%
Annual Average Efficiency Solar to Electricity	10%	11%	12.4%	12.4%	14%	15.5–17%
Total MWhe Output/Year	30 000	80 500	92 000	91 000	254 000	—
Project Price $US	$62 million	$96m	$124m	$116m	$230m	—
Price/kWe Installed $US	$4 500	$3 200	$4 130	$3 870	$2 875	—
Life Cycle Cost US¢/kWhe	24	—	12.5	11.5–12	7.5–8.5	6 expected

POTENTIAL APPLICATIONS OF PARABOLOIDAL DISH ARRAYS

The improving technology and economics of dish systems are pointing to early applications in many situations requiring high quality heat and high photon flux, leading to the mass utilization of solar energy in the near term, for example:

1. As indicated in Tables 1 and 2, the generation of high quality heat for industry and electricity generation suggests practicability of viable installations now.

2. Solar/Fossil Combinations

   With the recent commercial availability of gas turbine/steam turbine combined systems of high overall conversion efficiency (over 50% in large units), paraboloidal dish systems can make a major contribution to new power generation, allowing rapid response to changing load demands due to the quick startup of gas turbine units. For a suitably proportioned system of gas turbine/steam turbine with the steam turbine fed by steam from the gas turbine exhaust heat and also from a paraboloidal dish array sized to provide rated power on full sunshine, solar steam can provide more than half the annual energy. Such a system would produce less than 25% of the emissions from a coal-fired power station of the same output, with excellent economics in view of the low cost of solar-generated steam. This approach obviates the need to store solar energy.

   Solar/fossil combinations also allow the actual storage and transportation of solar energy via the solar gasification of coal, the product gas, syngas (CO and H_2) being produced without pollution via solar energy. Using syngas with a solar steam array/gas turbine/steam turbine combination enables coal also to be used more benignly, providing systems with quick startup and having accordingly peaking capabilities, as well as allowing effective solar storage — up to 70% of the incident solar energy being stored in the syngas. The gas can of course be transported, stored and used for process heat or as a town gas.

   Solar/fossil combinations have the major attraction that their early application is likely to be favoured because of the lack of major disruption to current energy infrastructures, facilitating changeover, within 3 or 4 decades, to renewable energy systems.

3. High Temperature Solar Thermal/Thermochemical/Photon Chemical Applications

Apart from strictly thermal applications, concentrated solar flux can be used to drive thermochemical and photon chemical reactions and processes, allowing a great number of products, including fuels and chemicals, to be produced. Such processes include:

- High Temperature Solar Chemistry:

  Conversion of sunlight and material resources into products, using concentrated sunlight as heat and/or using photon energy to facilitate reactions otherwise not practicable, allows solar thermal and photon energy to produce fuels and chemicals, and provides means for long term energy transport.

- Direct Absorption of Concentrated Solar Radiation — Photon Chemistry

  Radiance temperature of concentrated solar radiation occurs over a wide spectral range compared to flame temperatures of conventional sources, allowing effective heat transfer by absorption of solar energy, for example in volumetric and direct absorption receivers. The energetically useful fraction of direct sunlight can be high, eg 45% in a multistage thermochemical water-splitting process.

  Concentrated solar radiation has chemically unique properties beyond the capability to heat materials very rapidly to very high temperatures. Photoreactions form an important class of potential reactions, including conventional photochemical, photoelectrochemical, photoenhanced catalytic reactions, and hybrid reactions which combine photon and thermal reactions in a synergistic way, promising a major contribution to chemistry.

- Thermochemical Energy Conversion

  High temperature solar sources are attractive for thermochemical processes because they allow an efficient coupling of the chemical plant to the primary energy source. Second law of thermodynamics efficiency maximises in the temperature range 750°–950°C, which is a good match for solar heat sources. The first law thermal efficiency of thermochemical processes is in the range 40–50% which is also favourable.

  The development of efficient exothermic and endothermic cycles allows:

  - Storage and transport of solar heat.
    Solar power stations with thermochemical storage potentially can collect solar heat by concentrating collectors covering vast areas (hundreds of square kilometres) and conveying the heat (as chemical energy) to a central plant(s) for utilization. Storage can be effected within pipe networks connecting collectors to central plant, in aquifers and in old oil and gas wells under certain conditions.
  - Hydrogen production; oil shale retorting; coal gasification; destruction of hazardous wastes; steam reforming of methane and carbon dioxide; carbothermic reduction of metal oxides; pyrolysis of biomass; and many other processes.

4. The Production of Hydrogen

One of the major future means for energy storage and utilization is likely to be via the use of hydrogen. Current ways of producing hydrogen, by reforming carbonaceious materials, are not overly economical and produce CO_2.

- Hydrogen Production by Water Electrolysis

  Advanced electrolysis concepts over the past 5–10 years have ensured substantial increases in electrolysis efficiency and lowered plant costs. The main problem is the cost of electricity — usually 4¢/kWhe or more, giving an equivalent cost of some $12/GJ for the produced hydrogen. Since natural gas costs some $3/GJ thermal, to compete on a simple cost accounting basis, electricity costing less than 1¢/kWhe would need to be used in the electrolytic production of hydrogen.

- Hydrogen Production by Thermochemical Means

  Thermochemical systems promise economical production of hydrogen. For example, gas reactions based on carbon and sulphur cycles allow thermochemical water splitting at temperatures below 1000°C, and sulphur/iodine cycles are also highly efficient ($\approx$50%). Other thermochemical exchange reactions also exist and provide scope for further development.

- Hydrogen Production by Photochemical, Photoelectrochemical and Photobiological Systems

  Provide a rich field of approaches: electrolysis; plasmolysis; magnetolysis and magmalysis (see Roberts and co-workers 1984 and Schiavello 1985).

CONCLUSIONS

Solar thermal systems have only recently reached an effective level of development appropriate to commercial application. Demonstration plant and various installations, together with new research findings point to the potential for mass utilization, in the relatively near term. Support needs to be strong now to enable appropriate development to proceed not only with solar thermal power systems alone, but with solar/fossil combined systems which can provide the most practicable transition towards an all-renewable energy supply. The solar thermal/thermochemical/photon chemical and solar/fossil fuel paths have, in addition, a wide-ranging repertoire of means for achieving energy storage and in supplying heat, potential energy, energy-rich products (eg fuels and chemicals), and other materials.

Successful development of solar thermal arrays for the collection and concentration of energy at high temperatures and high photon flux densities seems set to facilitate changeover of our energy systems to benign sustainable forms in the next decades.

REFERENCES

Bilodeau E.A., D.N. Borton, W.E. Rogers and E.K. Inall. Design of the small communicty program SCSE No 2 at Molokai, Hawaii. Proc Joint ASME/JSME/JSES Thermal Engineering Conference, Honolulu, 22-27 March. In Solar Engineering 1987, ASME, 2, 810-814.

Carden P.O. and P.K. Bansal (1991). Optimization of steam-based energy transport in distributed solar systems. *Solar Energy*, in press.

Kaneff S. (1989a). Cost-effective paraboloidal dish solar heat and electricity generating systems. *Proc Annual Conference of Australian and New Zealand Solar Energy Society*, Brisbane, 30 Nov–2 Dec, Ed P. Jolly, Publ ANZSES, 6.1–6.6.

Kaneff S. (1989b). On the design of viable paraboloidal dish solar collector systems. In *Clean and Safe Energy Forever*, Proc 1989 Congress of International Solar Energy Society, Kobe, 4-8 Sept 1989, Eds Horigome, Kimera, Takakura, Nishino and Fuji, Pergamon Press, Oxford, 2, 1303–1307.

LUZ International Ltd (1989). Newsletter 5(1) and Publications FM16, FM20, Communications Department, April.

Roberts R., R.P. Ouellette, M.M. Muradax, R.F. Cozzens and P.N. Cheremisinoff (1984). *Applications of Photochemistry*, Technomic Publishing Company, Lancaster Pennsylvania USA, 101 pages.

Schiavello M. [Ed] (1985). *Photoelectrochemistry, Photoanalysis, and Photoreactions: Fundamentals and Developments*, D. Reidel Publishing Company, Dordrecht.

Williams T.A., D.R. Brown, J.A. Dirks and M.K. Drost (1987). Analysis of solar thermal concepts for electricity generation. *Proc Joint ASME/JSME/JSES Conference*, Honolulu, 22-27 March. In Solar Engineering 1987, ASME, 2, 779–785.

ANALYSIS OF A PLANO-CONCAVE HEAT PIPE RECEIVER

by

Youssef A. M. Elgendy Ph.D.

Georgia Institute of Technology

P.O.Box 36970

Atlanta, GA 30332

ABSTRACT

A dish advanced heat engine with heat pipe receiver has been shown by preliminary modeling to be a competitive technology for both space and earth. The proposed paper presents the design and the analysis of a plano-concave heat pipe receiver especially suited to Stirling, Ericsson, Brayton or similar modular heat engine systems. A significant contribution to the solar dish advanced power system is based on the design of the solar collector with a heat pipe receiver. The design of the collector should have high energy capture while maintaining low thermal losses and a small cost/effectiveness ratio.

The amount of energy that falls on the receiver has been determined. A simple formalism has been developed to analyze the optical performance of the paraboloidal concentrator. This research is based through the numerical determination of the distribution of the concentrated solar flux on the receiver, its optical efficiency, and the amount of thermal losses from the receiver.

KEYWORDS

Plano-concave receiver; heat pipe receiver; optical analysis; thermal loss; numerical analysis; shading.

INTRODUCTION

The proposed receiver has an external absorber rather than the conventional cavity receiver. The external receiver is a plano-concave surface comprising a flat disk in the focal plane and a cylindrical surface around the optical axis connected by a toroidal surface. A heat pipe couples the absorber to the heat engine.

A plano-concave heat pipe receiver would couple the absorber surface to an advanced heat engine with a closed evaporation-condensation loop as shown in Figure 1. High heat flux densities at the absorber can be accommodated without overheating the absorbing surface due to the minimal thermal resistance in the evaporating film. The excellent overall thermal conductance of the heat pipe would minimize the temperature difference between the absorber and the heat engine. This good thermal coupling would maximize the effective temperature of heat input to the engine and improve the conversion efficiency.

A plano-concave heat pipe receiver has been used to maximize the optical efficiency and minimize thermal losses. A cavity receiver may have slightly better absorptance; however, an external receiver can shade the concentrator less. More importantly, the external plano-concave design does not impede the flow of vapor to the engine and provides smooth return and distribution of the liquid.

Detailed optical analysis of the solar concentrator requires computation of the distribution of the concentrated solar flux on the surface of the receiver. An intensity distribution that accounts for both a non-uniform source and imperfect reflection with scattering has been selected to illustrate this model. The reported research is concerned with determining the performance of a typical plano-convex receiver. The overall optical efficiency and the concentrated flux distribution along with the thermal losses have been investigated. In computing this concentrated flux for the paraboloidal solar collectors, an analytical method using an infinitesimal pyramid method (IPM) has been presented [1]. This method is based on the established concepts of the flux integral and radiant intensity. IPM is implemented by simple geometrical constructions involving infinitesimal pyramids based on the reflecting surface. This technique is believed to be conceptually simple and straightforward to implement. The heat pipe receiver would allow the engine to be mounted behind the concentrator. This new design configuration would minimize shading of the reflector. Mounting the engine near the vertex of the paraboloid should also reduce the mass and moment of inertia of the system; consequently, less structural strength and steering torque would be required, beneficial characteristics for both space and terrestrial applications.

ANALYSIS

The basic concept is to subdivide the solar dish into infinitesimal rectangular elements, dA_c, as represented in Figure 2. The total energy per unit time leaving a reflecting element, dA_c, and incident upon a heat pipe receiving element, dA_r, as

$$d^4\dot{Q} = (\Lambda \, d\Omega) \cos\theta_r \, dA_r \qquad\qquad 1$$

Four integration limits are required to integrate Equation 1. Two integration limits for the receiving area and the other for the reflecting element.

The proceeding analysis has been generalized to consider a receiving point off the normal focus as shown in Figure 3. Receiving planes perpendicular, parallel, and curved to the optical axis has been specifically considered in a plano-concave receiver. The pertinent geometry is illustrated in Figure 4.

Illustrated in Figure 4 is a generally representative geometry for the three receiving surfaces of the plano-concave receiver. This refers to point F on the focal plane and a point P on the surface of the receiving element. The central ray vector is U, which is toward the focus, and the projection path, ρ, of a general ray. A general point, P, on the receiving surface is ξ units left of the nominal focus, F, and ς units toward the vertex.

In the coordinate system (x,y,z) originated at the reflecting element dA_c, the components of these vectors are:

$$U_x = U \sin \psi \cos \beta \qquad\qquad 2$$
$$U_y = -U \sin \psi \sin \beta \qquad\qquad 3$$
$$U_z = U \cos \psi \qquad\qquad 4$$
and
$$\rho_x = U_x - \xi = U \sin \psi \cos \beta - \xi \qquad\qquad 5$$
$$\rho_y = U_y = -U \sin \psi \sin \beta \qquad\qquad 6$$
$$\rho_z = U_z - \varsigma = U \cos \psi - \varsigma \qquad\qquad 7$$

The magnitude of vector ρ is

$$\rho = \sqrt{\rho^2_x + \rho^2_y + \rho^2_z} \qquad\qquad 8$$

The differential solid angle subtended by the reflecting element can be defined as follows:

$$d\Omega = dA_c \cos\theta_c / U^2 \qquad\qquad 9$$
$$dA_c = (U \sin\psi \, d\beta) [U \, d\psi / \cos(\psi/2)] \qquad\qquad 10$$

From the scalar product of vectors n and ρ

$$\cos\theta_c = n_c \bullet \rho / |\rho| \qquad\qquad 11$$

The components of the unit vector normal to the reflecting element at dA_c are:

$$n_x = \sin(\psi/2) \cos \beta \qquad\qquad 12$$
$$n_y = -\sin(\psi/2) \sin \beta \qquad\qquad 13$$
$$n_z = \cos(\psi/2) \qquad\qquad 14$$

then $\cos\theta_c = (U/|\rho|) [\sin\psi/2 (\sin\psi - (\xi/U) \cos\beta) + (\cos\psi - \varsigma) \cos\psi/2]$ 15

The solid angle can be written as

$$d\Omega = (1/\rho^2) [U^2 \sin\psi \, d\beta \, d\psi / \cos(\psi/2)] \cos\theta_c \qquad\qquad 16$$

The incident cosine at the three geometric receivers, shown in Figure 4, can be evaluated as follows

$$\cos\theta_r = n_r \bullet \rho / |\rho| \qquad\qquad 17$$

The vector n_r and the incident cosine for the flat surface, the straight tube, and the torus surfaces are defined as follow:

For A Flat Surface : $\cos\theta_r = -(\cos\psi - \varsigma/U)(U/|\rho|)$ 18

For A Straight Tube: $\cos\theta_r = -\rho_x / |\rho|$ 19

For Toroidal : $\cos\theta_r = (-1/|\rho|)[(U\sin\psi\cos\beta - \xi) \cos\phi_t + (U\cos\psi - \varsigma)\sin\phi_t]$ 20

The last equation represents the general case. The other two cases can be obtained by substituting the value of ϕ_t by $0°$ for the straight tube surface and by $90°$ for the flat surface.

OPTICAL EFFICIENCY

Optical efficiency is the requirement design parameter for the detailed simulation model for the receiver heat pipe. The amount of heat absorbed by the receiver consists of three parts. These parts are: the heat absorbed by the flat disk surface, the heat absorbed by the torus surface, and the heat absorbed by the straight tube surface. The proposed geometry for the plano-concave receiver is assumed to be symmetric. Therefore, the first integration of the double integration over the receiving element can be integrated simply over 2π. This means that the amount of solar energy absorbed by the receiver can be determined by integrating the flux distribution over $-\psi$ to ψ, $-\beta$ to β, and over the contour surface for the receiver as follows:

For Flat disk surface

$$\dot{Q} = \int_{R_p+R_t}^{R_d} q \, dA \qquad\qquad 21$$

For Tube surface

$$\dot{Q} = \int_{Z_o}^{Z_o+L_p} q \, dA \qquad\qquad 22$$

For Torus surface

$$\dot{Q} = \int_{0}^{\pi/2} q \, dA \qquad\qquad 23$$

By substituting the heat flux which is a function of concentration ratio, and the receiver area for the flat disk into the heat equation:

$$\dot{Q} = \int_{R_p+R_t}^{R_d} C_r \, G_{bn} \, F^2 \, (2\pi \ R/F \ dR/F) \qquad\qquad 24$$

The optical efficiency is defined as

$$\eta_{opt} = (\Sigma \dot{Q}) / A_c \, G_{bn} \qquad\qquad 25$$

where $\qquad \Sigma\dot{Q} = \dot{Q}_1 + \dot{Q}_2 + \dot{Q}_3 \qquad\qquad 26$

the subscript 1, 2, and 3 are referring to the flat plate, torus, and tube surfaces.

$$\eta_{opt} = 2\pi \left[\int_{R_p+R_t}^{R_d} C_{r1} R^* dR^* + \int_{R_p}^{R_p+R_t} C_{r2} \, X dX + \int_{Z_o}^{Z_o+L_p} C_{r3} \, R_p dZ \right] / 4 \, \pi \sin\psi_m / (1+\cos\psi_m)^2 \quad 27$$

The distribution of solar intensity for the flat disk and the torus receiver is shown in Figure 5 at 0.95 specular reflectance, 0.083246 flat disk diameter to focal length, 0.0077 heat pipe radius to focal length, 0.01 torus radius to focal length, 0.007 rad solar scattering parameter, and $60°$ rim angle. Note that the amount of the flux at the last point on the torus is exactly the same as that for the first point at the flat disk. Additionally, the flux at the first point on the torus is the same as that for the first point on the tube receiver. The flux distribution depends on the dimension of the plano-concave receiver.

Although the increasing receiver area increases the optical efficiency, it also increases the thermal heat losses. Note that the flux calculated is at the optimum rim angle ($60°$). The optical efficiency at the above parameters is 94.42 %.

The result of this numerical method has been compared with the local point source model and an excellent agreement has been obtained.

THERMAL HEAT LOSS FROM THE RECEIVER

The rate of useful heat gain was determined from a concentrator optical efficiency, η_o, and heat losses models, h_k, h_r, and h_c such that:

$$\eta_{dish} = Q_u / G_{bn} \, A \qquad\qquad 28$$

$$\eta_{dish} = \eta_{opt} - (Q_{cond} + Q_{rad} + Q_{conv}) / G_{bn} \, A$$

$$\eta_{dish} = \eta_o - (h_k + h_r + h_c)(T_r - T_a)/(C_r \bullet G_{bn}) \hspace{2cm} 29$$

where h_k, h_r, h_c have been defined. The variation of radiation heat loss coefficient, h_r, with receiver temperature is plotted in Figure 6. The magnitude of radiation loss is larger than magnitude of the convection loss coefficient at the same temperature range (600-1600 K).

For receiver design purposes, calculation of plano-concave receiver conduction and radiation heat loss is relatively straightforward, but calculating the convection loss is more difficult. While there is a correlating increase between receiver temperature and radiation loss, the convection loss remains small.

RESULT AND CONCLUSIONS

The reported research is concerned with determining the performance of a typical plano-convex receiver. The proposed paper has presented the overall optical efficiency and the concentrated flux distribution along with the thermal losses and the thermal conductance. In computing this concentrated flux for the paraboloidal solar collectors, an analytical method using an infinitessimal pyramid method (IPM) has been presented.

The plano-concave receiver is subject to radiation, convection, and conduction heat losses. An analytical model for the overall thermal conductance for a heat pipe receiver has be investigated. The thermal performance for the plano-concave receiver has been examined at various operating temperatures.

The preliminary results of the numerical optical analysis for the plano-concave receiver and the analytical analysis for the overall thermal conductance and heat losses for the heat pipe receiver indicate that a high system efficiency is achievable with existing paraboloidal solar dish technology. Good collection efficiency depends on high optical efficiency and low heat losses. The heat loss depends on operating temperature, the operating temperatures depend on the heat rate and the overall thermal conductance of the receiver. The combined optical, thermal loss, and thermal conductance models presented in the proposed paper are necessary to fully characterize the thermal performance of the plano-concave heat pipe receiver.

This numerical analysis technique is utilized in evaluating the concentration intensity. This may be useful in determining the optical efficiency required for the detail design model for the solar dish Ericsson thermal power system. The overall paraboloidal solar dish efficiency against receiver temperature at various concentration ratio is plotted in Figure 7. The effects of thermal heat loss from the receiver on the solar dish efficiency is highly magnified with increasing receiver temperature, at a given concentration ratio. The paraboloidal solar dish efficiency increases along with increase in the concentration ratio for a given receiver temperature. Therefore, the amount of solar energy which falls on a plano-concave heat pipe receiver has been calculated.

NOMENCLATURE

A = aperture area
A_c = reflector area is defined as $\pi/4\ D^2$
C_r = concentration ratio
F = the focal length and is used as a reference length.
G_{bn} = beam normal irradiance
$\dot{Q}$ = total energy
q = heat flux density ($\dot{Q}/A$)
Q_u = useful heat collection
R^* = R/F
T_r = receiver temperature
T_a = ambient temperatures
η_{opt} = optical efficiency
ρ = reflectance of the reflector
α = absorptance of the receiver
γ = interceptance
τ = transmittance of the cover
ψ = rim angle
σ = sun width (mrad)
Λ = radiant intensity

θ = incident angle at the receiver
Ω = infinitesimal solid angle subtended by
dA_c when viewed from dA_r.

ACKNOWLEDGMENT

The author would likes to thank Dr. Sheldon M. Jeter Associate professor in the School of Mechanical Engineering, Georgia Institute of Technology, for his constructive comments in reviewing this work.

REFERENCE

1. Jeter, S. and Y. A. Elgendy, "Simulation of a Dish Ericsson Solar Thermal Power System with Heat Pipe Receiver", Proceedings of the 1989 Annual Conference American Solar Energy Society, pp. 356-361, Denver, Colorado, June 19-22, 1989.

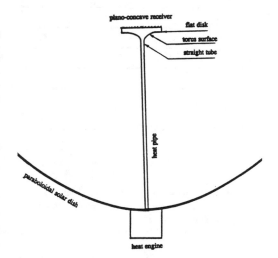

Figure 1 A concept for an external plano-concave receiver coupled to a heat engine by a heat pipe.

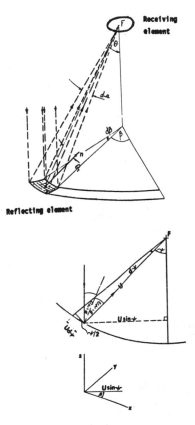

Figure 2 Basic relationship between a differential reflecting and receiver elements.

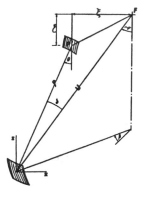

Figure 3 Generalized geometry for a receive element off focus.

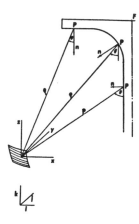

Figure 4 Simple geometry showing the normal vector, n, and the distance between the reflector and the receiver, ray ρ, at various positions for the plano-concave receiver.

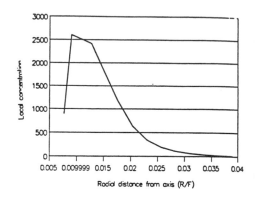

Figure 5 Intensity distribution on the surfaces of the plano-concave receiver.

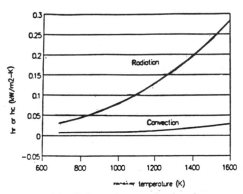

Figure 6 Radiation and convection heat loss coefficients at various receiver temperature for a plano-concave receiver.

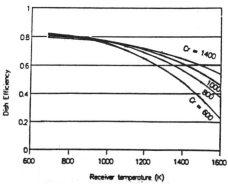

Figure 7 The effects of a receiver temperature on the thermal performance of a solar dish system, utilizing a plano-concave heat pipe receiver, at various concentration ratio.

SOLAR PLANT STUDY OF POWER GENERATION FOR GAS TURBINE

L. León, M. Toledo, E. Villanueva, F. Verduzco, P. Quinto and A. Sanchez

S.G.I. - E.S.I.M.E. - I.P.N. - U.P. Zacatenco Edif. 5
Col. Lindavista 07738 México, D.F.

ABSTRACT

This study shows the viability of electric generation of energy around 500 kW
using solar energy for a Gas Turbine operation. First of all, all of the thermodynamic aspects of the whole system are analyzed, taking into consideration that
the whole system is working only with solar energy. Then, taking into account
the solar energy incidence in Mexico City for the last year, the necessary area
was calculated to obtain enough energy, considering the optical efficiencies of
the receptors. In addition, two alternatives are suggested. The first one is
that the receiver's systems be of a modular type. This means a revolution
paraboloide system of 7m diameter each. The second alternative is that it is
recommended that the captation area of each paraboloide has to be formed of
flat, hexagonal mirrors of three different sizes in three levels on the same
antenna. The working fluid is atmospheric air. For determining the useful
and gained energy an analysis was made tracking the solar radiation until the
focal point of each antenna was reached. Then the optimum form and dimensions
of this focal point were determined.

KEYWORDS

Modular system; revolution parabolic antennas; flat hexagonal mirrors,
efficiencies.

INTRODUCTION

Solar energy shows great advantages over conventional energy forms. This is
the reason why, in recent years, there have occurred a notable growth of investigations related to the conversion processes of solar energy to thermal,
mechanical or electrical energy. Specifically, a proposal of solar energy to
electrical energy conversion is called the cycle Joule-Brayton.
The concept of concentration for parabolic discs is advantageous, given the
temperature range that is desired.
In a receptor-absorber system where high temperatures are obtained, that
proves to be independent of all conventional thermal sources. The system
proposed here changes the solar energy first to mechanical energy and later to
electrical energy. Production ranges normally from 300 kW to 500 kW, and, under
ideal conditions, it reaches a maximum of 800 kW.
This system includes a grou pf 105 collectors of 7 m diameter each, a control
system, and an energy conversion system. The whole system begins to operate
when the starter dispositive of the Turbine pulls the compressor which impels
and compresses the working fluid to the heat exchanger, (focal point of the
antenna) where this fluid is heated to nearly 1100K.

THERMODYNAMIC BEHAVIOR.

In the figure No. 1 it is shown the proposed system.
Where: (1) intake filter; (2) intake silencer; (3) compressor;
(4) recuperator; (5) absorber; (6) turbine; (7) and (8) blow-off
valve; (9) gearbox; (10) generator and (11) exhaust silencer.

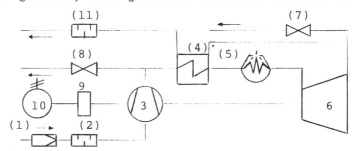

Fig. 1. Gas turbine proposed system.

Specifically, analyzing the open cicle configuration of Gas
Turbine, determines the dependence that exists between
thermodinamical efficiency (η_{th}) and the pressure relation (π),
for different temperatures where it is observed that the
efficiency is maximum for a unique value of (π).

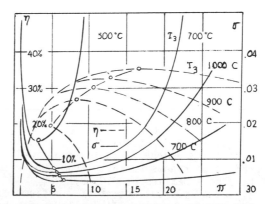

Fig. 2. Analysis (η_{th}) agaiast (π).

Now, doing the same analysis for the compressor, that has
variable velocity because the solar energy is variable then, the
caracteristic curves where the pressure relation (π_c) against the
reduced masic use (m_c) is charted, this is

$$\overset{\circ}{m}_c^* = \overset{\circ}{m}_c \times \frac{(T_{tot,1} / T_s)^{1/2}}{P_{tot,1} / P_s} \qquad (1)$$

To access the compressor, where Ts = 288.15 K and Ps = 1.013 bar are the standards to sea level condition and, Ttot and Ptot are the real temperature and real pressure respectivelly, then obtained the next figure.

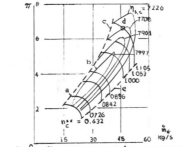

Fig. 3.° Compressor's curves.

The (a) curves shows the jumped of pressure that occured when reduced speed compressor is constant n^{**}. This is

$$n_c^{**} = \frac{n_c (T_{tot,1,d})}{n_{c,d} (T_{tot,1})} \qquad (2)$$

Where nc is the compressor speed and the "d" subindex design values. The efficiency curves of compressor are limited for the lines of "surge" and lines of "strike". Thinking in the last, we obtain the next design parameters:

TABLE 1 Design Parameters

CYCLE		
Electrical output at terminals	500	kW
Ambient temperature	15	°C
Ambient pressure	1.013	bar
Pressure ratio of compressor	6.8	-
Turbine inlet temperature	800	°C
Turbine mass flow rate	4.8	kg/s
EFFICIENCY		
Efficiency of compressor	77	%
Efficiency of turbine	86	%
Parabolic dish efficiency	80	%
Absorber efficiency	60	%
Generator efficiency	88	%
Gearbox efficiency	96	%

ABSORBER CONCEPT

For obtaining the ideal energy distribution radiant in the absorber, is neccesary to analvze equations of the posible receptor and the form itself of the system in whole, where we seek the optimal points of the system.

If we coursed of determining heat quantity incidents an the absorber Q_{Re}, we have

$$Q_{Re} = {}^{\pi}/_4 \; \rho_{pa} \; S \; (D^2 - D_a^2) \qquad\qquad (3)$$

Where: ρ = reflectivity; S = Total average radiation.

To optimize the system, is neccesary relate at thesame time size of paraboloide and absorber size, also with curvature angle, reflexion deviation and other coordinates system itself.

$$D_{a \; mim} / D = |1 - 2F/D \left[1 - (D/4F)^2 \right] tg \left[2 \; arctg \; (D/4F) + 16 \right]|$$
$$(4)$$

Where: F = Focal distance; F/D = Focal radio.

In base to the last equations we can obtain the form and dimension of absorber, this is shown in figure 4.

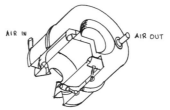

Fig. 4. Absorber.

MECHANICAL CONCEPTS

Because of the neccesity of a captation area of $5\;362.3\;m^2$ that is required for a generation of 500 kW, we suggest a modular system of 105 revolution parabolical antennas with 7 m of diameter each one.

For construction of these paraboloides, we advised that they will are assemble for structural supports and reinforce like it is shown in the figure 5.

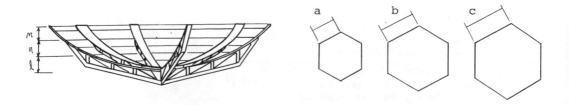

Fig. 5. Paraboloide's structure.

Fig. 6 Forms of the mirrors.

The structure is covered with sheet, later the mirrors are mounted which are flat and of first surface, They are hexagonals in three differents sizes, They are presented in figure 6.

The mirror types (a) cover until level l of the figure 5, the mirrors type (b) cover until level n and mirrors type (c) cover until level m of the same figure. For this, we neccesity 118 mirrors of type (a), 50 of type (b) and 197 of type (c). They are three screw each one for posterior part to permit their orientated toward the focal point.
The size of the mirrors have direct relation with the diameter of absorber opening.

For the design of the paraboloide support think that this most permit to parabolide a continue orientation respect to the sun. Finally, the paraboloide has the form like the figure 7.

Fig. 7.   Paraboloide with the support.

COMPARATIVE EFFICIENCIES

In the next step was determined the variation of yield of the system in fuction of the temperature to the turbine's access, like it is shown in the figure 8.

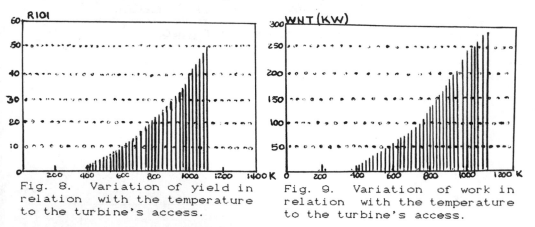

Fig. 8.   Variation of yield in relation with the temperature to the turbine's access.

Fig. 9.   Variation of work in relation with the temperature to the turbine's access.

Then doing the same   net work. with respect to the temperature, to the turbine's access figure 9.
Moreover, when next resolving an equations system we found that theoretically our system, as a whole, got a yield of 19.32 % for December until 28% in March. Our plant has the next appearance:

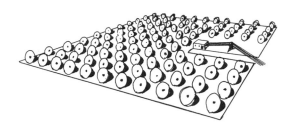

Fig. 10.   Plant of generation modular type.

CONCLUSIONS

The study presented here leads   us to obtain the global dimensions of the proposed system to take into account the particular conduct of all the elements of these; they are the gas turbine, absorber design, yield of the paraboloid and conversion system of energy. Specifically, it was demostrated that the efficiency is good for our intentions, which are to employ a modular system and flat mirrors in the paraboloids.

REFERENCES

K. Bammert and H. Lange "Part-Load Behavior of a Solar Heated and Fossil-Fueled   Gas   Turbine   Power   Plant" ASME Vol. 109 January 1987.
K. Bammert and J. Johannig. "Dynamic Behavior of a Solar Heated Reserver of a Gas Turbine Plant ASME Vol. 109 January 1987.
L. León and A. Sánchez "Parámetros Termodinámicos de una Instalación Solar por Turbina de Gas" ANES 1990.
L. León and F. Aguirre "Diseño del Banco de Pruebas de una Instalación   Solar   con   Paraboloide   de   Revolución" 1ª reunion nacional de estudiantes de posgrado Salca. Gto. 1991.

COMPARISON OF PREDICTED OPTICAL PERFORMANCE WITH MEASURED RESULTS FOR DISH CONCENTRATORS

Gary Jorgensen

Solar Energy Research Institute
1617 Cole Boulevard, Golden, Colorado 80401 USA

ABSTRACT

Several optical design tools have been developed at the Solar Energy Research Institute (SERI) during the past two years. These have been used extensively both in house and by industry to analyze dish concentrator systems and to optimize performance of such designs. The first program, OPTDSH, models single-element dish concentrators. The second code, ODMF, allows multifacet dish arrays to be modeled. The accuracy of performance simulations by these programs has been established by comparing predicted results with measured on-sun data.

ODMF evolved from SERI's High-Flux Solar Furnace (HFSF) design tool, SOLFUR, and in fact is a special case of SOLFUR in which the primary facet array is "on sun." Consequently, confirmation of the accuracy of SOLFUR would verify the results from ODMF as well. Furthermore, because OPTDSH can be viewed as a single-facet case of ODMF, determination of the precision of SOLFUR/ODMF would also substantiate OPTDSH. Thus, the approach to verifying the correctness of all three codes was to compare flux patterns as predicted by SOLFUR with those actually measured at SERI's HFSF.

Measured vs. calculated data have been compared on the basis of flux distribution (in terms of contour plots) and peak flux for both single-facet and multiple-facet cases. Agreement in measured vs. predicted peak flux values has been obtained within the uncertainty associated with the measurement/calibration process. Excellent agreement has also been demonstrated by comparing contour maps of measured vs. computed flux levels.

KEYWORDS

Optical performance modeling; measurement of optical performance; dish concentrators.

INTRODUCTION

A computer model named SOLFUR was written at the Solar Energy Research Institute (SERI) to serve as a design tool during the development of a High-Flux Solar Furnace (HFSF), also at SERI (Lewandowski, 1989). Because ODMF evolved from this solar furnace design tool and in fact is a special case of SOLFUR in which the primary facet array is "on sun," validation of SOLFUR would effectively validate ODMF as well. Furthermore, because OPTDSH (Balch and co-workers, 1991) can be viewed as a single-facet case of ODMF, validation of SOLFUR/ODMF would also validate OPTDSH. Thus, the approach to validating all three codes was to compare flux patterns as predicted by SOLFUR with those actually measured at SERI's HFSF.

All of these programs use a three-dimensional ray-trace procedure as described by Spencer and Murty (1962). Rays are generated at infinity either on a uniform (Cartesian) grid or in a random fashion. The intersection of each ray with the appropriate optical surfaces (heliostat, dish facet, target plane for SOLFUR; dish facet and target plane for ODMF; and dish surface and target plane for OPTDSH) is sequentially computed, and optical errors are incorporated into the reflected ray directions. The target plane is divided into a two-dimensional grid and a tally is kept of the number of rays that intersect each grid area. The concentration ratio (C_g) within each grid is then proportional to the ratio of the number of rays per grid area to the number of rays per dish area:

$$C_g = \rho_h * \rho_d * \left(\frac{N_g}{A_g}\right) * \left(\frac{N_d}{A_d}\right)$$ (1)

where ρ_h = the solar-weighted reflectance of the heliostat
ρ_d = the solar-weighted reflectance of the dish array
N_d = the number of rays that strike the dish array
A_d = the projected area of the dish array
N_g = the number of rays that intersect the·grid element
A_g = the area of the grid element.

Given the concentration ratio and the value of the direct solar irradiance (I, for example as provided by a Normal Incidence Pyrheliometer (NIP) reading, in W/m^2), the flux within a given grid (F_g) can be calculated as:

$$F_g = I * C_g$$ (2)

Calculated results have been compared to measured data on the basis of flux distribution (in terms of contour plots) and peak flux for both single-facet and multiple-facet cases. Agreement in predicted vs. measured peak flux values has been obtained within the uncertainty associated with the measurement/calibration process. Excellent agreement has also been demonstrated by comparing contour maps of measured vs. computed flux levels.

EXPERIMENTAL APPROACH

The basic approach to verifying the optical performance codes was to compare measured flux profile data obtained at SERI's HFSF with results calculated by SOLFUR in simulating the geometrical configuration and optical specifications of the various elements of the HFSF. Using a commercial beam analysis software package named BEAMCODE (1990), which has been integrated into the HFSF data acquisition and control system (Bingham and Lewandowski, 1991), the following flux profiles were measured:

- Two flux maps at the optimum target distance (i.e., at the focal plane) taken over a fairly short time period (several minutes or less between each measurement) to demonstrate the variability associated with the measurement process.
- Two flux maps each at a distance +8 cm (away from the dish array) and −8 cm (closer to the dish array) along the optical axis from the ideal target plane. This was carried out to verify the angular distribution of rays at the target; if a series of flux profile "snapshots" taken along the optical axis agrees well with predicted results, then confidence in the directionality of rays at the target plane is high.

TABLE 1   Measurement Parameters at SERI's High-Flux Solar Furnace, 8/3/90

Run/ Flux File Name	Time	# of Facets	Distance of Target Plate from Focal Plane (cm)	Peak Scale Factor	Direct Peak Normal Flux (kW/m²)	Concentration	
						Irradiance (W/m²)	Ratio (suns)
G1B	10:54	1	0	35.8	106	890	119
G2A	11:56	23	−8	34.6	1873	957	1957
G2B	11:58	23	−8	34.6	1868	957	1952
G2C	12:00	23	0	35.8	2263	954	2372
G2D	12:03	23	0	35.8	2249	954	2357
G2E	12:04	23	+8	37.0	2085	950	2195
G2F	12:05	23	+8	37.0	2091	950	2201

- A flux map for the central facet only (i.e., all other dish facets covered). This corresponds to a single-facet "on-axis" dish concentrator as modeled by OPTDSH.

Results of these experiments are presented in Table 1. NIP readings provide the direct-normal irradiance required to convert between flux and concentration ratios. The scale factors listed in Table 1 vary because the video camera remained stationary during the measurement process while the target flux plate (and its intercepted flux profile) was moved about the focal plane, resulting in differences in the projected size of the image at the camera sensor array.

Values for the following optical parameters were used as input to the SOLFUR model:

$\rho_h = 0.879$ = reflectance of heliostat facets
$\rho_d = 0.904$ = reflectance of dish facets
$\sigma_h = 0.1$ mrad = specularity of heliostat facets
$\sigma_d = 0.1$ mrad = specularity of dish facets
$s_h = 0.5$ mrad = slope error associated with heliostat facets
$s_d = 0.2$ mrad = slope error associated with dish facets

Heliostat and dish reflectances were measured after cleaning the mirror surfaces prior to the set of experiments discussed herein. The sun was modeled as a pillbox distribution having a characteristic size of 4.65 mrad. Additional design parameters required to model the HFSF are given by Lewandowski (1989) and by Lewandowski and co-workers (1990).

A common format was desirable in order to readily compare predicted results with measurements. Such an arrangement is provided by a program named SURFER (1989). This program allows three-dimensional information (for example, flux level as a function of x and y coordinates) to be graphically represented as topological contour maps (TOPO feature) and as three-dimensional isometric projections (SURF option). This program was used to accept both measured and calculated data.

RESULTS

A number of SOLFUR simulations were run corresponding to the measurements taken for the various configurations specified in Table 1. The results predicted by SOLFUR fluctuate based

TABLE 2   Simulation Results Predicted by SOLFUR*

Run/ Ray File Name	1-D # of Rays	# of Grids	# of Facets	Distance from focal plane (cm)	Peak Flux (kW/m²)
B05K0000	5000	16	1	0	86
B12K0000	12500	25	1	0	93
B25K0000	25000	35	1	0	98
B50K0003	50000	25	1	0	87
B50K0004	50000	50	1	0	96
B50K0006	50000	25	23	0	2091
B50K0005	50000	50	23	0	2306
B50K0008	50000	25	23	−8	1767
B50K0012	50000	35	23	−8	1852
B50K0007	50000	50	23	−8	1948
B50K0010	50000	25	23	+8	1817
B50K0011	50000	35	23	+8	1891
B50K0009	50000	50	23	+8	1986

*For comparison with the various measurement configurations enumerated in Table 1; the random number generator seed was held fixed for all runs.

on several factors including the number of rays traced and the size of the grid pattern used to tally rays in the target plane. The predicted peak flux is a strong function of the size of the grids used to compute concentration ratios (and hence flux levels). This can be seen from the results presented in Table 2. In general, for a large number of rays, the greater the number of grids the more accurate will be the predicted peak flux. The reason for this is that the grouping of rays will tend to be more uniform and jagged features of local peaks will be smoothed into a more rounded distribution.

Figure 1 is a contour map of the measured flux levels for 23 dish facets with the target plane located at the focal plane. The flux levels predicted by SOLFUR presented in Fig. 2 have been scaled such that the peak flux agrees with measured values. As suggested above, the calculated peak flux can be made to artificially agree with measured values (Table 1) by varying the number of grids used in the simulation (Table 2). The scaling process was within the uncertainty of measured peak flux levels reported by BEAMCODE; calibrated values are generally 7%–10% higher than expected based upon measured direct normal irradiance. Comparing measured results (Fig. 1) with calculated flux levels (Fig. 2) indicates excellent agreement.

Figures 3 and 4 present the measured and predicted contour maps for 23 dish facets with the target plane located −8 cm from the focal plane. Although some structure appears in the measured data (Fig. 3) that is not evident in the calculated pattern (Fig. 4), the spacing between lines of constant flux in both plots gives a superb match. The pattern associated with the measured data may be due to the fact that the primary dish array was aligned such that the centroids of each facet coincided at the focal plane. This is in contrast to the optical models where the aim points which are specified for each dish facet refer to the central ray rather than the centroids of the resulting flux patterns. Similar results are obtained when the target plane is positioned +8 cm from the focal plane. These results suggest that the angular distribution of rays at the focal plane is correctly predicted by the optical models.

To validate the single-facet model (OPTDSH), a simulation was run for just the central facet of the primary array. The measured data are presented in Fig. 5; the corresponding predicted results are given in Fig. 6. Outstanding agreement can readily be seen in terms of both the shape and the orientation of the contour lines.

CONCLUSIONS

The performance simulations of several optical ray-trace programs have been validated by comparing predicted results with measured on-sun data. Comparison of flux patterns measured at SERI's HFSF and those predicted by SOLFUR reveal excellent agreement for both peak values

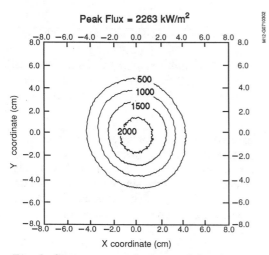

Fig. 1. Contour map of measured flux levels for 23 facets with target plane located at focal plane.

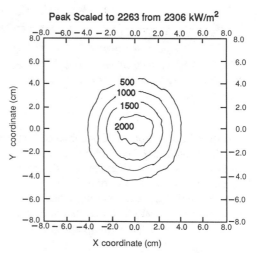

Fig. 2. Contour map of flux levels predicted by SOLFUR for 23 dish facets with target plane located at focal plane.

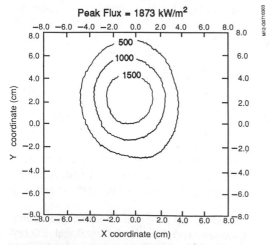

Fig. 3. Contour map of measured flux levels for 23 dish facets with target plane located −8 cm from focal plane.

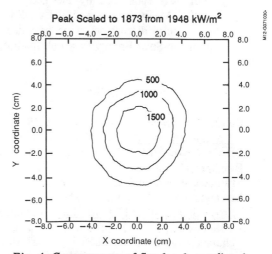

Fig. 4. Contour map of flux levels predicted by SOLFUR for 23 dish facets with target plane located −8 cm from focal plane.

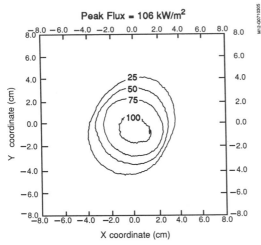

Fig. 5. Contour map of measured flux levels for single (central) dish facet with target plane located at focal plane.

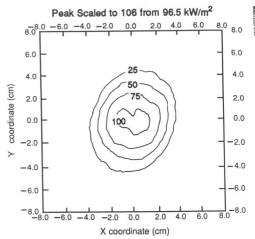

Fig. 6. Contour map of flux levels predicted by SOLFUR for single (central) dish facet with target plane located at focal plane.

and distribution profiles. Concordance between measured peak flux values and those predicted by the model has been obtained within the uncertainty associated with the measurement/calibration process. Comparison of contour maps of measured vs. computed flux levels demonstrates the validity of SERI's optical ray-trace codes for both the single-dish element case and the multifacet case. Strong evidence for the accuracy of the predicted angular distribution of rays at the focal plane has also been shown by comparing measured vs. predicted results for target planes offset on either side of the focal plane.

ACKNOWLEDGMENT

This work was sponsored by the U.S. Department of Energy under contract DE-AC02-83CH10093. Measurements at SERI's HFSF were performed by Al Lewandowski and Carl Bingham. Useful discussions were provided by Al Lewandowski, Tim Wendelin, and Meir Carasso of the Solar Energy Research Institute.

REFERENCES

Balch, C., C. Steele, G. Jorgensen, T. Wendelin, and A. Lewandowski (1991). Membrane Dish Analysis: A Summary, SERI/TP-253-3432, Solar Energy Research Institute, Golden, Colorado.

BEAMCODE 6.0 Rev 19 (1990). Big Sky Software Corp., Bozeman, Montana.

Bingham, C., and A. Lewandowski (1991). Data acquisition and control of SERI's high flux solar furnace. American Society of Mechanical Engineers, paper to be presented in Reno, Nevada.

Lewandowski, A. (1989). The design of an ultra-high flux solar test capability. Proceedings 24th IECEC Meeting, Washington, D.C.

Lewandowski, A., C. Bingham, J. O'Gallagher, R. Winston, and D. Sagie (1990). Performance characterization of the SERI high flux solar furnace. Presented at the International Energy Agency 5th Symposium on Solar High-Temperature Technologies, Davos, Switzerland.

Spencer, G. H., and M.V.R.K. Murty (1962). J. Opt. Soc. Am., 52, 672-678.

SURFER Version 4 Reference Manual (1989). Golden Software, Inc., Golden, Colorado.

SOLAR THERMAL ENERGY UTILIZATION

Efstratios Soubassakis and José G. Martín

Chemical and Nuclear Engineering Department
University of Lowell, Lowell, MA 01854, U.S.A.

ABSTRACT

The viability of a solar thermal system depends on the efficiency with which the intercepted solar energy can be converted to a useful form of energy such as electricity. At any concentration, the thermal collection efficiency decreases with temperature. Because the maximum efficiency for the conversion of thermal energy to work increases with temperature, there is a certain temperature at which a dynamic system can operate most effectively.

This paper proposes a model for estimating the effect of the thermal resistance of the absorber on solar thermal energy utilization. Two equations can be used to model a dynamic system: one for the maximum conversion efficiency and one to specify the constraint that, in the quasi-steady state, the net energy collected must be transferred to the coolant. These two equations define a constrained optimization problem in two variables, the temperature of the fluid and the temperature of the absorber. The method of Lagrange multipliers may be used to convert this problem to an equivalent unconstrained formulation. The nonlinear equations that describe the stationary point of the augmented Lagrangian function can be solved using available software.

For electrical generation, the results have interesting implications for collectors which are cooled by liquid-metals or by forced boiling liquids. The model can also be applied profitably to other high temperature solar applications, such as the manufacture of synthetic fuels and chemicals.

KEYWORDS

Solar concentration; collection efficiency; direct absorption; solar fuels; solar chemicals.

INTRODUCTION

The intensity of solar radiation is relatively low at the surface of the earth in comparison with most human needs and large areas of collectors are required to intercept that radiation. Because of costs, the viability of the solar option depends on the efficiency with which the intercepted energy can be converted to a useful form such as electricity. Since the spectrum of solar radiation reaching the earth may be approximated by that of a black body at 6200 K, it is in principle possible to convert this energy to work at high efficiency.

In a solar thermal system, the losses from the active absorber area increase with the surface temperature; at a given temperature, the losses are proportional to the active area, and lower losses, higher efficiencies, and/or higher temperatures are possible if the incident radiation is concentrated.

Because the maximum work obtained from the collected energy decreases with temperature, there is a temperature at which the system will operate most effectively. Suppose that the collected energy heats a coolant, which drives a heat engine. In the absence of thermal resistance between the active surface and the coolant, one can estimate the effect of concentration on the thermal collection efficiency and on the maximum possible (Howell and Vliet, 1982, Martín and Blanco, 1987).

Actual efficiencies fall short of the theoretical maximum, due to the intermittence of the solar source and material and conversion equipment limitations. Heat transfer through a solid wall imposes another major limitation; at high flux, there is a large temperature differential across the absorber.

EFFECT OF CONCENTRATION ON EFFICIENCY

The effect of solar energy concentration on the efficiency at which solar energy can be collected thermally and converted to work is discussed by Howell and Vliet (1982). Suppose that q W/m2 of solar energy are intercepted by collector of area A_c, so that a fraction (1-F) of the total power is lost. Of the total power incident on the collector area (q Ac F), a fraction G can be redirected and absorbed by a smaller area A_a. For a concentrating system based on mirrors, G may be written as $\gamma \alpha \rho$, where γ is the intercept factor, ρ is the reflectivity, and α is the absorptance of the active surface.

The absorber loses energy by radiation, conduction, and convection. Assume that the environment is at temperature T_e. The amount of the energy which can be extracted, Q_u, is the total energy absorbed minus the losses. In terms of the hemispherical emissivity of the absorber, ε, and an overall heat loss coefficient U to the environment, it is possible to define a <u>collector thermal efficiency</u>, η, as

$$\eta = (Q_u / F q A_c) \tag{1}$$
$$= G - [(\varepsilon \sigma A_a / Fq)(T^4 - T_e^4)](FqA_c) - [U(T - T_e)](FqA_c) \tag{2}$$

The expression for η may be rewritten in dimensionless form as:

$$\eta = G - [(a/F)(Z^4-1) + b/F (Z-1)](1/C) \tag{3}$$

where $a = \varepsilon \sigma T_e^4 / q$, $b = U T_e / q$, $Z = T / T_e$, and the concentration $C = A_c / A_a$.

For any T_e and solar flux, a is function of ε while b depends on U. T is a characteristic value for the absorber. Note that for constant G, ε, and U, η increases with C at any temperature.

If the collected energy is converted to work in a Carnot engine, with efficiency $\eta_t = 1 - 1/Z$, the maximum attainable <u>overall efficiency</u> W for conversion of the incident solar energy to work is

$$W = \{G - [(a/F)(Z^4-1) + b/F (Z-1)](1/C)\} [1-1/Z] \tag{4}$$

Since η_t increases with temperature while η decreases, the overall efficiency W is maximum for some optimum temperature. Making the derivative of (4) with respect to Z equal to zero, we have

$$Z^5 - (3/4)Z^4 + (b/4a)Z^2 - [(FGC + A + b)/4a] = 0. \tag{5}$$

The real solution of eq. (5) gives the optimum temperature for the system, that is, the one corresponding to the maximum efficiency. A designer aims at keeping the values of a and b as low, and F and G ($= \gamma \alpha \rho$) as close to one, as practical. For <u>constant</u> ε and U, the optimum temperature is higher the larger the term in the brackets, i.e., it increases with concentration.

THE EFFECT OF FINITE THERMAL RESISTANCE

The previous discussion was based on the assumption that the temperature of the absorber surface, T_a is a constant and equal to the temperature of the working fluid, T_f. In fact, hot spots will contribute disproportionately to the losses. Also, in most envisioned receivers, the absorber is a solid wall. Energy must be conducted through this wall into a coolant, and then is transported (directly or through storage tanks or heat exchangers) to a power conversion system or chemical processing plant. If $T_a = T_f$, no heat is conducted. The differential between the T_a and T_f increases with the thermal resistance. If the difference is high, the losses are high and the system is inefficient. This provides an incentive to incorporate coolants of exceptional heat transfer characteristics - liquid metals, for example - even when their chemistry or thermal storage properties introduce formidable challenges. The thermal resistance of the solid wall depends on its thickness and thermal conductivity. Once a good conductor compatible with the coolant is chosen, there is a strong incentive to make the absorber wall as thin as possible and this tends to compromise the receiver robustness and expected lifetime.

A model to quantify the effect of the thermal resistance of the first wall on system performance was first proposed by Soubassakis, White and Martín (1991). The model assumes that the heat loss coefficients are constant over all surfaces. The coolant, at T_f, drives a Carnot cycle. Losses along pipes or heat exchangers are neglected. Hence, the maximum conversion efficiency for the system is

$$W_{max} = \{G - [(a/F)(Z_a^4-1) + b/F (Z_a -1)](1/C)\}[1-1/Z_f], \tag{6}$$

where $Z_a = T_a / T_e$ and $Z_f = T_f / T_e$. $\tag{7}$

In the quasi-steady state the total net heat absorbed by the first wall, Q, which is equal to the total heat absorbed minus the radiation and convection losses, is transported to the coolant. In other words,

$$Q = q \, A_c \, F - [\, aqA_a \, (Z_a^4-1) + bqA_a \, (Z_a -1)], \tag{8}$$

$$= U'A_a \, (T_a - T_f) = U'A_a \, T_e \, (Z_a -Z_f) = b'q \, A_a \, (Z_a -Z_f), \tag{9}$$

where U' is an overall heat transfer coefficient between the surface and the coolant, and $b' = U'T_e/q$. From eqs. (8) and (9),

$$aZ_a^4 + (b + b')Z_a - b' \, Z_f - (CF + a + b) = 0. \tag{11}$$

Equations (6) and (11) define a constrained optimization problem in two variables, Z_a and Z_f. The method of Lagrange multipliers (Reklaitis, 1983) is used to convert this problem to an equivalent unconstrained formulation. The nonlinear equations that describe the stationary point of the augmented Lagrangian function are then solved using available software.

RESULTS

It is an interesting exercise to solve these equations for reasonable values of ε and U, for different values of U'. Figures. 1 to 4 show the results for ε = 0.5 and U = 18 (W/(m2·oK), for U' ranging from 100 W/(m2·oK) to infinity and C from 250 to 1500.

Figure 1 shows that W drops with Ta.and increases with C. More interesting, it shows the effect of U'. Suppose that for a hypothetical oil, under forced convection, U' = 100 W/(m2·oK). For C = 250, W is only about 7% - if the oil could tolerate 1000 C. At the same concentration, however, if the coolant was a liquid metal with U'= 50,000 W/(m2·oK), a system with C = 250 could reach an efficiency of 45% at a much lower temperature. The difference between Ta and Tf is shown in Fig. 4. For low resistance, Tf approaches Ta and any further increases in U' have a small effect on the system performance. We may be able to apply these results to some actual examples.

Sodium-Cooled Receivers

In many ways, sodium is an ideal coolant. It has a high thermal conductivity, k, and low vapor pressure Two sodium-cooled receivers were installed at the Central Receiver System of the International Energy Agency Small Solar Power System project in Almería, Spain (Kesselring and Selvage, 1986). One of these, the Advanced Sodium Receiver (ASR) was a 2.7 MWth external type that consisted of five panels arranged to form a rectangular absorber 2.85 m high and 2.78 m wide. Each panel consisted of 39 fourteen mm diameter vertical tubes, a bottom and top header and downcomer. Liquid sodium was pumped from the cold storage tank at 270 C into the bottom of the receiver panel, through each of the panels in series and out at the top of the central panel at 530 C.

For a liquid metal flowing in a circular tube of radius r_o, the convection heat transfer can be approximated by $h=4k/r_o$ (Incropera and DeWitt, 1985). The conductivity of sodium ranges from about 86.2 W/(m·°K) at 366 K to 59.7 W/(m·°K) at 977 K. For the ASR tubes at 500 C one can estimate $h \approx 37,000$ W/(m²·°K), decreasing with temperature. These are only approximations, but Figs. 1-3 indicate that for U'≥2,500 W/(m²·°K), the efficiencies are not a strong function of U'.

Because the intercept factor for diffuse radiation is very small for these systems, we take q to be the direct solar flux. For the plant design point conditions (solar noon at equinox), F=1, direct irradiance =920W/m² and T_e=25C. The average cosine of the heliostat field is 0.95 resulting in q=874 W/m². Also, ρ =0.911 and α =0.95 (DFVLR, 1984) and γ=0.92 (Schiel and Lemperle, 1985), so that G=0.8. Finally, U =18W/(m²·°K) (Jacobs and colleagues, 1985).

Consider a receiver operating at 550°C and C= 250. From Fig. 3, the receiver can collect thermal energy with an efficiency close to 70%. This is <u>lower</u> than the actual ASR efficiency; ε for that receiver was 0.3, lower than the one assumed for these calculations. From Fig. 2, it can be seen that this receiver could convert solar energy to work with a maximum conversion efficiency of about 45%. This is, of course, much higher than the best Almeria results.

It is possible to operate the receiver at T_f= 600 C by setting C=500. (At 600 C, the solubility of iron in sodium is about 1 ppm, and that of carbon is 1.8 ppm. This may lead to the mass transport of iron and the decarburization of steel, and also adds impurities to the sodium, raising its melting point). Since U' for sodium tends to decrease with temperature, Fig. 2 shows that the 50°C increase in the operating temperature for the coolant <u>decreases</u> the thermal collection efficiency. More pertinently, Fig. 2 shows that this sodium temperature increase <u>does not increase</u> the maximum work that can be theoretically generated by the system. Of course, the results are dependent on the assumptions.

Boiling Coolant

Consider a system which operates at relatively low T_f (≤500 C), C = 250, and is cooled by forced convection by high-temperature oil. For forced convection, U' ranges from 60 to 1,800W/(m²·°K). Assume U'= 100 W/(m²·°K). For ε = 0.5, and α = .9, Fig. 3 indicates that the maximum efficiency at which this system can collect solar energy is about 30%, close to 400 C. Figure 2 shows that the maximum overall theoretical efficiency for conversion to work by the system is about 17% - or lower, for higher C and higher temperatures. The performance of the system improves dramatically if the thermal resistance is lowered. Suppose that an increase in the temperature to 500 C somehow increases U' to 750 W/(m²·°K): this will increase the overall conversion efficiency to about 38%.

A second system demonstrated in Almeria, was the Distributed Collector System (DCS) (DFVLR, 1983). A major part of the DCS was made up of one-axis tracking cylindrical troughs cooled with an organic oil. For the DCS, C = 18, γ=0.91, ρ=0.9, α=0.85, and the cover transmittance is 0.93 so that G=0.7 (Martín and Carmona, 1983). From calculations based on actual experiments, ε=0.25 and U=4.6 W/(m²·°K) (Jacobs and Carmona, 1983). The plant achieved very low efficiencies of conversion (≤5%), due in part to the low U'. U' is very large with phase change (boiling); typical values range from 2500 to 100,000 W/(m²·°K). With boiling heat transfer, even a very low-concentration system such as the DCS may achieve an overall conversion efficiency of 26% at 340C.

Fuels and Chemicals

Electricity generation is only one possible high-temperature application of solar energy; the manufacture of synthetic fuels and chemicals is another. The manufacture of ethylene, the steam reforming of methane, shale retorting, and solar coal gasification have been identified as specially promising applications. The feasibility of these schemes depends on the design of solar receivers with very low thermal resistance, such as one based on direct absorption (Martín and Vitko, 1982).

Graphs such as Figs. 2 to 4 can be used to evaluate these schemes. Consider shale retorting, which requires a temperature of about 550 C. Fig. 3 shows that, even at high concentration (C = 750), the efficiency of thermal collection is less than 18% if the U is 100.W/(m²·°K) The reason for this low

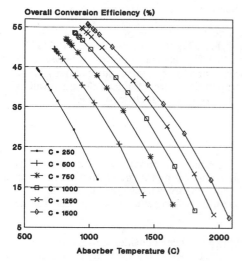

**Fig. 1 Effect of Concentration
on Absorber Temperature**

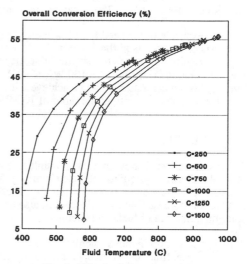

**Fig. 2 Effect of Concentration
on Fluid Temperature**

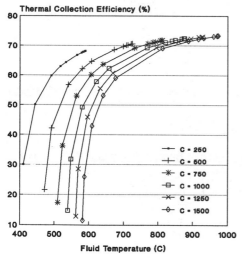

**Fig. 3 Effect of Concentration
on Thermal Collection Efficiency**

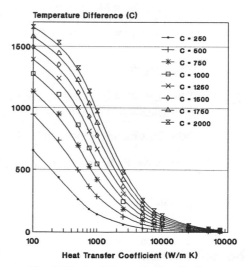

**Fig. 4 Effect of Concentration to the
Absorber to Fluid Temperature Difference**

efficiency is apparent from Fig. 4: with this high resistance, the absorber is about 1000 C hotter than the coolant. Even at C = 250, η can be made higher than 65% if U' is $\geq$ 1000 W/(m2•oK).

Another example of the need for a low thermal resistance is the production of ethylene by the pyrolysis of ethane, as in the reaction C_2H_6 + heat -> C_2H_4 + H_2 (Carling and others, 1981). The furnace required for the pyrolysis must operate at about 1100C. One can show that if one were to use a system characterized by U' = 100 W/(m2•oK), one would need a concentration of about 10,000. Even if this concentration could be achieved, it would be practically impossible to design a wall to withstand a power density of the order of 10 MW/m2 while operating at close to 4000C.

DISCUSSION

The results reported show that it is possible to quantify the effect of thermal resistance on the potential of a solar thermal system to generate electricity. They illustrate the improvements in plant performance that can be achieved in a system where the absorbed thermal energy is transferred by forced convection boiling. The formulation can also be used to make scoping studies of proposals for the solar manufacture of fuels and chemicals.

A major limitation in these calculations is that ε, U and U' have been considered to be constant. They are, of course, functions of temperature. Since much is known about the reflectivity of selective surfaces and about the dependence of the thermal loss coefficient and the thermal heat transfer coefficient on temperature, the authors expect to incorporate these effects in future work.

ACKNOWLEDGEMENT

The authors want to thank Messrs. Philip Delmolino and Harish Hande, of the Energy Engineering Program at the University of Lowell, who helped to prepare the graphs presented in this paper, and Professor John White, for his valuable suggestions.

REFERENCES

Carling, R. W, et al., "Solar Central Receiver Fuels and Chemicals Project Status Report", Sandia National Laboratories, SAND 81-8232, 1981.

CRS and DCS Construction Report, DFVLR, Germany, 1983.

Howell, J., Bannerot, R. B., and Vliet, G. C., "Solar Thermal Energy Systems", McGraw Hill, N. Y., 1982.

Kesselring, P. and Selvage, C., editors,"The IEA/SSPS Solar Thermal Power Plants; Vol.1, Central Receiver System", Springer -Verlag Berlin, 1986.

Martín, J. and Vitko, J. Jr., "ASCUAS: A Solar Concept Utilizing a Solid Thermal Carrier", Sandia National Laboratory Report, SAND 81-8005, 1981.

Martín, J. G. and Carmona, R., "Optical Losses", Proc. DCS Intl Workshop:The Fist Term, DFVLR, 1983.

Martín, J. G. and Blanco, M., "Terminal Concentrator for Solar Central Receiver", Proc. Intl. Solar Energy Society Congress, Hamburg, Germany,1987.

Reklaitis, G.V. and others., "Engineering Optimization", John Wiley and Sons, New York, 1983.

Schiel, W. and Lemperle, C., "Measurements and Calculations on Heliostat Field", IEA SSPS Final Evaluation Report, Vol. 1, 18.7. Kesselring, P. and C. S. Selvage, C.S., Editors, 1985.

Soubassakis, E., White, J. and Martín, J., "Conversion Efficiency in Solar Thermal Systems", accepted for presentation at ATHENS'91, Intl. Conf. on the Second Law.

2.23 Line-Focus Collectors

A NEW HIGH-CONCENTRATION TWO-STAGE OPTICAL DESIGN FOR LINE FOCUS SYSTEMS

M. Collares-Pereira[1], J. M. Gordon[2], A. Rabl[3] and R. Winston[4]

[1] Centro para a Conservaçao de Energia, Est. Alfragide, Alfragide, 2700 Amadora, PORTUGAL

[2] Center for Energy & Environmental Physics, Blaustein Institute for Desert Research, Ben-Gurion U., Sede Boqer Campus, ISRAEL, and the Pearlstone Center for Aeronautical Engineering Studies, Dep't of Mechanical Engineering, Ben-Gurion U., Beersheva, ISRAEL

[3] Centre d'Energétique, Ecole des Mines, 60 Blvd St.-Michel, 75272 Paris CEDEX 06, FRANCE

[4] Dep't of Physics and Enrico Fermi Inst., U. of Chicago, 5640 S.Ellis Ave., Chicago, IL 60637

ABSTRACT

A new high-concentration, two-stage optical design is proposed for line focus systems, with specific applications to parabolic trough solar collectors with tubular absorber. It can increase the flux concentration ratio by a factor of 2.5 relative to conventional designs, while maintaining the large rim angles that are desirable for practical and economical reasons. The second stage is comprised of asymmetric nonimaging (CPC-type) concentrators matched to first-stage off-axis parabolic mirrors. The second stage can be accommodated inside an evacuated receiver, allowing the use of first-surface silvered reflectors. The low heat loss of this design opens the possibility of producing steam at temperatures and pressures of conventional power plants, using only one-axis tracking. The improvement in conversion efficiency would be substantial.

KEYWORDS

Secondary concentrator; CPC; nonimaging optics; solar thermal; parabolic trough

INTRODUCTION

The efficiency of solar thermal power plants can be improved by raising collector temperature. The associated heat losses can be reduced by increasing concentration. But higher concentration implies smaller acceptance angle. Define geometric concentration C as the ratio of absorber area to aperture area, and acceptance angle 2δ as the angular region over which all rays incident on the collector aperture reach the absorber. For a given δ, the maximum attainable concentration is

$C_{max} = 1/\sin\delta$ in two dimensions (geometry of troughs), or

$C_{max} = 1/\sin^2\delta$ in three dimensions (geometry of dishes or cones)

[Welford and Winston, 1989]. This limit has been called the thermodynamic limit of concentration because it is a direct consequence of the second law of thermodynamics [Rabl, 1985].

For focusing devices, C is well below C_{max}. For a parabola with tubular absorber, $C = 2x_A/(2\pi r)$ $= C_{max} \sin(\phi)/\pi$, where $2x_A$ = aperture width, r = tube radius, and ϕ = rim angle $\angle$ AFO, which is at best (at $\phi = 90°$) a factor π below C_{max} (Fig. 1). For tracking concentrators, δ is typically around 0.4°-1.0°, for which we can approximate $\sin\delta \approx \delta$.

C_{max} can be reached with (nonimaging) CPCs (compound parabolic concentrators) [Welford and Winston, 1989]. The CPC is well suited for applications with large acceptance angle, but its depth-to-aperture ratio grows excessive at high concentration. For intermediate temperatures (200 to 400 C), a tracking one-stage 2-D CPC is usually impractical. One usually chooses parabolic troughs, despite the poor relation between δ and C.

A suitable combination of a focusing first stage with a CPC second stage can approach C_{max} [Rabl and Winston, 1976; Kritchman, 1981, 1982]. In Fig. 2 the first stage, or primary, is a parabola that directs all incident rays toward the entrance BB' of a secondary CPC (we refer to BB' as virtual absorber). The concentration of the primary is $C_1 = AA'/BB' = \sin\delta \cos\delta/\delta$ in the limit

$\delta \rightarrow 0$ at fixed ϕ. Having an acceptance half angle equal to ϕ, the CPC boosts concentration by a factor $C_2 = 1/\sin \delta$ to $C_{tot} = C_1 C_2 = \{\cos \delta \}/\delta$.

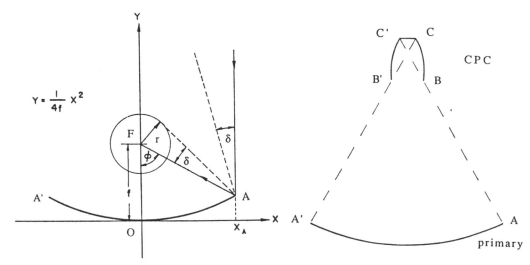

Fig. 1. Schematic of parabolic trough with focal point F, focal length f, aperture A'A and acceptance half angle δ. Equation of parabola is $y = x^2/4f$.

Fig. 2. Two-stage concentrator with parabolic primary and one symmetric CPC secondary. CPC entrance aperture BB' is centered in focal plane of primary. Dashed lines show acceptance angle of CPC is centered in focal plane of primary

Secondary concentrators have not been used in line focus systems because they offer little at large rim angles. For practical line focus systems one strongly prefers large rim angles (80°-120°) in order to improve mechanical stability. For a given aperture, a larger rim angle implies that the receiver is closer to the center of mass, which facilitates tracking and lessens the cost of receiver support structure. Hence the simple parabola of Fig. 1, with tubular absorber and rim angles around 90°, has become the favorite choice for solar applications at intermediate temperatures.

In this paper we present a new design, based on asymmetric CPCs [Mills and Giutronich, 1979; Mills, 1980]. One can boost concentration to within 90% of C_{max} while allowing rim angles as large as 90°. One can realistically increase the <u>flux</u> concentration by a factor of about 2.5 relative to a conventional design. Thus it may become possible to use parabolic troughs for producing steam at temperatures (500 to 550 C) and pressures required by conventional power plants with substantial improvement in solar-to-electric efficiency.

SECOND STAGE WITH ASYMMETRIC CPC

Overall concentration can be improved by using several adjacent, each facing a different portion of the primary and each with its own absorber. Some examples are shown below in Figs. 5-7. First consider a single asymmetric CPC secondary with flat absorber, collecting radiation from parabolic primary AA' (Fig. 3). In polar coordinates (ρ,ϕ) the parabola is $\rho = 2 f/(1 + \cos \phi)$, with ϕ measured from the parabola's axis and f = focal length; ρ is the distance from a point P = (ρ,ϕ) to focus F. The end points A = (ρ,ϕ) and A' = (ρ',ϕ') correspond to the upper and lower rim angles ϕ and ϕ'. The aperture width a of the primary is $a = 2 f \left[\frac{\sin(\phi)}{1+\cos(\phi)} - \frac{\sin(\phi')}{1+\cos(\phi')} \right]$.

The angle δ, as drawn in Fig. 3, is exaggerated for visibility. For small δ, b = BB' is

$$b = \frac{4\,f\,\delta}{\sin(\phi - \phi')} \sqrt{\frac{1}{(1+\cos(\phi'))^2} + \frac{1}{(1+\cos(\phi))^2} - \frac{2\cos(\phi - \phi')}{(1+\cos(\phi))(1+\cos(\phi'))}} \; .$$

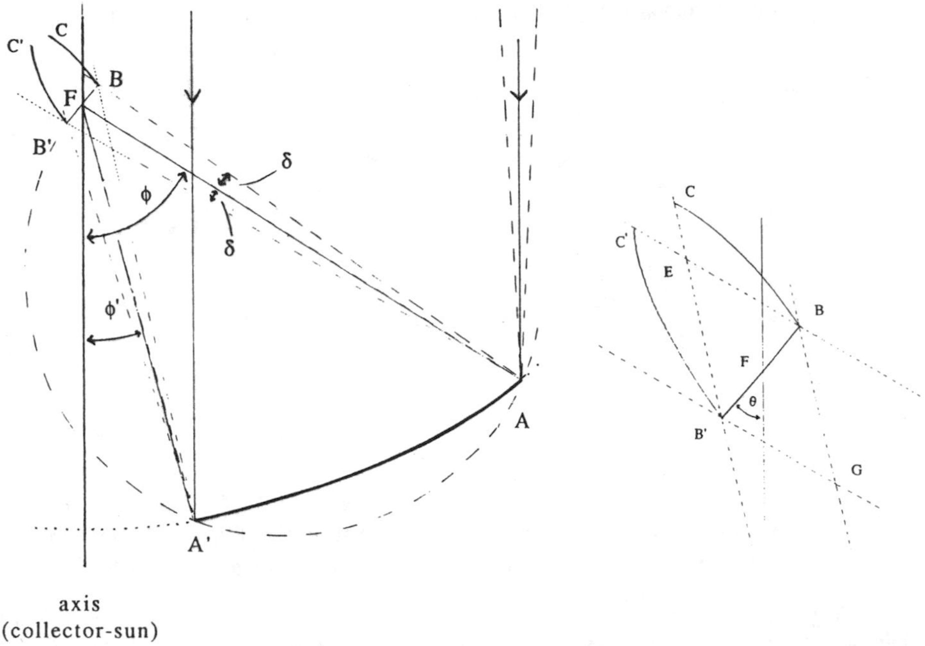

Fig. 3. Asymmetric two-stage concentrator. Primary is parabolic segment AA', secondary is asymmetric CPC with entrance aperture BB'. (a) Primary and secondary entrance aperture (for clarity the latter's size is exaggerated). (b) Close up of the secondary.

The tilt θ of BB' from the collector-sun axis can be determined from the relation (Fig. 3)

$$\cos(\theta + \phi) = (e'^2 - e^2 - b2)/(2\,b\,e) \quad \text{with}$$
$$e = EB = 2\,\delta\,\rho'/\sin(\phi - \phi') \quad \text{and} \quad e' = EB' = 2\,\delta\,\rho/\sin(\phi - \phi').$$

One can then obtain the concentration ratio of the primary

$$C_1 = \frac{a}{b} = \frac{\left[\dfrac{\sin(\phi)}{1+\cos(\phi)} - \dfrac{\sin(\phi')}{1+\cos(\phi')}\right]\sin(\phi-\phi')}{2\,\delta \sqrt{\dfrac{1}{(1+\cos(\phi'))^2} + \dfrac{1}{(1+\cos(\phi))^2} - \dfrac{2\cos(\phi - \phi')}{(1+\cos(\phi))(1+\cos(\phi'))}}} \; .$$

In the limit of small δ, the CPC has a concentration ratio $C_2 = b/c = [\cos(\theta+\phi') - \cos(\theta+\phi)]/2$, where $c = CC' = $ CPC exit aperture. The total concentration C_{tot} can then be expressed as

$$C_{tot}/C_{max} = (a/c)/(1/\delta) = C_1\,C_2/(1/\delta) = \frac{\cos(\frac{\phi-\phi'}{2})\,[\cos(\frac{\phi+\phi'}{2}) + \cos(\frac{\phi-\phi'}{2})]}{[1 + \cos(\frac{\phi+\phi'}{2})\cos(\frac{\phi-\phi'}{2})]} \; .$$

The section BC is a parabola whose axis is parallel to extreme ray A'B' and whose focus is the end point of the other parabolic section of the CPC; furthermore C' lies on the continuation of extreme ray AB. Analogous statements hold for B'C'. Straightforward but tedious algebra yields the positions of C and C', as well as the equations of the parabolic segments BC and B'C'.

NUMBER OF CPCS, CHOICE OF RIM ANGLES AND RECEIVER DESIGN

To choose rim angles for the individual primary segments, we plot C_{tot}/C_{max} vs. ϕ for several ϕ' values in Fig. 4. This form is convenient when the primary is cut off below a rim angle $\phi'=3°$ due to shading by the receiver. At a focal length of 1.25 m, this rim angle corresponds to the shade cast by a receiver of radius 6.5 cm - a good initial guess for the size of the resulting receivers.

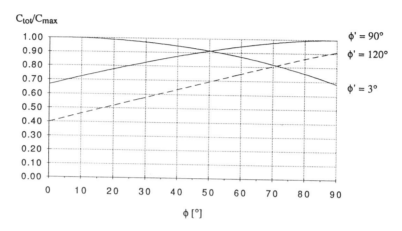

Fig. 4. Concentration ratio of two-stage concentrator, in the form C_{tot}/C_{max} vs. lower and upper rim angles ϕ and ϕ'. Equation is invariant under interchange of ϕ and ϕ'.

If the secondary is to contain only one CPC on each side (2x1), the curve $\phi'=3°$ shows C_{tot} for the rim angle ϕ shown on the abscissa. With $\phi=90°$ this design reaches $C_{tot}/C_{max}= 0.68$, for total rim angle 90°; at $\phi=60°$ it would be 0.87. Two CPCs on each side can do even better. To choose the intermediate rim angle, note that the results in Fig. 4 are symmetric under exchange of ϕ and ϕ'. Thus the $\phi' = 90°$ curve indicates the concentration of the outer collector portion if the total rim angle is 90°. Taking ϕ of the abscissa as the intermediate rim angle, the $\phi' = 3°$ curve shows the concentration of the inner collector portion and the $\phi' = 90°$ curve shows that of the outer portion. The highest overall concentration corresponds to the intersection of the two curves.

For a total rim angle of 90° the intersection is at $\phi = 50°$, with $C_{tot}/C_{max} = 0.91$, with 2x2 CPCs. The corresponding results for 120° are $C_{tot}/C_{max} = 0.81$, with 2x2 CPCs and an intermediate rim angle 71°. For higher concentration, one can increase the number of CPCs but with increasing secondary complexity and reflective losses; hence no more than 2 CPCs on each side are analyzed.

Since different portions of the collector are optically independent, one has some freedom to avoid both spatial interference and shading. The primaries cannot belong to a single parabola, otherwise the virtual absorbers would overlap at the same focus. This problem is avoided if the primaries are displaced horizontally, just enough to accommodate the CPCs. Similarly, for a design with 2x2 CPCs, the outer two primary portions are placed slightly closer toward the sun than the inner two.

To enable the CPCs to illuminate different portions of the tube perimeter, without leaving any gap and without obstructing each other, we use a patchwork of light guides composed of: (a) involutes of tubes; (b) circular segments (involutes of straight lines); and (c) straight parallel segments. Position the CPCs as close as possible to the absorber tube and to each other, each with the correct tilt. The solution tries to minimize the size of the glass tube that will enclose the receiver and to minimize reflective losses in the light guides (Figs 5-7). Each portion of the tube circumference is connected to its CPC by means of light guides composed of the above 3 elements.

For instance, in Fig. 6 the length of $D_1D_1' = C_1C_1'$, and the connection is made with involutes D_1C_1 and $D_1'C_1'$. The CPC for the outer portion, $B_2'B_2C_2C_2'$, is linked to the portion D_2D_2' of the absorber tube by means of a circular light guide from C_2 to S, followed by straight parallel segments from S to R and from C_2' to D_2', and by the involute for the tube from R to D_2.

PRACTICAL CONSIDERATIONS

Placing the secondary reflectors inside an evacuated glass tube permits the use of bare silver. Secondary reflective losses are then only a few percent [Rabl, 1977; Rabl, 1985]. To minimize glass envelope diameter, it should not be concentric with the absorber. Both tube diameter and reflector depth can be reduced by truncating the CPC. Cutting off about 1/2 the CPC sacrifices only 10% in concentration. Truncated aperture $B_{2t}B_{2t}'$ should be parallel to B_2B_2' (and likewise for the other CPC). The primary should be positioned to aim at $B_{2t}B_{2t}'$. Since the upper CPC is larger than the lower one, the glass tube is reduced only by truncation of the former.

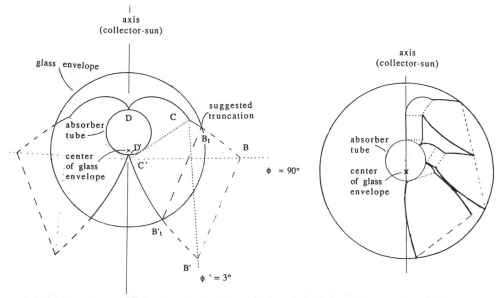

Fig. 5. 2x1 secondary; asymmetric CPCs; $\phi' = 3°$ and $\phi = 90°$. Subscript "$_t$" indicates suggested truncation points. Involute DCD'C' is light guide from CPC to absorber tube. Small dashes indicate extreme rays.

Fig. 7. Secondary of Fig. 6 with upper CPC replaced by 2 CPCs of same shape but 1/2 the size to reduce glass tube diameter (left half not shown) CPCs are truncated. Dotted lines show transition between light guides from CPC to absorber.

Dotted lines show transition between different parts of the light guide from CPC to absorber. For the truncation in Fig. 6, the glass tube diameter is around 6.5 times the diameter of the absorber tube. For example, for a parabolic trough with primary aperture width 5.0 m and absorber tube diameter 6.4 cm, the design of Fig. 6 achieves 2.57 times the concentration of the conventional design. The absorber tube diameter becomes 2.5 cm, requiring a glass tube diameter of 16.3 cm. By replacing the upper CPC by two parallel CPCs of 1/2 the size (Fig. 7), the light guides become more complex and reflection losses increase, but the diameter of the glass tube is reduced to 4.5 times the absorber tube, i.e., 11.3 cm.

To estimate the possible increase in operating temperature for the same heat loss, suppose heat loss is proportional to T^5 (allowing emissivity to increase as T). Doubling concentration would boost T by $2^{0.2} = 1.15$; one could go from 400 C to 500 C. Hence the new design opens the possibility of operating conventional power plants with collectors that need only one-axis tracking. That would represent a dramatic improvement in the conversion efficiencies of such solar power plants.

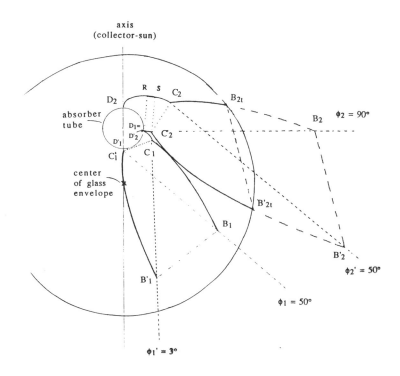

Fig. 6. 2x2 secondary; asymmetric CPCs; total rim angle=90°; mirror image left half not shown.

SUMMARY

We have presented new designs of secondary concentrators for parabolic troughs with tubular absorbers. Total concentration can be increased by a factor of 2-2.5 relative to conventional designs. The extra absorption losses in the secondary amount only to a few percent if silver inside an evacuated tube is used. Compared to the conventional design the secondary requires somewhat larger glass tubes, but with acceptable diameters.

REFERENCES

Kritchman, E. (1981). Applied Optics 20, 3824.
Kritchman, E. (1982). Applied Optics 21, 870-873.
Mills, D.R. and Giutronich, J.E. (1979). Solar Energy 23, 85-87.
Mills, D. R. (1980). Solar Energy 25, 505-509.
Rabl, A. (1977). International Journal of Heat and Mass Transfer 20, 323-330.
Rabl, A, Goodman, N.B. and Winston, R. (1979). Solar Energy 22, 373-381.
Rabl, A. (1985). Active Solar Collectors and Their Applications. Oxford University Press.
Rabl, A. and Winston, R. (1976). Applied Optics 15, 2880-2883.
Welford, W. T. and Winston, R. (1989). High Collection Nonimaging Optics. Academic Press.

SECOND GENERATION OF ALUMINUM FIRST SURFACE MIRRORS
FOR SOLAR ENERGY APPLICATIONS.

R. Almanza*, R. Soriano* and M. Mazari**

*Instituto de Ingeniería and **Instituto de Física.
Universidad Nacional Autónoma de México, Ciudad
Universitaria, 04510, México, D.F., México.
FAX (5) 548-3044

ABSTRACT

A second generation of aluminum first surface mirrors for solar energy applications is in progress. The development of the first generation of aluminum
first surface mirrors was a first step, but as they were built with tungsten
filaments, the aluminum reacted with the tungsten, and some pinholes were observed on the mirrors. When these mirrors were exposed to the environment, some
corrosion was detected and started just in the holes. To eliminate such holes,
an electron gun was built and is being used to evaporate Al and SiO without any
contamination nor pinholes. The behavior of this second generation of mirrors
under a cycling environmental test chamber is being studied. The initial tests
are reported in this paper.

KEYWORDS

First surface mirrors; front surface mirrors; solar materials; aluminum mirrors;
electron gun.

INTRODUCTION

Since the reflectivity of a solar concentrator as well as its mean life plays a
vital role in the performance of the concentrator, the R & D of mirrors are important for thermal applications of solar energy, at high temperatures, mainly.
The thermal energy produced with these systems can be used directly or converted
into electricity.

Different types of solar mirrors are being developed. Second surface mirrors
that use silver as reflecting material are the most common ones employed in parabolic troughs and heliostats; these devices are the most popular concentrators
in large area plants for electricity generation, such as the Luz SEGS (Kearney,
1990; Jaffe, 1987) and the "Solar One" (López, 1988), located in the Mojave
desert in California, USA. However, these mirrors are extremely heavy and
must be sopported by expensive framings. Another problem is that after few
years some corrosion appeared on such mirrors. Furthermore, the glass used is
a special material known as white glass which has a low absorption and a low
content of iron oxide.

The first surface mirrors are an alternative that is being developed. For example, the sol-gel mirrors (Ashley, 1988) are one of them; another one are the
aluminum first surface mirrors presented in this paper.

The following are some of the several advantages that the aluminum first surface mirrors have over second surface mirrors:

1. It is not necessary to have a low absorption glass substrate, and therefore the iron oxide contaminant is not important either.
2. The protection of the aluminum is automatically achieved.
3. A higher reflectance is obtained.
4. A better adherence to the substrate and to the front surface is acquired.
5. Fewer corrosion problems are expected due to point four.

A first generation of aluminum first surface mirrors has been developed (Almanza, 1989) by using the following technique: A glass substrate of 30 x 30 cm was used with a previous chemical cleaning. An oxygen glow discharge of 3 KV and 250 mA occurred inside the vacuum evaporator and lasted 20 minutes. After this cleaning, the aluminum was evaporated through a tungsten filament. In order to protect the aluminum from corrosion and abrasion, a Si_2O_3 film was deposited by means of a reactive evaporation (Hass, 1982). This process was carried out by thermally evaporating SiO with a high pressure of oxygen and low rates of evaporation, $p=10^{-4}$ Torr of oxygen and 4 $\text{Å}/s$, respectively. Using this technique after the aluminum evaporation, some holes were observed. If the same tungsten filament is used several times, the holes increase progressively. This is because the Al reacts with the W (Glang, 1970). These mirrors were exposed to the environment during 600 days, and after one year some corrosion was detected and started just in the pinholes. To eliminate such holes, an electron gun was built in order to evaporate the different film layers of the mirror without any contamination.

DEVELOPMENT OF SECOND GENERATION OF ALUMINUM FIRST SURFACE MIRRORS

An electron gun was built and is being used in the vacuum evaporator, Fig. 1, to manufacture the Al and Si_2O_3 layers that constitute the mirror. This electron gun makes possible to evaporate just Al without any contamination. In order to do that an NB crucible is used. A graphite crucible is utilized for the evaporation of SiO. The purity of the Al was of 99.99%, and the one of the SiO was of 99.89%. An electron beam of about 60 mA was necessary to evaporate the Al, while for the SiO was of about 100 mA. The SiO was transformed in Si_2O_3 by a reactive evaporation in the presence of oxygen and at a pressure of 10^{-4} Torr (Hass, 1982).

According to Hass (1982), a higher specular reflectance can be obtained if the SiO film is changed to Si_2O_3.

The substrate selected was a commercial glass (soda-lime), the only one produced in Mexico. All the substrates tested have been of 10 x 15 cm with a 3 mm thickness.

Four chemical methods were used to clean the substrates: 1) An HF solution at 0.5% at an ambient temperature was used; 2) A mechanical polish with CeO_2; 3) An ultrasonic cleaning method: 20 minutes in a KOH solution, 20 minutes in distilled water and 20 minutes in isopropyl alcohol; 4) The rinsing in a cromic mixture. For methods 1,2 and 4 a final cleaning was made at the end of the process by a rinsing with distilled water. After all these processes the substrates were dried in a vacuum glass container at a temperature of 150°C for about 2 or 3 hours.

Fig. 1. Electron gun used for the manufacture of the mirrors.

TESTS OF THE MIRRORS

Two performance techniques have been selected to study the degradation of the mirrors: Specular reflectance measurements and surface analysis by an optical photomicroscopy. The specular reflectance was measured with a solar spectrum reflectometer (SS) from Devices & Services Co. The main characteristics of this device are: a resolution of 0.001 units, a repeatability of ±0.003 units, and an accuracy of ±2%.

Figures 2a and 2b show the micrographs (100X) of a mirror built with a tungsten filament (Fig 2a) and a mirror built with the electron gun (Fig 2b). Both show one difference. The first one has pinholes and the other one has not.

Fig. 2a. Micrograph (100X) of a mirror manufactured with a
tungsten filament.

Fig. 2b. Micrograph (100X) of a mirror manufactured with
an electron gun.

Two tests have been made in an environmental test chamber for accelerated
weather evaluations. Tables 1 and 2 show the results obtained after different
conditions of humidity, temperature and time. Table 1 shows aluminum mirrors
with different cleaning techniques under 50°C and 80% of relative humidity.
The mirrors were inside the environmental chamber during four hours every day
for one and two weeks. Table 2 shows another set of mirrors under 70°C and
40% of relative humidity conditions. These mirrors also were in the chamber
during four hours every day for one and two weeks.

TABLE 1   Tests of First Set of Mirrors.
Temperature 50°C & relative humidity 80%

Mirror number	Average specular reflectance at 7 milliradiance at the beginning	Specular reflectance after one week	Specular reflectance after two weeks	Cleaning technique
53	.906	.875	.815	ultrasonic
54	.905	.889	.877	ultrasonic
55	.900	.854	.658	cromic mixture
56	.910	.861	.666	cromic mixture

TABLE 2   Tests of a Second Set of Mirrors
Temperature 70°C & relative humidity 40%

Mirror number	Average specular reflectance at 7 milliradians at the beginning	Specular reflectance after one week	Specular reflectance after two weeks	Cleaning Technique
1	.892	.855	.847	cromic mixture
2	.894	.852	.848	cromic mixture
43	.852	.820	.813	ultrasonic
45	.810	.793	.782	HF
46	.802	.780	.779	HF
47	.817	.793	.789	CeO_2
48	.845	.827	.821	CeO_2
51	.849	.845	.843	CeO_2
74	.902	.897	.886	ultrasonic
75	.824	.817	.813	ultrasonic

CONCLUSIONS AND SUGGESTIONS

A this stage of the development, in which it   used an environmental test chamber to evaluate the degradation of aluminum first surface mirrors, it is possible to affirm that the cleaning process of the substrate is an important aspect that must be considered for the mean life of a mirror. Under high humidity conditions, as it is shown in Table 1, the ultrasonic cleaning is the most effective one. In Table 2 it can be seen that the cleaning with HF gave a low reflectivity since the beginning. However, with the other methods under the same conditions does not seem to exist any difference.

When these mirrors were examined under the microscope, some corrosion was detected in the mirrors that were under high humidity conditions (80%).

The next step in this project is to compare the manufacturing of these mirrors using another manufacturing technique, such as sputtering, with a magnetron, to try to compare which technique is more advisable for the manufacturing of aluminum first surface mirrors for solar energy applications.

REFERENCES

Almanza R., F.Muñoz, and M.Mazari (1989). Development of Aluminum First Surface Mirrors for Solar Energy Applications, Clean and Safe Energy Forever Vol.3, Pergamon Press.

Ashley C.S., T.R.Scott, and A. K. Mahoney (1988). Planarization of Metal Substrates for Solar Mirrors, Sandia National Laboratories, CONF-88 04114-4.

Glang R., (1970). Vacuum Evaporation, in Handbook of Thin Film Technology (Maissel L. and Glang R. ed.) McGraw Hill.

Hass G., J.B. Heany, and W.R. Hunter (1982). Reflectance and Preparation of Front Surface Mirrors for Use at Various Angles of Incidence from the Ultraviolet to the Far Infrared, Physics of Thin Film Vol.12, Academic Press.

Jaffe D., S.Friedlander, and D. Kearney (1987). The Luz Solar Electric Generating Systems in California, Advances in Solar Energy Technology Vol.1, Pergamon Press.

Kearney D., D.Jaffe, and L.Daniel (1990). Design Aspects of an 80 MW Solar
  Electric Plant, Proceedings of the 1990 Annual Conference ASES, Austin, Texas.
López C.W. (1988). Solar One-A Success Story, Southern California Edison,
  Fourth International Symposium on Research, Development and Applications of
  Solar Thermal Technology, Santa Fe, New Mexico.

LOW EMISSIVE TiN$_x$O$_y$-Cu-SOLAR SELECTIVE ABSORBERS FOR HIGH TEMPERATURE APPLICATIONS

B. Röhle, M.Lazarov, R.Sizmann

Sektion Physik, Ludwig-Maximilians-Universität München,
Amalienstr. 54, D 8000 München 40, Germany

ABSTRACT

We investigated TiN$_x$O$_y$ films on Cu which show an emittance $\epsilon(250\ °C) \leq 0.05$ when properly prepared. Titanium nitride films were deposited by activated reactive evaporation (ARE) on copper at room temperature and at temperatures up to 400 °C. An analysis shows that inspite of a low oxygen partial pressure $(P_{O_2}/P_{N_2} \leq 10^{-3})$ the coatings contain more than 10% oxygen. Nitrogen partial pressure P_{N_2}, plasma current I_{are} and evaporation rate r_D determine the composition of the coating. This, together with the substrate temperature T_S, influence the selectivity of the coatings. These preparation parameters were investigated to find their influence on the selectivity of the absorber. Best results are achieved when P_{N_2} is above $1 \cdot 10^{-4}$hPa and the plasma current above 0.5A. The reactive gas should contain 1–5% oxygen.

KEYWORDS

selective absorber; TiN$_x$O$_y$; sheet resistance; reactive evaporation, non-concentrating collector; process heat

INTRODUCTION

Nonconcentrating evacuated collectors can in principle provide process heat with temperatures above 250 °C. The efficiency is controlled by the emittance, whereas the absorptance is less important for reaching elevated temperatures (Lazarov, 1990). Although the research on selective coatings in the last two decades has been promising, low emitting coatings are at present not available.

Sputtered zirconium- and titaniumnitride tandem absorbers on silver have been reported by (Blickensderfer, 1977) to exhibit highly selective properties. We prepared TiN$_x$O$_y$ films on copper by activated reactive evaporation (ARE) (Plenk, 1990) with similar results. The ARE process is governed by four parameters: nitrogen partial pressure P_{N_2}, plasma current I_{are}, deposition rate r_D and substrate temperature T_S. As a fifth parameter we found that oxygen as well as oxygen containing gases can improve the optical selectivity.

These parameters show different importance on the selectivity of the coatings. Concerning low cost preparation of large areas of selective absorbers, it is important to know the sensitivity of these process parameters on performance of the coatings. We investigated TiN$_x$O$_y$ films produced under various conditions within a range of each parameter to find the best results.

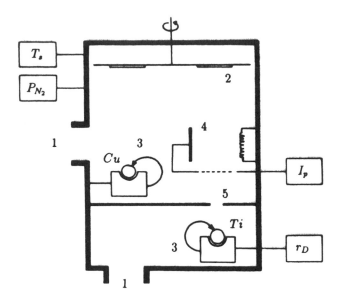

Fig. 1. UHV-chamber:
(1) turbomolecular pumping unit, (2) sample holder, (3) e-gun, (4) catode 15×10 cm^2, at a distance of 15 cm of the anode, (5) 30mm$\oslash$ hole in separator shield.

EXPERIMENTAL

Sample Preparation

TiN$_x$O$_y$-Cu tandem absorbers were prepared in a UHV-chamber shown schematically in Fig. 1. The chamber is separated into two compartments, each being evacuated by a turbomolecular pumping unit. In the upper level, copper is evaporated on glass substrates at pressures of $5 \cdot 10^{-8}$ hPa using a Varian 980-001 e-gun. After deposition of copper the substrate is moved above a 30 mm $\oslash$ hole in the compartment seperating shield. Titanium is then evaporated in the lower level by another e-gun, and deposited onto the substrate through this hole. The upper level is floated with pure nitrogen to attain a working pressure of $2.5 \cdots 20 \cdot 10^{-4}$ hPa. The differential pumping maintains the pressure in the lower level at 10^{-5} hPa to ensure good conditions for the e-gun evaporation. The nitrogen atmosphere is activated by a plasma in the upper level. A tungsten filament supplies electrons which are accelerated to a cathode at $+50 \cdots +250$V. Along their paths to the cathode they ionize part of the nitrogen gas. The plasma current I_p can be measured as the difference between the current after the cathode and the current difference before and after the filament. I_p is then characteristic for the degree of the ionized gas molecules. The activated gas reacts with the titanium and forms a TiN$_x$O$_y$- film on the copper substrates.

The substrate holder is kept at a subtrate temperature T_S between 20 °C and 400 °C. It can hold 10 samples. During deposition the film thickness d_s is monitored with a quartz crystal sensor (XTC Leybold Heraeus).

The pressure was measured with a frictional pressure gauge (SRG-2, MKS Company) ensuring an accuracy of 1% of the working pressure.

Sample Classification

During deposition we measured the following parameters: deposition rate r_D, substrate temperature T_S, plasma current I_p and nitrogen partial pressure P_{N_2}. The DC conductivity of the deposited film was measured employing the symmetrical van der Pauw set up (van der Pauw, 1958). This method allows measuring resistivity of a growing film. The measuring apparatus is described in (Röhle, 1991). For quick optical qualification of the samples the near normal spectral reflectance was measured with Beckman DK 2A Spectrometer for wavelength $\lambda = 0.3 \cdots 2.5 \mu m$ and with a Beckman IR 4260 spectrometer in the wavelength range $\lambda = 2.5 \cdots 50 \mu m$. The error in these measurements is less then 2%. Since the absorber shows low emittancies $(0.03 \cdots 0.08)$, this error leads to very high uncertainties in the emittance. Accurate determination of the hemispherical emittence was performed by calorimetric measurements (Brunotte, 1990), see Fig. 2. The film composition was measured using heavy ion elastic recoil detection (ERD) (Assmann,1991).

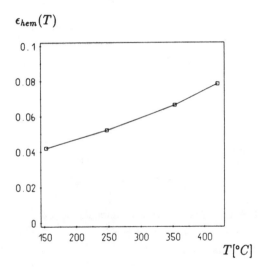 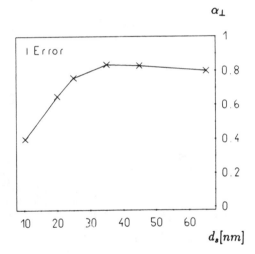

Fig. 2: Calorimetric measurement of the hemispherical emittance of TiN_xO_y-Cu absorber (Brunotte, 1990)

Fig. 3. Near normal solar absorptance versus film thickness.

RESULTS

Influence of the Film Thickness

Infrared emittance and solar absorptance depend on film thickness. To determine the influence of the process parameters on emittance and absorptance for coatings prepared under different conditions it is necessary that the thickness of the coatings be the same. This was possible within 5%. In Fig. 4 we present the near normal spectral reflectance versus wavelength of TiN_xO_y-Cu absorbers of different TiN_xO_y film thicknesses, prepared under identical conditions. The influence of increasing thickness is to shift parallel the cut-off-wavelength towards the infrared, as is seen in a semilogarithmic plot. Simultaneously the IR-reflectance becomes lower which increases the emittance. In the solar wavelength region interference minima and maxima can be observed which change their position with growing thickness. This causes a weak maximum of the absorptance at approximatelly 35 nm, see Fig. 3.

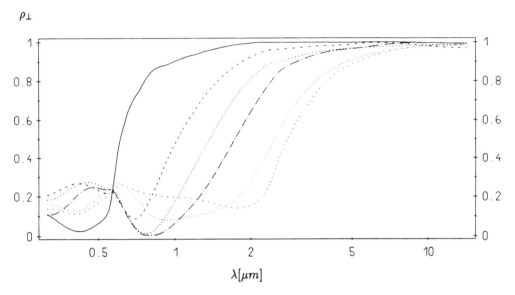

Fig. 4. Near normal reflectance of TiN_xO_y-Cu absorbers versus wavelength for different TiN_xO_y thicknesses : 10 nm (————); 20 nm (——); 25 nm (— —); 35 nm (——— -); 45 nm ($\cdots$) and 65 nm (- - -). All samples were prepared at $P_{N_2} = 9 \cdot 10^{-4}$hPa, $T_S = 100°C$, $I_{are} = 0.5A$ and $r_D = 0.1$nm/s.

Influence of the Pressure and Plasma Current

In addition to the location of the cut-off-wavelength λ_c, the selectivity is determined by the slope of the reflectance at λ_c. Coatings with steep slopes are produced when the working pressure is above $1 \cdot 10^{-4}$ hPa and the plasma current is above 0.5A, see Fig. 5. Under these conditions TiN_x with $x \simeq 1$ is formed developing a fcc-lattice (Sundgren, 1982). A further increase of P_{N_2} or I_p does not change the performance drastically, see Fig. 6.

Furthermore, the slope is influenced by the substrate temperature T_s and the H_2O partial pressure. Preparation of samples with H_2O partial pressure of about $3 \cdots 5 \cdot 10^{-5}$ hPa show a better performance than those prepared under clean conditions, i.e., with $P_{H_2O} < 10^{-5}$

hPa and those with high H_2O partial pressures. This can be seen from Fig. 6, where the near normal α/ε is plotted versus the plasma current for two different H_2O partial pressures. The absorptance α and emittance ε were derived from reflectance measurements at room temperature. The optical quality is influenced by the evaporation rate. Above a deposition rate of 0.3 nm/s higher nitrogen pressures and plasma currents are necessary.

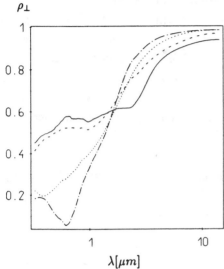

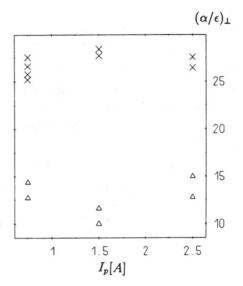

Fig. 5: Near normal reflectance of TiN_xO_y-Cu absorbers versus wavelength for working pressures: $1 \cdot 10^{-7}$hPa (———); $5 \cdot 10^{-5}$ (—); $1 \cdot 10^{-4}$ (— —) and $9 \cdot 10^{-4}$ (— — -). All samples were prepared at $T_S = 100\,°C$, $I_{are} = 0.5A$ and $r_D = 0.1nm/s$.

Fig. 6: Near normal α/ε measured at room temperature versus plasma current (α and ε were derived from reflectance measurements). Above 0.5 A no correlation with the performance can be observed. ($\times$) = samples prepared with P_{H_2O} of $3 \cdot 10^{-5}$ hPa; ($\triangle$) = P_{H_2O} lower than $1 \cdot 10^{-6}$ hPa. The value of α is between 0.65 and 0.75.

Influence of the Substrate Temperature

Best performance was observed with coatings which were prepared at substrate temperatures at $T_s = 80 \cdots 120\,°C$. However, these coatings are not thermally stable. The performance after tempering at 400 °C for 16 hours is not distinguishable to that of samples prepared at $T_s = 250 \cdots 350\,°C$.

DISCUSSION

We examined the process parameters of TiN_xO_y-Cu absorbers prepared with ARE. The cut-off-wavelength can be tuned in the range of $0.6 \cdots 2.0\mu m$ in a simple way by varying the film thickness. For good selectivity a N_2 working pressure of more than $1 \cdot 10^{-4}$ hPa is needed with a H_2O background pressure in the range of $3 \cdots 5 \cdot 10^{-5}$ hPa. The plasma current density should exceed a certain level. During deposition it is preferable to keep the substrate temperature above 150 °C to ensure good IR-reflectivity of the copper layer.

ACKNOWLEDGMENT

We are grateful to Bundesministerium für Forschung und Technologie, Bonn, Federal Republic of Germany, for funding and support.

REFERENCES

Assmann W. (1991). *to be presented at the DPG Meeting 91 in Münster*

Brunotte, A., F. Liebrecht, M. Lazarov and R. Sizmann (1990). *Proceedings of the 7.Intern. Solar Forum 90, Frankfurt*, 2:963–967.

Blickensderfer, R., Deardorff D.K., Lincoln R.L. (1977). *Solar Energy*, 19:429–432.

Plenk, J.,M.Lazarov, F.Liebrecht and R.Sizmann (1990). *Proceedings of the 7.Intern. Solar Forum 90, Frankfurt*, 2:957–962.

Lazarov, M., T.Eisenhammer and R.Sizmann (1990). *Proceedings of the 7.Intern. Solar Forum 90, Frankfurt*, 2:968–973.

Röhle, B. (1991). Master's thesis, LMU München.

Sundgren, J.-E. (1990). *PhD Thesis, Department of Physics, Linköping Institute of Technology, Linköping Sweden.*

van der Pauw L.J. (1958). *Philips Res. Repts.*, 13:1 –9.

INFLUENCE OF THE SUBSTRATE-TEMPERATURE OF
SPUTTERED Au-MgO-FILMS DURING THEIR FABRICATION
ON THEIR SPECTRAL REFLECTANCE AND THERMAL
STABILITY

O. Gutfleisch and L.K. Thomas

Institut für Metallforschung - Metallphysik -
der Technischen Universität Berlin,
Hardenbergstr. 36, 1000 Berlin 12, Germany

ABSTRACT

Solar selective surfaces have been prepared by sputtering Au_{40} $(MgO)_{60}$ cermet films on Ni at various substrate temperatures between 20 and 475 °C. The half-space reflectance was measured with normal incidence of light in the wavelength range $0,37 \lesssim \lambda \lesssim 20$ µm shortly after the preparation and after annealing the samples for 500 h at 400 °C in an Ar-atmosphere. The structure of the films was investigated by Scanning Electron Microscopy. The stability of the spectral reflectance and of the structure of the films is increased by higher substrate temperatures during preparation.

KEYWORDS

Solar selective surface, cermet-film, sputtering, substrate temperature, normal spectral reflectivity, thermal stability, high temperature annealing.

INTRODUCTION

Au-MgO-cermet films on metallic substrates show spectral selective absorption in the range of solar radiation (Fan, 1978; Gittleman, 1977; Mazière-Bezes, 1982; Thomas, 1989). Because of the oxidation resistance of the components, it may be expected that this material is a model substance to be used for photothermal energy conversion at a temperature of 400 °C. This investigation was done to test the influence of the temperature of the metallic substrate during the fabrication of the cermet films on the stability of the spectral reflectance and on the structure after an anneal at 400 °C for 500 hours in an Ar-atmosphere.

EXPERIMENTS

The films were produced by the method of sputtering (Vossen, 1978) with a Perkin-Elmer 3140-6J system (13,56 MHz). The targets were disks with a diameter of 60 mm, the composition was Au_{40} $(MgO)_{60}$. They have been pressed and sintered from the powder of the components by Demetron GmbH, Hanau, Germany. With steady state sputtering conditions the composition of the cermet is the same as the target. This has been tested with these samples by microprobe analysis. The substrates were various Ni-discs (diameter 10 mm, thickness 1 mm), which were positioned symmetrically on a copper block (diameter 65 mm) with an overall plane surface. This block was bolted to the water-cooled substrate table. One Ni-disc in the center was in contact with a thermocouple which was pressed against it on the one side. The other side, in the same way as the other discs, was exposed to the plasma for the deposit of the film. Within the copper block a resistance heating coil was mounted. Because of the electric contacts no bias-sputtering was possible. Different temperatures of the substrate could be adjusted by combined variations of the sputtering power P, the water flow of the cooling system and the current in the heating element. - The Ar-pressure was 4 Pa, the distance between target and substrate was 40 mm. The thickness of the sputtered films was measured with the interference of light on a wedge produced by covering a section of the substrate during the sputtering process. Since the thickness influences strongly the position of the reflectance-edge between 1 and 4 μm, this position itself can be used for the estimation of the thickness. The structure was observed by means of a scanning electron microscope. The measurement of the reflectance was done at room temperature with normal incidence of light. All light reflected symmetrically into a 156° cone was detected in the wavelength range $0,37 \lesssim \lambda \lesssim 20$ μm. The interference pattern of the reflectance curves at smaller wavelengths gave additional information on the thickness.

The heattreatment was done by heating the samples for 500 hours at 400 °C in a sealed quartz-tube filled with Ar, the pressure being about 1 atm.

RESULTS

The temperature of the samples during sputtering was not constant. Within about 45 minutes an equilibrium temperature was reached, see. Fig. 1. Starting from room temperature with increasing sputtering power the temperatures 60, 140 and 220 °C were reached. With different constant currents in the heating coil, the final settings of the temperature were 300°, 400° and 475 °C.

The structure of the cermet being formed consisted of columns with diameters in the range of 100 nm, see Fig. 2. The diameters increase slightly with higher temperatures during the sputtering process. Each column consists of about 30 nm Au-particles in a MgO-matrix. This has been shown earlier (Thomas 1989).

Figure 3 shows the spectral reflectance of samples with the film thickness 0,15 μm, prepared with T_{max} = 300, 400 and 475 °C. Within the experimental scatter the values do not differ. The main edge of the reflectance is between 1 and 2 μm. The effect of the heat treatment on the spectral reflectance is shown in Fig. 4. Here we observed a shift of the main absorption edge to smaller wavelengths. The films made at high temperatures (370 → 475 °C), however, show a smaller shift in comparison with the films made at low temperature (20 → 150 °C). So although the films show no strong difference in reflectance and microstructure when prepared at various tempera-

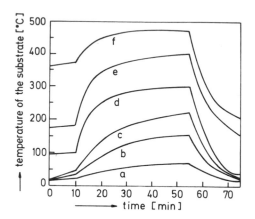

Fig. 1. Temperature change of the substrate during
preparation of the cermet-films. Sputtering
power P: curve a: 200 W, b: 400 W and
c: 600 W (no heating of the sample); P was
constant 250 W with continuous constant heating
of the substrate; temperature before start of
sputtering: curve d: 95 °C, e: 180° C and
f: 370 °C.

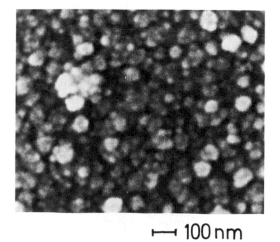

⊢⊣ 100 nm

Fig. 2. Structure of the surface of $Au_{40}(MgO)_{60}$
cermet-films.

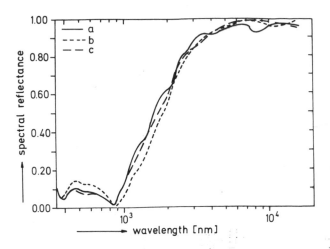

Fig. 3. Spectral reflectance of $Au_{40}(MgO)_{60}$ cermet-
films on Ni with the thickness 0,15 µm
measured after preparation of the samples
without heat treatment P = 250 W, U = 1250 V,
t_{sp} = 45 min. Preparation at various substrate
temperatures: curve a: 100 → 300 °C, b: 180 →
400 °C, c: 370 → 475 °C.

tures, after the heattreatment the films made at higher temperatures show a smaller shift of the absorption edge. In the long wavelength range the reflectance decreases for about 5 %. This has been
the general result with all films, also made with various thicknesses. Figure 5 shows reflectance curves for films with three
different thicknesses.
The shift of the reflectance edge as a function of the temperature
of preparation is shown in Fig. 6.

DISCUSSION

Although sputtering is being used for the production of thin films,
the details of the nucleation and growth processes of the films and
their dependence on the parameters of sputtering are not known in
detail (Mattox, 1989). Our results show that independent of sputtering power and Ar-pressure during sputtering, the temperature of
the substrate is the most decisive parameter for the stability of
the films at higher temperature. However, sputtering power may com-
plement the effect of temperature. High sputtering power, together
with low substrate temperatures, results in stable films in a similar
way as low sputtering power with high temperature. We only measured
the temperature of the substrate but not directly the temperature
of the growing cermet. This surface layer hit by the impinging ions
apparently is influenced by high sputtering power in the same way
as by high temperature. A high temperature during deposition causes
conditions, which, with low temperature, are created later at a high
temperature anneal.

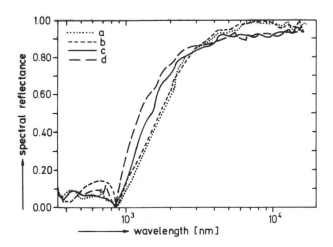

Fig. 4.   Spectral reflectance of $Au_{40}(MgO)_{60}$ cermet-
films on Ni. Film thickness 0,15 μm. Sputtering
parameters: Temperature range 20 → 150 °C:
$P = 500$ W, $U = 1700$ V, $t_{sp} = 30$ min. Temperature
range 370 → 475 °C: $P = 250$ W, $t_{sp} = 45$ min.

$T_{substrate}$	as prepared	after heat treatment 400 °C, 500 h
20 → 150 °C	a	d
370 → 475 °C	b	c

A measure for the stability of the cermet-films is the position of
the main reflectance-edge. A shift of this edge to smaller wavelengths corresponds to smaller thickness values. The high temperature anneal apparently reduces the effective thickness of the
films. The structure of the films, as shown in Fig. 2, contains
voids. The density of the cermet after production at low temperature is likely to be smaller than the equilibrium value. At high
temperatures voids should diffuse to the surface and disappear.
This results in a smaller effective thickness which in turn causes
a shift of the reflection edge to smaller wavelengths. The structure of the cermet after high temperature anneal is about the same
as shown in Fig. 2, but with somewhat larger diameters of the columns forming the structure. By the diffusion of voids and growing
diameter of the columns the density increases and moves closer to
equilibrium. Under equilibrium conditions no variation of the optical properties should be observed. A high temperature during fabrication already results in a structure of the cermet which is
closer to equilibrium.

One problem for the change of the optical properties is the diffusion from the substrate into the cermet film. With Cu as the substrate metal we observed a strong diffusion into the cermet during
the preparation (Thomas, 1989). This was measured by microprobe
analysis. With Ni as the substrate no Ni could be found in the
cermet after the preparation of the sample.

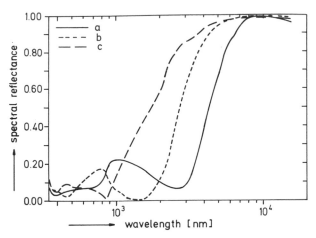

Fig. 5.   Spectral reflectance of $Au_{40}(MgO)_{60}$ cermet
films on Ni with various thicknesses.
a: 0,3 μm (T_{substr} 100   300 °C, t_{sputt} = 90 min)
b: 0,2 μm (T_{substr} 180   400 °C, t_{sputt} = 60 min)
c: 0,15 μm (T_{substr}370   475 °C, t_{sputt} = 45 min)
sputtering power P = 250 W.

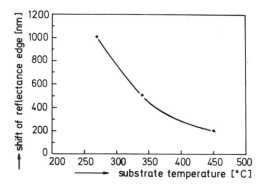

Fig. 6.   Difference of the position of the reflectance
edge before and after heattreatment as a function
of the medium substrate temperature during the
preparation of the films (P = 250 W, t_{sputt} = 45 min)

The roughness of the films is smaller than the wavelength of
the light. So the surface can be considered to be smooth with
respect to optical reflection properties. Therefore for the
calculation of the reflectance a simple three layer model-air,
cermet, Ni-substrate - can be applied. The optical properties
of the cermet can be calculated according to effective medium
theories (Landauer, 1978). Al calculation of this kind according to the model of Bruggemann (1936) has been done with
agreement between measurement and calculation.

CONCLUSION

$Au_{40}(MgO)_{60}$-cermet films on Ni-substrates which have been prepared by sputtering at substrate temperature up to 475 °C do not
change by an anneal at 400 °C for 500 h in an Ar-atmosphere. For
good stability the temperature of the preparation of cermet layers
should be in the same range as the temperature of the planned use.

REFERENCES

Bruggeman, D. A. (1936). Ann. d. Phys., 25, 645-672.
Fan, J. C. C. (1978). Thin Solid Films, 54, 139-148.
Gittleman, J. I., B. Abeles, P. Zanzucchi and Y. Arie, (1977).
  Thin Solid Films, 45, 9-18.
Landauer, R. (1978), in J. C. Garland and D.B. Tanner (Ed.),
  Electrical Transport and Optical Properties of Inhomogeneous
  Media, AIP Conf. Proc. No 40, pp 2-45.
Mattox, D. M. (1989) in E. Broszeit, W. D. Münz, H. Oechsner,
  K.T. Rie and G.K. Wolf (Ed.), Plasma Surface Engineering,Vol. 1,
  DGM Informationsgesellschaft mbH, Oberursel, pp 15-34.
Mazière-Bezes, D. and J. Valignat (1982), Solar Energy Mat., 7,
  203-211.
Thomas, L. K. and Tang Chunhe (1989), Solar Energy Mat., 18,
  117-126.
Vossen, J. L. and J. J. Cuomo (1978), in J. L. Vossen and
  W. Kern (Ed.), Thin Film Processes, Acad. Press, New York,
  Chap. 2, pp 11-73.

PARABOLIC TROUGH OPTICAL EFFICIENCY VERSUS ASSEMBLY TOLERANCES. ANALYTICAL AND EXPERIMENTAL APPROACHES.

J.I.Ajona; E.Zarza

Instituto de Energias Renovables. CIEMAT
Av. Complutense 22. 28040 Madrid. Spain

ABSTRACT

In order to design a thermal solar collector for a given application, it is necessary to search for an optimum compromise between optical quality and heat losses. This paper deals with the optical characteristics of a parabolic trough collector (PTC), as a result of the attainable mirror quality, absorber positioning errors, tracking accuracy and structure twists. We present an experimental approach to characterize the optical behaviour of PTC fields.
To calculate the optical efficiency and the intercept factor as a function of the incident angle, we have developed an analytical tool based on the approach made by Jeter(1987) properly modified to take into consideration not perfectly configured collectors. For evaluating large collector fields it is not feasible to use a laser scanner in order to see whether or not it is necessary to adjust the mirrors, absorbers, etc. To separate the sources of optical errors once you know the optical efficiency, we have prepared a diagnostic procedure based on our analytical approach to calculate the optical efficiency and distant observer techniques, with a video camera and image digitalization.

KEYWORDS

Solar Energy; parabolic trough collector; solar concentrators; optical characteristics; distant-observer techniques

INTRODUCTION

A perfect parabolic trough (PTC) can be defined by its concentration ratio, the acceptance half-angle and the rim angle. Errors in mirror shape, receiver collocation and tracking degradate the attainable performance due to tolerances in manufacturing and installation impossed. as a result of economic constrains. Our concerns are to determine, in the design stage, how the tolerances act on the degradation of the performance in order to fix them in a good compromise, and to prepare a tool for analysing the optical errors present in a PTC field.
In order to study optical errors in PTC, we can sort them into · random and non-random errors. Random errors can be represented by normal distributions and give rise to spreading of the reflected energy distribution. Non-random errors are deterministic and provoke changes in the mean direction of the reflected beam (tracking errors, receiver positioning,...). Examples of random errors are microscopic imperfections in the mirrors, slope errors in the parabola profile (macroscopic errors), and the non-random errors averaged in a large field. As a whole they are characterized by a zero mean and a standard deviation σ_{op}. When the half-width of the sun angular distribution is smaller than σ_{op}, it is possible to consider it as a normal distribution with zero mean and standard

deviation σ_{sun} and to combine the effects of random errors and sun angular distributions in one effective source with zero mean and σ_{tot} as standard deviation (fig. 1)

$$\sigma^2_{tot}(\theta_i) = \sigma^2_{sun}(\theta_i) + \sigma^2_{opt}(\theta_i) \quad (1)$$

where θ_i is the incident angle on the aperture.When we have deterministic errors, one convenient way of dealing with them is to calculate the concentrated flux density distribution on the receiver of the PTC as a first step to calculate its effect on the collector optical efficiency.

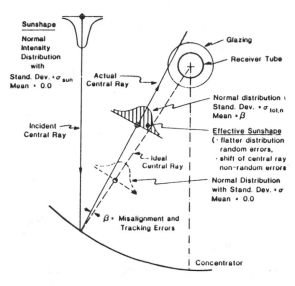

Modelling of potential cal errors in PTC's. (Guven, 1986)

Fig.1.

ENERGY FLUX DENSITY ON THE RECEIVER

Jeter (1987) developed a method to calculate the energy flux density on the receiver of a perfectly configurated PTC without deterministic errors. We have extended his method to take into consideration the effect of various deterministic errors and the influence of the receiver glazing. As a result we can calculate:
- Local concentration ratios on any point on the receiver.
- Optical efficiencies as a function of incident angle.
- Expected observations with distant-observer techniques.

Following Jeter (1987), the energy flux that an infinitesimal element, dA_r, of a concentrating surface reflects onto an element dA of the receiver located at certain point P is given by:

$$dQ = I*d\Omega*cos\theta*dA \quad (2)$$

where:
I = radiant intensity.
$d\Omega$ = differential solid angle, subtended by reflecting element.
$cos\theta$ = incident cosine at receiver.

and the flux density

$$d^2q = \frac{dQ}{dA} = I*d\Omega*cos\theta \quad (3)$$

For PTC the cylindrical symmetry hints that a first integral of eq (3) can be derived. Considering an effective source with normal angular distribution with zero mean and standard deviation $\sigma_t(\theta_i)$ and doing $\varepsilon = 4\sigma_t(\theta_i)$, Jeter (1987) obtained

$$dq = \frac{I_b}{\sigma_{tot}*\sqrt{(2*\pi)}}*F(\psi)*exp\left\{-\frac{\delta^2_o}{2*\sigma^2_t}\right\}*d\psi*cos\theta_o \quad |\delta_o|<\varepsilon' \quad (4.a)$$

$$dq = 0.0 \qquad\qquad\qquad |\delta_o| > \varepsilon' \qquad (4.b)$$

where for a PTC without deterministic errors and with all the random errors included in σ_t, we have:

$$F(\psi) = F_1(\psi) * \cos^2 \Phi_o \qquad F_1(\psi) = \frac{u_{xy} * \cos(\psi/2 - \delta_{xy})}{\rho_{xy} * \cos(\psi/2) * \cos \Phi_o}$$

$$u_{xy} = \frac{f}{\cos^2(\psi/2)} \quad ; \quad \rho_{xy} = \sqrt{R_a^2 + u_{xy}^2 - 2 * u_{xy} * R_a * \cos(\beta - \psi)} \qquad (5)$$

$$\delta_{xy} = a\sin\left\{ \frac{R_a * sen(\beta - \psi)}{\rho_{xy}} \right\} \quad ; \quad \begin{array}{l} \cos(\Theta_o) = \cos \Phi_o * \cos \delta_{xy} * \cos(\beta - \psi) - \\ \quad - \sin \delta_{xy} * \sin(\beta - \psi) \end{array}$$

u_{xy}, ψ = polar coordinates of the parabola.
δ_{xy} = projection on the XY plane of δo.
δo = skewness angle, angle between the reflected rays toward P and towards the focus.
Φ_o = arctg (tg(Θi)/cosδxy, angle between the central ray of the pencil towards P and XY plane.
Θ_o = incident angle of the central ray of the pencil on P.

When the receiver is not centered in the focus we have to replace ρ_{xy} by ρ'_{xy} and δ_{xy} by δ'_{xy}:

$$\rho'_{xy} = \sqrt{\rho^2_{xy} + r^2 + 2 * r * u_{xy} * sen(\psi + \alpha) - 2 * r * R_a * sen(\beta + \alpha)} \qquad (6.a)$$

$$\delta'_{xy} = \psi - \psi' + arcsen\left\{ \frac{R_a * sen(\beta - \psi')}{\rho'_{xy}} \right\} \qquad (6.b)$$

the new position of the receiver center is obtained from.

$$u'_{xy} = \sqrt{(u^2_{xy} + r^2 + 2 * r * u_{xy} * sen(\alpha + \psi)} \quad ; \quad \cos(\psi') = (u_{xy} * \cos \psi + r * sen \alpha)/u'_{xy} \qquad (7)$$

When we have a tracking error ε_s, we obtain dq calculating δo with $\delta''xy$ instead of $\delta'xy$.

$$\delta''_{xy} = \delta'_{xy} - \varepsilon_s \qquad (8)$$

When the mirror halve is rotated an angle τ around the parabola vertex, its effect on dq is the same than that of a collector with the mirrors properly positioned but with the receiver displaced to a point given by the polar coordinates (r', α') and with a tracking error $\varepsilon's$. We consider $\tau > o$ ($\tau < o$) if the mirror end approach (separate) the parabola axis.

$$r' = abs(2 * f * sen(\tau/2))$$

$$\alpha' = \left\{ \begin{array}{ll} \tau/2 & \psi < 0 \ y \ \tau < 0 \\ \pi + \tau/2 & \psi < 0 \ y \ \tau > 0 \\ -\tau/2 & \psi > 0 \ y \ \tau > 0 \\ \pi + \tau/2 & \psi > 0 \ y \ \tau < 0 \end{array} \right. \qquad \varepsilon'_s = \left\{ \begin{array}{ll} \tau & \psi < 0 \\ -\tau & \psi > 0 \end{array} \right. \qquad (9)$$

Combining flexion errors (τ), with receiver displacement and tracking errors :

$$u'_{xy} = \sqrt{(u^2_{xy} + r_p^2 + 2 * r_p * u_{xy} * sen(\alpha_p + \psi)} \quad ; \quad \cos(\psi') = (u_{xy} * \cos \psi + r_p * sen \alpha_p)/u'_{xy}$$

$$\rho'_{xy} = \sqrt{\rho^2_{xy} + r_p^2 + 2 \ast r_p \ast u_{xy} \ast sen(\psi + \alpha_p) - 2 \ast r_p \ast R_a \ast sen(\beta + \alpha_p)} \tag{10}$$

$$\delta'_{xy} = \psi - \psi' + asin[R_a \ast sen(\beta - \psi')/\rho'_{xy}] \; ; \; \delta''_{xy} = \delta'_{xy} - \varepsilon_s - \tau \ast \psi/abs(\psi)$$

where:

$$r_p = \sqrt{r_{px}^2 + r_{py}^2} \; ; \; \alpha_p = atg(r_{py}/r_{px}) \; ; \; r_{px} = r \ast cos\alpha + r' \ast cos\alpha' \; ; \; r_{py} = r \ast sen\alpha + r' \ast sen\alpha'$$

so the elements to calculate **dq** take the form:

$$F_1(\psi) = \frac{u_{xy} \ast cos(\psi/2 - \delta'_{xy})}{\rho'_{xy} \ast cos(\psi/2) \ast cos\Phi_o}$$

$$cos(\theta_o) = cos\Phi_o \ast cos\delta'_{xy} \ast cos(\beta - \psi) - sin\delta'_{xy} \ast sin(\beta - \psi) \tag{12}$$

$$sin\delta_o = cos\theta_i \ast sin\delta''_{xy} \; ; \; sen\Phi_o = \left\{ \frac{tg\theta_i}{\sqrt{(tg^2\theta_i + cos^2\delta''_{xy})}} \right\}$$

δ_o and Φ are both parameters of the incident beam and consequently functions of δ'', so they incorporate the effect of the tracking error. θ_o and $F_1(\psi)$ are parameters of the reflected beam, they do not care the beam origin and do not depend on tracking error, so they are function of δ'.

We also have to take into consideration the mirror reflectivity, ρ_m, the transmisivity of the cover, τ_c, and the absortivity of the receiver coating, α_r. An additional factor to calculate **dq** is the effect of shadowing of the mirror by the receiver. Defining, d_{im}, as the impact parameter of an incident ray, we get

$$d_{im} = u'_{xy} \ast sen(\psi' - \delta') \tag{13.a}$$

$$dq_s = \begin{cases} \rho_m \ast \tau_c(\theta_o, \beta) \ast \alpha_r(\theta_o) \ast dq(\beta, \psi) & dim > R_a \\ 0.0 & dim < R_a \end{cases} \tag{13.b}$$

As a compensating effect the receiver is directly illuminated in its upper part.

LOCAL CONCENTRATION AND OPTICAL EFFICIENCY

The radiant flux may be computed by integration of ec (13) with ψ from $-\Psi_m$ to Ψ_m where Ψm is the rim angle and adding the direct illumination term so

$$q_s(\beta) = \int_{-\Psi_m}^{\Psi_m} dq_s(\beta, \psi) \tag{14}$$

$$q_c(\beta) = \begin{bmatrix} q_s(\beta) & -\pi/2 < \beta' < \pi/2 \\ q_s(\beta) + I_b \ast cos(\theta_i) \ast cos(\varepsilon_s) \ast cos(\pi - \beta) \ast \{\tau\alpha\}(\theta_o, \psi, \beta) & 3 \ast \pi/2 > \beta' > \pi/2 \end{bmatrix}$$

where $\beta' = \beta - \varepsilon_s$. Local concentration ratio on the receiver are obtained with Ib=1. Instead of integrating in ψ as in (14) we can integrate the flux on the receiver, in β, to obtain:

$$q_m(\psi) = \int_0^{2\pi} dqs(\beta, \psi) \tag{15}$$

To calculate radiant flux per unit aperture area we have to divide (15) by u_{xy}:

$$q_a(\psi) = q_m(\psi)/u_{xy}(\psi) \qquad (16)$$

Equation (16) represents the contribution of each element of aperture area to the illumination of the whole receiver. With this new interpretation we can reverse the path of the rays and say that if we put an observer far enough looking at the collector with a certain "incident" angle, he will see the receiver image covering the apertura area. Each element of the aperture will take a colour which depends directly on its contribution to the absorbed energy in the receiver when the sun stays in the same relative position. For a perfect PTC with a black receiver and no cover, a distant observer will see the aperture with a uniform black colour. Optical errors of the collector, non perfectly black receivers and cover transmitances degradates the optical efficiencies and as a result a distant observer will see a grey image of the receiver. The grey level of a particular element on the aperture will be proportional to its local optical efficiency obtaine from Eq (17) taking Ib = 1. This method is a way of predicting analitycally the experimental procedure recomended by Wood (1981) to determine optical errors in PTC fields, as an alternative to the laser scanner (Orear, 1978). It will be the tool we are using to experimentally characterize, by means of a video recorder and image digitalization, the new ACE20 PTC prototype we have just built in collaboration with an Spanish enterprise, Abengoa S.A.

To obtain the collector optical efficiency η_o we have to integrate eq (14) in all receiver area, and divide Q by the maximum available flux Q_M :

$$Q = \int_0^{2\pi} \{q(\beta)*dA_a\} = 2*\pi*R_a^**\langle q\rangle \qquad (17)$$

$$Q_M = I_a*W = Ia*2*\pi*R_a*C$$

$$\eta_o = Q/Q_M = \langle q\rangle/(I_a^**C)$$

where W is the aperture width, C the geometric concentration ratio, Ra the receiver radius $\langle q\rangle$ the averaged flux on the receiver and Ia the available radiation on the aperture.

As representative examples of what can be done with the procedure outlined here, in figure 2 we show ,(for a colector defined by W=2.62m, Ra=21mm, C=19.8, Φ_M = 86.14, without cover, $\rho m=1$, $\alpha r=1$)^M , the local concentration ratio for $\theta_i=0$, as a function of tracking error with $\sigma t=6.2$mrad and without any other deterministic errors. Figure 3 shows the influence of the tracking error (epsi) on the local optical efficiency for the colector defined before but with $\rho_m=0.93$; $\alpha_r=0.93$; cover radius Rc=3.25/3cm (glass), extintion coefficien for the cover=4m-1, and refraction index = 1.52, without receiver or mirror displacement (r=0,$\alpha=0,\tau=0$).
In Figure 4, we have varied the receiver displacement (r=1,2,3,4cm, $\alpha=45$, $\tau=0$, $\varepsilon_s=0$) and in figure 5 the

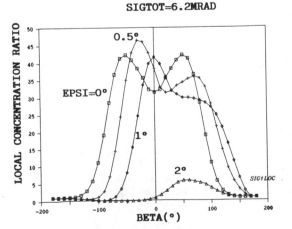

Fig.2.   Local concentration ratio

mirror flexion (r=0cm, ε_s=0, $\tau_{left}=\tau_{right}$=0.25,0.5,1). Figure 6 combine receiver displacement with tracking errors.

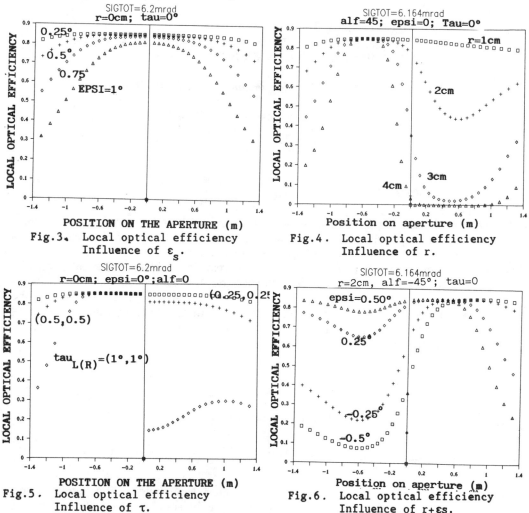

Fig.3. Local optical efficiency Influence of ε_s.

Fig.4. Local optical efficiency Influence of r.

Fig.5. Local optical efficiency Influence of τ.

Fig.6. Local optical efficiency Influence of r+εs.

EXPERIMENTAL DETERMINATION OF OPTICAL ERRORS

In order to apply this analytical tool for analysing the errors that occur in a PTC, we have to compare the measured local optical efficiency obtained by digitalization of a video image of the collector aperture with the expected results as a function of the set of optical errors. After some preliminary tests, we are quite confident of the aplicability of our method for large PTC fields.

References:
Jeter S.M. (1987). Analytical Determination of the Optical Performance of Practical Parabolic Trough Collectors from Design Data. Solar Energy, 1, 11-21.
Orear L. (1978). Sensitivity of Slope Measurements on Parabolic Solar Mirrors to Positioning and Alignement of the Laser Scanner. SAND 78-0700.
Wood R.L. (1981). Distant-Observer Techniques for Verification of Solar Concentrator Optical Geometry. UCRL-53220
Guven H.M. (1986). Determination of error tolerances for the optical design of parabolic troughs for developing countries. Solar Energy, 36, 535-550.

APPLICATION EXPERIENCE AND FIELD PERFORMANCE OF SILVERED POLYMER REFLECTORS[*]

Paul Schissel, Gary Jorgensen, and Roland Pitts

Solar Energy Research Institute[**]
1617 Cole Boulevard
Golden, Colorado 80401

ABSTRACT

The solar-weighted hemispherical reflectance of unweathered silvered acrylic mirrors exceeds 92%, and specular reflectance into a 4-milliradian, full-cone, acceptance angle is greater than 90%. Comparison of outdoor and accelerated tests suggests that the protected silver can resist corrosion for the five-year life that is the current goal. An installation of parabolic troughs has been cleaned monthly for two years, and reflectance is regularly returned to within a few percent of the initial reflectance values. In the presence of moisture the silver/acrylic bond can delaminate to form a maze of tunnels and destroy specular reflectance. Proper edge preparation and protection delays the initiation of tunnels.

KEYWORDS

Silvered polymer mirrors, weathering, durability.

INTRODUCTION

For most applications silver is the reflective material of choice. The hemispherical reflectance of freshly deposited silver weighted over the solar spectrum (0.3–3.0 µm) is greater than 97%. A transparent layer is required to protect the silver from abrasion, soiling, and corrosion. An acrylic polymer with ultraviolet (UV) absorbers (to inhibit UV-photon-activated degradation) can be used. The solar-weighted hemispherical reflectance of new, unweathered silvered acrylic material exceeds 92%.

The composite mirror is shown schematically in cross section in Fig. 1. The performance goals for silvered polymer films are a five-year life with a specular reflectance greater than 90% into a 4-mrad, full-cone, acceptance angle. The optical goals for unweathered mirrors have been met, and current emphasis is on durability in the environment (Susemihl and Schissel, 1987).

---

[*]This work was supported by the U.S. Department of Energy (DOE) under Contract No. DE-AC02-83CH10093.
[**]A U.S. DOE facility.

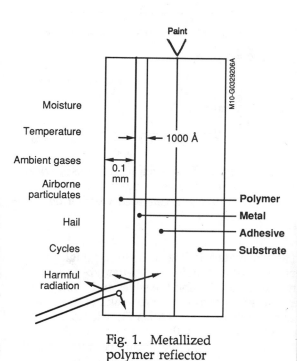

Moisture

Temperature

Ambient gases

Airborne
particulates

Hail

Cycles

Harmful
radiation

Paint

1000 Å

0.1
mm

M10-G0329206A

Polymer

Metal

Adhesive

Substrate

Fig. 1. Metallized
polymer reflector
construction

Fig. 2. Solar-weighted hemispherical
reflectance vs. months of outdoor
exposure, silvered polymer mounted
on aluminum

Fig. 3. Parabolic troughs, Industrial
Solar Technology, Denver

Experimental mirrors are tested in accelerated weathering tests. Laboratory results have
led to a series of production materials called ECP 300 followed by ECP 300A and
ECP 305 from the 3M Company. The production materials also are undergoing
accelerated weathering tests and outdoor tests near Denver, Albuquerque, Miami,
Phoenix, and other sites.

Silvered acrylic mirrors lose reflectance principally in the following ways: (1) trace
impurities in the mirror materials and/or the atmosphere can cause corrosion of the
silver; (2) soil coats the mirror surface; and (3) delamination at the polymer/silver
interface can decrease specular reflectance.

SILVER CORROSION

Outdoor performance is site dependent. Denver's climate is relatively mild, and test
samples continue to maintain reflectance after years of exposure (Fig. 2). In Fig. 2 we
have plotted hemispherical rather than specular reflectance because it is more convenient
to measure, and results show that the two types of reflectance degrade similarly because
of corrosion near the polymer/silver interface. Approximately 700 m² of ECP 300A were
installed four years ago by Industrial Solar Technology (IST) on parabolic troughs near
Denver. The film has resisted corrosion of the silver and has withstood monthly
noncontact spray cleaning that regularly recovers reflectance above 90% (Fig. 3). Other
sites like Phoenix are harsher (Fig. 2).

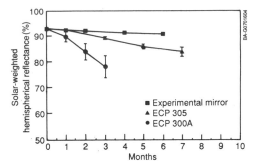

Fig. 4. Solar-weighted hemispherical reflectance vs. months in the Weather-Ometer accelerated weathering chamber

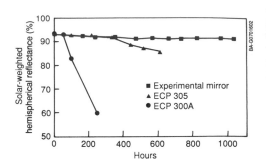

Fig. 5. Solar-weighted hemispherical reflectance vs. hours in the solar-simulator accelerated weathering chamber

We have investigated improved silvered acrylic materials (Fig. 1) including high-purity polymers, more effective UV stabilizers, and coated polymers better able to resist scratches and to reject soil. Interlayers in front of and behind the silver and paint layers on aluminum or stainless steel substrates can retard corrosion. We evaluate new materials in accelerated laboratory tests in our Weather-Ometer (Atlas Electric Co.). In the process, samples are illuminated with a xenon arc lamp with filters to match the solar spectrum, and are maintained at 60°C in air at 80% relative humidity (Schissel and Neidlinger, 1987). Figure 4 compares the durability of reflectance in the Weather-Ometer for three sample types. ECP 300A, an earlier production material from the 3M Company, is now superseded by ECP 305. An experimental mirror that was fabricated at the Solar Energy Research Institute (SERI) is also shown in Fig. 4.

The experimental mirrors are now so stable that we also use a more accelerated test. The collimated beam from a solar simulator concentrates the near-UV radiation to more than ten suns while the samples are maintained at 80°C in air at 75% relative humidity. Figure 5 again compares the three types of mirrors and demonstrates the better performance of the newer materials. In Fig. 5 the corrosion of ECP 300A and ECP 305 has been slowed because the mirrors are mounted on painted aluminum substrates, which are known to provide better durability than unpainted aluminum for these mirror types. The data of Figs. 2, 4, and 5 suggest that the newer films can resist corrosion of the silver to meet the durability goal of five years.

SOILING AND CLEANING

Noncontact cleaning using only a medium-pressure spray of deionized water is effective, according to tests at Industrial Solar Technology (1989) (Fig. 3). The tests, conducted in Brighton, Colorado, cleaned 560 m^2 of parabolic troughs that use the silvered polymer mirror type ECP 300A. The specular reflectance was measured periodically, including immediately before and after noncontact cleaning, using a Devices and Services Model 14R portable reflectometer. Figure 6 shows the specular reflectance immediately after washing at 650 nm wavelength and at a larger full-cone acceptance angle (25 mrad) because of the curvature of the troughs. The troughs had been operating for about four months when the monthly cleaning test began. As Fig. 6 shows, the specular reflectance is regularly returned to about 93% after cleaning.

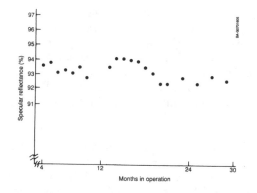

Fig. 6. Average specular reflectance at 650-nm wavelength and 25-milliradian acceptance angle immediately after noncontact spray cleaning of a 560-m^2 installation of parabolic troughs successively over a two-year period. Reflective film ECP 300A.

Between the monthly washings reflectance fell at rates between 0.1% and 0.5% per day depending on weather conditions, with an average rate of 0.26% per day. Unweathered mirrors have a specular reflectance of about 97% at 650 nm, so that a reflectance loss of about 4% occurs that is caused by tenaciously held soil and is not retrievable by noncontact cleaning. Contact (abrasive) cleaning returns reflectance nearly to initial values, but soiling rates increase so that the average reflectance is not improved much during longer-term tests.

The study by Industrial Solar Technology also indicated the following preliminary conclusions regarding soiling and cleaning: (1) Detergents or higher-pressure sprays do not improve cleaning. (2) Stowing troughs face down minimizes soil accumulation. Soiling depends upon collector position; test samples that are held in a fixed position are of little value for predicting performance of operating systems. (3) The 560-m^2 field could be washed in two hours using 120 gallons of water.

Laboratory and outdoor studies have identified factors such as hardness, smoothness, hydrophobicity, and low surface energy that may influence soiling (Cuddihy and Willis, 1984). Hardcoats do make silvered polymer mirrors more resistant to scratching. Temperature-cured hardcoats also weather well, but low curing temperatures that are limited by the polymer slow the curing and may increase costs. UV-cured hardcoats are equally scratch resistant. They cure rapidly but they have not weathered well.

Studies have also indicated that detergents or surface coatings (e.g., fluorine-containing layers) improved the performance of photovoltaic modules (Cuddihy and Willis, 1984). Fluorine-containing molecular monolayers can be stable and can alter the performance of mirrors (Shutt and others, 1989). Other research (Baum, Cross, and Hilyard, 1990) also finds that surface coatings can improve the performance of mirrors, but the effects are lost after continued cleaning cycles. An acceptable coating that resists soiling or allows better maintenance of reflectance during cleaning cycles has not yet been identified.

TUNNELING

ECP 305 mirrors can have a sporadic failure mode termed "tunneling," which usually occurs when the mirror is exposed to high humidity. During tunneling the polymer separates from the silver in a characteristic pattern. The tunnels are usually about 1 in. wide and are separated by about 3 in. A maze of tunnels can meander over the complete mirror surface if the tunnels are not repaired as they initiate. If water is allowed to puddle on a mirror, the mirror might fail within a few days. However, a parabolic trough installation in Denver by Industrial Solar Technology has not had any tunneling problems after more than one year of operation. Tunneling is believed to begin when stresses induced at the silver/polymer interface by differential thermal and hygroscopic expansion overcome the weak adhesion between the polymer and silver. The adhesion between the polymer and silver is weak initially and is further weakened upon exposure to moisture.

We use two laboratory tests to evaluate tunneling. The first exposes mirrors to moisture by immersing them in a bath of tap water at room temperature. The second procedure is a cyclic test that alternates the mirrors from a tap-water bath (23°C) to a dry oven (60°C). The water bath tests have identified the following variables that influence tunneling.

Substrates

Experiments show that tunneling occurs more readily when ECP 305 is mounted on aluminum or stainless steel substrates rather than on painted aluminum or glass substrates.

Edge Preparation and Protection

Tunneling virtually always initiates at edges. How the edges are cut affects tunneling. Microscopic examination of edges reveals cut-induced flaws in the brittle polymer which depend on the cutting method. Razor cuts visually are poor and they perform poorly in the water bath. A heated knife or a laser beam that melts through the polymer yields better edges, probably because local melting anneals the flaws. After cutting, the edges are protected. The standard procedure recommended by the 3M Company is to tape the edges with an aluminized polymer tape. We are investigating other more effective edge protection methods such as Tedlar tapes.

Adhesion

If the adhesion between the polymer and silver can be significantly increased, tunneling could be avoided not only from edges that are formed during fabrication, but also from damage (hail, vandalism) that occurs internal to edges. We are investigating optically clear adhesive layers, which, preliminary experiments show, delay tunneling when the mirrors are exposed to the water baths.

The effects of some of these variables are shown in Fig. 7. In Fig. 7 the percent of surviving samples is plotted versus days in the water bath for three constructions. Razor-cut samples that are not protected by edge tape tunnel very quickly. Samples cut with a heated knife and not edge taped resist tunneling longer but finally fail. When razor-cut samples are edge taped with a Tedlar tape, there is significant improvement. Other samples were cut with a heated knife and edge taped with Tedlar tape and are totally intact after 120 days in water.

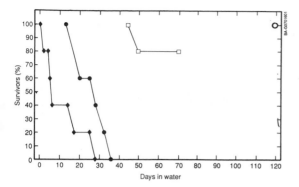

Fig. 7. Percent of intact samples vs. days in a water bath for razor-cut samples with no edge tape (◆), for samples cut with a heated knife and without edge tape (●), and for razor-cut samples that are edge taped with Tedlar tape (□). Samples (○) cut with a heated knife and edge taped with Tedlar tape are all intact after 120 days in water.

CONCLUSIONS

Silvered acrylic mirrors such as ECP 305 can have a high specular reflectance and, based upon accelerated laboratory tests, can plausibly be expected to maintain optical performance by resisting silver corrosion for the five-year life that is the current goal. Silvered acrylic mirrors that have been cleaned monthly for two years have maintained reflectance within a few percent of initial values. In the presence of moisture the silver/acrylic bond is weak and can delaminate. Delamination is delayed or prevented when the edges are properly formed and protected or when interlayer adhesion is enhanced as exhibited by some new laboratory samples.

ACKNOWLEDGMENTS

The authors thank B. A. Benson of the 3M Company for supplying test materials and weathering data. Industrial Solar Technology kindly supplied results of their cleaning studies. We thank Cheryl Kennedy and Yvonne Shinton of SERI for sample preparations and optical measurements.

REFERENCES

Baum, B., S. Cross, and G. Hilyard (1990). Protective Treatments for Membrane Heliostat Mirrors. Springborn Laboratories, Enfield, Connecticut. Prepared for the Solar Energy Research Institute, Contract No. HX-0-19028-1.

Cuddihy, E. F., and P. B. Willis (1984). Antisoiling Technology: Theories of Surface Soiling and Performance of Antisoiling Surface Coatings. Jet Propulsion Laboratory, Pasadena, California. DOE/JPL-1012-102 (DE 85006658).

Industrial Solar Technology (1989). Soiling Rates and Cleaning Techniques for ECP 300, Final Report. Prepared for the Solar Energy Research Institute, Contract No. HL-9-19017-1.

Schissel, P., and H. Neidlinger (1987). Polymer Reflector Research During FY 1986. Solar Energy Research Institute, Golden, Colorado. SERI/PR-255-3057.

Shutt, J. D., D. J. O'Neil, P. M. Hawley, D. K. Dawson, and J. A. Kearns (1989). Antisoiling Solar Reflector Research. Georgia Institute of Technology, Atlanta, Georgia. Prepared for the Solar Energy Research Institute, Contract No. XX-7-07029-1.

Susemihl, I., and P. Schissel (1987). Specular reflectance properties of silvered polymer materials, Solar Energy Materials, 16, 403.

2.24 Detoxification and Materials

Solar Power Station with Thermochemical Storage System

J. Kleinwächter*, M. Mitzel*, M. Wierse**, R. Werner**,
M. Groll**, B. Bogdanovic***, B. Spliethoff***, A. Ritter****

   * Bomin Solar GmbH & Co.KG, Lörrach
  ** Institut für Kerntechnik und Energiewandlung e.V. Stuttgart
 *** Max-Planck-Institut für Kohlenforschung, Mülheim a.d. Ruhr
**** Max-Planck-Institut für Strahlenchemie, Mülheim a.d. Ruhr

Parabolic concentrators with Stirling generators have been identified as a very promising option to generate electricity in
regions without public grid. A thermal heat store guarantees
constant power output during periods of cloud covering and after
sunset. This heat store is based on the reversible chemical
reaction of active magnesium hydride. The system will be operated between 300°C and 480°C. The first two lab models will contain 20 kg of magnesium powder with a storage capacity of about
12 kWh.

Introduction

Thermal energy can be stored as sensible heat, latent heat or in
a thermochemical way. Thermochemical heat storage is based on a
reversible chemical reaction. Heat dissociates a compound into
its components. Their recombination releases the same quantity
of heat again. The components of a solid-gas-reaction can be
easily separated. Metal hydride / metal-systems meet these requirements. A very high hydrogen storage capacity (theoretically
8.3 wt.%, referred to the magnesium) at a high temperature level
(about 300°C to 500°C) can be achieved with the magnesium-hydri-
de/magnesium (MgH_2/Mg) - system / 1 /.

Since early 1989, Bomin Solar (Lörrach), the Max-Planck-Institu-
tes für Kohlenforschung and für Strahlenchemie (Mülheim/Ruhr)
and Institute für Kerntechnik & Energiewandlung e.V. (Stuttgart)
are cooperatively developing a solar energy station with thermochemical store, under contract to the German Ministery for Research and Technology (BMFT).

Functional Description

The solar power station is schematically shown in **Fig. 1.** The
solar radiation is concentrated by a Fix-Focus-concentrator
(**Fig. 2**). The excentric sector of a paraboloid of revolution
produces a stationary fixed focus / 2 /. For that reason it is

possible to place the whole energy conversion unit (receiver, thermal energy store, Stirling engine, generator) fixed on the ground, thus its weight is of secondary importance. It is not neccessary to track it with the mirror. The concentrator surface itself is divided in six segments. Caused by the stretched-membrane-construction a thin aluminium covered fluorpolymer-film is used as a reflector. The film is pneumatically formed into the accurate shape. In this way a light-weight concentrator can be built.

A cavity receiver, located in the focal plane in front of the collector, minimizes the thermal losses. For the heat transfer from the receiver to the Stirling engine and the MgH_2/Mg - store the heat pipe technology was chosen. Thereby a high heat transfer coefficient and a very small temperature difference between the heat source and the heat sink can be realized. Heat pipes also tolerate the varying heat flux at the receiver surface. That is the reason for low thermal stresses and long life of the system.

During daytime the thermochemical store is charged, in parallel to the electricity generation. Therefore a hermetically sealed Stirling generator unit is used. With the other part of the solar energy the MgH_2 is dissociated

$$MgH_2 + 75 \text{ kJ / mol } H_2 \quad --> \quad Mg + H_2.$$

The released hydrogen gas can be stored in a pressure vessel; it can also be stored as a hydride in an appropriate low temperature metal alloy.

During periods of cloud covering and at night time the hydrogen of the secondary store recombine with the magnesium releasing the same amount of reaction heat.

$$Mg + H_2 \quad --> \quad MgH_2 - 75 \text{ kJ / mol} H_2$$

The discharge temperature depends on the pressure level of the hydrogen store. The characteristics of the two above-mentioned variants are shown in Fig. 3 and Fig. 4 / 3 /.

During the desorption of hydrogen from the low temperature metal hydride useful cold can be produced. As it is shown in Fig. 4 the temperature in the MgH_2/Mg store is at about 320°C while the temperature of the low temperature metal hydride is at about - 10 °C. (Fig. 4 shows the reaction bed temperature.)

Design of the System   / 3 /

The receiver / storage unit with the integrated Stirling heater head is shown in Fig. 5. The main component is the large cylindrical primary heat pipe. It is about 1 m long and 0,5 m wide. In the lower part 14 storage containers filled with MgH_2 are mounted. Alltogether the store will be filled with 20 kg of the nickel-doped magnesium powder. The corresponding storage capacity is expected to be about 12 to 15 kWh.
Concerning the safety, the fabrication and the mounting the containers are inserted in jacket tubes. The cavity receiver is welded to the side of the heat pipe. Caused by the melting point of the storage material the maximum temperature is limited to about 480°C. The corresponding decomposition pressure of MgH_2 is to 8.8 MPa / 1 /.

A smaller secondary heat pipe above the storage bottles contains the Stirling heater head. Thus the large heat pipe is protected against damages of the Stirling engine. Potassium is suited for the heat pipe working fluid at this temperature range. Each of the two heat pipes is designed to transfer a maximum thermal load of 6 kW. The aperture area of the used Fix-Focus-concentrator amounts to 6,5 m². This results in a receiver thermal input of 5,5 kW, assuming an optical efficiency of 85 % and a solar radiation of 1 kW / m². The maximum electric power output of the Stirling generator is expected to about 1 kW. On the one hand the efficiency of the solar power station depends on the insolation, on the other hand on the operating mode and the working conditions. The expected efficiency is about 12 % for electricity generation without thermal energy storage. Including the storage the efficiency amounts to about 15%.

References

/ 1 / Bogdanovic, B; Ritter, A; Spliethoff, B;

Active MgH$_2$ - Mg - Systems for Reversible Chemical Energy Storage, Angew. Chem. Int. Ed.Engl. 29 (1990), 223

/ 2 / Kleinwächter, J;

Paraboloidsysteme, Seminar: solarthermische Stromerzeugung 3. und 4. September 1987, Lampoldshausen.

/ 3 / Groll, M; Wierse, M; Werner, R;

Magnesium-Hydride For Thermal Energy Storage and Small Scale Solar-Thermal Power Station, Int. Symposium on Metal-Hydrogen Systems, Bauff, Alberta, Canada, September 2nd - 7th, 1990.

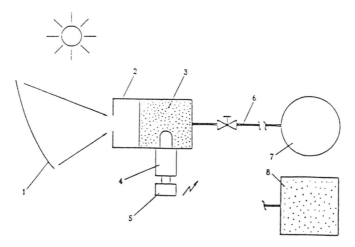

1 : Fix-focus concentrator	5 : Electricity generator
2 : Solar receiver	6 : H$_2$-piping
3 : MgH$_2$/Mg-store	7 : H$_2$-pressure vessel store (variant 1)
4 : Stirling engine	8 : Low temperature metal hydride store (variant 2)

Fig.1: Schematic of the planed solar power station with thermal energy store /3/

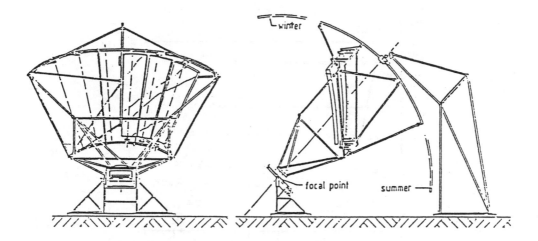

Fig.2: Fix-focus concentrator (Bomin Solar)

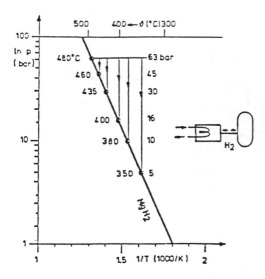

Fig.3: Characteristics of the combination MgH$_2$-store and pressure vessel /3/

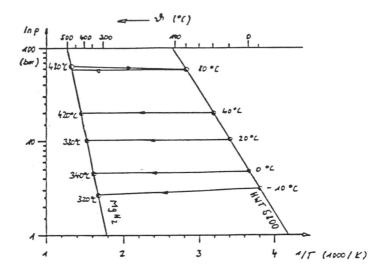

Fig.4: Characteristics of the combination MgH₂/Mg-store and a low temperature metal alloy/hydride-store /3/

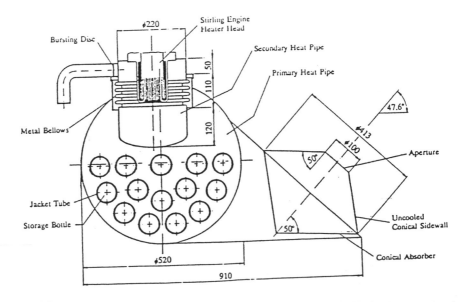

Fig.5: Receiver / storage unit /3/

SOLAR THERMAL DECOMPOSITION OF $CaCO_3$ ON AN ATMOSPHERIC-OPEN CYCLONE REACTOR

A. Imhof, C. Suter and A. Steinfeld[1]

Paul Scherrer Institute, CH-5232 Villigen-PSI, Switzerland

ABSTRACT

The solar thermal decomposition of $Ca\,CO_3$ is thermodynamically examined. A new scheme for making a high-temperature solar receiver-reactor to conduct such process is proposed. It consists of a cyclone gas-particle separator that has been modified to let concentrated solar radiation enter the cavity through a windowless (atmospheric-open) aperture. A prototype was fabricated and its heat transfer characteristics were investigated using a 15 kW electric heating source. The new design offers high energy transport efficiency and permits continuous feeding of reactants and separation/removal of products.

KEYWORDS

Solar reactor, solar receiver, solar fuels, solar energy storage, cyclone reactor, calcium carbonate, calcium oxide.

INTRODUCTION

The production of calcium oxide by thermal decomposition of calcium carbonate using concentrated solar energy is an attractive candidate for solar thermochemical conversion, in which concentrated solar radiation is used as an energy source of high temperature process heat to drive this highly endothermic reaction. Previous studies on the decarbonation of calcite using solar energy (Badie and co-workers, 1980, Salman and Kraishi, 1988) have shown that the process is technically feasible and high degree of chemical conversion can be achieved. Preliminary experimentation at our solar furnace, in which samples of $Ca\,CO_3$-powders were exposed to direct high solar fluxes, resulted in complete decomposition to CaO and CO_2 (Durisch, 1990). Various solar reactor types, among them rotary kiln and fluidized beds, have been tested, and thermal efficiencies of about 15% were measured (Flamant and co-workers, 1980). Of particular interest is the cyclone gas-particle separator, which has been suggested as a solar receiver-reactor to conduct gas-solid heterogeneous chemical reactions (Villermau, 1979, Jacobs and Gunther, 1975). The cyclone configuration provides a practical technique for continuous feeding of reactants and *in-situ* separation and removal of products. In a previous paper (Imhof, 1990), a new scheme for making an atmospheric-open cyclone reactor for the calcite decarbonation process has been proposed. In the present work we examine how to effect such process thermodynamically, and we describe in detail the fabrication and testing of a prototype solar receiver-reactor.

---

[1] To whom correspondence should be addressed

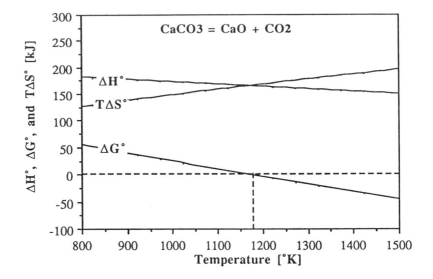

Figure 1: Variation of $\triangle H°$, $\triangle G°$ and $T \triangle S°$ with temperature for reaction [1].

THERMODYNAMIC ANALYSIS

The reaction of interest is characterized by the equation:

$$Ca\,CO_3 = CaO + CO_2 \tag{1}$$

In order to drive this reaction at a given temperature and pressure, the amount of energy that must be added to the system is the enthalpy change $\triangle H°$ of the reaction. Application of the second law of thermodynamics to determine whether a given reaction can possibly occur is straightforward. We check to see whether the entropy generation is positive (or if the Gibbs free energy change $\triangle G°$ is negative), and if so, the reaction is thermodynamically favorable. At room temperature,

$$\triangle S_{universe} = \frac{-\triangle G°}{T} = -0.45\,kJ/K < 0$$

which means, according to the 2nd. law, that reaction [1] cannot occur spontaneously. Heat addition alone is not sufficient to effect the transformation. An amount of $\triangle G°$ of *free* energy is required, for example in the form of electrical work. The total energy requirement, however, is at room temperature:

$$\triangle H° = \triangle G° + T \triangle S° = 178\,kJ$$

where $T\triangle S°$ is the amount of energy that must be supplied as process heat for the completely reversible reaction. To decompose $Ca\,CO_3$ into CaO and CO_2 without the need for electrical work, the system should be brought to a temperature at which $\triangle G° \le 0$ in order to make the reaction [1], as written, go spontaneously to the right. The variation of $\triangle H°$, $\triangle G°$ and $T\triangle S°$ with temperature at 1 atm is shown in Fig. 1, for the temperature range 800 K-1500 K (Data taken from Janaf Thermochemical Tables, 1985, J. Phys. Chem. Ref. Data, 1982, Smyth and Adams, 1923). The slope of these curves are:

$$\left(\frac{\partial\triangle H}{\partial T}\right)_p = \triangle Cp \quad ; \quad \left(\frac{\partial\triangle G}{\partial T}\right)_p = -\triangle S$$

The $\triangle Cp$, the difference between specific heat capacities of products and reactants, is small.

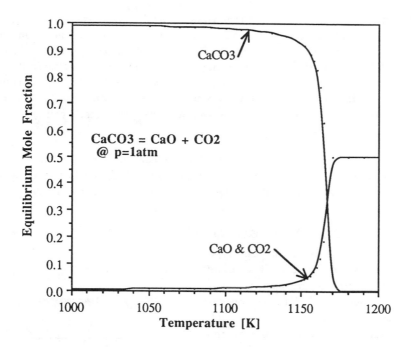

Figure 2: Equilibrium composition of $CaCO_3$ at 1 atm.

But the variation of ΔG with temperature depends on ΔS of the transformation, which is a large positive number. It can be observed in Fig. 1 that $\Delta H°$ is almost constant with temperature, while $\Delta G°$ decreases. Therefore, as the temperature is raised, the fraction of electric energy required to decompose $CaCO_3$ decreases. For example, at 800 K and 1 atm:

$$\Delta H°(800\,K) = 182\,kJ \quad ; \quad \Delta G°(800\,K) = 55\,kJ$$

Thus the production of 1 gram-mole of CaO from $CaCO_3$ would ideally require at 800 K and atmospheric pressure the input of a total of $182\,kJ$ of energy, $55\,kJ$ to be in the form of electric energy. At the same time the system would take up $127\,kJ$ of process heat from the surroundings. At 1170 K and 1 atm, $\Delta G° = 0$. At temperatures above 1170 K, if we simply permit the reaction proceed without doing any useful work, we would have to supply the $\Delta H°$ of the reaction. The equilibrium composition of the system at 1 atm was calculated using *FLAME* computer code (Gordon and McBride, 1985) and is plotted in Fig. 2 as a function of temperature for the range 1000 K-1200 K. It can be seen that $CaCO_3$ starts to thermally decompose at around 1100 K. At 1180 K the only species in equilibrium are CaO and $CO_2(g)$. If we were to conduct the reaction in the presence of air, which for this specific reaction acts as an inert gas, then the decomposition of $CaCO_3$ will take effect at a lower temperature. The reason for that is explained in the following paragraphs. Since CO_2 is the only gaseous species that participates in the reaction, then the equilibrium constant, k_p, is equal to the partial pressure of CO_2, given by:

$$k_p = p_{CO_2} = \frac{n_{CO_2}}{n_{CO_2} + n_{air}} p$$

where n_{CO_2} and n_{air} stand for number of moles of CO_2 and air respectively, p_{CO_2} for CO_2 partial pressure, and p for total pressure in atmospheres. As we add more air and keep the temperature and total pressure constant (the volume certainly increases), then n_{air} increases. But since k_p is constant, n_{CO_2} should also increase and the equilibrium is displaced in such way to keep k_p constant. That is, the equilibrium of equation [1] is displaced to the right. For example, if $CaCO_3$-particles are fed using air as carrier gas, same mass flow rate for reactants and air (mole fraction: $x_{CaCO_3} \approx 0.23, x_{air} \approx 0.77$), then the equilibrium curves of Fig. 2 are shifted in such way that in the abscissa the temperatures are about 100 K less. The situation can be further improved if we work at lower pressures, recalling:

$$\Delta G(T, p) = \Delta G^{\circ}(T) + RT \ln p_{CO_2};$$

where R is the gas constant, and p_{CO_2} the partial pressure of CO_2 in atmospheres. When $p_{CO_2} < 1$, the second term on the right becomes negative. ΔG switches to a negative sign at a lower temperature, meaning that $CaCO_3$ thermally decomposes at a lower temperature.

SOLAR RECEIVER-REACTOR DESIGN

A scheme of the receiver-reactor prototype fabricated is shown in Fig. 3. It consists of a conventional cyclone gas-particle separator that has been modified to let concentrated solar energy enter the cavity through an aperture. It is designed for use at the PSI facilities that include a 55 kW McDonnel Douglas dish and a 17 kW solar furnace (described in Ries and Schubnell, 1990). Its main body is a truncated conical cavity, made of heat-resistant steel (DIN 1.4841, max. use temp. 1373 K), 80cm height, 17° cone opening angle, major circular opening of 54cm-diameter, 3mm thickness. The walls have been insulated with a 12cm-thick layer of Cerablanket-800 (47 % Al_2O_3, 53 % SiO_2, max. use temp. 1533 K), and Wacker-WDS (65 % SiO_2, 15 % FeO, 16 % TiO_2, max. use temp. 1123 K) by Schneider Dämmtechnik, CH-8401 Winterthur, Switzerland. The remainder of the cavity is formed by two concentric cones that form a conical gap for the gas exhaustion. The upper cone contains a 17cm-diameter circular aperture and is made of aluminum oxide casting Gopelit-18 (Al_2O_3 and TiO_2, max. use temp. 1973 K), manufactured by TKT, CH-8240 Thayngen, Switzerland. The lower cone joins the main body

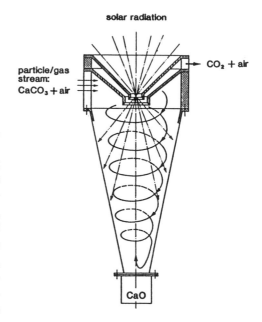

Fig. 3. Receiver-reactor scheme.

through a concentric cylinder that has a small tangential slot through which the particle/gas stream is being fed.In the cyclone separator the carrier gas (air) forms a strong and confined vortex with a high tangential velocity which gives a high centrifugal force to the entrained particles. Therefore, the particles are thrown to the cavity walls where they swirl in a downward spiral path until they descend to the bottom. Incoming concentrated solar radiation enters the cavity through the aperture and strikes directly the particles and the cavity walls, undergoes multiple reflections, and is being redistributed. Some of this radiation will be absorbed by the insulated walls. As their temperature rises, they will emit more and more diffuse radiation, which may undergo multiple reflections until it is absorbed either by the particles or by the cavity walls, or eventually escapes through the aperture. Other mechanisms of heat transfer include forced convection between the cavity walls and the gas stream, and between particles and gas stream, convection losses through the aperture, and conduction losses through the insulation. Energy transferred to the fluidized $CaCO_3$ particles is used to raise their temperature and to drive their decomposition reaction. CaO produced is collected and removed at the bottom. CO_2 gas evolved is exhausted, together with the air stream, through the exhaustion channel.

In order to gain some insights on the heat transfer characteristics of such apparatus, and to investigate practical problems associated with its use, we conducted a set of experiments using an electric heating source. The electric element is an inconel steel (DIN 1.4435) cylindrical coil, and delivers about 15 kW. It was positioned along the cavity axis. Thermocouples type K were placed on various locations inside the cavity. Other parameters measured were pressure, temperature, and volume flow rate of inlet air (no particles), outlet air temperature, power supplied to heater, and temperature of insulation outer walls.

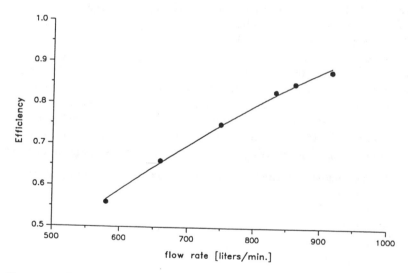

Figure 4: Heat transfer efficiency to air flow as a function of the inlet air flow rate. Input power is 13125 W.

The efficiency of heat transferred to the air as a function of the inlet air volumetric flow rate is plotted in Fig. 4 for steady state conditions and 13125 W power supplied. The maximum efficiency observed was 88 % for 916 liters/min.; power lost through the walls by conduction was about 730 W. The average thermal resistance of the insulation was estimated to be about $1 \, m^2 \, °C/W$. The system has a large thermal inertia due to the large heat capacity of the massive insulation, which is an inconvenience for start-up operations, but an advantage when transients of solar energy occur. In order to reduce heat losses by convection and leakage of air/particles through the windowless aperture, a suction unit is to be implemented at the air outlet duct. Such arrangement was experimentally studied on a smaller scale model, using a cold particle/gas stream of approx. 2 kg of $CaCO_3$ particles per 1 kg of conveying air; particle diameter less than $10\mu m$. The air/particle leakage through the aperture could be effectively controlled and practically eliminated. The fluid mechanics inside the reactor have not been yet well characterized, and for that reason the convection heat transfer coefficient has not been evaluated. Tangential velocities at the wall boundaries, measured at room temperature with a hot-wire anemometer, ranged between 2 m/s at the lower portion of the conical cavity to 10 m/s at the upper portion for a flow rate of 900 liters/min.. A theoretical simulation model is now under development, and experimentation on our solar furnace will start this summer.

In summary, this type of configuration offers several advantages: (1) It is atmospheric-open, thus eliminating the need for windowed reactors. Windows for solar receivers are expensive, brittle, require careful mounting, often fail to withstand high solar fluxes/high temperatures, and reduce the solar energy absorption efficiency. Therefore, when possible, a windowless reactor is a preferred alternative, and it was shown that the addition of air to the system has a positive effect, because the equilibrium is displaced to favor the products CaO and CO_2. (2) It is a cavity-type enclosure which can capture solar radiation efficiently. The apparent absorptance of a cavity, defined as the fraction of energy flux emitted by a black-body surface stretched across the aperture that is absorbed by the cavity walls (Lin and Sparrow, 1965), usually exceeds the surface absorptance because of multiple reflections among the cavity walls. For solar radiation coming from the solar concentrator, confined to a cone whose apex angle is the rim angle of the concentrator, one can design a conical cavity with an apparent absorptance close to 1, even for highly reflective cavity walls. (3) Direct incidence of concentrated sunlight at the reaction site, therefore providing an efficient means of energy transport and, under proper conditions, photochemical enhancement. In addition, heat absorbed by the cavity walls is efficiently being transferred by forced convection to the carrier gas. (4) Reactants are continuously fed; products are being separated and are continuously removed. (5) The particle residence time can be adjusted to meet kinetics requirements by changing geometric proportions, mass flow rate, and loading ratio.

Acknowledgements. This work is being funded by the Swiss Federal Office of Education and Science.

REFERENCES

Badie, J. M., Bonet, C., Faure, M., Flamant, G., Foro, R., and Hernandez, D. (1980). *Chem. Eng. Sci. 35*, 413-420.

Durisch W. (1990). *Proc. 5th Symp. Solar High Temperature Technologies*, Switzerland.

Flamant, G., Hernandez, D., Bonet, C., and Traverse, J. (1980). *Solar Energy 24*, 385-395.

Gordon, S., and McBride, B. J. (1985). *NASA SP-273*, NASA, Washington, D.C.

Imhof, A. (1990). *Proc. 5th Symp. Solar High Temperature Technologies*, Switzerland.

Jacobs, J., Gunther, R. (1975). *Int. Chem. Eng. Symp. Series 43*, Harrogate.

JANAF Thermochemical Tables (1985). National Bureau of Standards, 3rd. edition.

J. Phys. Chem. Ref. Data (1982), Vol. 11, Suppl. 2.

Lin, H. S., and Sparrow, E. M. (1965). *J. Heat Transfer*, 299-307.

Ries, H., and Schubnell, M. (1990). *Solar Energy Materials 21*, 213-217.

Salman, O. A., and Kraishi, N. (1988). *Solar Energy 41*, No. 4, 305-308.

Smyth, Adams (1923). *J. Am. Chem Soc. 45*, 1167-1184.

Villermau, J. (1979). *Entropie 85*, 25-31.

PSA's WORK IN SOLAR PHOTOCATALYTIC WATER DETOXIFICATION

J. Blanco*, S. Malato*
M. Sánchez**, A. Vidal**, B. Sánchez**

*Plataforma Solar de Almería, Box 22,
04200 Tabernas (Almería), Spain
**Ciemat, Instituto de Energías Renovables,
Av. Complutense 22, 28040 Madrid, Spain

ABSTRACT

A Solar Photocatalytic Detoxification Loop has been developed at the Plataforma Solar de Almería for the destruction of toxic organic compounds dissolved in water at low concentrations. For this purpose, two-axis solar-tracking parabolic trough collectors originally designed to concentrate and transform solar radiation into thermal energy have been adapted to this solar chemical application, in which the solar ultraviolet band activates a chemical reaction in the presence of a semiconductor-type catalyst. In this paper, loop engineering is described.

KEYWORDS

Solar water detoxification; solar chemical engineering; ultraviolet band concentration; photocatalytic oxidation; semi-conductor catalyst.

INTRODUCTION

The relatively recent proliferation of photocatalytic process applications, together with an ever growing worldwide interest in environmental matters, have led to the birth of solar photocatalysis, a very new application of solar energy, practically unsuspected up to just a few years ago. This photochemical reaction takes place in the presence of a semiconductor-type catalyst excited by the ultraviolet band of the spectrum concentrated in the collectors and which provokes the jump of valence band electrons to the conduction band. These mobile electrons may then readily be used for instigating any of various types of reactions. Specifically, and undergoing the most spectacular development, is the one used for solar photocatalytic detoxification, in which the existing electronic cloud produces in the presence of oxygen some strongly oxidizing radicals which, between 30 and 85°C, are capable of completely oxidizing almost any organic compound, including the great majority of halogenated organic compounds, solvents, pesticides, etc., from ppm to ppb levels. With this reaction toxic residues dissolved in water at low concentrations are not simply transformed or transfered from one medium to another, but totally mineralized into carbon dioxide, water and diluted concentrations of simple inorganic acids, and are thus really eliminated. This process has already been widely evaluated in the laboratory (Fox, 1983; Okamoto, 1985; Pelizzetti, 1986; and others) and a detoxification loop has now been set up at the Plataforma Solar de Almería (PSA), a solar research center in Southern Spain belonging to the Spanish Research Institute, CIEMAT (Centro de Investigaciones Energéticas, Medioambientales y Tecnológicas), for large-scale testing and development of a pre-industrial process technology. The project is sponsored by the European Communities, within the "Access to Large-Scale Scientific Installations Program" and is a joint effort by several European institutions.

LOOP DESCRIPTION

The loop consists of 12 two-axis tracking collectors, developed by the German company, M.A.N. and initially designed to convert direct solar radiation into heat. These 12 collectors are connected in series, in such a way that at any given moment any one of them may be eliminated or modified with a by-pass valve on each. Fig. 1 shows a diagram of the detoxification loop.

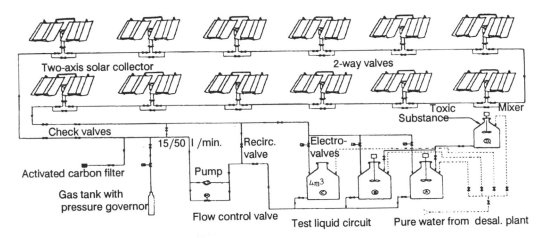

Fig. 1.   Detoxification Loop Diagram

There are 4 tanks for loop inlet and outlet and pure water from a Desalination Plant, also located at the Plataforma Solar de Almería, which evaporates water by a multieffect system using solar energy and obtaining water with conductivity between 5 and 10 μsiemens/cm, is initially collected in these tanks. The substance whose descomposition is to be tested in the loop is added to this water and disolved with mixers. Later the catalyst is added at the desired ratio. The most commonly used catalyst for this application is titanium dioxide (Degussa P25), normally used as a paint pigment and which has the additional advantage of low cost. Recent results (Ollis, 1985; Matthews, 1987; Alpert, 1990) indicate that the laboratory optimun ratio is around 0.1% TiO_2. In the test the final mixture is circulated through the loop. Residence time, or time of each test liquid volume unit is exposed to radiation in the absorber, is computer controlled and can be fixed between 10 and 30 minutes, and more with recirculation. There are different process parameters (pressure, temperature, flowrate, pH, oxygen concentration, direct and global UV radiation) continually measured by on-line instrumentation and transmitted to the DAS (data adquisition system) which sends the data to computer control.

Polyethylene pipes and tanks are used as this material behaves well in the presence of low concentrations of chemicals agents. Total loop volume is 838 liters. The computer-controlled flowrate can be varied between 15 and 50 liters/minute depending on UV radiation available in order to make comparable tests with the same absorbed radiation. The nominal loop aperture area of 384 m^2 consists of 12 collectors, each of which has four 4,50 x 1,81 m parabolic troughs with a 70.5° aperture angle and concentrating ratio of 20. In order to study the process, samples may be taken from different parts of the loop for laboratory analysis with Heraeus LiquiTOC total organic carbon (TOC) analyzer and Hewlett-Packard 1050 gradient high performance liquid chromatograph (HPLC) with variable wavelength UV-Visible detector.

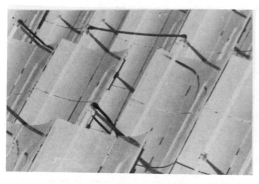

Fig. 2.   View of Sun-tracking Collectors

REFLECTIVE SURFACE

One of the most difficult engineering tasks was the improvement of the reflective surface. Original mirror (5 mm glass back silvered with black antisplintering protective coating) reflectivity measured in the interval between 295 and 385 nm, with Pelkins Elmer λ-9 fUV/VIS/NIR spectrum photometer with diffuse reflection integrating sphere, was found to be rather low: 48.6%. This value (and all other reflectivity and transmitivity percentages) was obtained by integrating the resulting spectral curve and comparing it with the ASTM standard solar spectrum for the interval [295 - 385] nm. The reason for predefining this specific interval, is that this is the band measured by UV radiation instrumentation, thus all parameters and performances have the same spectral band reference in order to be compatibles.

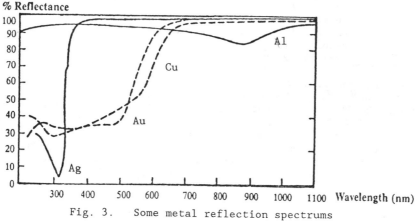

Fig. 3.   Some metal reflection spectrums

As is well known, aluminium is the best surface for UV radiation, therefore a 30μ copolymer film of teflon and polyethylene (Hoechst Hostaflon), later aluminized by vacuum deposition, thus simultaneously acting as shield and support for the aluminium, was the chosen solution, since the film remain on the outside of the reflective surface. This film (1.75 gr/cm³ specific weight, 230°C maximum working temperature and 275°C melting point) has high performance UV transmission (84%), is highly resistant to degradation outdoors and provides a practically first-surface aluminium reflective element. Reflectivity after aluminization was 83%. As the aluminized side had to be glued to the original glass mirror, the last step was to make a plastic composite by adding 100μ polyester film over the aluminium side for easier handling, glueing, and possible later disassembly from the original mirrors if required. Final surface reflectivity after glueing was 78%.

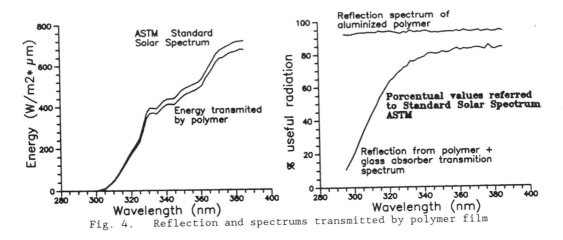

Fig. 4. Reflection and spectrums transmitted by polymer film

ABSORBER TUBES

The fotochemical reaction takes place in the collector absorber tubes which are thus the chemical reactors of the loop. As the collector aperture plane is always perpendicular to the solar rays, the UV rays are reflected by the parabolic trough onto the focus where the contaminated water to be detoxicated circulates through a 56 mm inner diameter borosilicate glass. The combined reactor length is 216 m with a 484 liter capacity. The 2 mm thick tube wall transmits 85% of the UV radiation. Total optical collector performance is defined by the following factors:

UV available radiation	100%	[100%]
Construction factors	91%	[91%]
Sun tracking error	92%	[84%]
Hemispherical film reflectivity	83%	[69%]
Specular film efficiency	93%	[65%]
Glass absorber transmission	85%	[55%]

Therefore, only 55% of direct UV radiation is really usable. Assuming radiation between 295 and 385 nm on the earth's surface to be 34.57 W/m^2 (ASTM standard) and aperture area of 384m^2, total loop UV is 13,275W, which, with 55% performance yields a real power of 7300W. Since reactor capacity is 484 liters (1.26 liters/m^2), the usable intensity is 15.08 W-UV/liter of test mixture. The titanium dioxide semi-conductor catalyst absorbs 87% of UV radiation (between 295 and 385 nm and refered to the ASTM solar spectrum), but the absorptivity in the rest of the spectrum is very low, causing the suspension to appear bright white in colour.

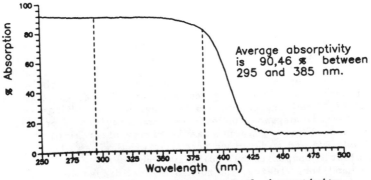

Fig. 5. TiO$_2$ catalyst spectral absorptivity

The nominal usable intensity for a photocatalytic detoxification reaction then is 13.25 W-UV/liter of test mixture. For the measurement of UV radiation, an Eppley TUVR ultraviolet radiometer and an International Light SUD400 direct UV sensor have been installed at the Plataforma Solar de Almería. The DAS inputs UV radiation readings to the computer from which the amount of energy required in the test mixture can be set with the keyboard. The control program then calculates the time of residence and flowrate according to the radiation available. Flowrate is again adjusted as necessary during the test according to the variations in radiation to keep preset UV wattage received by each unit of volume constant. Typical radiation curves for spring and autumn days in Almería are shown in Fig. 6.

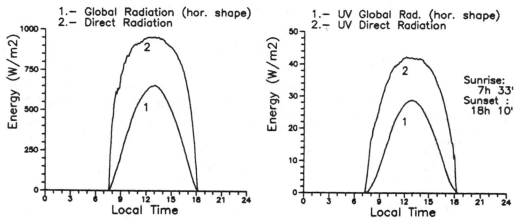

Fig. 6. Direct and global radiation curves on 30th Oct. 1990

CONCLUSIONS AND OBJETIVES

The loop described is the initial configuration and is designed as a large-scale test facility available to photochemical research groups and scientific institutions for the testing the viability of projects already performed in the laboratory. From this point of view, the detox loop will be a living project continually modified and perfected, in which sensor configuration may be varied or new sensors added as needed for individual tests or substances. Engineering and handling of the installation will be perfected in order to make possible commercial applications which could attract the participation of private enterprise.

In a first phase the decomposition testing will only affect one susbtance at a time so that in a second phase the reaction can be approached with mixtures of various toxic ingredients. First phase test plan includes the following substances:

Organic compounds
- Chlorophenols (e.g., p-chlorophenol)
- Haloaliphatics (e.g., tetrachloroethylene)
- Pesticides (e.g., bentazon)
- Ionic surfactants

Heavy metals
- Chromium (VI)
- Mercury (II)
Other compounds
- Cyanides

The substances selected above are typical water contaminants and part of the wide range of agricultural pesticides which are appreciably contaminating the water systems of many European and other countries. The main objectives of the project are the following:

- Development of the technology necessary for pre-commercial application by 1993 and use of the detox loop as a demonstration plant for the transfer of the technology to private enterprise.

- Determination and quantification of the process parameters in order to optimize the several subsystems and hardware components, including determination of optimum reactor type and scale factor for each application.
- Find an optimum catalyst fixation system in the reactor (absorber) and solve catalyst handling problems. Development of new catalysts allowing the use of a wider radiation spectrum band than the ultraviolet only must also be considered.
- Study and modelling of the different chemical reactions. Kinetic models describing their evolution will be essential to future process design.

Participating research institutions in the European Community will be carrying out this ambitious program in close collaboration with the Plataforma Solar de Almería within the next two years.

ACKNOWLEDGEMENT

The authors would like to express their gratitude to the PSA maintenance team: J.M. Molina, M. Carreño, M. Hernández, and M. Flores for their invaluable help, as well as the close collaboration of G. García, J.A. González, and J. Aranda. Thanks are due to our colleagues at the Instituto de Energías Renovables including M. Romero, and J. Herrero. The authors would also like to thank D. Fuldauer and C. Montesinos who assisted in the preparation of the paper. Special recognition is due to A. Sevilla for his encouragement and continued support.

REFERENCES

Ahmed, S., and D. F. Ollis (1984). Solar photoassisted catalytic decomposition of the chlorinated hidrocarbons trichloroethylene and trichloromethane. Solar Energy, 32, 597-601.

Alpert, D. J., J. L. Sprung, and J. E. Pacheco (1990). Sandia National Laboratories' work in solar detoxification of hazardous wastes. Proceedings 5th IEA Symposium, Davos, Switzerland.

Barbeni, M., E. Pramauro, and E. Pelizzetti (1986). Photochemical degradation of chlorinated dioxins, biphenyls, phenols and benzene on semiconductor dispersions. Chemosphere, 15, n° 9, 1913-1916.

Fox, M. A. (1983). Organic heterogeneus photocatalysis: chemical conversions sensitized by irradiated semiconductors. Acc. Chem. Res., 16, 314-321.

Matthews, R. W. (1987). Solar water purification using photocatalytic oxidation with TiO_2 as a stationary phase. Solar Energy, 38, n° 6, 405-413.

Kamat, P. V., and N. Dimitrijevic (1990). Colloidal semiconductors as photocatalysis for solar energy conversion. Solar Energy, 44, n° 2, 83-98.

Okamoto, K., and colleagues (1985). Heterogeneus photocatalytic decomposition of phenol over TiO_2 powder. Bull. Chemical Society Japan, 58, 2015-2022.

Ollis, D. F. (1985). Contaminant degradation in water; heterogeneus photocatalysis degrades halogenated hydrocarbon contaminants. Enviromental Science and Technology, 19, n° 6, 480-484.

Pacheco, J. E., and C. E. Tyner (1990). Enhancement of processes for solar photocatalytic detoxification of water. Proceedings ASME International Solar Energy Conference, Miami, Florida.

Pelizzetti, E., and colleagues (1989). Photocatalytic degradation of bentazon by TiO_2 particles. Chemosphere, 18, n° 7-8, 1437-1445.

Serpone, N., and E. Pelizzetti (1989). Photocatalysis fundamentals and applications, Wiley-Interscience, John Wiley & Sons.

Tributsch, H. (1989). Feasibility of toxic chemical waste processing in large scale solar instalations. Solar Energy, 43, n° 3, 139-143.

Yamagata, S., and colleagues (1988). Photocatalytic oxidation of alcohols on TiO_2. Bull. Chemical Society Japan, 61, 3429-3434.

DETERMINATION OF ACCURACY OF MEASUREMENTS BY SERI'S SCANNING HARTMANN OPTICAL TEST INSTRUMENT

Gary Jorgensen, Tim Wendelin, and Meir Carasso

Solar Energy Research Institute
1617 Cole Boulevard, Golden, Colorado

ABSTRACT

SERI's Scanning Hartmann Optical Test (SHOT) instrument is routinely used to characterize the surface of candidate dish concentrator elements for solar thermal applications. An approach was devised to quantify the accuracy of these measurements. Excellent reproducibility was exhibited and high confidence was established in the absolute error related to individual characterizations.

The SHOT instrument was designed to allow the surface figure of large optical test articles to be accurately specified. Such test articles are nominally parabolic shapes with an f/D ratio (where f = focal length and D = aperture diameter) in the range of 0.5–1.0. Recent modifications of SHOT have extended the characterization range out to about f/D = 3.0.

A series of experiments was designed to investigate and quantify the uncertainties associated with optical characterizations performed by SHOT. This approach involved making a series of measurements with an arbitrary test article positioned at a number of locations transverse to the optical axis of SHOT.

KEYWORDS

Optical measurements; optical characterization; surface figure; large optics; dish concentrators.

INTRODUCTION

A reliable and accurate means of quantifying the surface figure of large-aperture dish concentrators is required to allow optical performance to be predicted. Such a capability would allow comparison of candidate prototype designs, suggest improvement of fabrication techniques during the manufacturing process, and provide quality control of mass-produced modules. An instrument has been developed at the Solar Energy Research Institute (SERI) and has been used to test a variety of concentrators fabricated for the U.S. Department of Energy's Solar Thermal Program (Wendelin, Jorgensen, and Wood, 1991).

The standard test configuration is shown in Fig. 1. A laser beam emanating from the center of a screen is sequentially directed toward a user-specified number of points located on the test

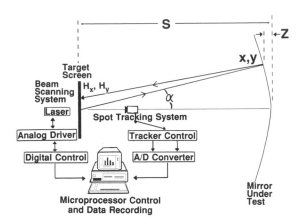

Fig. 1. Standard SHOT test configuration.

article. To minimize the time required to process each point, a regularly spaced test grid pattern is used as a first-approximation sampling scheme. To avoid aliasing the data by any periodic features that may be present whose size or spatial frequency is of the order of the sampling grid, each point is located randomly about its nominal regular grid coordinate.

The intersection of the reflected spot with the screen is detected by a phase-locked quad-cell tracking unit. By knowing the distance from the screen to the vertex of the test article (s, typically chosen to be slightly greater than 2f) and the direction angles (α_x and α_y) of the scanned laser beam, the position at which the laser beam intersects the surface can be estimated and the slope of the surface (ϕ_x and ϕ_y) can be determined from the measured position (H_x and H_y) of the return spot on the screen.

A Zernike monomial equation (Malacara, 1978) is used to represent the surface of the test article. This representation is non-axisymmetric and the various terms of the expansion can be related to standard optical errors such as tilt, coma, etc. Measured data points are used to fit a surface equation expressed as

$$z(x-\Delta_x, y-\Delta_y) = \sum_{i=0}^{k} \sum_{j=0}^{i} B_{ij} * (x-\Delta_x)^{i-j} * (y-\Delta_y)^j \qquad (1)$$

where Δ_x and Δ_y are the transverse distances (decenter) a given test article is displaced from the optical axis ($\alpha_x = 0$ and $\alpha_y = 0$) of the scanning laser in the x and y directions respectively. A best fit least-squares approach is used which minimizes the root mean square (RMS) of the sum of the differences between the measured and calculated slopes. These residuals are given by

$$r_x = \phi_x - \frac{\partial z}{\partial x} \quad \text{and:} \quad r_y = \phi_y - \frac{\partial z}{\partial y} \qquad (2)$$

The RMS residual slope is defined as $\qquad \sigma = \sqrt{\dfrac{\sum\limits_{m=1}^{n} (r_x^2 + r_y^2)}{n}} \qquad (3)$

with n being the number of measured data points. The magnitude of σ is an indication of how well the monomial representation defines the surface of the test article; it compares the fit curve to the measured slope data.

In practice, the procedure is to align the test article with the optical axis of the scanning laser, measure s and D, specify the number of data points desired and the designed focal length (f) of the (assumed parabolic) test article, and acquire data. Given the order of the fit (k), the number of desired iterations (I) to be performed during the fitting process, and the offset positions (Δ_x and Δ_y), the surface equation (1) can be fit to the data. Results from SHOT measurements can be input to the OPTDSH program (Balch and co-workers, 1991) to obtain optical and thermal performance information.

Ideally, the number of iterations and the order of the fit should be chosen to maximize the significance of regression (as determined by an F-test). If too many terms are included in the fit, the confidence in higher-order terms may be low. Such higher-order terms may be applied to fitting experimental errors in the data rather than contributing to a description of "true" surface shape. Typically, σ is monitored until convergence (in terms of changes in σ after each iteration during the fitting process) is obtained.

When the fitting process has been carried out, the amount of tilt of the test article with respect to the optical axis will be given by

$$\theta_x = \tan^{-1}(B_{10}) \quad \text{and:} \quad \theta_y = \tan^{-1}(B_{11}) \tag{4}$$

The focal lengths in the x (horizontal) and y (vertical) directions are given respectively by

$$f_x = \frac{1}{4B_{20}} \quad \text{and:} \quad f_y = \frac{1}{4B_{22}} \tag{5}$$

EXPERIMENTAL APPROACH

A series of experiments was carried out to investigate and quantify the uncertainties associated with optical characterizations performed by SHOT. The ideal way to validate an instrument such as SHOT would be to obtain a standard test article having a surface figure that is well known a priori. A measured characterization could then be carried out using SHOT, and the results could be compared with the known standard surface. Unfortunately, no such standards exist having the applicable size and geometry required. Consequently, an alternate means was devised to verify the operation of this instrument in an absolute sense. This was accomplished by characterizing the surface figure of a particular test article a number of times without changing any of the test parameters, and then comparing the results with those for the same test article moved transverse to the optical axis. By moving the test article orthogonally to the optical axis a known distance, the degree of internal consistency could be derived for the measurement system. Although the fitting process allows an unknown surface to be represented in terms of the offset parameters Δ_x and Δ_y, the measurement system independently characterizes the surface of each new test article. The measurement system does not discriminate between one test article positioned exactly along the optical axis and another that has been shifted relative to that axis. Therefore if an identical test article is characterized both on axis and at some displaced position, and if the same surface is predicted to within the accuracy of the measurement/repositioning process, then the surface figure as predicted by SHOT is correct and valid. A summary of these experiments and their results is presented in Table 1.

The test article used for these experiments was a fiberglass composite membrane dish having a 3-m diameter and a designed parabolic shape with f/D = 0.6 at a stabilization pressure (required to maintain the desired parabolic shape during operation) of about 1900 Pa. For each experiment (Run #) the decenter terms (Δ_x and Δ_y) were determined as the values that minimized the root mean square (RMS) of the residuals between measured slopes and a second-order least-squares fit. For Runs 1–5, the values of Δ_x and Δ_y which were used during the fitting process were the

TABLE 1   Results of SHOT Verification Experiments

Run #	Δ_x (cm)	Δ_y (cm)	B_{20}	B_{22}	σ^* (mrad)	# of Pts.	P (Pa)	Remarks
1	−0.99	1.63	0.1274	0.1318	5.018	2000	1900	1st replication
2	−0.96	1.58	0.1272	0.1316	4.930	2000	1900	2nd replication
3	−0.95	1.61	0.1273	0.1316	5.023	2000	1900	3rd replication
4	−1.02	1.57	0.1271	0.1315	4.871	1000	1900	½ # of points
5	−0.96	1.57	0.1273	0.1314	4.917	4000	1900	2X # of points
6	−17.35	0.88	0.1265	0.1313	5.025	2000	1900	Moved 15.24 cm to left
7	−24.62	1.17	0.1259	0.1309	4.857	2000	1900	Moved 22.86 cm to left
8	13.98	1.02	0.1269	0.1313	5.080	2000	1900	Moved 15.24 cm to right
9	−1.77	0.81	0.1269	0.1313	4.903	2000	1900	Re-centered (0 cm)
10	−0.88	0.92	0.1269	0.1313	4.876	2000	1750 to 2250	Moved 18.29 cm forward
11	−0.89	0.88	0.1269	0.1315	4.933	2000	1900	Moved 18.29 cm forward
12	−1.01	1.09	0.1258	0.1307	5.378	2000	1900	Re-centered; $R=R_{Dish}$
13	−1.14	0.70	0.1259	0.1308	5.225	2000	1900	Relocated tracker
14	−0.18	1.34	0.1266	0.1315	5.141	2000	1900	Moved SHOT and test article to new lab

*   All RMS residual slopes are after fifth iteration of a fourth-order fit. The tracker calibration coefficients were the same for experiments 10-12; for all other runs the tracker calibration procedure was explicitly performed.

averages obtained from the first three "identical" replications. The monomial terms related to the focal length in the x (horizontal) and y (vertical) directions are also tabulated, along with the RMS residual slopes for a fourth-order fit and five iterations. Additional distinguishing information associated with each run is also provided in Table 1.

A figure of merit for comparing various experimental effects was defined as the RMS difference between the monomial coefficients used to describe the surface for different runs. Tilt terms (B_{10}

TABLE 2   Surface Figure of Merit for Various Experimental Effects

Comparison Run #'s (μ,ν)	ε (x10^5)	Effect Being Computed
0, 1	0.533	Average for replications vs. replication #1
0, 2	0.455	Average for replications vs. replication #2
0, 3	0.375	Average for replications vs. replication #3
0, 4	1.448	Decrease number of points by half
0, 5	1.129	Double number of points
0, 9	2.160	Move and reposition test article
0,14	2.553	Relocate SHOT and test article
9, 6	1.906	Move 15.24 cm left
9, 7	4.747	Move 22.86 cm left
9, 8	1.350	Move 15.24 cm right
9,11	0.559	Move 18.29 cm forward
9,12	5.319	Sample closer to outer perimeter
9,13	5.291	Move tracker
11,10	0.721	Variation of nominal stabilization pressure

and B_{11}) were not included because they are not integral to the surface being tested but rather are an artifact of positioning. The figure of merit that was used is

$$\varepsilon = \sqrt{\frac{\sum_{i=2}^{k=4} \sum_{j=0}^{i} [B_{ij}(\mu) - B_{ij}(\nu)]^2}{\eta}} \tag{6}$$

where the number of coefficients excluding the two tilt terms, η, is given by

$$\eta = \frac{[(k+1)(k+2) - 2]}{2} \tag{7}$$

and $B_{ij}(\mu$ or $\nu)$ are the monomial coefficients associated with Run #'s μ or ν. The average monomial coefficients from the first three experiments are denoted as $B_{ij}(0)$. Key comparisons that isolated various experimental effects are summarized in Table 2.

RESULTS

The effect that exhibited the largest uncertainty was that of sampling a greater number of data points near the outer perimeter of the test article (Table 2, comparison 9,12). This is the region where it is most difficult to maintain a desired surface figure during fabrication.

Very close agreement was found between the calculated lateral displacements and the measured position of the test article (Δ_x vs. movement reported in "Remarks" in Table 1). Errors associated with small movements transverse to the optical axis (±15.24 cm, comparisons 9,6 and 9,8) are

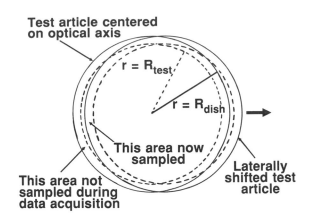

Fig. 2. Perimeter area is sampled when test article is moved laterally.

close to those related to simple repositioning of the test article (comparisons 0,9 and 0,14). For larger movements (22.86 cm, comparison 9,7), an appreciable area along half of the outer perimeter of the test article is moved into SHOT's scannable range (Fig. 2). Data points from this region can introduce errors as discussed above. Based upon these results, the description of the laterally shifted surface of the test article is identical (to within the uncertainties related to other experimental factors) to the description of the unshifted surface. This result verifies the SHOT measurement process. Random variation in stabilization pressure (roughly 10%–20% for the particular test article under consideration) and longitudinal movement of the test article along the optical axis introduce minor uncertainty into the measured surface figure. Such effects are barely more significant than the inherent error associated with the measurement process itself (as represented by the first three entries of Table 2).

CONCLUSIONS

An approach was devised to quantify the accuracy associated with optical characterization of the surface of dish concentrators by SERI's SHOT instrument. This approach was independent of the details of the surface figure of the test article. Excellent reproducibility was exhibited and high confidence was established in the absolute error related to individual measurements.

ACKNOWLEDGMENT

This work was sponsored by the U.S. Department of Energy under contract DE-AC02-83CH10093. Doug Powell of SERI assisted in performing the series of experiments presented in this paper.

REFERENCES

Balch, C., C. Steele, G. Jorgensen, T. Wendelin, and A. Lewandowski (1991). Membrane Dish Analysis: A Summary, SERI/TP-253-3432, Solar Energy Research Institute, Golden, Colorado.

Malacara, D. (1978). Optical Shop Testing. John Wiley and Sons, New York.

Wendelin, T., G. Jorgensen, and R. Wood (1991). SHOT: A method for characterizing the surface figure and optical performance of point focus solar concentrators. Proceedings 13th Annual ASME Solar Energy Conference, Reno, Nevada.

GAS-PHASE SOLAR DETOXIFICATION OF HAZARDOUS WASTES: LABORATORY STUDIES

Mark R. Nimlos, Thomas A. Milne
Solar Energy Research Institute, 1617 Cole Blvd.,
Golden, CO 80401

J. Thomas McKinnon
Center for Combustion Research, University of Colorado,
Boulder, CO 80309

ABSTRACT

The Solar Energy Research Institute (SERI) in coordination with the Center for Combustion Research is investigating the potential for using solar energy for the destruction of hazardous materials. In support of this program, bench-scale laboratory experiments are being pursued to help unravel the chemistry and photochemistry of various solar detoxification processes. Key to these studies is the use of SERI's molecular beam mass spectrometer (MBMS), which allows real-time detection of virtually all products of gas-phase reactions. The MBMS has been used to monitor reactants, intermediates, and products for gas-phase reactions at atmospheric pressure and temperatures from ambient to 1000°C. In this paper, three solar waste destruction processes will be discussed: 1) Steam reforming over a Rh catalyst using solar energy to provide direct heating of the catalyst surface. 2) High-temperature gas-phase decomposition of wastes using the visible and infrared portions of the solar spectrum to provide the heat and the ultraviolet portion of the spectrum to initiate or enhance the destruction. 3) Low-temperature heterogeneous processes using solar energy to excite photoactive solids (such as semiconductors) leading to the decomposition of gas-phase wastes. Steam reforming studies are being done in conjunction with Sandia National Laboratories, where preliminary field tests have been carried out using a solar concentrator. Destruction efficiencies, catalyst deactivation and intermediates have been determined. These tests are being used to evaluate and plan experiments at Sandia. The MBMS results concerning the other processes listed above are being used as an aid to SERI's own field tests. Kinetic and photochemical parameters for both starting materials and intermediates are determined; these are used to evaluate and predict field tests. In addition to the MBMS results, absorption properties and chemical kinetics models are being evaluated at the Center for Combustion Research.

KEYWORDS

Hazardous waste destruction; solar energy; steam reforming; oxidation; TiO_2;

heterogeneous; semiconductor; chloronaphthalene; rhodium.

MOLECULAR BEAM MASS SPECTROMETRIC DETECTION

SERI's molecular beam mass spectrometer (MBMS) is an apparatus that allows the detection of species in high-temperature (up to about 1000°C), ambient pressure, gas-phase reactors. It is particularly well suited for studying waste destruction, as a wide variety of species (such as condensible and highly reactive compounds) can be directly monitored in real time. A schematic diagram of the MBMS is shown in Figure 1 and the reader is referred elsewhere (Milne and Soltys, 1983; Evans and Milne, 1987a, 1987b) for a detailed discussion of the experimental hardware.

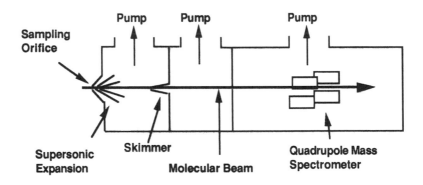

Fig. 1. Schematic representation of molecular beam mass spectrometer (MBMS)

Hot gases are drawn through a small orifice (0.01 in.) into a differentially pumped vacuum system where the molecules are ionized and detected with a mass spectrometer. The adiabatic cooling that occurs in the expansion from atmospheric pressure to high vacuum increases ion signals and quenches further reaction. In the past, the apparatus has been used to sample products from such complicated processes as biomass pyrolysis and combustion.

CATALYTIC STEAM REFORMING

The idea of solar-driven steam reforming hazardous wastes over a rhodium catalyst is an offshoot of studies of solar energy storage utilizing CO_2 reforming of methane (Richardson, Paripatyodar, and Chen, 1988; Keehan and

Richardson; 1989). Preliminary field tests were conducted at Sandia National Laboratories of the catalytic steam reforming of trichloroethylene, and laboratory studies were conducted at the University of Houston (Richardson, 1990) to measure the kinetics in the catalytic steam reforming of 1,1,1-trichloroethane. At SERI, exploratory studies have been done using the MBMS to monitor destruction levels and the formation of intermediates for a variety of compounds. Figure 2 shows a plot of destruction of methylchloride (CH_3Cl) as a function of temperature for steam reforming and pyrolysis (thermal destruction in the absence of a catalyst). The residence times for the pyrolysis experiments were about 3.7 s. while for the steam reforming experiments the residence time was about 1.0 s.

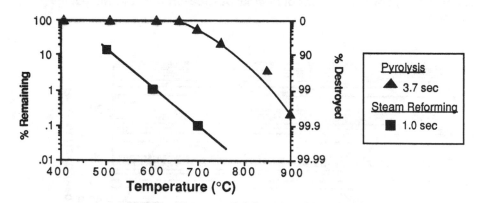

Fig. 2. Destruction of CH_3Cl as a function of temperature for pyrolysis and catalytic steam reforming

Other compounds tested over the Rh catalyst include chlorinated and nonchlorinated aromatics and aliphatics as well as acetonitrile. All of these compounds showed high levels of destruction in the catalytic steam reforming environment. In addition to high destruction levels, the catalytic steam reforming experiments also produced a minimum of undesirable intermediates. In all the tests run, the only intermediate identified was dichloroethylene formed from the partial conversion of trichloroethane. As models for phosphorous- and sulfur-containing wastes, dimethyl methylphosphonate and 3-methylthiophene were tested. These compounds were found to rapidly deactivate the catalyst.

HOMOGENEOUS GAS-PHASE DESTRUCTION

The use of a solar-driven gas-phase process for the destruction of hazardous wastes was first investigated by Graham and Dellinger (1987) under a subcontract from SERI. In those initial studies, high levels of destruction of several compounds were demonstrated when hot gases were irradiated with ultraviolet light from the solar region of the spectrum. Laboratory experiments at SERI have utilized the MBMS to monitor destruction levels and product formation in high-temperature photodestruction processes. For these experiments, an argon ion laser with high power output at roughly 300 nm and 330 nm was used as a simulation of solar ultraviolet radiation. Tube furnaces were used to heat the gases.

Chloronaphthalene was used as a model compound in laboratory tests because this compound has a fairly high absorption cross-section in the solar spectrum. Figure 3 shows the results of thermal destruction with and without the laser radiation.

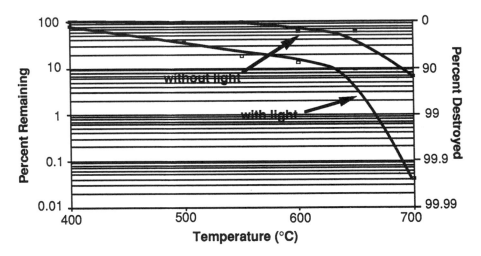

Fig. 3. A plot of the destruction of 1-chloronaphthalene in 80% He and 20% O_2 as a function of temperature with and without ultraviolet irradiation by the argon ion laser

The products from both the thermal and the photothermal decomposition seem to be similar and the total amount appears to be about the same for comparable levels of destruction. However, for comparable levels of destruction, in the presence of ultraviolet light there are fewer chlorinated products produced. This is because photothermal destruction causes preferential scission of C-Cl bonds relative to purely thermal destruction. The products detected were mostly substituted single ring systems (benzaldehyde, ethynylbenzaldehyde, ethynylbenzene, etc), and a mechanism has been developed to explain the observed products.

For compounds that do not directly absorb solar radiation, research is being directed at discovering photoinitiators that could be added in small quantities to gas-phase mixtures containing hazardous waste. These compounds would have to absorb solar ultraviolet light and decompose to form radicals. These radicals would than initiate chain reactions to enhance the thermal destruction of the waste. To maximize the length of chain reactions initiated by these substances, temperatures need to be high, and thermally stable compounds are necessary. Formaldehyde is an initiator that has been considered because it is known to decompose in solar light to form CHO, H atoms, H_2, and CO. Chemical kinetics modeling of the thermal oxidation of trichloroethylene (TCE) and methane showed that including a step with the photodecomposition of formaldehyde could lead to significant destruction of the TCE or methane. Experiments are planned to verify this modeling.

HETEROGENEOUS PHOTOCHEMISTRY

First explored by Dibble and Raupp (1988, 1990), this gas-phase process would utilize the heterogeneous photochemistry that occurs when semiconductors are illuminated by uv radiation. This type of process has been studied extensively for the detoxification of aqueous solutions, but little has been done for gas-phase wastes. In experiments with the MBMS, tests have been performed with TiO_2 at temperatures from ambient up to 250°C. As Dibble and Raupp reported, high levels of destruction of TCE were seen when the TiO_2 was irradiated with ultraviolet light. However, unlike their experiments, the intermediates pentachloroethane (C_2HCl_5), dichloroacetylchloride (C_2HOCl_3), chlorine (Cl_2), and phosgene (CCl_2O) were seen. The reason for this difference is unclear, but it may be attributed to difference in contact time with the catalyst or it may be a result of the different analytical techniques. In any event, by increasing the contact time or increasing the temperature, the formation of these intermediates seems to be diminished.

CONCLUSIONS

Several processes that show promise for the use of solar energy in the detoxification of hazardous compounds are being investigated at SERI. However, there are a number of issues that need to be addressed. These include catalyst lifetime and range of applicability, as well as issues concerning real waste mixtures and engineering problems concerning scaling up. Experiments conducted at the SERI MBMS facility have demonstrated its capability in the investigation of some of these issues.

ACKNOWLEDGMENTS

We would like to thank Barry Dellinger, Jim Richardson and Dan Blake for their help in this work.

REFERENCES

Dibble, L. A., and G. B. Raupp (1988). Proceedings of the Arizona Hydrological Society, 1st. Annual Symposium, 221.

Dibble, L. A., and G. B. Raupp (1990). Catalyst Lett., 4, 345.

Evans, R. J., and T. A. Milne (1987a). Energy and Fuels, 1, 123.

Evans, R. J., and T. A. Milne (1987b). Energy and Fuels, 1, 311.

Keehan, D., and J. T. Richardson (1989). "Carbon Monoxide Rich Methanation Kinetics on Supported Rhodium and Nickel Catalysts," SAND88-7149, Contract Report for Sandia National Laboratory, Albuquerque, NM.

Graham, J. L., and B. Dellinger (1987). Energy, 12, 303.

Milne, T. A., and M. N. Soltys (1983). J. Anal. Appl. Pyrol., 5, 93.

Richardson, J. T., S. A. Paripatyodar, and J. C. Chen (1988). AIChE Journal, 34, 743.

Richardson, J. T. (1990). Private communications.

RELATIVE COST OF PHOTONS FROM SOLAR OR ELECTRIC SOURCES FOR PHOTOCATALYTIC WATER DETOXIFICATION

Craig S. Turchi and Harold F. Link

Solar Energy Research Institute
1617 Cole Boulevard
Golden, CO 80401 USA

ABSTRACT

A heterogenous photocatalyst such as titanium dioxide allows the near-ultraviolet photons available in the solar spectrum to be used for chemical oxidation reactions. Applications include the purification of contaminated drinking water supplies and the destruction of organic contaminants in industrial waste water. The overall treatment cost for such systems is governed by the capital and operating cost of the light-generating or -collecting equipment. It is estimated that a nonconcentrating solar collector can supply near-ultraviolet photons less expensively than either a concentrating solar collector or a typical electric lamp system.

KEYWORDS

Photocatalysis; solar detoxification of water; titanium dioxide; water treatment; waste treatment

INTRODUCTION

Photochemical methods for the detoxification of waste water are being explored and marketed as viable competitors to traditional water purification techniques. As more research has focused on the area of photocatalysis, an important question has been: what is the most cost effective means of providing the required near-ultraviolet (near-UV) photons that drive the photocatalytic process? In general, the most expensive component in photocatalytic systems is the light-producing or light-collecting equipment. Consequently, the cost of building and/or operating this equipment is responsible for most of the overall photocatalytic treatment cost. In this paper we investigate the relative cost of generating near-UV photons from electric lamps compared with the cost of collecting solar near-UV photons using two different types of solar collectors.

COMPARISON OF ELECTRIC AND SOLAR PHOTON COSTS

A simple procedure was used to estimate the cost per photon from hypothetical solar and lamp treatment systems. Estimates of the solar resource were made for two different locations: Livermore, California, representing a medium-insolation site, and Albuquerque, New Mexico, representing a high-insolation site. A key assumption in this analysis is that all near-UV photons (wavelengths between 300 and 385 nm) are equivalent for catalyst excitation.

Estimate of Lamp Costs

The lamp system was assumed to be powered by medium-pressure mercury arc lamps. These lamps were chosen for the following reasons:

- High fraction of near-UV output
- Available in high wattage
- Reasonable lifetimes (2000-3000 hours, warranted for 1000)
- Commonly used in industrial and non-catalytic oxidation systems

The values used to generate the photon cost for a representative medium-pressure mercury lamp are shown in Table 1.

TABLE 1 Cost of Photons from a Single Medium-Pressure Mercury Lamp

Conditions		
Lamp input power	2400	Watt
Lamp cost	220	$
Lamp useful life	2000	hours
Avg output over life	85	% of initial
Near-UV output	25	% of input
Operating time	7000	hours per year
Electricity	0.05	$/kWh
Capital cost	1	$/input Watt
Maintenance cost	3	% of capital
Photon Production		
(using an average	3.4×10^{24}	per operating hr
energy of 3.39 eV/ph)	2.4×10^{28}	per year
Annual Costs		
Lamp replacement	771	$
Electricity	841	$
Maintenance	72	$
Capital × FCR (13%)	312	$

Lamp near-UV output and operational lifetime were based on manufacturers' and literature data.

Estimation of Solar Resource

A limited amount of information is available on the amount of near-UV flux available in sunlight (Riordan, Hulstrom, and Myers, 1990). For purposes of this study, we use a procedure based on a model that projects spectral characteristics of sunlight as a function of atmospheric conditions (Bird and Hulstrom, 1982). This BRITE model estimates both direct-normal and global-horizontal radiation levels. Direct-normal light consists of the direct-beam radiation (propagating directly from the solar disk) that strikes a surface oriented normal to the beam). Global-horizontal light consists of both direct-beam and diffuse radiation (scattered by the atmosphere) that strike a horizontal surface.

For this study we consider trough-type concentrating solar collectors which use single-axis tracking systems to capture the direct-beam radiation. The direct-beam radiation available to a trough is characterized by the direct-normal estimates after accounting for incidence angle losses. Nonconcentrating systems typically do not track the sun but can capture diffuse light in addition to the direct beam. For this study we consider nonconcentrating systems with a horizontal orientation such as a solar pond or raceway.

To estimate direct-normal insolation, we first estimated *near-UV* flux as a function of air-mass[1] by curve-fitting projections of clear-sky, direct-normal insolation made with the BRITE model. Similarly, a relationship was established for the *full-spectrum*, direct-normal flux as a function of air-mass for clear-sky conditions. These relationships were then used to estimate flux as a function of time of day using formulas that calculate air mass for any given date, time, and location (Kasten and Young, 1989).

Clouds and haze were assumed to affect near-UV flux at the same rate as full-spectrum flux. Their impact was accounted for by first comparing the predicted, *clear-sky*, full-spectrum insolation with estimated, *actual*, full-spectrum insolation at the site (SERI, 1981). The resulting "cloudiness" factor was then applied to the predicted, *clear-sky*, near-UV insolation to estimate actual near-UV insolation at the site. These calculations were repeated to estimate the near-UV resource at each site for every month of the year. These insolation levels were then averaged and used to calculate the cost per photon for trough-type solar concentrators.

Global-horizontal near-UV was estimated using the same procedure as that used to estimate direct-normal near-UV. The resulting insolation levels were used to estimate the cost per photon for nonconcentrating solar collectors. Comparison of model-predicted insolation with a limited number of daily near-UV measurements has shown that the BRITE model reasonably predicts the solar resource.

Figure 1 shows how these procedures predict solar flux levels for June 15 at Albuquerque, New Mexico, under clear-sky conditions. Note that, despite its lack of tracking, the nonconcentrating collector receives 80% more photons than the trough concentrator. This performance is due to the significant level of atmospheric scattering of near-UV photons.

Table 2 displays the estimation of solar photon capture for a nonconcentrating collector at the two locations. The difference in average photon capture rates for the two locations is primarily due to different cloudiness factors --- clear-sky insolation levels are within 5% despite the difference in latitude.

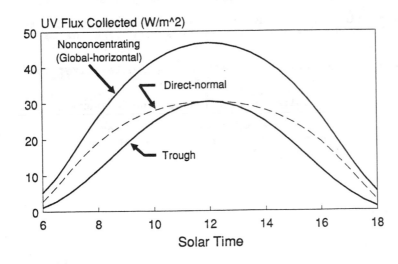

Fig. 1. Estimated near-UV flux available at Albuquerque, New Mexico, in mid-June and the flux captured by nonconcentrating and trough solar collectors.

---

[1]   One air mass designates the amount of air through which sunlight passes to strike a surface at sea level when the sun is directly overhead. Air mass increases proportionally to the secant of the sun's zenith angle for angles less than 75°.

TABLE 2   Estimated Capture of Solar Global-Horizontal (GH) Photons with a Nonconcentrating Collector

Month	Livermore, CA			Albuquerque, NM		
	Daily Clear-Sky GH UV 10^{25} ph/m^2	Cloud Factor	Daily GH UV	Daily Clear-Sky GH UV 10^{25} ph/m^2	Cloud Factor	Daily GH UV
Jan.	0.069	0.39	0.027	0.078	0.76	0.060
Feb.	0.102	0.43	0.044	0.112	0.71	0.080
March	0.150	0.45	0.067	0.159	0.68	0.108
April	0.201	0.48	0.097	0.206	0.67	0.137
May	0.238	0.49	0.117	0.241	0.66	0.160
June	0.253	0.51	0.129	0.254	0.66	0.168
July	0.248	0.56	0.139	0.249	0.55	0.136
Aug.	0.219	0.53	0.116	0.223	0.60	0.133
Sept.	0.172	0.55	0.095	0.179	0.65	0.117
Oct.	0.118	0.52	0.061	0.128	0.70	0.089
Nov.	0.078	0.45	0.035	0.088	0.80	0.070
Dec.	0.060	0.41	0.025	0.071	0.71	0.050
Annual avg (10^{25} ph/m^2-day)			0.079			0.109

Conditions:
    Operating days: 292 per year
    Collector efficiency: 80%

Estimates of Levelized Cost of Photons

We estimate capital cost for the lamp system as $1 per input watt. This number is consistent with quotes from the manufacturer, Canrad Hanovia Inc. (O'Reilly, 1990). Capital cost for the trough-type solar collector is estimated at $200/m^2, listed as current technology in the Solar Thermal Technology Program's Five Year Plan 1986-1990. The cost for a nontracking, nonconcentrating collector is estimated at $50/m^2 based on work done at Lawrence Livermore National Laboratories (Parsons, 1985).

Annual operating and maintenance (O&M) costs for the lamp system are given in Table 1. O&M costs for the solar systems were taken as 3% of the capital cost.

Finally, the cost comparison was made by calculating the levelized cost for each system, defined as

Levelized cost = (FCR x capital + O&M)/(annual photon output)

The fixed charge rate (FCR) was set at 13% -- consistent with U.S. Environmental Protection Agency estimates for conventional water treatment systems (Adams and Clark, 1989).

The results of this analysis for the baseline parameter values are shown in Table 3. Note that the cost per photon in a nonconcentrating system is significantly less than that of either a trough concentrator or a medium-pressure lamp system.

Sensitivity to Baseline Conditions

The impact of various changes in the baseline conditions is shown in Table 4. The first two changes from baseline conditions did not change the ranking of the three systems. Change (3) represents a substantial reduction in solar concentrator cost. This change places trough and nonconcentrating collectors at an equal cost and thereby demonstrates the impact of the diffuse component of near-UV sunlight. Even at equivalent capital costs, a nonconcentrating collector has a lower cost per photon.

TABLE 3   Cost per 10^{25} Photons for Different Sources

Baseline Conditions	
Fixed charge rate (FCR)	13%
Electricity cost	$0.05/kWh
Solar trough collector cost	$200/m^2
Solar nonconcentrating collector cost	$ 50/m^2
System availability	80%
Useable photons	<385 nm
Location	Albuquerque, New Mexico

Costs	
Medium-pressure mercury lamp	$0.84
Solar trough concentrator	$1.64
Solar nonconcentrating collector	$0.30

TABLE 4   Sensitivity Analysis ($/$10^{25}$ photons)

	Medium Pressure Lamp	Solar Trough Alb.	Liver.	Solar Nonconcen. Alb.	Liver.
Baseline Values (see Table 3)	0.84	1.64	2.40	0.30	0.45
Changes from Baseline					
(1) Electricity $0.10/kWh	1.20	1.64	2.40	0.30	0.45
(2) FCR: 7%	0.78	1.02	1.50	0.21	0.30
(3) Solar trough: $50/m^2	0.84	0.41	0.60	0.30	0.45
(4) Useable photons: 300-385 nm only	1.21	1.64	2.40	0.30	0.45
(5) Useable photons: all <415 nm	0.73	0.80	1.16	0.16	0.24

Alb.:   Albuquerque, New Mexico (high-insolation site)
Liver.:   Livermore, California (medium-insolation site)
FCR: Fixed Charge Rate

Change (4) reflects the uncertain response of the catalyst to photons with wavelengths shorter than 300 nm. If these short-wavelength photons do not excite the photocatalyst, lamp costs would increase significantly.

Finally, (5) depicts the relative advantage to be gained by a solar system if more easily activated photocatalysts, such as rutile TiO_2 or tungsten trioxide, prove to be effective.

Tables 3 and 4 depict the relative *photon cost* for three systems. The relative *water treatment cost* for these systems will depend not only on this photon cost but also on the efficiency with which the photons are used and the cost of equipment specific to each design. For example, if the photocatalyst is expensive, the smaller photoreactor associated with a solar concentrator might be preferable. The catalyst would have to be more expensive than current estimates would indicate. Presently the photoreactor tube (including catalyst) comprises less than 10% of the treatment cost in a trough-based solar detoxification system (Link and Turchi, 1991).

There is less certainty about the efficiency of the photon use. No conclusive performance tests have been completed with photoreactors designed for nonconcentrating systems. However, literature data suggest that one measure of photoreactor efficiency, the quantum yield, may decrease with increasing light concentration (Egerton and King, 1979; Okamoto and co-workers, 1985). If confirmed in practical solar detoxification systems, this behavior would increase the relative advantage of nonconcentrating systems over trough-type concentrators.

Finally, in addition to medium-pressure mercury lamps, we estimated costs for a high-pressure mercury lamp system and a fluorescent lamp system using a method similar to that depicted in Table 1. These costs were $1.09 per 10^{25} photons and $2.59 per 10^{25} photons, respectively.

CONCLUSIONS

Because of their ability to capture both direct-beam and diffuse sunlight, nonconcentrating solar systems can collect more solar near-UV photons for less cost than concentrating systems. Additionally, for chemistry that can operate with near-UV photons, such as the TiO_2-photocatalyst system, a nonconcentrating solar collector can acquire photons less expensively than electrically generated photons. Naturally, total treatment cost depends on additional variables such as photocatalyst cost, photoreactor efficiency, pumping cost, and land requirements. Current research and development efforts are investigating these variables for both concentrating and nonconcentrating systems.

ACKNOWLEDGMENTS

The authors recognize that part of this analysis followed the method developed by Jane Diggs of Sandia National Laboratories, and we appreciate her assistance.

REFERENCES

Adams, J. Q. and R. M. Clark (1989). Cost estimates for GAC treatment systems. In *Organics Removal by Granular Activated Carbon*, American Water Works Association, Denver, CO, pp. 91-98.

Bird, R. and R. Hulstrom (1982). Extensive modeled terrestrial solar spectral data sets with solar cell analysis. SERI/TR-215-1598, Solar Energy Research Institute, Golden, CO.

Egerton, T. A., and C. J. King (1979). The influence of light intensity on photoactivity in TiO_2 pigmented systems. *J. Oil Col. Chem. Assoc.*, **62**, 386.

Kasten, F. and A. T. Young (1989). Revised optical air mass tables and approximation formula. *Applied Optics*, **28**, 4735-4738.

Link, H. F. and C. S. Turchi (1991). Cost and performance projections for solar water detoxification systems. Presented at American Society of Mechanical Engineers International Solar Energy Meeting, Reno, Nevada, March 1991.

Okamoto K., Y. Yamamoto, H. Tanaka, and A. Itaya (1985). Kinetics of heterogeneous photocatalytic decomposition of phenol over TiO_2 powder. *Bull. Chem. Soc. Japan*, **58**, 2023.

O'Reilly, J. (1990). Canrad Hanovia Area Representative, Wichita, Kansas, personal communication.

Parsons, R. E (1985). Shallow solar ponds: economical water heaters for industry. *Energy and Technology Review*, UCRL-52000-85-4, Lawrence Livermore National Laboratories, Livermore, CA.

Riordan, C.J., R.L. Hulstrom, and D. R. Myers (1990). Influences of atmospheric conditions and air mass on the ratio of ultraviolet to total solar radiation. SERI/TP-215-3895, Solar Energy Research Institute, Golden, CO.

SERI (1981). Solar radiation energy resource atlas of the United States. SERI/SP-642-1037, Solar Energy Research Institute, Golden, CO.

2.25 Solar Heat

DEVELOPMENT OF VOLUMETRIC AIR RECEIVERS

M. Böhmer*, M. Becker*, M. Sanchez**

* Deutsche Forschungsanstalt für Luft- und Raumfahrt e.V., Köln
** CIEMAT-IER, Plataforma Solar de Almeria, Tabernas

ABSTRACT

Volumetric air receivers are believed to be an alternative in the generation of electric energy from solar radiation as well as for the production of solar process heat, which can be used for many chemical processes. Though substantial development work has to be performed, volumetric receivers show clear advantages compared with tube receivers, mainly because of the three dimensional absorber geometry instead of the quasi-two dimensional form in the the case of the tube receivers. Volumetric receivers are light-weight, they show less infrared radiation losses; convection losses are lower, cracks or other failures in the structure do not lead to dangerous accidents and the heat transfer medium air is cheap, has an unlimited temperature capability, is not dangerous in case of leakages and anywhere available. However, of highest importance is the fact that flux densities up to a few MW/m^2 are tolerable. This is almost one magnitude higher than achievable with tube receivers.

During the last years a lot of different volumetric receiver concepts have been tested on the Plataforma Solar de Almeria in a 200 kW scale. Metallic absorber structures as well as ceramic ones, designed and built from wires, foils or foam were investigated. The present paper shows essential results of the different tests and compares and evaluates the findings.

KEYWORDS

Receiver, solar thermal energy, efficiency, material, solar tower, volumetric effect, solar power.

Though volumetric receivers have been investigated in the early eighties at different research institutes of the United States, a systematic investigation was initiated by the Sulzer Company and the SOTEL-Consortium in 1985 in Switzerland (Fricker, 1983, 1986). This finally led to first volumetric receiver tests in Almeria in 1987-88 under the roof of the IEA-SSPS-Project with participation of Germany, Spain and Switzerland.

In contrast to tube receivers, where the incoming solar energy heats the outer tube walls & is conducted to the inner surface of the tube and thus heats the cooling medium inside. this tube by enforced convection, in a volumetric receiver a quite different concept is followed: an absorbing structure of a specified porosity is installed in a certain volume within the receiver body. The concentrated radiation penetrates throughout the depth of the structure and is absorbed within the structured surfaces. At the same time air, acting as the heat transfer medium, is drawn through the structure in the same direction as the radiant energy, thus flowing around the absorbing structures and heated up convectively. The solar irradiation and the heat extraction occurs on the same surface simultaneously, whereas in the case of a tube receiver irradiation and the conversion into heat occurs on different surfaces.

As already mentioned above, the first volumetric receiver test was proposed by the Swiss Sulzer Company. The absorber consisted of ring-shaped wire mesh elements, made of stainless steel wires of 0,4 mm diameter and 1,65 mm mesh size. The absorber was installed into a receiver housing which also was used for the most of the following tests until today. This so-called "volumetric receiver testbed" is shown in Fig. 1. Testbed, absorber design, test procedure and results are explained in detail in SSPS TR 2/89 (1989). The calculated efficiency of that first wire mesh absorber was lower than expected. The reason for that low efficiency was the fact that some distortion occurred between the wire mesh rings, so that an uncontrolled air flow led to an increase in radiative losses and over-heating of some wire mesh parts. Therefore, a second absorber, made of coiled wire knitting, was installed during the test period. This so-called "Sulzer wire knitting receiver" showed a much better behavior than the first design. Figure 2a shows an efficiency of about 85% at low temperatures and even 75% in the maximum reached temperature range between 650 and 680°C.

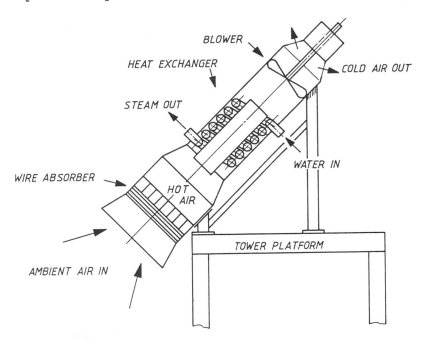

Fig. 1. Schematic View of the Volumetric Receiver Testbed on the SSPS-Tower

The next test, also using the Sulzer volumetric receiver testbed was proposed by the SANDIA National Lab., Albuquerque, in cooperation with the University of Colorado. The absorber was made from ceramic foam (the material being aluminum oxide), with a porosity of 80% and 20 pores per inch. The thickness of this ceramic foam was 30 mm.

As can be seen from Fig. 2b the so-called "SANDIA Ceramic Foam Receiver" showed a somewhat lower efficiency over the whole temperature range compared to the wire knitting receiver. The reason for this lower-than-expected efficiency is explained by the investigators namely with the fact that a lot of pores or micro-pores were blocked by the Pyromark paint that was used to increase the absorptivity. Furthermore the optical density of the material was too high. In addition some lost area existed because of ceramic cement used to join the 17 pieces of the absorber. Details are reported by Chavez and Chaza (1990).

Freudenstein and von Unger (1990) reported on the results of the "Interatom Metal Foil Receiver CATREC", which was built from modules designed for automotive catalytic converters. Figure 2c shows a relatively wide scatter band of the efficiencies in the lower temperature range up to 500°C around 80%. The absorber is designed of metal foils with a thickness of 0.05 mm, built with 62 channels per cm^2. Because of manufacturing reasons the absorber had to be put together from 5 pieces, which, with an increasing number of tests, formed gaps and led to a considerable amount of losses in the efficiency. With that the wide scatter band can be explained, but in principle for lower temperature ranges this absorber can be expected to work efficiently.

In order to reach higher temperatures the metal foil absorber can be replaced by a ceramic foil absorber. Such a test was performed in Almeria in 1989/90 with the "DLR/CeramTec Ceramic Foil Receiver", the results of which were reported by Becker, Böhmer, Cordes (1991). This nearly 1 m diameter absorber was designed from an array of .8 mm foils and 3 x 3 mm rods produced of SiSiC (infiltrated siliconcarbide). The absorber showed efficiencies in the order of 85 - 90% up to temperatures of about 500°C, where the efficiency started to decrease, but even at the maximum reached air temperature of 800°C the efficiency was still 60% (see Fig. 2d).

In parallel to these tests with pressureless, open air receivers in the volumetric receiver testbed two other research groups investigated volumetric receivers with windows. DLR-Stuttgart investigated a ceramic foam absorber with quartz glass window which is designed to work under high air pressures or even with other gases. Atlantis GmbH designed a ceramic grid absorber, which also worked with quartz glass tubes, but at atmospheric pressure. This receiver was tested at SANDIA in 1990. The results of these receivers cannot be compared directly with the results of the open atmospheric receiver tests. They are reported in the literature. A short summary of the tests is given in SSPS TR 2/91 (1991).

In principle it can be stated that with different designs of wires, foams and foils and different materials (metals or ceramics) volumetric receivers can be designed, which produce air up to 800°C at high efficiencies. It could be shown that the absorbers did not fail during the tests, even if the material temperatures exceeded 1000°C by far. During some tests even cracks in the materials occurred, parts broke off from the structure, but no change in the performance or even efficiency of the absorbers was detected. All of these absorber tests were first of their kind and it can be expected that with better manufacturing or improved design an increase in air temperature and efficiency is easily possible. Compared to tube receivers of the same power range, the volumetric receivers are cheap, show an easy design and are much smaller in size.

As planned between the participants of the Almeria test series and within the SSPS project Task III (Receiver Technology) a workshop on the status and the potential of volumetric receivers took place at DLR-Cologne, February 13 to 15, 1991. the most important recommendations of the experts convened were (SSPS TR-2/91):

- for small scale investigative field tests perform studies of relation between solar flux distribution and aperture surrounding flux resisting shield

- for a larger scale (2.5 MW) system tests specify the requirements, evaluate the facility capacities and design the experiment such that the absorber can be exchanged later on

- for computer modelling determine all absorber properties and "low-Reynolds number" heat transfer coefficients, carefully perform sensitivity analyses and produce a total program package for general evaluation.

The potential of volumetric receivers is clearly directed towards easy and inexpensive operation of heat transfer units that function in an efficient and safe way at flux densities up to the order of MW/m^2. In respect to closed cycles (high pressures, use of chemically reacting fluids) the long-term behavior of quartz windows has to be investigated in future and special knowledge and careful determination of interactions between radiation/absorption processes with any chemical (e.g. catalytic) action has to be developed.

Table 1 Plataforma Solar de Almeria Volumetric Absorber Test Results (1987 - 1990)

ABSORBER		SULZER WIRE MESH	SULZER WIRE KNITTING	IA METAL FOIL	SANDIA CERAMIC FOAM	DLR CERAMIC FOIL	CONPHOEBUS CERAMIC FOIL (PROVISIONAL)
Diameter	mm	875	875	875	875 - 5%	875	700
Peak Flux	kW/m^2	960	757	844	824	840	917
Average Solar Flux (Solar Flux/ area)	kW/m^2	265	218	254	246	240	255
Material Temperature	° C	---	---	1070	1350	1320	1238
Maximum air Temperature	° C	780	680	570	730	782	788
Thermal Efficiency at 550 °C air Temperature	%	68	79	73	65	80	70
Thermal efficiency at maximum air Temperature	%	46	75	69	54	59	60

Table 1 represents the most essential date of six volumetric test series performed at the Plataforma Solar de Almeria from 1987 to the end of 1990. All test results concern an open cycle when atmospheric air was aspirated. The information addresses the diameters, flux densities, material and air temperatures and efficiencies.

References

Becker, M., Böhmer, M., and Cordes, S. (1991), DLR/CeramTec Volumetric Ceramic Foil Receiver. SSPS TR-1/91, DLR, Köln, 1991

Chavez, J., and Chaza, C. (1990). Results of the Testing of a Porous Ceramic Absorber for a Volumetric Air Receiver. Presented at the 5th Symposium on Solar High Temperature Technology, Davos/CH, 1990

Freudenstein, K., and v. Unger, E. (1990) In Becker, M. and Böhmer, M. (Eds.). Volumetric Metal Foil Receiver CATREC - Development and Tests. SSPS TR-1/90, DLR, Köln

Fricker, H.W. (1983). A proposal for a novel type of solar gas receiver. International Seminar on Solar Thermal Heat Production, DLR-Stuttgart, October 13-14, 1983

Fricker, H.W. (1986). Tests with a small volumetric wire receiver. In Becker, M. (Ed.), Solar Thermal Central Receiver Systems, Vol. II. Proceedings of the Third International Workshop, Konstanz, 1986, Springer Verlag.

SSPS TR-2/89 (1989) Report of the Wire Pack Volumetric Receiver Tests Performed at the Plataforma Solar de Almeria, Spain, in 1987 and 1988 (SSPS Task VII - First Experiment), DLR, Köln

SSPS TR-2/91 (1991). Böhmer, M., and Meinecke, W. (Eds.) Proceedings of the Volumetric Receiver Workshop, Köln, Feb. 13th - 15th, 1991.

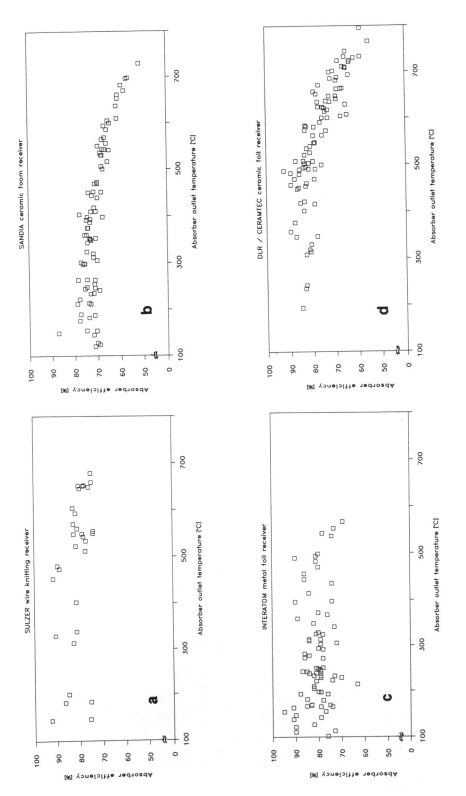

Fig. 2. Efficiencies of different absorbers vs. air outlet temperature.
a: metal wire absorber, b: ceramic foam absorber, c: metal foil absorber,
d: ceramic foil absorber

PREDICTION OF RADIATIVE PROPERTIES OF COAL PARTICLES
SUITABLE FOR ABSORPTION OF SOLAR ENERGY IN GAS

J. Oman, P. Novak

Faculty of Mechanical Engineering
University of Ljubljana, Slovenia, Yugoslavia

ABSTRACT

Black gas is an ideal absorber of solar energy. Such ideal conditions can be approached by adding absorptive particles to gas. Together with a high range of absorptivity, the particles must have minimal capability of reflection of solar radiation. On the basis of Mie's theory, projections of the optical properties of individual small black and brown coal, anthracite and coke particles are illustrated and compared. A backscattering efficiency factor was introduced to determine the solar radiation scattered backwards in an hemispherical angle towards the source of radiation. The results of analysis show which material and particle size are most appropriate for the absorption of solar radiation in gas particle suspension.

KEYWORDS

Solar energy; particle-gas receiver; small particles; coal; reflection; absorption;

INTRODUCTION

The process of heating gases to high temperatures by concentrated solar radiation can be performed by adding particles which absorb solar radiation to the gas. This process was first described by Hunt (1978). To minimize the energy loss in the initial layer of the absorber, the added particles must have, in addition to high absorption, as low as possible an ability for reflecting solar radiation back towards the source of radiation. It is necessary to answer questions as which size and material of particle fits those conditions, how large is the amount of scattered radiation which can be considered reflected and how to minimize it. Although the particles of pulverized coals cannot be spherical in shape, Blokh (1988) and Viscanta (1981) found that for randomly oriented particles, the radiative properties are not very sensitive to shape. The particles discussed in this paper are considered as homogeneous spheres. The optical properties of spherical particles can be predicted using the approach suggested by Mie, clearly explained by Hulst (1957) and Kerker (1969). Experiments on the absorption of solar radiation applied to particle-gas suspension, performed by Oman (1989), indicate that acetylene carbon black particles produced by pyrolytic decomposition have the predicted properties and are very suitable in terms of optical properties. However, the major problem is the high cost technology of carbon black particle production. So in the view of the lower production expenses, coke, black and brown coal, anthracite, etc.,are of interest as a particle mixture for absorbing highly concentrated solar radiation. Nevertheless, there is another possibility worth attention: to carry out the process of gasification of coals in particle-gas solar receivers by means of highly concentrated solar radiation.

THEORY

Spectral properties

From the results of the Mie theory, the dimensionless efficiency factors for scattering and extinction are given as suggested by V.D.Hulst (1957):

$$K_{sc\lambda} = \frac{2}{x^2} \sum_{n=1}^{\infty} (2n+1)[|a_n|^2 + |b_n|^2] \tag{1}$$

and

$$K_{ex\lambda} = \frac{2}{x^2} \sum_{n=1}^{\infty} (2n+1)[Re(a_n + b_n)] , \tag{2}$$

respectively, where the size parameter $x = \pi D/\lambda$, D the diameter of the particle and λ the wavelength of the incident electromagnetic radiation. The scattering coefficients a_n and b_n are expressed in terms of spherical Bessel functions. The efficiency factor for absorption is expressed indirectly:

$$K_{ab\lambda} = K_{ex\lambda} + K_{sc\lambda} . \tag{3}$$

A sphere with diameter D and geometrical cross section $A_c = \pi D^2/4$ will intercept power $I_0 K_{ex\lambda} A_c$, from the incident radiation of intensity I_0. The part absorbed and the part scattered are found by replacing $K_{ex\lambda}$ with K_{ab} and $K_{sc\lambda}$ respectively. When consider the intensity scattered in different directions, the angular distribution function should be defined. Kerker (1969) proposed the expression:

$$G_u = \frac{2}{x^2} [i_1(\theta) + i_2(\theta)] \tag{4}$$

as an angular efficiency function for unpolarized incident radiation. Herein belong the complex scattering amplitudes $S_1(\theta)$ and $S_2(\theta)$ (Kerker, 1969) which determine the intensity functions $i_1(\theta) = |S_1(\theta)|^2$ and $i_2(\theta) = |S_2(\theta)|^2$. The efficiency for scattering can be now calculated from the expression:

$$K_{sc\lambda} = \frac{1}{x^2} \int_0^{\pi} [i_1(\theta) + i_2(\theta)]\sin\theta \, d\theta. \tag{5}$$

The plane through the center of the particle, perpendicular to the incident radiation, divides all space around the particle into the hemispherical angle $(2\pi)^+$ in the direction of propagation of incident radiation (forward side), and the hemispherical angle $(2\pi)^-$ in the direction towards the source of radiation (backside). Similary, the right side of the equation (6) is divided into two parts, which yields:

$$K_{sc\lambda} = K_{od\lambda} + K_{nap\lambda} \tag{6}$$

The first term on the right side is the efficiency factor for reflection and may be calculated from:

$$K_{od\lambda} = \frac{1}{x^2} \pi \int_{\frac{\pi}{2}}^{\pi} [i_1(\theta) + i_2(\theta)]\sin\theta \, d\theta . \tag{7}$$

If $I_{0\lambda}$ is the spectral intensity of radiation incident on a particle with geometrical cross section A_c, then the particle will scatter $K_{od\lambda} A_c I_{0\lambda}$ power backwards into the hemispherical angle $(2\pi)^-$.

Complex Refractive Index, Solar Spectral Irradiance

The complex refractive index which characterizes the interaction of electro-magnetic radiation with a substance is set out in the form $m = n - n'i$. The values of the real and imaginary part act as functions of the wavelength. In our calculations we used the data on complex refractive indices for various fuels that was collected by Blokh (1988). The radiative properties of four different carbonaceous materials, anthracite, black and brown coal and coke were examined. When using the ASTM classification the rank of coals taken into account is as follows: anthracite from group 2; black coal: -bituminous, -high volatile A; brown coal: group -subbituminous C with 50% volatile matter in the moisture -and ash-free mass.
Data for solar spectral irradiance on the earth's surface in the wavelength interval 0.3 μm $\leq \lambda \leq$ 2.5 μm were taken from ISO (1987) and have been arranged so as to enable Simpson's composite integration. Global solar irradiance over all included wavelengths of such simulated solar radiation is 750.4 W/m^2 when integration wavelength subintervals are 0.05 μm. Figure 1. presents the spectral irradiance used in calculations of global properties.

Global Properties

The power that individual particles scatter into the hemispherical solid angle $(2\pi)^-$ from the incident solar radiation is calculated by the equation:

$$P_{od} = A_c \int_{\lambda_1}^{\lambda_2} I_{0\lambda} K_{od\lambda} \, d\lambda . \tag{8}$$

The power which a particle scatters in all directions P_{sc}, and the power that particle absorb P_{ab}, is found by replacing $K_{od\lambda}$ in eq.(10) by $K_{sc\lambda}$ and $K_{ab\lambda}$, respectively. The power that a particle physically intercepts; i.e. the power contained in the incident beam of solar radiation having the same cross-sectional area as the particle is:

$$P_{in} = A_c \int_{\lambda_1}^{\lambda_2} I_{0\lambda} \, d\lambda . \tag{9}$$

The integrals in this section are calculated by the application of composite Simpson's rule for individual particle sizes from $\lambda_1 = 0.3$ μm to $\lambda_2 = 2.5$ μm with an integration subinterval $\Delta\lambda = 0.05$ μm.

RESULTS AND DISCUSSION

The calculations show that the absorptivity of carbonaceous particles increases quickly with particle diameter. At wavelength 0.3 μm, coal particles with diameter around 0.5 μm are most absorptive. At longer wavelengths, the maximal absorptivity is for the largest particles and differ greatly with the particle substance, in large intervals of diameter, as can be seen from figure 2.

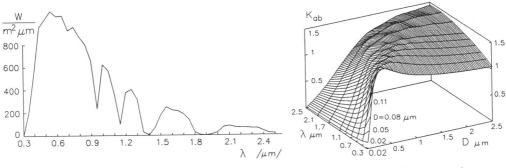

Fig. 1. Spectral solar irradiance.   Fig. 2. $K_{ab\lambda}$, brown coal.

The power reflected by a particle in solid angle $(2\pi)^-$ varies with the substance, diameter of particle and wavelength of incident radiation. The interaction is clear when the plot of angular efficiency function G_u is drown in the angle with borders $\pi/2$ and $3\pi/2$. The integral of function G_u with borders $\pi/2$ to $3\pi/2$ is proportional to the power that the particle reflect in the hemispherical angle $(2\pi)^-$, i.e. the values of G_u indicate the amount of reflected power. Figure 3. presents the angular distribution of scattered radiation, for different diameters and four substances at wavelength 0.5 μm, around which the intensity of solar radiation is highest.

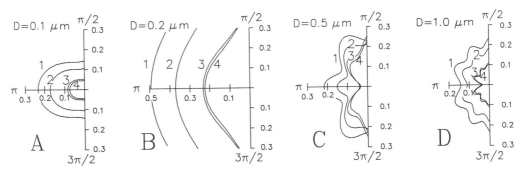

Fig. 3. The angular efficiency function at λ = 0.5 μm. Substances are:
1- coke, 2- anthracite, 3- black coal, 4- brown coal.

It is evident that the maximal ability of reflection exist as a function of particle diameter and wavelength of incident radiation. This is clear when the values of reflection efficiency are presented as functions of diameters and wavelengths for individual substance, fig. 4., where four points indicate the situations presented in fig. 3. in the case of anthracite particles. When increasing the diameter of particles, three intervals exist. The first, where the reflection by unit of geometrical cross section of particles is minimal, for instance at point A, the second, where maximal reflectivity appears (point B), and the third interval of diameters, where the reflection by unit of geometrical cross section is small (points C, D).

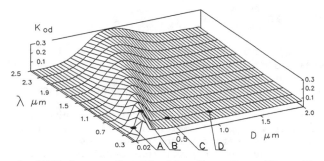

Fig. 4. Efficiency of reflection $K_{od\lambda}$ for anthracite.

At the same wavelength the location and maximal values of reflection efficiency differ with particle substances.

TABLE 1. The Power of Solar Radiation Absorbed, Scattered and Reflected by Carbonaceous Particles.

D	P_{in} $\cdot 10^{10}$	$\dfrac{P_{ab}}{P_{in}}$ $\cdot 100$	$\dfrac{P_{sc}}{P_{in}}$ $\cdot 100$	$\dfrac{P_{od}}{P_{in}}$ $\cdot 100$	$\dfrac{P_{ab}}{P_{od}}$	$\dfrac{P_{ab}}{P_{in}}$ $\cdot 100$	$\dfrac{P_{sc}}{P_{in}}$ $\cdot 100$	$\dfrac{P_{od}}{P_{in}}$ $\cdot 100$	$\dfrac{P_{ab}}{P_{od}}$
μm	W	%	%	%	–	%	%	%	–
		COKE:				ANTHRACITE:			
0.04	0.009	9.5	0.23	0.11	171	17.6	0.18	0.09	205
0.06	0.021	31.1	1.2	0.57	54	27.8	0.91	0.43	64
0.08	0.038	44.3	3.8	1.7	25	39.2	2.86	1.30	30
0.10	0.059	59.0	9.1	3.9	15	51.7	6.68	2.89	18
0.12	0.085	74.2	17.1	6.8	11	64.8	12.9	5.10	12
0.22	0.28	129.5	70.1	18.9	7	118.1	59.5	14.9	8
0.27	0.43	142.3	90.4	20.1	7	133.0	79.9	16.2	8
0.32	0.60	149.1	104.3	19.7	8	142.3	95.1	16.1	9
0.37	0.81	152.4	113.4	18.7	8	147.5	106.0	15.4	10
2.05	24.8	113.5	133.6	9.7	12	117.5	129.8	7.8	15
2.14	27.0	112.4	133.4	9.7	12	116.0	129.8	7.8	15
2.23	29.3	111.5	133.4	9.7	12	115.0	129.7	7.8	15
		BLACK COAL:				BROWN COAL:			
0.04	0.009	9.7	0.09	0.045	224	10.4	0.08	0.04	273
0.06	0.021	15.1	0.45	0.21	70	16.2	0.4	0.19	85
0.08	0.038	20.1	1.4	0.64	33	22.4	1.3	0.57	39
0.10	0.059	27.4	3.4	1.4	19	29.1	2.9	1.3	23
0.12	0.085	34.3	6.6	2.6	13	36.2	5.8	2.3	15
0.28	0.46	88.5	63.2	9.6	9	89.9	55.5	8.4	10
0.30	0.53	93.6	70.2	9.7	10	95.0	62.4	8.5	11
0.32	0.60	98.3	78.0	9.7	10	99.6	68.9	8.5	12
0.34	0.68	102.5	84.7	9.6	11	103.8	75.1	8.4	12
2.05	24.8	124.9	123.1	4.49	28	124.9	121.1	3.95	32
2.14	27.0	123.9	122.6	4.43	28	123.9	120.9	3.92	32
2.23	29.3	122.9	122.2	4.38	28	122.9	120.7	3.88	32

CONCLUSIONS

The article presents calculated values by means of which the radiative properties of individual anthracite, black and brown coal and char particles can be
predicted. The data of solar spectral irradiance were used to calculate the
power that individual particles absorb, reflect and scatter from incident
solar radiation.
The calculations show that the efficiency factor for absorption and absorptivity rapidly increases with the diameter of particles and remains high for all
substances and particle diameters taken into account.
The scattering efficiency and the power that particle scatter in all direction
also rapidly increases with diameter but primarily as forward scattering. Such
a kind of asymmetric scattering is characteristic of most carbonaceous
substances. The power scattered by particles in a forward direction is not
very significant in our case. More important is the power scattered by the
particle back i.e. reflected power.
In connection with reflected radiation by carbonaceous particles, three
significant intervals of particles diameters exist. The first is the interval
of the smallest particles, up to 0.1 μm, where reflection is minimal, yet
absorption remains sufficient for all substances. In the second interval
particle diameters, up to 1.5 μm, the absorption and also reflection is
maximal (at coke particles up to 20 % of incident solar power). In the
interval of particle diameters greater than 2.0 μm, the reflection is
significantly smaller again, but only for black and brown coal. The results
presented on Table 3 show, that in third interval brown coal particles will
reflect 3-4%, anthracite almost 8% and char particles around 10% of the power
of incident solar radiation. When using coke or anthracite particles the high
efficiency of a gas-particle absorber can be attend only through mixing
particles with modal size around 0.1 μm or less. But when mixing brown coal
particles, very high efficiency can be attend also, when the size of particles
is 2 μm or more.
It is obvious that in the examined interval of diameters, black coal and
especially brown coal particles are optically much more suitable than
anthracite or char particles as absorptive matter in a gas-particle absorber
of solar radiation.

REFERENCES

Hunt, A.J. (1978). Small Particle Heat Exchangers. Lawrence Berkeley Lab.,
Univ. of California, LBL-7841, UC 62.

Hulst, V.D. (1957). Light Scattering by Small Particles. J.Willey, New York.

Kerker, M. (1969). The Scattering of Light. Acad.Press, New York.

Blokh, A.G. (1988). Heat Transfer in Steam Boiler Furnaces. Hem. Pub. Corp.,
New York.

Viscanta, R., A. Ungan (1981). Prediction of Radiative Properties of Pulverized Coal and Fly-Ash Polydispersions, ASME Paper, No.81-HT-24.

Oman, J., (1989). Direct Absorption of Solar Radiation with Solid Particles in
Gas. Dr.thesis, Faculty of Mechanical Engineering, Ljubljana.

ISO/DP 9839 (Draft Proposal), (1987). Solar Energy; Terrestrial Solar Spectral
Irradiance.

SOLAR THERMAL WATER PUMP

A.Venkatesh

Department of Mechanical Engineering
Indian Institute of Technology, Madras,India

ABSTRACT

The operation of a solar thermal water pump is explained briefly with the help of schematic and p - v diagrams. The performance of the pump is predicted through heat transfer analysis of the flat plate collectors and the thermodynamic analysis of the cycle of events. The predicted values are compared with the experimentally determined values for three discharge heads of 10 m, 7.5 m and 5 m and the agreement between the two is found to be good.

KEYWORDS

Flat plate collectors, collector analysis, water pump, non-conventional pump, N - pentane.

INTRODUCTION

The pump discussed in this paper operates on the following principle: Vapor of a low boiling liquid, generated through flat plate collectors provides the motive power for lifting water, while the condensation of the vapour and the subsequent decrease in pressure provides for suction of water into an un-conventional pump (Rao and Rao, 1976; Sudhakar and co-workers, 1980). A small unit, with flat plate solar collectors of 3 m^2 exposed area, operating on this principle, with N-pentane as the working sub-stance, built in the laboratory, is subjected to experiments. For an assumed set of parameters number of cycles per day, mass of water lifted per day and the overall efficiency are evaluated by the thermodynamic analysis of the system. The predicted values are compared with the experimentally determined values.

SYSTEM DESCRIPTION

Figure 1 shows the schematic of the solar thermal water pump. Laboratory grade N-pentane is heated in the flat plate collectors by thermosyphon flow. When the pressure in the separation tank reaches the pressure corresponding to the discharge head of the pump, valves 1 & 2 are opened slowly. Pentane vapour separating in tank S enters vessel A containing water. Consequently water in vessel A is gradually transferred to vessel B which contains air initially at atmospheric pressure. Air in vessel B gets compressed. When the pressure of air in vessel B reaches the

pressure corresponding to the discharge head of the pump the
compressed air displaces water in vessel C to the overhead tank
D, thereby effecting the required pumping. When water in vessel C
is completely pumped out, valves 1 & 2 are closed. At this instant vessel A contains pentane vapour, vessel B contains water
and compressed air and vessel C contains compressed air. Valve 3

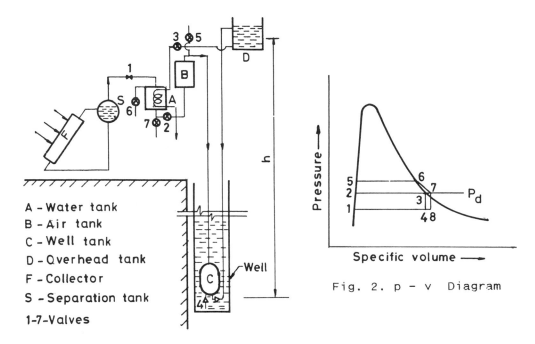

A - Water tank
B - Air tank
C - Well tank
D - Overhead tank
F - Collector
S - Separation tank
1-7 - Valves

Fig. 1. Schematic of the pump.

Fig. 2. p - v Diagram

is now opened to the required extent to allow water from the
overhead tank D to flow through the cooling coils in vessel A. As
pentane vapour starts condensing, the pressure in vessel A begins
to decrease and eventually reaches the saturation pressure of
pentane corresponding to the condensing temperature. During this
period water in vessel B returns to vessel A. This reduces the
pressure of air in vessels B and C, as a consequence of which
water from the well gets into vessel C through the one way valve
4. The system is now ready to perform the next cycle. In the p -v
diagram shown in Fig.2, 1-2-3-4-1 shows the processes undergone
by pentane in the first cycle of pumping water. During the
period of condensation, pentane in the collectors is being continuously heated by solar energy. Hence, at the start of the
second cycle the temperature, and hence the pressure of pentane
(states 5 and 6) will be higher than those at the beginning of
the first cycle. Valves 1 & 2 are opened slightly at this instant. Pentane vapour at state 6 in the separation tank S expands
to state 7 to reach the pressure corresponding to the discharge
head. The process of lifting water and the subsequent condensation of pentane follow the same pattern as before. The process
undergone by pentane during this cycle is shown by 2-5-6-7-8-1.
Similar sets of events occur in the subsequent cycles which

continue as long as pentane in the collectors can get heated up.

ANALYSIS OF THE PUMP

To begin with it is assumed that the collector and the separation tank S (Fig.1) contain m kg liquid pentane at a temperature of t_1°C (Fig.2). t_1 is assumed to be equal to the ambient temperature, t_a. At this instant the system pressure will be equal to the saturation pressure of pentane corresponding to temperature t_1. The collector is exposed to sun. Solar radiation incident on the collectors is assumed to vary sinusoidally from sunrise to sunset with the maximum intensity occurring at solar noon. The temperature, and thus the pressure of pentane gradually increase as it gets heated up by solar energy. During this period the heat transfer is essentially transient in nature. Heat transfer to liquid pentane is equal to solar energy absorbed by the collector minus the sum of thermal losses from the collector and the increase in internal energy of the collector components. With the knowledge of the collector characteristics, the meteorological data and the thermodynamic and transport properties of pentane as a function of temperature, and using the technique employed by Venkatesh and Sriramulu (1989), the heat transfer to pentane, and thus, the rise in pentane temperature at step interval of 0.01 hours is evaluated with the help of a computer. The pentane pressure at each step is evaluated from (Reid and Sherwood,1958),

$$p = \{exp.[15.833 - 2477.07/(T - 39.94)] \times 1.013/760\} \qquad (1)$$

where p is in bar and temperature T is in Kelvin.
When the pressure reaches the value that corresponds to the discharge head of the pump (say 2 bar abs. for a discharge head, h = 10 m) valves 1 & 2 are opened slightly. At this instant, in the separation tank, S, the liquid pentane is at state 1 and pentane vapour is at state 2 (Fig.2). Pentane vapour slowly travels to vessel A to do the job of pumping the water as explained earlier. An assumption made is that pentane vapour travels slowly to vessel A such that liquid pentane in the separation tank does not flash, and the vapour is generated in the tank at a steady temperature of t_2. The mass of pentane vapour required in the first cycle of events, m_1, is calculated using the relation,

$$m_1 = V/v_3 \qquad (2)$$

where V is the volume of pentane vapour required and v_3 is the specific volume of pentane vapour at state 3. Theoretically V is equal to volume of water transferred from vessel A to vessel B. Assuming air in vessel B gets compressed isothermally V is related to the discharge head, and the volumes of vessels A, B and C by

$$V = V_B(1 - p_d^{-1}) + V_C \qquad (3)$$

The time required to generate m_1 kg of pentane vapour is calculated as that during which heat transferred, Q_1, to pentane in the collector is given by,

$$Q_1 = m_1 \times h_{fg} \qquad (4)$$

where h_{fg} is the enthalpy of vaporisation of pentane corresponding to temperature t_2. At the end of pumping of water from Vessel C, valves 1 and 2 are closed and valve 3 is opened partially such that cooling water passes through the cooling coils in vessel A at a required rate. Pentane vapour at state 3 is condensed to reach state 1. At the end of condensation, the condition in vessels A,B and C would be the same as those at the start of the first cycle and hence the system will now be ready to perform the next cycle.

During the time taken for the condensation in the first cycle, which is calculated theoretically using established methods relevant to shell and tube condensers, the temperature and pressure of remaining pentane in the collector and separation tank, $(m - m_1)$ kg, would have gone up because of its continuous heating by solar energy. The heat transfer required to raise the temperature of this mass of pentane from t_2 to t_5 (Fig.2) is evaluated using the technique referred to. Assuming that at the start of the second cycle, 5 is the state of liquid pentane and 6 is the state of vapour, the pressure of pentane, p_6, will be higher than that corresponding to discharge head, p_2. It is assumed that pentane vapour expands adiabatically from p_6 to p_2 and enters the vessel A at state 7. The mass of vapour required for this cycle, m_2, is given by,

$$m_2 = V/v_7 \qquad (5)$$

where v_7 is the specific volume of pentane vapour at state 7. The time required to generate this mass of vapour is evaluated as in the case of the first cycle. The process of lifting water from the well and the subsequent condensation of pentane vapour follow similar steps as explained earlier. The analysis is carried out for all subsequent cycles until such time that the temperature of pentane in the collectors keeps increasing. The analysis reveals that the pentane temperature keeps increasing until around 14.00 hours.

V_w, the volume of water lifted per day by the pump is given by,

$$V_w = V_c \times N \qquad (6)$$

where N is the number of cycles per day.
η, the overall efficiency of the pump is given by,

$$\eta = V_w \rho_w g \ h/H_{tot} \qquad (7)$$

where ρ_w is the density of water and H_{tot} is the total solar radiation incident on the collectors during the period of working of the pump.

EXPERIMENTAL SET-UP

The experimental set-up consists of tube-in-sheet type, single glazed, copper flat-plate collectors of exposed area 3 m^2. The absorber sheet is 0.4 mm thick and, while the riser tubes (six per collector of 1 m^2 area) are 12.5 mm in diameter, the headers are 25 mm in diameter. The separation tank S is a mild steel cylinder, 15 cm in diameter and 3 m long. The vessels A,B & C are fabricated out of 3-mm thick mild steel plates. While V_C, the volume of vessel C is 15 liters, V_A and V_B are 50 and 60 liters respectively, and these are the design values for a discharge head of 10 m. Calibrated Bourdon pressure gauges are used to measure the pressure in the separation tank S and vessel A. Calibrated copper-constantan thermocouples, in respective thermometric wells, are used to measure the temperature of pentane in S and the inlet and outlet temperatures of water passing through the cooling coils. The cooling coil is made out of 6.5 mm diameter copper tubes and is 5 m long in 6 coils. The average diameter of the coils is 25 cm.

The collectors and the separation tank are filled with 27 kg of N-pentane. The collectors are exposed to solar radiation from 8.00 hours. The pressures and temperatures of pentane are measured at frequent intervals of time. The intensity of global radiation is continuously recorded using a Kipp & Zonen solarimeter. The pumping of water is carried out, more or less, on the lines explained in theory, until around 14.00 hours.

RESULTS AND DISCUSSION

Figure 3 shows the variation in number of cycles per day, both theoretical and experimental, with discharge head for an assumed maximum intensity of solar radiation at noon of 1000 W/m^2 and an ambient temperature of 30°C. It can be seen from Fig.3 that the number of cycles per day increases with a decrease in the discharge head. As the discharge head decreases, the hydraulic work

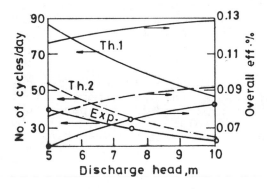
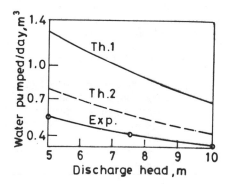

Fig. 3. Effect of discharge head on   Fig. 4. Effect of head on
      no.of cycles and efficiency.   water pumped/day.

decreases. To meet the decreased hydraulic work, the volume of pentane vapour required to pump a given amount of water per cycle decreases. As the volume of pentane vapour required decreases, the time required to generate this vapour and to condense it in each cycle decreases. This increases the number cycles that could be performed in a given interval of time. There is, however, a large difference between the theoretical (as shown by curve marked 'Th.1') and experimental values. This difference is mainly because of the practical difficulties that exist in performing experiments strictly according to theory. In theory it is assumed that pentane, after condensation in each cycle, would continue to float on water in vessel A. However in experiments the condensed pentane will have to be essentially removed as frequently as possible. The liquid pentane cannot be taken back to the separation tank because of adverse pressures. When the condensed pentane floating on water increases beyond 1 litre, it is drained out to a storage vessel. It is found that it takes nearly ten minutes to drain the pentane each time. During this period the pump is out of action. This requirement reduces the total operating time of the pump and thus actual number of cycles is small. Contrary to the assumption made in theory, in practice the pump starts operating only when the pressure of pentane in the separation tank is atleast 60 % more than the discharge head. This causes a delay in the start of the first cycle which reduces further the number of cycles. When these experimental constrains are included in theory and the number of cycles is re-evaluated, the variation follows the curve marked 'Th. 2' in Fig. 3. It can now be seen that agreement between predicted and the observed values is good.

Figure 4 shows that the total quantity of water lifted per day increases with a decrease in the discharge head. It may be pointed out here that the total quantity of water lifted per day depends only upon the number cycles per day, as the quantity of water lifted per cycle is fixed at 15 liters in theory. In practice water lifted in each cycle is slightly less than 15 litres.

Figure 3 also shows the variation in the overall efficiency of the pump with discharge head. The efficiency decreases with a decrease in discharge head. This is because , although the number of cycles per day increases with a decrease in the discharge head, the hydraulic work required to lift water per cycle decreases. Although the absolute values of efficiencies are very small, there is a good agreement between the predicted and measured value.

CONCLUSIONS

Using heat transfer analysis of the collectors and the thermodynamic analyses of the cycles of events, the performance of a solar thermal water pump of the kind explained in this paper can be predicted. For a given intensity of solar radiation, while the number of cycles per day and the volume of water pumped per day increase with a decrease in discharge head, the overall efficiency decreases. The nature of variation of experimental results agree well with that of predicted results. However, there is a difference in the measured and predicted magnitudes. This difference is mainly due to the fact that, for practical reasons, the pump can operate only for about 30 to 40 % of the theoretical time of operation of the pump. When the practical restrictions are incorporated in the theory, the agreement between the predicted and experimental results is found to be fairly good.

REFERENCES

Rao,D.P., and K.S.Rao (1976). J.Solar Energy, 18, p.405.
Sudhakar,K., M.Muralikrishna, D.P.Rao, and A.K.Srivastava
    (1980). J.Solar Energy, 24, p.71.
Venkatesh,A., and V.Sriramulu (1989). J.Energy, 14, p.23.
Reid,R.C., and T.K.Sherwood (1958). The Properties of Gases
    and Liquids. McGraw - Hill Book Co., Inc.

DOUBLE DIAPHRAGM SOLAR POWERED WATER PUMP

M.R.Amor[*], J.K.Raine[**], A.S.Tucker[**]

[*]Postgraduate student, [**]Senior Lecturers
Department of Mechanical Engineering, University of Canterbury,
Christchurch, New Zealand.

ABSTRACT

To utilise solar power for small scale water pumping in developing countries, the trade-offs between mechanical simplicity, maintainability and performance are critical. This paper presents the basis of theoretical analyses to optimise dimensions and improve the performance of a novel double-diaphragm, solar-thermal powered water pump operating on a modified Rankine cycle. In finding the most desirable operating temperature and cycle, it is noted that the degree of cloudiness has little effect on the optimum operating temperature. From exergy analyses it is found that the conversion from solar radiation to stored energy in the working fluid is the major irreversibility followed by the irreversibilities brought about by the processes within the actual operating cycle. By computer simulation, optimum designs for different operating conditions are found.

KEY-WORDS

Solar power; water pump; appropriate technology; solar water pumping; multi-variable optimisation; exergy analysis.

INTRODUCTION

The thermodynamic utilisation of solar radiation for water pumping is an inherently sensible proposition given that a solar water pump's output will be higher in hot sunny weather, when demand for water is also high. An analysis of the economics of such water pumping systems leads to the conclusion that these systems are potentially feasible in areas where there is no readily available alternative form of power other than solar radiation. (eg. in developing countries or isolated locations). Such pumps must suit local conditions, taking into account not only the required pumping head and climatic conditions of the locality, but also the pump should meet the following criteria to be of practical use:

- The unit should perform an equivalent amount of work to that of the hand pump which it would most likely replace.
- It should be a low maintenance device having few moving parts, able to be easily maintained by local poorly skilled technicians and thus fit into a Village Level Operation and Maintenance (VLOM) scheme as set out by the World Bank for similar devices (Arlosoroff, 1987).
- The unit should rely on appropriate technology to enable local manufacture and utilise a non-CFC working fluid.
- The working fluid should preferably work at above atmospheric pressure so that any leaks are to the atmosphere, ensuring no deterioration in performance due to system leakages.

The design and construction of a solar water pump to meet the above conditions at an appropriate price would have many advantageous applications in the developing countries, providing potable water with little or no associated human labour.

OPERATING PRINCIPLE

The apparatus described here uses a plain hydrocarbon working-fluid, n-pentane, common to both the solar collector and pump unit. The pentane is vaporised in the solar collector effecting an 100-fold increase in volume for a small increase in temperature. This increase in volume and pressure is the driving force for the displacement pump. The motion of the displacement pump impels an amount of the second fluid (water), which it is desired to pump from a certain depth below the pump unit to a certain height above the pump unit. Upon the displacement pump reaching its full extension the vapour in the pump is exposed to a condenser, causing a reduction in pressure and drawing the displaced unit back to its initial position. In the initial position the displacement unit is once again exposed to the high pressure vapour from the solar collector. To complete the cycle the condensed liquid is returned to the solar collector via a small return pump. The thermodynamic cycle undergone by the working-fluid is a modified Rankine cycle.

DESCRIPTION

The whole system has been designed to draw water up 6.5 metres and deliver it at up to 1.5 metres above the pump unit. The pump unit comprises two separate chambers arranged so that an increase in volume of one chamber effects an increase in volume of the second chamber. The ratio of volume increase of the lower (vapour) chamber (see Figure 1) to upper (water) chamber can at the design and production stages be altered so as to produce optimal pressures and temperatures for the working fluid. In choosing the design temperature, consideration is given to the head of water to be pumped, ambient conditions and incoming solar radiation (related to latitude and cloudiness).

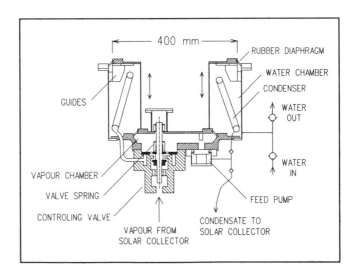

Fig. 1. Double diaphragm water pump

The pump unit is controlled by a two-way valve (Stubbs, 1980) which toggles between its two operating states, linking the vapour chamber directly to either the solar collector (boiler) or the condenser situated in the water chamber. The valve's change from one position to the other is controlled directly by the pump's displacement. Initially the displacement of the lower diaphragm compresses a valve spring. Before the spring is fully compressed, however, a stop is reached which causes the valve to be unseated from its stable position, where it was being held by the differential pressure across the seated portion of the valve. The valve is then delivered to its opposite position by the spring. Once in the new position, the valve is held in place by the spring until the pressure in the vapour chamber has changed sufficiently for the valve to be held in place by differential pressure. As the diaphragm returns towards its initial position, a similar spring and stop action toggles the valve back to its initial position. This in effect makes the valve end biased according to the pump diaphragm's displacement.

The solar collector is of the plate and tube variety with the tubes 90% full of the pentane working fluid. The collector utilises a transparent polycarbonate cover which exhibits similar spectral properties to glass. The polycarbonate cover has twice the thermal resistance of glass due to multiple air gaps running the length of the cover, effectively providing a double baffle for conductive-convective heat losses.

Most designs for such water pumps conform to a single diaphragm arrangement where the induction of the vapour causes the ejection of the water and vice versa. These units operate with essentially the same pressure on either side of the diaphragm. For such units to draw water up from a well, the diaphragm unit can be placed at the bottom of the well and a high pressure working fluid used to force the water to the top of the well. This method distances the solar collector from the pump, makes the pump unit hard to get at for maintenance and looses a significant amount of energy between the pump unit and the solar collector. A second option is to use a vapour at well below atmospheric pressure in a pump at ground level to suck the water up to ground level. The low pressure units are subject to leaks of air into the system, deteriorating the performance of the pump and require more frequent purges of the working fluid. The first system can be used at the top of the well by using a large spring to store the energy between successive parts of the cycle (Sharma, 1980). For such systems to support different water pumping heads, changes in spring force, working fluid, or position of the pump head must be made in order to achieve optimum system performance.

THEORETICAL ANALYSIS

The simulation of a solar-thermal device's performance over a period of time is complicated by the interrelationships of many of the fundamental variables. To achieve the optimal dimensions of the solar water pumping system, a multi-variable optimisation computer program was developed. The optimisation program operated in conjunction with a program simulating the steady state operation of the solar water pumping unit. Figure 2 and Fig. 3 are flow diagrams representing the optimisation routine and simulation program respectively. The theoretical basis of the simulation is an energy balance over one operating cycle
i.e.:

solar radiation * collector efficiency = thermal losses of pump unit + work output/pump efficiency.

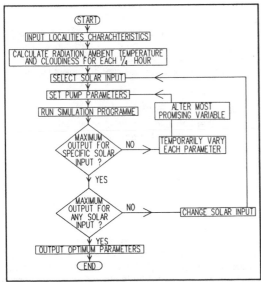

Fig. 2. Optimisation flow diagram

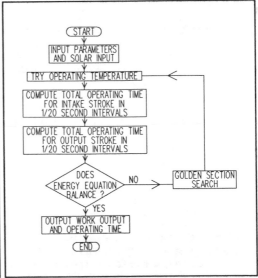

Fig. 3. Simulation flow diagram

Although the simulation program finds the steady state operating point for the given conditions, the transient nature of solar radiation is taken into account by the inclusion of a calculated period of down time which allows for both the possibility of clouds and the increase in operating temperature of the whole unit as the morning progresses. The two optimisation routines, one operating within the other, provide the possibility for the optimum pump unit to be found. The first routine indirectly manipulates the working temperature of the system giving the ability to compare over a one day period: a low collector temperature (low cycle efficiency) unit operating over a long period of time, with a high collector temperature (high cycle efficiency) unit operating over a short period of time. The second optimisation routine nested within the first finds the optimum pump parameters, based on daily water output, by temporarily varying each of the variables, then permanently altering the most promising variable. This optimisation method was chosen above gradient methods because the iterative simulation program does not yield sufficiently accurate changes in output for small changes in input variables (basic to most gradient methods).

Given the operating temperatures and pressures an analysis of the working cycle can be undertaken by means of first law and exergy (availability) analyses. The main purpose of using an exergy analysis on the solar water pumping unit's operating cycle is that the exergy analysis highlights the areas which hold the greatest potential for improvements in performance.

RESULTS AND DISCUSSION

Initial runs of the optimisation programme produced large values for the variables condenser volume and condenser area, as would be expected. These variables were then held to a realistic size and the final results for a particular collector area were optimised only in terms of the stroke, diameter and volume ratio. The increase of stroke (diaphragm length) was limited by the maximum pressure that could be withstood across the exposed area of cupped rubber diaphragm. An increase in stroke also elevates the heat losses in the pump unit across the uninsulated diaphragm and detrimentally increases the dead volume in the pump unit's vapour chamber (due to the deflection of the diaphragm). An increase in the diameter ratio between water and vapour chambers gives a relatively lower area of uninsulated diaphragm, beneficially increases pumping cycle time, but detrimentally increases the dead volume. An increase in volume ratio induces higher working fluid temperatures which add to thermal losses through the uninsulated diaphragm and also lower the collector's efficiency, but give higher possible pumping cycle efficiencies. An increase in volume ratio also makes the pump unit cycle slower due to more water being induced with less pressure being available to accelerate the water into and out of the pump's water chamber.

Given New Zealand summer conditions, drawing water from 6.5 m below the pump and delivering 1.5 m above the pump, using a 3 m^2 solar collector, the optimal conditions for the pump's operation was radiation at 700 W/m^2 and pump dimensions of 0.25 m diameter lower diaphragm, an upper to lower diaphragm diameter ratio of 1.67 and 30 mm diaphragm travel. For a cloudless day an output of 11 m^3/day has been predicted.

The inclusion of a period of down-time to allow for the effect of cloudy periods was only of major consequence in the results for the simulations which operated at low levels of direct solar radiation. The down-time allows for cloudy periods by calculating the temperature to which the system would fall after a certain period of cloudy conditions and then computes the period of time necessary for the direct solar radiation to re-heat the system up to operating temperature. In this scenario, systems which have higher operating temperatures and greater thermal masses would lose more energy than systems with lower thermal mass and lower temperature. In the final analysis this mode of operation was not of major consequence to the results due to the following: on the cessation of direct radiation the system performs several pumps until the pressure in the solar collector is not sufficient to displace the lower diaphragm of the pump. At this point there is no vapour (energy) being extracted by the pump unit and thus the solar collector will go into temporary stagnation conditions. For cloudy intervals the diffuse radiation can rise from its normal level with no cloud cover of approximately 1/10 of the direct component, to between 1/7 and 2/5 of the global radiation depending on cloud cover (Lestrade, 1990). For a collector operating at stagnation conditions under such diffuse radiation, the stagnation temperature is close to if not higher than the unit's operating temperature in direct sunlight, and thus the solar collector will stagnate at just below operating temperature. On the recommencement of direct solar radiation the system will start up with only a small time delay.

The main benefit of this design over the numerous small solar water pump designs is the possibility for the one pump unit with minor changes in the diameter of the upper diaphragm to work efficiently for many different pumping heads, using only the one working fluid (generally it is necessary to choose the working fluid once the pumping head is known, or alternatively use large springs to support a non-optimal fluid). The optimisation program was run through several different head configurations, the optimal pump parameters were found and the output for an ideal day was plotted against pumping head of water (Fig. 4). The effect of varying the water delivery height above the pump unit was minimal because this portion of the cycle is dependent upon the condenser temperature (related to the temperature of the water being pumped) and has no direct bearing on the collector's temperature or pressure. The change in well depth altered the optimum volume ratio and diameters to a fair extent, in general the operating pressure and temperatures increasing with decreasing well depth. The best results are represented by the peak in the m^4 versus head graph (Fig. 4). The m^4 or hydraulic energy equivalent performance (where $m^4 = m^3$ (volume of water pumped) * m (pumped head)) is a measure of work output of a pump unit. To perform the same amount of work over different head conditions the ideal device would have a constant hydraulic energy equivalent line.

The drop in energy equivalent for the 9.5 m pumping head (8m suction) is due to cavitation arising in the intake during the suction stroke. This reduction in performance is not caused solely by the water being displaced by vapour, but also by the vapour pressure during cavitation limiting the accelerating pressure and thus the water's velocity into the water chamber. It is possible for the vapour cavity to collapse at the end of the pump intake stroke, lifting the foot valve and replacing with water what would otherwise be volume occupied by vapour. This is possible because of the slow motion of the diaphragm and slight pauses at the top and bottom of each stroke. The drop off in hydraulic energy equivalent for the shallower well depths is due to higher water flow rates causing the parasitic losses to be much greater, while fluid friction factors remain essentially constant (ie. same foot valve and pipe diameter). This can be overcome by using larger diameter pipes, fittings and valves where appropriate.

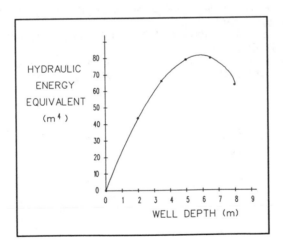

Fig. 4. Pump performance

A theoretical exergy analysis was carried out for two variants of the double diaphragm pump. The first operating with no irreversibilities other than those due to the ideal processes within the cycle. The second for comparison included the effects of: accelerating the fluid into and out of the pump, dead volume and thermal losses. The reference condition (dead state) for these analyses was chosen as pentane fluid at $17^{\circ}C$ and one atmosphere pressure. This reference state was chosen as it represents the minimum temperature to which energy can be transferred and the pressure at which the system is in mechanical equilibrium with the atmosphere. The inefficiencies within the processes are small when compared to the irreversibilities of the processes themselves. For example the unresisted expansion process accounts for 14% of the total exergy lost in the cycle. In contrast the effect of accelerating water into and out of the pump contributes by only 10% to the exergy loss within its process and by 2% to the total exergy loss in the cycle.

Most of the developments in the solar field have gone into making high efficiency, high power output, lower cost units using readily available technology. For an average of $120m^4$ and greater per day, Stirling or Rankine cycle engines powered by two axis tracking collectors (Halcrow, 1983) have a financial advantage over flat plate collector systems but for smaller systems of $40m^4$ the cost of producing the complex miniature tracking mechanisms and small power systems are well outweighed by the simplicity of the lower temperature flat plate systems.

CONCLUSION

The pump design proposed in this paper holds possibilities for efficient usage under many different pumping head applications. The economic viability of such units is yet to be proved as to date there has been no mass manufacture of such small scale solar thermal devices. To further the understanding of this unit and its anomalies a prototype unit has been constructed and is now under test.

ACKNOWLEDGEMENTS

The authors would like to express their gratitude to: The Institute of Professional Engineers New Zealand (IPENZ) who provided a scholarship for M.Amor to study this topic. and The University of Canterbury for the provision of a research grant to cover the expense of prototype production.

REFERENCES

Arlosoroff, S., G. Tschannerl and co-workers (1987). Community Water Supply (The Handpump Option), The World Bank. pp. 7-29.

Halcrow, Sir W. and Partners (1983). Small-scale Solar-powered Pumping Systems: The Technology, Its Economics And Advancement, The World Bank. pp. 6.1-6.12

Lestrade, J. P., B. Acock and T. Trent (1990). The Effect Of Cloud Layer Plane Albedo On Global And Diffuse Insolation. Solar Energy, Vol. 44, 115-121.

Sharma, M. P. and G. Singh (1980). A Low Lift Solar Water Pump. Solar Energy, Vol. 25, 273-278.

Stubbs, H. P. (1980). Vapour Operated Motor. Australian Patent No. A1-54,705/80.

PERFORMANCE OF A LIQUID PISTON ENGINE FOR LOW HEAD PUMPING

Pio C. Lobo & Manoel P. Martins

Centro de Tecnologia, Universidade Federal da Paraiba, 58059 João Pessoa. Pb, BRAZIL

ABSTRACT

The liquid piston pump with air as the working fluid is investigated for local construction to irrigate small remote rural landholdings or vegetable gardens. A model was built and tested based on the "Fluidyne" proposed by West (1971). Results are presented and further investigation seems warranted.

KEYWORDS

Liquid piston pumps, Stirling engine, small pumps, experimental measurement, air engines, solar energy, heat engines.

INTRODUCTION

The main energy source in the semi-arid Northeast of Brazil is sunshine, only exploited for drying. As a result small scale agriculture is very dependent on rainfall. Even where water is available near the surface, it is uneconomical to transport energy for pumping over large distances. The advantage of solar energy in these cases is that it arrives free at the point of use, so the scale of demand does not affect cost of supply. Several small solar engines have been proposed for this application (Ginell,1979; Taylor, 1979; West, 1971; Murphy, 1979). The most efficient use solid piston or turbine expanders and run on the Rankine and Stirling cycles. Photovoltaic cells driving electric pumps are also widely used (Rios, 1981). However, small landowners have low purchasing power and little access to skilled labour. So systems requiring large investments or sophisticated maintenance and repair are ruled out. A viable pumping system would be fabricated, or at least assembled, and repaired with inexpensive materials and unskilled farm labour. It is believed that this objective can be met by the liquid piston pump, based on the Stirling Cycle, with integral water pump, proposed by West (1971) with a single phase thermodynamic working fluid (air), and studied by Murphy (1979) with a two phase fluid (an organic refrigerant). With liquid pistons, cylinder finish, roundness and uniformity are not critical. The investigation was directed to West's "Fluidyne" pump. As a preliminary step, a model was built with standard hydraulic pipes and fittings. They are oversized for the application, and will be optimised later.

PRINCIPLE OF THE ENGINE

The sketch in Fig.1. serves to describe the operating principle. The working fluid (air) is contained in the connecting pipe and regenerator by the two water column liquid pistons. Above one column, the hot space communicates with the heat source. Above the other is the cold space, cooled by the surroundings. On oscillation of the water columns, if most expansion occurs at higher pressure in the hot space and compression at lower pressure in the cold, more work is done in the hot space than supplied in the cold since, for air, the pressure x specific volume product is proportional to temperature. Hence net work is done by the air, part of which can be usefully employed in pumping.

EXPERIMENTAL APPARATUS AND INSTRUMENTATION

The test model is sketched in Fig.1. Two 51mm (2") diameter vertical cast iron pipes, designated the hot and cold cylinders respectively, terminate in tees with two vertical legs.

The bottom tees are connected by a 64mm (2.5") diameter horizontal PVC pipe to form a "U" containing the displacer water column. Glass observation windows on axial slits in the vertical pipes permit viewing water levels. A 32mm (1.25") diameter plastic water pipe, designated the output column, connects the lower (vertical) leg of the hot cylinder bottom tee to the pumping tee. The other two legs of the pumping tee are connected to the inlet and outlet water pumping lines. A 19mm diameter copper pipe from the outlet leg of the pumping tee to the connection for the outlet hose, passes concentrically inside the cold cylinder between the vertical legs of the cold cylinder, cooling the water and the cold space. The horizontal legs of the top tees are connected by a 12mm diameter rubber hose.

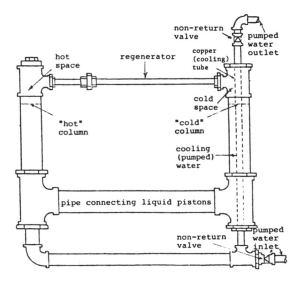

Fig. 1.   Liquid piston pump model

It contains the regenerator, consisting of 40g of 40μm diameter 3 mm mesh galvanised iron wire netting. The working fluid (air) is contained in the upper part of the hot and cold cylinders, connecting pipe and regenerator by the displacer water columns. To facilitate measurement of input power, the hot space is heated by a 100W electrical element in a ceramic sheath, in a 19mm diameter concentric copper well. The cold space is cooled by wall conduction to the surroundings. On oscillation of the water columns, water is induced at the bottom of the output column, through a non-return valve to avoid loss, and pumped through the copper section of the outlet pipe to the output hose.

Ambient temperature was measured but not controlled. Due to lack of instrumentation, electrical power to the heater was not monitored, except for measurement of direct heat losses during the determination of dynamic characteristics (see below). Maximum output (and efficiency) was attained when the peak water level just touched the bottom of the heater tube. Water inlet and pumping heads were varied, and measured with metre scales to a resolution of 1mm. Inlet head was measured between the static column level in the cold space and the inlet reservoir free surface. The pumping head between the level of the outlet pipe exit and the inlet reservoir free surface. Electrical heating power was evaluated from potential read on a voltmeter and current on an ammeter. Mean air temperature in the hot and cold spaces was determined from the e.m.f. of copper-constantan thermocouples read on a potentiometer calibrated against a mercury-in-glass thermometer with a resolution of 0.1C and an uncertainty estimated at 0.2C. Mean pump flow was obtained from the time, measured on a chronometer with a resolution of 0.2s, to fill a 500 cm^3 measuring jar with uncertainty estimated at 0.1cm^3. All dynamic measurements could not be simultaneous due to limitations in instrumentation. Air pressure in the hot and cold spaces was measured by pressure transducers calibrated against water column manometers with an uncertainty of 2mm water head (about 20Pa). The volume of the hot and cold spaces was determined as a function of water level by adding measured volumes of water and noting the corresponding level. Air volume was obtained by subtraction. Instantaneous water levels were measured on a frame-by-frame playback of a videotape of their motion, filmed through the glass windows in the hot and cold cylinders. Readings were taken at intervals of one thirtieth (1/30) of a second as measured on the camera clock. The oscilloscope pressure-time trace was filmed simultaneously, and time values corresponding to pressure maxima and minima noted. Since oscilloscope resolution was insufficient for accurate measurement of indicated power from the pressure-volume diagram, the pressure-time trace was recorded on a Campbell 21MX datalogger with an internal clock, at intervals of 50 ms. The time values corresponding to pressure maxima and minima were used to synchronize pressures and volumes and plot the pressure-volume diagram.

RESULTS AND OBSERVATIONS

All measurements were made in the steady state, defined by the cyclic repetition of state properties in time, attained in about ninety minutes. Control variables were ambient temperature, heater electrical power input, initial column heights, (pumped) water inlet and outlet heads. Output variables are mean pumped flow rate, pumping frequency, air space pressures and water levels. Time averaged, or "static", results are presented in Table 1.

TABLE 1. Time Average Readings and Results

Inlet head -23 cm

Pumping head	cm	78.5	83	90	99	105	112	118	130	143	149		
Time for 500 cm^3	s	15	15	15	19	23	27	33	47	47	52		
Flow rate	cm^3/s	33.3	33.3	33.3	26.3	21.7	18.5	15.2	10.6	10.6	9.6		
Thermal efficiency	%	0.26	0.27	0.29	0.26	0.22	0.20	0.18	0.14	0.15	0.14		

Inlet head   3 cm

Pumping head	cm	67	74.5	87.5	95	110	119	124	131	141	148		
Time for 500 cm^3	s	12	14	16	16	22	24	27	32	36	42		
Flow rate	cm^3/s	41.7	35.7	31.3	31.3	22.7	20.8	18.5	15.6	13.9	11.9		
Thermal efficiency	%	0.27	0.26	0.27	0.29	0.25	0.24	0.23	0.20	0.19	0.17		

Inlet head 37 cm

Pumping head	cm	64	73	80	89	94	101	107	115	120	132	143	151	161
Time for 500 cm^3	s	13	14	15	16	17	19	20	25	25	34	46	59	76
Flow rate	cm^3/s	38.5	35.7	33.3	31.2	29.4	26.3	25.0	20.0	20.0	14.7	10.9	8.5	6.6
Thermal efficiency	%	0.24	0.26	0.26	0.27	0.27	0.26	0.26	0.23	0.24	0.19	0.15	0.12	0.10

Inlet head 68cm

Pumping head	cm	77	81	92	103	113	125	133	143	153	160		
Time for 500 cm^3	s	13	14	17	19	20	26	32	43	57	66		
Flow rate	cm^3/s	38.5	35.7	29.4	26.3	25.0	19.2	15.6	11.6	8.8	7.6		
Thermal efficiency	%	0.29	0.28	0.27	0.27	0.28	0.24	0.20	0.16	0.13	0.12		

2150

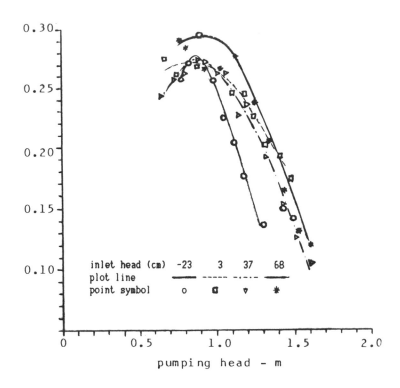

Fig. 2. Overall efficiency vs. pumping head

Net pumping head (output - input) is plotted against efficiency for input heads
between -230 mm and 680 mm in Fig. 2. Significant scatter of data points may be
largely due to errors in electrical power input as a result of variations in
mains voltage (assumed constant). Maximum pumping head is about 1.6 m and
corresponding thermal efficiency about 0.1%. The maximum thermal efficiency is
between about 0.27% and 0.29%, which is the average for this engine using air
as a working fluid. With "perfect" insulation in the hot space, regenerator and
connecting tube, heat input could be reduced by 25% so thermal efficiency would
rise to between 0.34% and 0.36%. The influence of inlet head on pump
performance over the range of 30 to 680 mm seems slight. At a negative inlet
pumping head (-230 mm) scatter was large.

Peak water levels at maximum output and efficiency overran the top of the
window. Air mass was increased to lower water levels and allow observation of
the peaks. As a result, maximum pumping output power was halved. Under these
conditions, at an ambient air temperature of 29.8 C, pumped water flow was 1.1
dm3/min (0.018 dm3/s) at a head of 1 m, equivalent to a reversible pumping
power of about 0.18 W. Air temperatures were about 87 C in the hot space and 39
C in the cold space. Dynamic air pressure and water level readings with
corresponding volumes are described as functions of time in Table 2. Cycle
frequency was 0.75 Hz. The "indicator diagram" or pressure-volume trace appears
in Fig. 3. The indicated work, from the measured area of the pressure-volume
diagram, was 0.53 J per cycle; and indicated power, the product of cycle work
and frequency, about 0.4 W. To estimate direct heat losses (mainly conductive)
to the surroundings, electrical power supplied to the heater was measured, with
the pistons at rest and average air temperature the same as during operation.
The results, described in Table 3, indicate that these losses are about 27 W.

TABLE 2. Dynamic Readings and Results

Time s	Water level mm hot space	cold space	Volume cm3 hot* space	cold* space	total	Pressure kPa hot space	cold space	mean
0	45	3	469	400	869	32.5	31.0	31.7
50	45	7.5	469	407.5	876.5	31.4	30.8	31.1
100	44	12	468	415	883	30.2	29.5	29.9
150	43.5	17.5	465	424	889	28.8	27.7	28.3
200	40	24	461	435	896	27	26.2	26.6
250	37.5	30.5	457	445.5	902.5	25.2	24.1	24.6
300	33	38	450	458	908	23.2	21.9	22.6
350	30	43.5	445	467	912	20.9	19.8	20.3
400	28	50	441	478	919	19	17.7	18.4
450	24	57	435	489	924	17.5	16	16.7
500	21	61	430	496	926	15.8	14.4	15.1
550	19.5	62	427	498	925	14.8	13.4	14.1
600	18	62	425	498	923	13.9	12.6	13.2
650	17	61	423	496	919	13.4	12.6	13
700	17	58	423	491	914	13.5	12.8	13.2
750	19	51	427	479	906	14.4	13.7	14
800	21	46	430	471	901	15.4	14.6	15
850	22.5	40	432	461	893	16.8	16.5	16.7
900	25	34	436	451	887	18.8	18.3	18.6
950	28.5	27.5	442	441	883	20.6	20.5	20.5
1000	31	21	446	430	876	22.8	22.5	22.6
1050	34.5	15	452	420	872	24.8	24.8	24.8
1100	38	10	458	412	870	27.2	27.0	27.1
1150	39.5	7.5	460	408	868	29.1	29.0	29.1
1200	43	4	466	402	868	30.9	30.9	30.9
1250	44.5	3	469	400	869	31.8	31.8	31.8
1300	45	3	469	400	869	32.4	31.8	32.1

*including part of the connecting hose; regenerator volume neglected

TABLE 3. Estimation of Static Irreversible Heat Losses

Pump Condition	Heater input Tension V	Current A	Power W	Thermocouple measurements cold space e.m.f. mV	temp. C	hot space e.m.f. mV	temp. C	mean temp. C
Operating	220	0.49	108	1.66	38.5	13.20	87.0	62.7
Stationary	104	0.26	27	0.44	33.0	14.5	92.5	62.7

Because of the limitation on minimum air volume due to column height restrictions described above, and because the available equipment prevented automatic data acquisition and processing, pressure-volume diagrams were not obtained at other operating conditions. This will be attempted after suitable modifications to equipment and instrumentation.

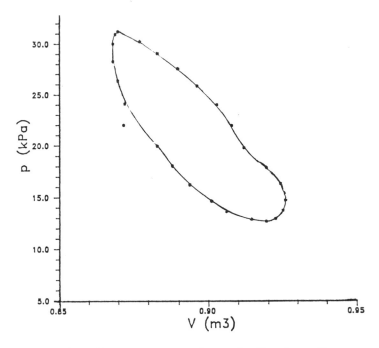

Fig. 3. Pressure - volume indicator diagram

CONCLUSIONS

A model was designed for assembly by unskilled labour with minimal training. It was built with standard hydraulic pipes and fittings and other easily available materials. Its performance is comparable to that of laboratory units using the same working fluid and thermodynamic limits. The pressure-volume trace is similar to other published diagrams. It is planned to study design changes to lower material costs. Loss reduction and the utilization of two phase working fluids will be investigated as means of increasing thermal efficiency.

ACKNOWLEDGEMENTS

The first author holds a research fellowship from the National Council for Scientific and Technological Development (CNPq - Proc. No. 30.1555/76-EM)

REFERENCES

Ginell, W. S., J. L. McNichols, Jr. and J. S. Cory (1979). Nitinol heat engines. Mech. Engng., 101, n 5, 28-33.
Murphy, C.L. Review of liquid piston pumps and their operation with solar energy (1979). ASME publication - 79-Sol-4.
Rios, M. (1981). An economic evaluation and experiments on application of photovoltaic systems in remote sites. Sandia National Laboratories.
Taylor, G. (1979). Free piston Stirling engines increase solar/thermal efficiency. International Power Generation, July/August 1979,
West, C. (1971). The Fluidyne engine. Electronics & Applied Physics Publication - U.K.A.E.A. Atomic Energy Research Group, Harwell, Berks., England.

FLOURESCENT SOLAR CONCENTRATORS - EXPERIMENTAL RESULTS

V.L. Fara   and R. Grigorescu

Physics Dept., Polytechnical Institute of Bucharest
Splaiul Independentei 313, Bucharest, Romania

ABSTRACT

A physical characterization is presented for classes of selected flourescent substances, with special emphasis on their concentration in the support. The influence of climatic factors was analyzed from two points of view: spectral and structural. Two types of geometrical models of fluorescent solar concentrators have been obtained: plates and cylinders.

KEYWORDS

Fluorescent solar concentrators; absorption spectra; emission spectra; climatic tests; solar cells.

PRESENTATION OF THE FLUORESCENT SUBSTANCES

The designing of fluorescent solar concentrators (FSC) requires careful study in both the selection of dyes and in the area of support - solid or liquid (Goetzberger and Greubel, 1977).

The main factors influencing the phenomenon of fluorescence are the molecular structure of the dye and the properties of the solvent. For this reason, the selection of a certain substance and of the appropriate solvent is essential for the realization of fluorescent solar concentrators. The fluorescent dyes must have the following properties (Rane and colleagues, 1984)

- existence of at least one absorption band in the
  visible range;
- a high value of the fluorescence efficiency;
- a minimum overlap of absorption fluorescence bands, in order to
  minimize absorption losses;
- a high photostability;
- a good solubility in the substances selected as support
  (PMMA - methyl polimetacrilate).

It is to be noticed that fluorescent dyes of benzentronic types have reduced overlaps of absorption and emission (fluorescent) bands. In order to improve the interaction between molecules of the dye and PMMA, one can introduce into the molecular structures some branched alkyl groups.

On this basis, the following dyes have been selected: ORACET 6 GF, ORACET ROSA, Rhodamine 6 GF. In all cases the support is PMMA; compounds with oxazolic ring (POPOP,TOPOT and XOPOX with Xilen as solvent); methyl blue, for which the substrate used was methanol, some compounds from pyrylium salts class.

ABSORPTION AND EMISSION SPECTRA

The physical characterization of the fluorescent substances selected requests both the study of the influence of the concentration for various experimentally obtained absorption spectra, and the comparative analysis of the absorption and emission spectra (Fara and coworkers, 1986, 1989).

Figure 1 presents as examples absorption spectra for different concentrations of the dye ORACET 6 GF. One can notice an increase of absorption in UV, without it being essentially modified in the visible. The concentration increase appears as important for achieving a high fluorescence efficiency, without allowing, however, a wider coverage of the solar spectrum.

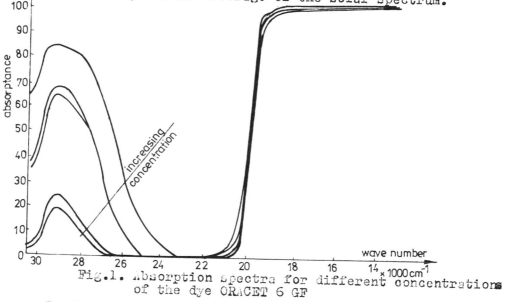

Fig.1. Absorption spectra for different concentrations of the dye ORACET 6 GF

In figure 2 one can see the emission spectra ($E_\lambda = f(\lambda)$) compared to the absorption ones ($\alpha_\lambda = f(\lambda)$), experimentally obtained with PMMA samples doped with ORACET ROSA and ORACET 6 GF. One has to emphasize that fluorescent emission takes place in a narrow spectral range (520÷560 nm for ORACET 6 GF). There is a very good overlap of emission for ORACET 6 GF and absorption for ORACET ROSA. The combination of the two does not essentially affect the emission spectrum of ORACET ROSA except only for a more pronounced self - absorption.

QUANTUM EFFICIENCY OF FLUORESCENCE

We have calculated quantum efficiency of the considered fluorescent dyes, relatively to B Rhodamine. The relative quantum efficiency can be defined as

$$R_i \equiv \frac{W_i}{W_{Rh}} = \frac{\mu_i}{\mu_{Rh}} \frac{s_i}{s_{Rh}} \frac{c_{Rh}}{c_i} \frac{I_{Rh}}{I_i} \frac{\int_{400}^{800} \frac{y_i(\lambda)}{M(\lambda) F(\lambda)} d\lambda}{\int_{400}^{800} \frac{y_{Rh}}{M(\lambda) F(\lambda)} d\lambda} \tag{1}$$

where: W_i, W_{Rh} stand for the energies emitted in visible range (400 ÷ 800 nm); μ_i, μ_{Rh} are the molar weights; S_i, S_{Rh} are the recorder scale factors; I_i, I_{Rh} are the exciting intensities; $J_i(\lambda)$, $J_{Rh}(\lambda)$ are the experimental emission spectrum intensities; c_i, c_{Rh} are the solution concentrations; $F(\lambda)$ is the photomultiplier contribution; $M(\lambda)$ is the monochromator contribution. All these sizes are related with the "i" fluorescent dye and the B Rhodamine, respectively. We have obtained the best results for 2, 4, 6 – triphenyl pyrylium oxalate (R = 328.7) and for 2, 6 – diphenyl pyrylium perchlorate (R = 145).

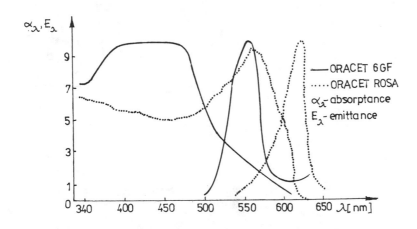

Fig.2. Emission and absorption spectra experimentally obtained with PMMA samples doped ORACET ROSA and ORACET 6 GF

INFLUENCE OF CLIMATIC FACTORS

We have considered samples both in the form of PMMA-plates doped with ORACET ROSA and ORACET 6 GF and as solutions (compounds with an oxazolic ring in xilen and with methyl blue in methanol).

The climatic tests undertaken have been: thermal cycling (in the range -5°C to +55°C) for 48 and 96 hours, respectively; the action of dust and the influence of solar radiation (both for 250 hours). The modification of absorption spectra of the samples (for a PMMA plate doped with ORACET 6 GF) compared with the witness sample is illustrated in Fig.3.

The analysis of the microstructure of the fluorescent plates surface under testing was performed with a scanning microscope (Stereoscan 180). We have obtained both topographic and scaning cathode luminescence images. Examination of the photographs (cf. Fig.4) gives information on the sample surface and the homogeneity of the luminous centers. One notices a relatively uniform dopping with fluorescent substances, the main defects observed being of mechanical nature.

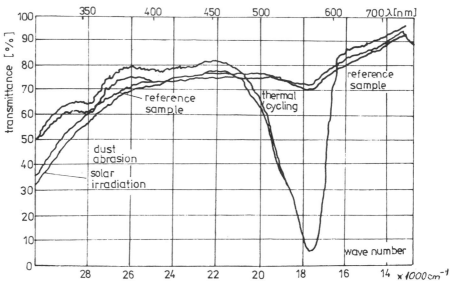

Fig.3. Modification of absorption spectra of the
climatic tests undertaken samples compared
with the reference sample

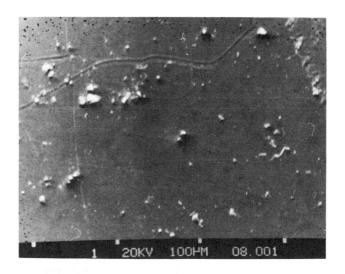

Fig.4. Topographic image of a fluorescent plate
surface, which was performed with a
scanning microscope

EXPERIMENTAL MODELS OF FLUORESCENT SOLAR CONCENTRATORS (FSC). UTILIZATION TRENDS

To obtain solar concentrators has involved two steps: obtaining a dye solution at a predetermined concentration and dopping the PMMA support with an organic dye. The studies performed led us to the consideration of two kinds of FSC: a plate and a cylinder. The main technological difference in building up the two geometrical models comes from the way the samples are cooled. The polymerization of plates is performed in a drift of hot air, whereas that of cylinders is done in water with controlled temperature. Our samples were plates of 150 x 400 x 6 mm and cylinders with diameters $\emptyset$ = 8 mm and maximum height of 350 mm. The wavelength of the fluorescent emission was 420 $\div$ 480 nm for the oxazolic compounds, 518 nm for ORACET 6 GF and 620 nm for ORACET ROSA.

The main advantages of solar installations with photo-voltaic attached to FSC are :

- the fluorescent systems do not require orientation and, at the same time, benefit of diffuse solar radiations;

- the conversion efficiency is aditionally increased by modification of the spectrum of the absorbed solar radiation and re-emission towards solar cells, in the blue-green range to oranger-red range, in which monocrystalline Si cells have an optimal spectral response;

- FSC do not lead to superheating of the solar cells since they are spectral concentrators and the energy concentration factor ranges between 2 and 10.

REFERENCES

Goetzberger A., W.Greubel (1977). Appl. Phys. 14, p.123.

Fara, V.L., R.Grigorescu (1986). Proc.Int.Congress on Renewable Energy Sources, p.1296, Madrid.

Fara, V.L., R.Grigorescu, I.A.Dorobanţu (1989), Applied Optics in Solar Energy III, Praga.

Rane, R., A.Harnish, K.H.Drexhage (1984). Heterocycles, 21, p.167.

2.26 Solar Heat Posters

SOLAR DRYING FOR CANDY FRUITS

Tang Yucheng*, Zheng Ruipei*, Yang Tiengzhu**,

*Energy Research Institute, Hebei Academy of Sciences,
Shijiazhuang, Hebei Province, China.
**Lianyungang Solar Energy Research Institute,
Lianyungang, Jiangsu Province, China.

ABSTRACT

This paper reports on a double air flow greenhouse type solar dryer which used for drying candy fruits.

KEYWORDS

solar drying; greenhouse type; candy fruits.

INTRODUCTION

The Zhongton village, Hebei province, China. is a fruitcultural area. Total harvest yield of apple, pear, apricot, jujube, plum etc over 10^7 Kg. one year. These fruits in addition to suppling market fresh fruit, will be produced into candy fruits (or preserved fruits) which is a kind of traditional Chinese sweet food. In the process of candy fruits, production must include drying to cut down moisture for long time preserving. Since 1984, studies have been done utilizing solar energy to dry candy fruits to replace the traditional coal fire drying house.

DRYING CHARACTERISTIC OF THE CANDY FRUITS

In drying process, the moisture level of material is important data.
Because of variance in type of candy fruits and because the process of candy fruits production is unstable, we have measured the moisture content of differential sort candy fruits repeatedly.
The average data is shown in Table 1.

2162

TABLE 1. THE AVERAGE MOISTURE DATA

sort of candy fruits	moisture content (% H.B.)		amount of removed moisture (Kg/1 ton finished product)
	before drying	after drying	
candy plum	32.0	18.0	206
candy apricot	44.2	20.4	426
candy apple	43.1	21.6	378
candy jujube	37.1	20.8	259
candy pear removed peel and first boil with sugar	45.3	35.4	181
candy pear removed peel and second boil with sugar	26.9	20.4	89
candy pear with peel and first boil with sugar	54.1	38.3	344
candy pear with peel and second boil with sugar	24.2	20.9	44
candy melon strip	16.6	14.5	25

The candy fruits semiproduct are high viscosity materials which contain large amounts of sugar.In sugar molecules there are many oxy-group link-up strong water molocules.   So there is difficulty in the drying process for removing moisture from candy fruits. Particularly, diffusion speed of inner moisture will control the all drying speed. When the drying temperature rises, the moisture diffusion is excited and the drying rate increases clearly. We have made a series experiment about candy apple drying at differential temperature and in a natural vent air condition. The typical candy apple constant temperature drying curves as shown in Fig.1. From Fig.1. Show eight drying hours   (nearly equal to one sunshine day solar drying) when the drying temperature at $60^\circ C$ or $80^\circ C$, the removed moisture from the candy apple 39% or 117% more than at $40^\circ C$. In addition, the candy apple produced in the same batch was cut to differential size, one weight 50g. and another 10g.. These samples got dry at $60^\circ C$ constant temperature and natural vent condition respectively. Figure 2 shows the constant temperature drying curves. After 8 hr. drying, the removed moisture from small samples (10g.)=41% more than big ones (50g). Increased flowrate of air that brushed the surface of the candy fruit also raised drying rate, but its effect smaller than tne above two options.

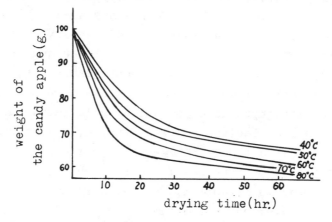

Fig.1. The temperature effect on candy apple drying rate

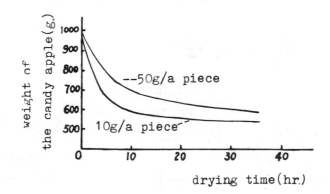

Fig.2. The sample size effect on candy apple drying rate

The removed moisture from samples (10g)=41% more than big ones (50g.) Increasing flowrate of air that brushes the surface of the candy fruit also can raise drying rate, but its effect smaller than the two options.

DESIGN AND BUILT DRYER FOR CANDY FRUITS

Quite a lot of articles on solar drying technique had been presented

by early investigators. (B.K.Huang, 1983; A.S.Mujumdar, 1987;
B.Norton, 1987; M.A.Zaman, 1989 and others) Linked with our practical
task, a double air flow greenhouse type solar dryer has been
designed and built. It was a novel style dryer which was situated in
Zhongton village, Hebei province, China. This unit was mainly used
for drying candy fruits. The production capacity of the solar dryer
was 500 tons/a year. The south surface glass area for collecting
solar energy was 303 m², the floor area of the drying house was
307 m², and inner volume was 542 m³. Total investment for this dryer-
about U.S. $12,000. The double flow channel exhaust system has been
adopted in this unit for drying the materials more uniformly. As
Fig.3 shows the air input hole and outlet hole can be alternated with
each other. When the electric fan was running, the fresh air was
down drawn from down air vent-pipe.   On the other vent model
that electric fan stops running. The air which flow through drying
house take an opposite direction. The wet air was exhausted by
natural vent. The over chimney became an outlet duct. Fresh air flows
into the drying house by down vent-pipe. Such a changeable vent model
as mentioned above, particularly suitable for solar cryer, was set
up in the rural area where electricity power supply was unstable.

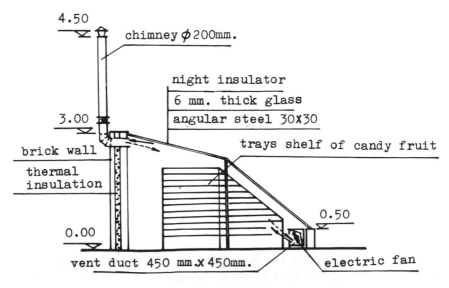

Fig.3. The scheme of solar dryer with double
flow channel exhaust system

RESULT OF PRACTICAL OPERATION

The solar dryer had been built in Septemper 1984. Ever since that time it already has dried thousands of tons of varied sorts of candy fruits sucsessfully: i.e.,   candy apple, apricot, jujube, plum, melon etc. The practical operation performance were observed and monitored. The main result as summarised in Table 2.

Table 2.

items	data	notes
highest temperature inside dryer	70° C	June--August
maximum temperature difference for inside and outside dryer	35° C	June--August
remove moisture ability for unit south glass area	$1.16 Kg/m^2 \cdot day$	candy apple septemper measured
solar efficiency for drying process	16.9-29.9%	June--October
term of a drying process	1--3 days 4--7 days	in summer and autumn in winter and spring
building plane area	$307 \ m^2$	
south glass area	$303 \ m^2$	
inside volume of the dryer	$541 \ m^3$	
designed loadage	6000 Kg.	
ventilation	$6289 \ m^3/hr.$	
thermal energy consumption	145130 Kcal/hr.	

These practical performance has already shown that utilizing solar energy drying candy fruits not only saves   the conventional fuels but also avoids environmental polution by comparison with the traditional sun drying in open air, or coal fire drying method. The candy fruits obtained from this solar dryer have all the better quality. In a national match with same production, these candy fruits won the gold prize.

ACKNOWLEDGEMENT

The authors would like to express their gratitude to the Sciences and Technology Commission of Hebei province, China. Thanks to their financial support this project was successful.

REFERENCES

B.K.Huang and M.Toksoy, (1983). Energy in Agriculture, 2. 115-136

A.S.Mujumdar,(1987). Solar Energy Utilization,(Turkey) 3. 630-655

B.Norton and P.D.Fleming(1987). Physics and Technology of Solar Energy. 1. 447-466

M.A.Zaman and B.K.Bala (1989). Solar Energy, 42. 167-171

A NEW CONCEPT ON DESIGN OF SOLAR WATER-PUMPING THROUGH OPEN HEAT PIPE

A. Hadji Saghati

Iran University of Science and
Technology Tehran, 16, Iran

ABSTRACT

In this approach the principle of heat pipe has been applied to design
for solar water-pumping system. The water to be pumped is heated
and evaporated through solar radiation, and vapor is moved toward
condenser section where it is condensed in an air-cooled condenser
slightly tilted towards water storage tank. The vertical pipe(open
heat pipe), connecting the evaporator and condenser sections together,
is insulated to prevent condensation during vapor transfer. This
paper describes a simple theoretical approach to analyse the per -
formance of this system.

KEYWORDS

Heat pipe; solar water-pumping; evaporator; condenser; Humphry pump;
Stirling engine.

INTRODUCTION

In many rural area in Iran, grid connected electricity is unavailable.
Thus water pumping depending on electricity is inaccessible. Other
types of pumps like Humphry pump using methane from bio-gas or stirling engine can also been taken into consideration as prime movers
for water pumping system, however, they suffer from problems arising
after the design process such as manufacturing, maintenance high
preliminary and operating expenditures, which altogether present
barriers for this adaptation. In present paper a new system based
on evaporation and condensation, with the advantages of reliability,
lack of maintenance and operation cost, has been designed.

SYSTEM DESCRIPTION

A schematic diagram of the solar water pump is shown in Fig,1. It
consists of heat-pipe with distilled water as operating fluid. The
heat-pipes are attached to an iron sheet. The absorber unit is
painted non-reflecting black to enhance its efficiency. The absorber
section is covered with double-glazing common window glass. The
condenser section of the heat-pipes is immersed in insulated water
tank. The heat absorbed by the heat-pipe is transferred to the water
and causes it to vaporize. This generated vapor moves along an
insulated vertical pipe(open heat pipe) and condenses in an air-cooled
condenser, and the condensed water flows to storage tank.

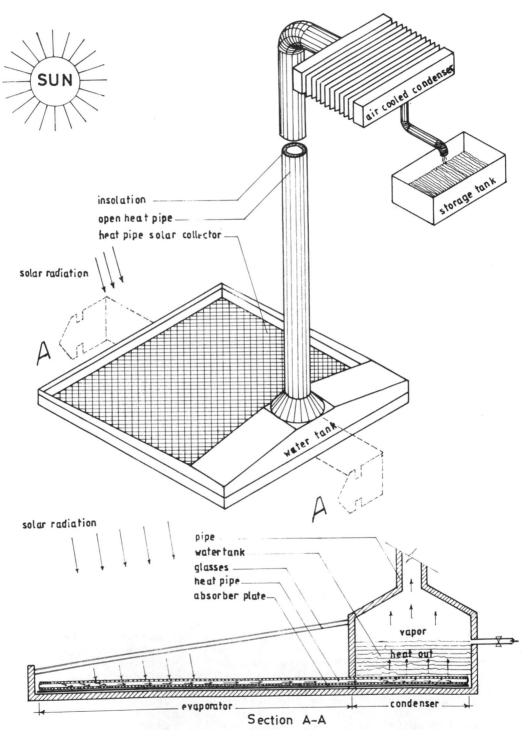

SUN

insolation

open heat pipe

heat pipe solar collector

solar radiation

air cooled condenser

storage tank

water tank

A

A

solar radiation

pipe
water tank
glasses
heat pipe
absorber plate

vapor

heat out

evaporator

condenser

Section A-A

Fig.1. Solar water pumping through heat pipes.

THEORY

To facilitate the theoretical analysis, the following assumptions
are made:
(i) The heat conduction through the covering glass can be ignored.
(ii) The air temperature in the environment is assumed to be the
     same as the surrounding area of the collector.
The energy in the collector received from the solar system is distributed into two parts:
(1) Useful or gained energy, and(2) heat loss.

Thus, the energy balance of the collector may be written as the
following formula:

$$I . A .(\mathcal{T}.\mathcal{K}) = Q_U + U_L . A (T_P - P_a)$$ (1)

and the vapor mass flow-rate in the condenser section can be shown
as:

$$\dot{m}_v = Q_u / (h_{fg} + C_p . \Delta t)$$ (2)

The relation of the maximum heigh of vapor flow(L) and the vapor
pressure drop (ΔP) and the diameter of the pipe(D),is obtained
from:

$$\Delta P = \rho . g . L + \Upsilon . \frac{L}{D} . \frac{V^2}{2}$$ (3)

The following are the symbols of the aforementioned formulas:

I   = Solar radiation intensity
A   = Solar collector area
Q_u = Useful heat gain
U_L = Overall heat-loss coefficient
T_P = Plate temperature
T_a = ambient temperature
L   = maximum Lengh (height) of the pipe

RESULT

The variations of the mass flow-rate of vapor ($\dot{m}_v$) and the solar radiation intensity (I) obtained from the equations (I) and (2)
can be illustrated in Fig. (2) as follows:

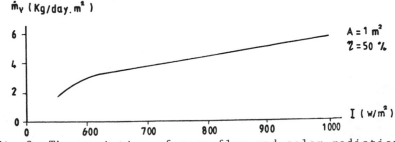

Fig.2. The variation of mass flow and solar radiation.

Vapor pressure drop along the vertical pipe, obtained from the equation(3), and varying with the height of the pipe(L) is shown in Fig. (3) as illustrated below:

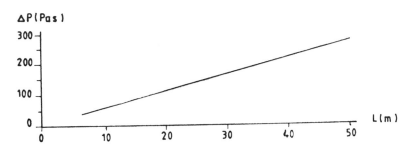

Fig.3. The variation of (ΔP) in relation to the height of pipe(L).

CONCLUSION

The proposed system has the main advantages of :

(i)   having no moving parts
(ii) having an independent external power
(iii) requiring no maintenance
(iv) being easy to install
(v)   being a silent operator
(vi) distilling water if it is salted

REFERENCES

Azad, E., F. Bahar, and Moztarzadeh (1987), Solar water heater using gravity-assisted heat pipe.
J. Heat Recovery Sys. & chapter, 7, 343-350.
Dunn, P.D., and D.A. Reay (1976), Heat pipes, pergoman-press, Oxford, Chap.1,

SOLAR CONTINUOUS CORN DRYER IN FLUIDIZED BED

E. Azad[*] and J. E. Mahallati

Iranian Research Organization for Science and
Technology
P. O. Box 15815-3538 Tehran, Iran

ABSTRACT

In design of solar dryer the combination of two new techniques, fluidized bed
for drying granular materials for continuous operation and heat pipes as thermal
energy transfer device for solar flat plate collectors, water-to-air heat exchanger and air-to-air heat recovery system to utilize waste exhaust moist air, are
employed. Thermal model of solar dryer in conjuction with thermal energy storage
and auxilary system for constant air temperature entering the dryer is given in
detail.

KEYWORDS

Solar dryer; heat pipe; fluidized bed; heat exchanger; heat recovery; storage.

INTRODUCTION

To conserve the world's limited resources of fossil fuels and increasing the
price of oil, the use of solar energy to replace conventional heat sources will
be more feasible.

The solar fluidized bed dryer was developed under a contract from Iranian Research Organization for Science and Technology. The system employs combination of
two new techniques, heat pipes and fluidized bed, for drying 1000 Kg. of corn
per day in particular but can be used for many types of granular materials.

SYSTEM DESCRIPTION

Figure. 1, shows a continuous multi-stage cascade solar dryer. The hot air stream
introduced at the base of the bed through the perforated plate and moist material
is fluidized by the hot air stream. Intimate contact between air and moist granular material results instantaneous evaporation.

In this system the corn will pan from section to section when the moisture content has reached a predetermined value. The final moisture content in each
section is governed by adjusting the partition plate in order to vary residence
time.

In this system an air-to-air heat pipe heat recovery, as shown in Fig. 2, is

---

[*] Energy consultant

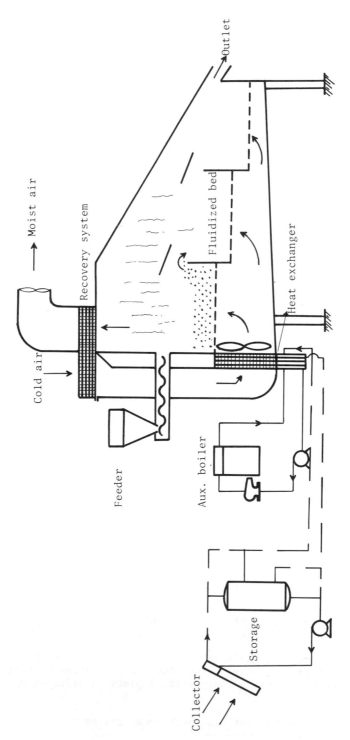

Fig. 1. Flow diagram of solar corn dryer

utilized in the process to preheat the feed air for the dryer from waste exhaust moist air (Azad, 1985).

Fig. 2. Air-to-air heat pipe heat recovery

The preheated air is further heated in water-to-air heat pipe heat exchangers. A detailed description and thermal performances of heat recovery and heat exchanger are given elsewhere (Azad, 1985). Hot water to heat exchangers are supplied through heat pipe solar collectors and auxiliary system.

PROCESS HEATING SYSTEM

Flow diagram for process heating system is shown in Fig. 1. The collector field is in series with load. The process requires fluid at temperature $T_p = 70$ C and return temperature from the process is the inlet temperature T_i of the collector field. The collector area is just large enough to match the load under peak insolation, then no energy will ever be wasted. The heat is supplied to the load when collector outlet temperature is greater than process temperature, and if it is not, then the heat is supplied through auxiliary system, and collector is used as preheater for the process or heat thermal energy storage. This heating system ensured continuous heat supplied at constant process temperature.

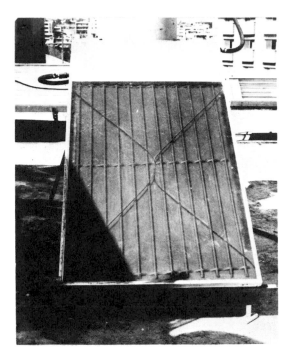

Fig. 3. Heat pipe solar collector.

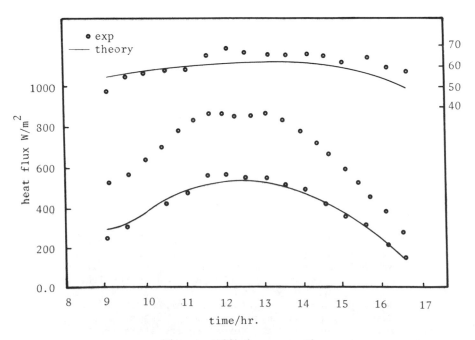

Fig. 4. Efficiency vs time.

THEORY

The main integral characteristics of fluidized bed are: the minimum fluidizing air velocity, V_{mf}, the bed pressure drop at minimum fluidizing velocity, ΔP_{mf}, and the average bed porosity, E_{mf}. Pressure drop can be obtained from Ergum's and Anderson's corrolation (Erdesz, 1986).

$$\Delta P_{mf} = H \left\{ 150 \frac{(1-E_{mf})^2}{E_{mf}^3} \cdot \frac{\eta}{d_p^2} V_{mf} + 1.75 \frac{(1-E_{mf})}{E_{mf}^3} \cdot \frac{P}{d_p} V_{mf}^2 \right\} \tag{1}$$

Minimum fluidizing air velocity can be obtained by using the following empirical formula of Kunii-Levenspiel (1969).

$$\frac{d_p V_{mf} \rho_g}{\mu} = \left[(33.7)^2 + 0.0408 \frac{d_p^2 \rho_g (\rho_s - \rho_g) g}{\mu^2} \right]^{0.5} - 33.7 \tag{2}$$

The energy needed to evaporate the water content in product is

$$Q = m_w h_{fg} \tag{3}$$

Where

$$m_w = w_i (M_i - M_f)/(100 - M_f) \tag{4}$$

Collector area is evaluated by

$$A = \frac{Q}{I \eta_{sys}} \tag{5}$$

Where

$$\eta_{sys} = \eta_c \cdot \eta_h \tag{6}$$

Figure. 3, shows heat pipe solar collector developed by Azad (1987) and its efficiency, η_c, can be obtained from Fig. 4.

CONCLUSION

Using heat pipe in the system has the following advantages:
  i) Redundancy: The heat pipes are independent of one another; failure of one has no effect on the others.
 ii) High thermal performance.

ACKNOWLEDGEMENT

The authors wish to acknowledge the support of this work by Iranian Research Organization for Science and Technology.

NOMENCLATURE

A = solar collector area (m^2).
d_p = particle diameter (m).
h_{fg} = latent heat of vaporization (KJ/Kg).
H = bed height (m)
I = solar insolation (KW/m^2)
M_i, M_f = initial and final moisture content (gr. H2O/gr. w. b)
ρ_g, ρ_s = gas and solid density (Kg/m^3)
E_{mf} = porosity (-)
η = efficiency (-)
η_c = collector efficiency (-)
η_h = heat exchanger effectiveness (-)
η_{sys} = system efficiency (-)

REFERENCES

Azad, E., F. Bahar and F. Moztarzadeh (1985), Design of water-to-air gravity-assisted heat pipe heat exchanger. J. Heat Recovery Systems. Vol. 5, No. 2, pp. 89-99.

Azad, E., F. Mohamadieh and F. Moztarzadeh (1985), Thermal performance of heat pipe recovery system. J. Heat Recovery Systems. Vol. 5, No. 6, pp. 561-570.

Azad, E., F. Bahar and F. Moztarzadeh (1987), Solar water heater using gravity-assisted heat pipe. J. Heat Recovery Systems. Vol. 7, No. 4, pp. 343-350.

Erdesz, K., A. S. Mujumdar and D. U. Ringer, Conventional and vibrated fluidized beds. Drying'86. Vol. 1, pp. 169-176.

Kunii, D., and O. Levenspiel (1969), Fluidization Engineering, John Wiley & Sons, Inc. New York.

COMPUTER-AIDED THERMODYNAMIC ANALYSIS OF A MINIMUM MAINTENANCE SOLAR PUMP

A. Brew-Hammond, J. Roullier and D. Appeagyei-Kissi

Department of Mechanical Engineering
University of Science and Technology, Kumasi, Ghana

ABSTRACT

The Minimum Maintenance Solar Pump (MMSP) is a solar-thermal pumping system which operates on a diurnal cycle with solar heating and nocturnal cooling/suction. A thermodynamic analysis of the MMSP has yielded an expression which is used with the aid of a micro-computer to predict the performance characteristics of the MMSP. The predictions compare favourably with available experimental results and indicate that it is imperative for high temperatures well above $80^{\circ}C$ to be obtained in the MMSP if pumping is to be achieved at heads of practical significance.

KEYWORDS

Solar pump; unconventional technology; nocturnal cooling; thermodynamic analysis.

INTRODUCTION

The Minimum Maintenance Solar Pump (MMSP) is one of many solar-thermal water pumping systems which employ "unconventional technology in order to be suitable for construction and maintenance in less developed regions of the world" (Spencer, 1989). The MMSP, shown schematically in Fig. 1, consists of a metal tank (or drum) painted black and placed inside an insulating box with a transparent cover to enable solar heating of the drum. It was first suggested and tested in France by Bernard (1983). Since then several prototypes of the MMSP have been constructed and tested with varying degrees of success (Brew-Hammond, 1988). This paper seeks to identify some of the critical requirements for successful operation of the MMSP.

THERMODYNAMIC ANALYSIS

The MMSP operates on a diurnal cycle with solar heating and nocturnal cooling/suction. During the day the MMSP tank absorbs solar radiation and heats up. This causes air in the tank to expand and escape while residual water in the tank evaporates and helps to evacuate the air. At night the tank cools off and condensation of water vapour takes place, creating a vacuum inside the tank and drawing in water from the source. In order to develop an expression for the pump discharge a detailed thermodynamic analysis of the diurnal cycle is necessary.

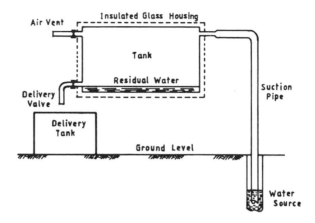

Fig. 1. A schematic diagram of the Minimum
Maintenance Solar Pump (MMSP).

The thermodynamic system comprises a mixture of moist air in equilibrium with saturated liquid. In the morning when the cycle begins (state 1), there would usually be a small quantity of water in the tank, left over from a previous day's operation, and the level of water inside the suction pipe would equal that for the source. As the tank absorbs solar radiation and heats up, the pressure of the residual air increases and forces the water inside the suction pipe downward until the pipe is completely empty of water (state 2), as shown in Fig. 2. The pressure inside the tank is given by (see nomenclature at end of paper)

$$P_{t,2} = P_{at} + \rho g h_b \qquad (1)$$

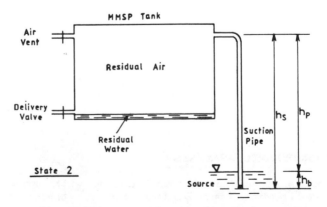

Fig. 2. A sketch of the thermodynamic system at state 2
(i.e. when maximum system pressure is attained.)

As further heating takes place the pressure inside the tank remains constant but the temperature of the tank continues to increase, depending on the solar radiation regime, until the maximum average tank temperature is attained (state 3). Assuming no significant change in the volume of residual water, the volume of residual air is given by

$$V_{ra,3} = V_t - V_{rv} + V_s \tag{2}$$

and the mass of residual air can be obtained from the perfect gas relation,

$$m_{ra,3} = \frac{P_{ra,3} V_{ra,3}}{R_a T_{t,max}} \tag{3}$$

From the Gibbs-Dalton Law of Partial Pressures the pressure inside the tank is the sum of the partial pressures of the residual air and the water vapour such that

$$P_{t,3} = P_{t,2} = P_{ra,3} + P_{wv,3} \tag{4}$$

As the diurnal cycle proceeds (i.e. towards evening) the temperature of the tank reduces, causing the total system pressure to decrease. Water from the source then starts to rise in the suction pipe. If the temperature drops low enough the water will rise to the top of the suction pipe and flow into the tank until the minimum (night-time ambient) temperature is attained (state 4) and the tank is nearly filled with water, as shown in Fig. 3. In this state the mass of air inside the tank is the same as that at state 3 and can again be obtained from the perfect gas relation,

$$m_{ra,4} = \frac{P_{ra,4} V_{ra,4}}{R_a T_{t,min}} \tag{5}$$

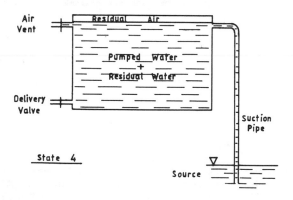

Fig. 3. A sketch of the thermodynamic system at state 4
(i.e. at the end of the suction process.)

The partial pressure of the residual air is obtained by applying the Gibbs-Dalton Law again to give

$$P_{t,4} = P_{ra,4} + P_{vv,4} \qquad (6)$$

and the total pressure in the tank is given by

$$P_{t,4} = P_{at} - \rho g h_p \qquad (7)$$

When the air vent and delivery valve are opened in the morning the pumped water flows out of the tank and leaves behind the residual water while any water inside the suction pipe simply flows back to the source. The volume of water pumped is therefore given by

$$V_p = V_t - (V_{rw} + V_{ra,4}) \qquad (8)$$

An expression for the final volume of residual air can be obtained by equating Eqns. (3) and (5), substituting for the partial pressures of the residual air from Eqns. (4) and (6), then for the total pressures inside the tank from Eqns. (1) and (7) and finally for the volume of residual air at state 3 from Eqn. (2) to give

$$V_{ra,4} = \frac{T_{t,min}(P_{at} + \rho g h_b - P_{vv,3})}{T_{t,max}(P_{at} - \rho g h_p - P_{vv,4})}(V_t - V_{rw} - V_s) \qquad (9)$$

The above expression can now be substituted into Eqn. 8 and, with the assumption that the suction pipe volume is negligible, the terms can be simplified and rearranged to yield

$$\frac{V_p}{V_t - V_{rw}} = 1 - \left[\frac{T_{t,min}}{T_{t,max}}\right]\left[\frac{P_{at} + \rho g h_b - P_{vv,3}}{P_{at} - \rho g h_p - P_{vv,4}}\right] \qquad (10)$$

It is worth noting that the term on the left-hand side of Eqn. (10) is the discharge ratio based on the maximum possible amount of water that can be pumped and the product on the right-hand side is also a dimensionless parameter involvong non-dimensionalised temperature and pressure.

APPLICATION

An interactive computer program was written for the solution of Eqn. (10). A subroutine within the program used a formula developed by Goff (1977) to calculate the partial pressures of water vapour. The minimum average tank temperature, assumed to be equal to the minimum night-time ambient temperature, was used to calculate the partial pressure of water vapour at state 4 while the temperature attained at the bottom of the tank, assumed to be equal to the temperature of the residual water, was used to calculate the partial pressure of water vapour at state 3.

In order to validate the model, discharge ratios obtained from experiments conducted at Brace Research Institute, Canada, were plotted against experimental values of the dimensionless parameter for temperature and pressure. A linear regression on the experimental data yielded a correlation coefficient of -0.9 and compared favourably with Eqn. (10), as shown in Fig. 4.

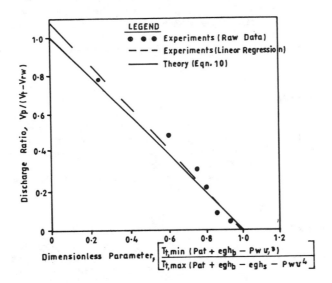

Fig. 4. A graph of discharge ratio vs. dimensionless
parameter for the MMSP.

The model was then used to determine the performance
characteristics of the MMSP for a given minimum tank temperature
of 25°C and a below-water suction pipe length of 1m, with the
additional assumption that the tank is heated uniformly so that
the maximum average tank temperature and the maximum temperature
attained at the bottom of the tank are the same. The results,
presented in Fig. 5, show that the discharge of the MMSP decreases
rapidly with decreasing maximum tank temperature and, therefore,

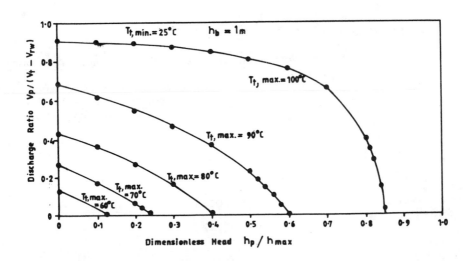

Fig. 5. Performance characteristics of the MMSP
predicted by Eqn. (10).

relatively high temperatures (i.e. above 80°C) are required for the system to operate successfully at suction heads of about 4m and above. Since the MMSP is essentially a non-concentrating solar collector, it becomes imperative that special care is taken during the system design to ensure that temperatures well above 80°C are attained.

CONCLUSION

An expression has been developed for computing the pump discharge of the MMSP. This expression gives results that compare favourably with available experimental data. The predicted performance characteristics of the MMSP indicate that high tank temperatures are necessary if pumping is to be achieved at heads of practical significance.

NOMENCLATURE

g - gravitational acceleration   R - specific gas constant
h - height/head   T - absolute temperature
m - mass   V - volume
p - pressure ρ - density of water

Subscripts

a   - air   rw - residual water
at - standard atmosphere   s - suction pipe
b   - below-water portion of   t - tank
      suction pipe   wv - water vapour
max - maximum   2 - state 2
min - minimum   3 - state 3
p   - pump/pumped   4 - state 4
ra - residual air

REFERENCES

Bernard, R. (1983). A low-maintenance solar pump. Appropriate Technology, 10 (1), 14 - 15.

Brew-Hammond, A. (1988). Development of a minimum maintenance solar pump. Journal of the University of Science and Technology, 8 (1), 7 - 12 (Reproduced as the Selected Article in ASSET, 10 (2), 19 - 22.)

Goff, J.A. (1977). Saturation pressure of water on the new Kelvin scale. In ASHRAE Handbook 1977 Fundamentals, ASHRAE, New York. p. 5.2.

Spencer, L.C. (1989). A comprehensive review of small solar-powered heat engines: Part III. Research since 1950 - "Unconventional" engines up to 100kW. Solar Energy, 43 (4), 211 - 225.

SOLAR ASSISTED HIGH PERFORMANCE AGRICULTURAL
DRYER

L.Imre, G.Hecker and L.Fábri

Technical University Budapest
1521 BME. Budapest, Hungary

ABSTRACT

A solar assisted lucerne dryer of 1000 tons/year drying capacity
has been developed and established at a stock-breeding coopera-
tive farm in Hungary for producing hay.
For all the year round utilization,hot water is produced by a flat
plat collector field of 900 m^2 surface area integrated into the
roof structure of a drying-storing barn.
Hot water is used for preheating the drying air in water-air heat
exchangers. In the idle periods of drying,hot water is used for
technological purposes.
The drying space is divided into four cells. Each cell has its fan
and its heat exchangers. Solar energy collected can optionally be
distributed among the cells.
A microprocessor is used for the direction and control of the drying process.

KEYWORDS

Solar drying; hay production; complex indirect system; microprocessing control.

INTRODUCTION

Stock-breeding farms need hay of good quality for feeding animals.
Conventional open-air drying methods can not fulfill the quality
requirements because of the losses of internal substances during
drying lucerne on the swath. Artificial drying methods   using
oil for heat production can meet the quality requirements, however, their operation is expensive and, the product is of high
price, - not considering the disadvantageous pollution effects of
the atmosphere.
Using solar drying for preservation of forages is principally reasonable since the requirements of the drying process to be realized are in harmony with the characteristic features of   solar
irradiation; the harvesting and drying season is in summer
and, for saving the internal substances of the dried product,low
temperature drying is needed. To increase the economy of the solar
system three main aspects should be considered. The first one is
to increase the energy effectiveness of the drying process by reducing the exit energy losses of the dryer. It is possible by feeding the wet material into the dryer by fairly thin layers. A new

layer can be fed in after having reached the 1st critical moisture content. The 2nd method for increasing the energy effectiveness of the dryer is the use of intermittent drying in the falling rate period of the drying process /Imre, Molnár and Farkas, 1983/. These two methods can be realized by dividing the drying space into separate cells. The drying processes in the individual cells should be directed shifted in time. The 3rd aspect is based on the fact that the solar energy collected should also be utilized in the idle periodes of drying, i.e. between the harvesting periods of lucerne.   All year round utilization is possible in farms which need hot water for technological purposes when using indirect solar system with flat-plate water medium collectors. The water tank of the hot water system of the farm can be used as a heat storage of the solar system. A complex solar-assisted dryer and hot water production system with microprocessing control has been established in a stock-breeding farm in Hungary /Imre, Farkas and Gémes, 1986/. The brief description of the system and the long term experiences are presented.

DESCRIPTION OF THE SYSTEM

Drying space is arranged in a barn of 54x16 m^2 surface area /Fig.1: 1/ which serves also for storing the hay. Drying capacity of the system is 1000 t/year, the storing capacity is 800 t/year. The single glazed, water medium flat-plate collector field of 900 m^2 surface area /2/ is integrated into the roof structure of the barn. The barn is divided into 4 cells of 12x16 m^2 surface area /3/. Each cell has its fan /4/ of 10^5 kg/h nominal mass flow rate of air. Maximum possible layer thickness of hay is of 7 m. Each fan has its water-air heat exchanger /5/ for preheating the air. Preheated air is introduced below the static bed and, the moist air leaves the barn at the openings below the roof /6/.

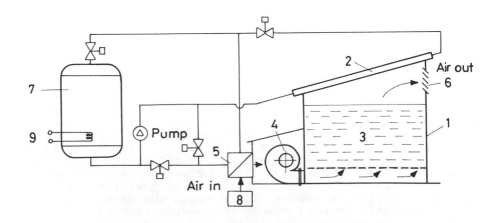

Fig. 1. Simplified scheme of the system.

In the middle of the barn an additional space is established for the feeding in system which distributes the wet lucerne equally in the cells by pneumatic transportation.

Hot water produced by the collector can optionally be distributed among the cells. The collector can be connected to the water tank of 100 m^3 volume capacity /7/ of the hot water system of the farm. It serves also as heat storage. As auxiliary energy source thermo-generator /8/ operating with natural gas can be used outside the fan boxes. For emergency situations in hot water supply electrical heating /9/ is built into the tank.

CONTROL OF THE OPERATION

The system has 5 main modes of operation /Imre, 1987/. The actual one is selected according to the technological requirements and to the meteorological conditions.

The 4-cells arrangement offers the possibility to realize various drying programs shifted in time in the different cells for achieving the most effective utilization of the solar energy collected and, to fulfill the actual requirement of the drying process. For the direction and control of the system a microprocessor has been designed and constructed. The control algorithm is based on a numerical analysis of the lucerne drying process /Imre and co-workers, 1983/ and of the operation of the complex system /Imre, 1987/. For the direction and control actions the following parameters are measured:

- actual value of the solar irradiation;

- ambient temperature and relative humidity;

- inlet and outlet temperature and relative humidity of the dry-ing air in each cell;

- temperature of the lucerne under drying in every layer of 1 m thickness in each cell;

- water inlet and outlet temperature of the collector;

- water temperature distribution in the layer-type storage tank.

The direction system produces all information about the actual operational characteristics of the system for the operator and after data processing it forms commands for the drying operations in the individual cells.

Signals are also formed for the operator when some disorder occurs in the operation.

LONG-TERM OPERATION

The complex solar-assisted drying and hot water supply system has been established in 1987. This year the experimental operation was performed. The regular operation started in 1988. To determine the performance and the main operational characteristics long-term monitoring was realized in 1988 and 1989. Main results and con-clusions are summarized as follows.

2186

Energy aspects

The annual average value of the solar energy used for drying was
127 MWh. Solar energy available for technological hot water production was 144 MWh.
Auxiliary air heater was needed in 1988 and produced of 12,1 MWh
/about 10 per cent/. In 1989 auxiliary energy was not used. Electrical energy consumption for driving fans was annually 112,5 MWh
in average.

Hay production

Lucerne has been harvested 3 times in the summer season. Mass of
hay fed in was 796 t/year /dry mass: 428 t/.
In Fig.2. the typical operational characteristics of the cell Nr.1
is presented. In the diagram the inlet, outlet and ambient air
temperatures /t/, the absolut moisture content of the inlet and
outlet air /x/ is given in the function of the time /from 7 a.m.
to 17 p.m. during the day/. On the axis of time the intervals of
the "fan in operation" is presented /with 2 breaks/.

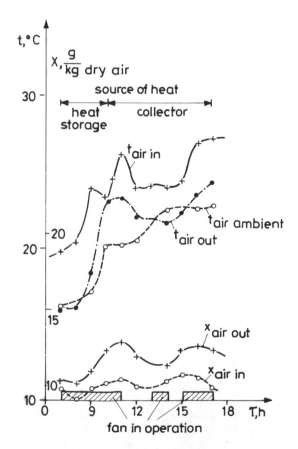

Fig. 2. Typical operational characteristics.

In Fig.3. the main material characteristics are given in the function of the time for the drying of one layer. At τ = 0 the carotin, the protein and the moisture content of lucerne before cutting is given. Harvest was timed early in the morning. Lucerne was dryed on the swath. 2nd and 3rd values are given after 6 and 12 hours open-air drying on the swath, resp.
Feeding in has been realized after 12 hours. Preservation by drying in the cells needed 72 hours.
It can be concluded that during drying the most important internal values of the material was practically equal to the values of feeding in.

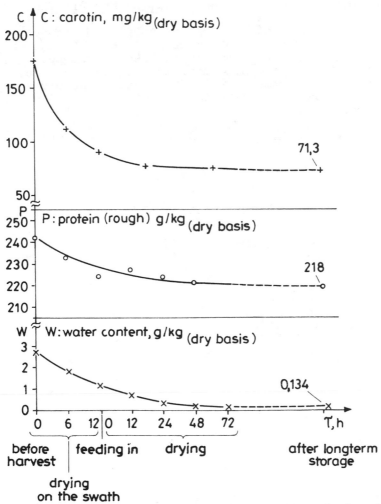

Fig. 3. Main material characteristics during drying.

Conclusions

Technically, the operation of the system was without any problem. Hay produced is of high quality with green colour and good smell. Milk production of the farm has been increased and deaths of calves /usually in winter time/ has been eliminated. As a result

of the high quality of hay the land area needed for feeding the
animals has been decreased.
In the Hungarian economy situation the payback time of the solar
system is about 6-7 years calculated on the basis of the substituted energy.

REFERENCES

Imre L., I. Farkas, L.I. Kiss and K. Molnár /1983/. Numerical
    analysis of solar convective dryers. In R.W. Lewis, K.A. John-
    son and W.R. Smith /Ed./, Numerical Methods in Thermal Problems,
    Vol.3. Pineridge Press, Swansea, pp. 957-967.
Imre, L., K. Molnár and I. Farkas /1983/. Some aspects of the in-
    termittent solar drying of lucerne. Proc. Int. Conference of
    Solar Drying and Rural Development, Bordeaux, pp. 93-101.
Imre, L., I. Farkas and L. Gémes /1986/. Construction, simulation
    and control of a complex and integrated Solar Drying System.
    In A.S. Mujumdar /Ed./, DRYING'86, Hemisphere, New York,
    pp. 676-684.
Imre, L. /1987/. Solar drying. In A.S. Mujumdar /Ed./, Handbook
    of Industrial Drying, Marcel Decker Inc. New York. Chap.11.
    pp. 357-417.

MODELIZATION OF A SOLAR FISH DRYER WORKING BY NATURAL CONVECTION

Michel FOURNIER, Yves MAURISSEN

IMP-CNRS
Perpignan University - France

ABSTRACT
A solar fish dryer has been built in Comores Island, to dry 30kg of fish. A mathematical model to simulate its behaviour has been written : It permits to optimize the design of the dryer.

KEY WORDS
solar dryer- natural convection- mathematical model

INTRODUCTION

The engineering of devices qualified as rustic (LE GOFF 1987) is that they can be handled by small scale production units, entrusted to developping countries and artisans with little technical skill.

Appliances should be designed to be easily repared or copied by local craftsmen, with available materials. They should be chosen by rational methods, by modelization and optimization of technical systems recently devel o ped by specialists in Chemical engineering. A devise will be that much easier to   put to use by a village craftsman if its preparatory theoretical study and its optimization have been carefully worked out.

In brief, for a device to be rustic, the preceding research shouldn't be.

DESCRIPTION OF THE DRYER

The dryer that has been studied is the chimney type. It's a derivative of the SKELTON's dryer (1985). Its content   is 30 kg of fresh fish. The dryer is raised so as to be out of reach of rodents, particularily numerous on the island. This makes it possible for the air to enter from below the collector and it is possible to use a porous absorber that is more efficient than a simple metal sheet.

The model is shown in the following figure.

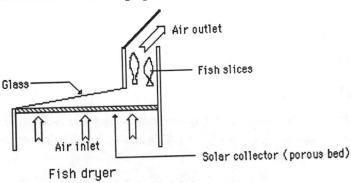

Fish dryer

MODELIZATION

1) The principle and working order diagram.

The working order of this dryer has been entirely simulated by computer, which has made it possible to optimize its design. For this, the dryer has been divided into two parts, clearly visible in the diagram:

1 - the solar collector with a porous absorber in a horizontal position;

2 - the drying box where the fish filets are hung, that is covered by a chimney;

The repeated calculations used for the modelization are as follows:

- the outside air enters the collector with a temperature T_e and a flow rate G_i

- the air is heated in the collector,

- the air heats the fish from which it recovers the water in cooling.

- Tm, the average air temperature is calculated and the difference $T_m - T_e$ is responsable for the air flow. One deducts a new air flow rate G_{i+1} with which one recommences a new calculation series and so on until the mesurements converge.

2) Simplifying hypotheses.

- the inertia of the thermal system is negligeable: so the calculations will be those that follow balanced states.

- in the beginning of the experiment, the produce to be dried is considered to have a regular internal humidity W_o.

3) The drying box.

The drying phenomenon consists in a double thermal process : under the action of water content gradient, the water inside the product migrates (dW_2/dt) and reaches the surface where it evaporates due to the drying air flow; the evaporating water flow is called dW_1/dt. Obviously, these two simultaneous phenomena should be such that the quantity of water exchanged is the same. One should have in a given time interval : $\Delta W_1 = \Delta W_2$

This is the condition that defines the system.

a) dW_1 calculation

The drying kinetic may be represented as :

$$\frac{dW_1}{dt} = -\alpha \left(p_s \; A_w - p_v\right)$$

p_s = saturation pressure

p_v = partial pressure of water vapor

A_w = Superficial water activity (Aw=1 for free water)

α = Coefficient depending on physical and geometrical properties of the product and air

This kinetic depends on water activity, therefore on temperature and on surface moisture content W_s of the product. This dependance is represented by the sorption curves of the product.

b) dW_2 calculation

To determine the internal water migration flow, the produce is considered as a plate, whose surface moisture content W_s, varies with time. This time is divided into intervals $\Delta\tau$, during which W_s remains constant .

Thus :

$$W_s = W_s(\tau) \qquad \text{for} \qquad \tau - \Delta\tau \leq t < \tau$$

$$W_s = W_s(\tau + \Delta\tau) \qquad \text{for} \qquad \tau \leq t < \tau + \Delta\tau$$

$$W_s = W_s(\tau + 2\,\Delta\tau) \quad \text{for} \qquad \tau + \Delta\tau \leq t < \tau + 2\,\Delta\tau$$

....

Moisture content in a plate

Let $W(x,t)$ the moisture content at any time t and any point x in the plate ; at the end of an interval (for instance, at $t = \tau$), it is possible to develop the difference $W(x,\tau) - W_s(\tau)$ into Fourier's serie:

$$W(x,\tau) - W_s(\tau) = \sum_i A_i \sin\left(\frac{(2i-1)\pi x}{L}\right)$$

with

$$A_i = \left[w_i(\tau) - W_s(\tau)\right] \frac{4}{\pi(2i-1)}$$

Then the average moisture content is given by the equation [*] :

$$\overline{W}(\tau) - W_s(\tau) = \sum_i \left[w_i(\tau) - W_s(\tau)\right] \frac{8}{\pi^2(2i-1)^2}$$

At the beginning of the next time interval ($t = \tau$), the surface moisture content is $W_s(\tau+\Delta\tau)$:

$$W(x,\tau) - W_s(\tau + \Delta\tau) = W_s(\tau) - W_s(\tau + \Delta\tau) + \sum_i A_i \sin\left(\frac{(2i-1)\pi x}{L}\right)$$

This expression may be written as :

$$W(x,\tau) - W_s(\tau + \Delta\tau) = \sum_i \left[w_i(\tau) - W_s(\tau + \Delta\tau)\right] \frac{4}{\pi(2i-1)} \sin\left(\frac{(2i-1)\pi x}{L}\right)$$

where appear new Fourier's coefficients.
The evolution of this system is ruled by FICK's equation.

$$D_m \frac{\partial^2 W}{\partial x^2} = \frac{\partial W}{\partial t}$$

with D_m the diffusion coefficient of water.
- the initial condition for $t = \tau$ is the expression of $W(x, \tau) - W_s(\tau + \Delta\tau)$ above,
- the boundaries conditions are :
$W(0,t) - W_s(\tau + \Delta\tau) = 0$ and $W(L,t) - W_s(\tau + \Delta\tau) = 0$ when $t > \tau$

According to DE VRIENDT (1984), the solution is:

$$W(x, t) - W_s(\tau + \Delta\tau) = \sum_i \left[w_i(\tau) - W_s(\tau + \Delta\tau) \right] \frac{4}{\pi (2i-1)} \sin\left(\frac{(2i-1)\pi x}{L}\right) e^{-(2i-1)^2 \frac{\pi^2}{L^2} D_m (t-\tau)}$$

letting

$$w_i(\tau + \Delta\tau) - W_s(\tau + \Delta\tau) = \left[w_i(\tau) - W_s(\tau + \Delta\tau) \right] e^{-(2i-1)^2 \frac{\pi^2}{L^2} D_m \Delta\tau}$$

so, the average value in $t = \tau + \Delta\tau$, is written:

$$\overline{W}(\tau + \Delta\tau) - W_s(\tau + \Delta\tau) = \sum_i \left[w_i(\tau + \Delta\tau) - W_s(\tau + \Delta\tau) \right] \frac{8}{\pi^2 (2i-1)^2}$$

We find again the equation[*], which gives the average moisture content, but with τ transformed into $\tau + \Delta\tau$. To study the next interval, it is only necessary to perform again the process.

The difference between two successive average moisture contents, separated by $\Delta\tau$, leads to obtain ΔW_2. At each step, using the two last equations (in addition of that giving dW_1/dt and those expressing heat and mass transferts), we seek the value of W_s, such as we obtain : $\Delta W_1 = \Delta W_2$.

For $t = 0$, we have $w_i = Wo$ (with Wo initial moisture content).

4) Particular cases

a - Phase I: In the beginning of the drying process, whilst the surface of fishes is very damp, this evaporation surface determines the drying speed, this depends on the heat exchange coefficient and the air flow rate.

The water activity is $A_w = 1$ so the solution is given by the only equation expressing dW_1/dt.

b - Phase II (or diffusional phase): It takes place when the surface is dry. Then the process rate is a function of D_m. This coefficient takes different values according to the nature of the bonds product-water. The drying speed is that much faster if the product temperature is high. We have taken two experimental values depending on the advancement of the drying process (JASON 1965).

In phase II , we have $W_s \approx W_e$ (the water content balance W_e is fixed by the condition of the drying air), consequently only subsist the equation leading to the calculation of dW_2/dt.

RESULTS OF SIMULATION

As the drying speed depends essentially on the temperature, we have noticed that it was not worth planning a collector surface greater than ten square meters, which is sufficient for obtaining an adequate air flow rate in order to avoid a temperature drop during the evaporation phase. This is interesting because Phase I only lasts for a short time .

In the same way, raising the chimney increases the air flow and reduces the temperature, so this is not necessarily a means of increasing the drying speed during Phase II.

- The most sensitive variable in the length of the drying operation is the thickness of the fish slices.

- The fat content of the fish (JASON 1965) plays a negative role in the choice of fish to be dried (influence on D_m): cod, for exemple is a good choice.

- Finally, in order to avoid the rehumidification of the fish during the night, the air flow circuit in the drying box is closed. This complete stop in the drying process encourages a uniform distribution of the water content in the product. The model takes this phenomenon into account.

EXPERIMENTAL RESULTS

- The comparison of fish drying operation times was satisfactory: a load of tuna filets about three centimeters thick was dried in 5 days with steady weather conditions (maximum outdoor temperature difference during the day 5°C; maximum insolation at midday over 1000W/m^2), whilst on open air trays, in the same conditions, the drying time is estimated at 8 days.

- The tent-dryer, experimented in Senegal by SY(1979), permit to dry whatever fish in 8 days. This dryer is low-cost, and, above all, protect the fish against the dust.

- The M5003 model built in Seychelles Island by THEMELIN in 1984 is more expensive: it permit to dry about 13kg in 5.5 days.

- In our case, the efficiency of the solar collector turned out to be relatively weak: a better distribution of the coconut fibre should improve the air-absorber heat exchange coefficient, and so the efficiency of the system.

CONCLUSIONS

Amongst the solar fish dryers existing in the world, we have chosen a type for the Comores Islands perfectly adapted to the local climatic and economic conditions.

The mathematical modelization has made it possible to simulate the working order of the dryer : it will be possible to simulate drying operations with regard to the different geometrical variables, permitting the optimization and improvement or to make changes in the design of the apparatus with regard to the needs so as to obtain the greatest efficacity.

BIBLIOGRAPHY

DE VRIENDT   A. Transmission de la chaleur - Conduction vol I, tomeII, p.110. Gaetan Morin ed., CHICOUTIMI, Quebec, Canada, 1984.

JASON A. C.   Effects of fat content on diffusion of water in fish muscle - J. Sci. Food Agric.,vol 16 , 1965.

LE GOFF P.   Colloque sur le Génie des Procédés - NANCY- France, Sept 1987.

SKELTON A.   Institut BRACE Université Mc Gill -1 Ave Stewart park - Ste Anne de Bellevue H9X1CO - QUEBEC - Canada, 1985.

SY A. Y.   Rapport d'expérimentation sur le séchoir tente solaire - ITA, BP 2765, DAKAR-HANN, Sénégal, 1979.

THEMELIN A.   Rapport de mission aux Seychelles Avril 1984 CEEMAT- GERDAT - BP 5035 34032 MONTPELLIER CEDEX, France, 1985.

SOLAR DRYING OF SAND FOR EGYPTIAN GLASS INDUSTRY

Mansour A. Mohamad[*] and Salah M. Khalil[**]

* National Research Centre, Solar Energy Dept,
Cairo, Egypt.
** Egyptian Petroleum Research Institute, Nasr City,
Cairo, Egypt.

ABSTRACT

The present study is made to investigate the possibility of energy saving in the glass industry, which is a heavy energy consumption one, using solar energy. The theoretical analysis showed that the drying of sand after washing is the suitable part to share by solar energy   It is technically viable to use solar energy to reduce the moisture content of washed sand from 11% to 5% using hot air from solar collectors at temperatures less than 80 $^{\circ}$C. This will lead to energy saving of about 360 MJ/ton of glass product. As the Egyptian present production is about 200 000 ton glass per year, so the possible national saving in energy in this sector due to utilization of solar energy would be approximately 72 000 000 MJ/year. Three methods of drying were tried. The results showed the possibility of drying the sand by all of them to a moisture content less than 5%. The results showed also that a period of two days is sufficient to reach this level of moisture content.

KEYWORDS

Solar   drying;   sun   drying;   natural   air   circulation;   flat reflectors; energy conservation; sand drying ; glass industry.

INTRODUCTION

Utilization of solar energy in industrial process heat, especially in processes involving low level temperatures is one of the most active options in the field of energy conservation. The specific energy consumption in the Egyptian glass industry is about 50 GJ/Ton of product, the international figure reaches about 20 GJ/Ton, so the Egyptian glass industry suffers from excessive energy consumption. In the process of sand drying in glass industry, the sand is dried from about 11% to about 5% moisture content. This can be achieved using solar kilns and would lead to considerable fossil fuel saving. Saving fossil fuel consumption in glass industry is not only important from the economical point of view, but also reduces the environmental pollution problems.

THE EGYPTIAN GLASS INDUSTRY

The glass industry started in Egypt during the forties, developed to cover most of the local needs through the fifties, and started exportation in the early seventies. The total production has been raised from about 21 thousand tons in 1954 and to about 200 thousand tons in 1990. Glass making incorporates a number of operations; mainly, preparation and handling of raw materials, (e.g. crushing, sieving, washing, and drying), melting, forming, annealing and inspection. The primary sources of energy are fuel oil, gas oil, and gaseous fuels. Fuel oil is used in the melting process, while gas oil or gaseous fuels are used in the sand drying and annealing processes. The melting process is performed in the furnace at temperatures up to 1300 $^{\circ}$C., while the annealing process is performed at temperatures ranging between 500 to 1000 $^{\circ}$C. The drying process of sand, is the point of interest, which takes place during the preparation of raw materials. In this stage, sand is washed by using a strong stream of water, then is left over night, when the moisture content falls to about 11%. Sand is then fed to a rotating dryer where it is subjected to the heat liberated by direct firing of gas oil or gaseous fuel to come out almost dry at about 110 $^{\circ}$C (Salah, 1983).

The idea of .the present research is based on the following ;

1. It is of great importance that sand is thoroughly mixed with other ingredients to have a good quality glass. To achieve this, water up to 4-5% is usually added to the sand, if it is dry, mixed with it before adding the other ingredients. Wet sand grains are capable of becoming a sort of nucleus to which soda, calcium carbonate and other constituents adhere.

2. Drying sand from 11% to 5% moisture content using solar energy can be achieved with hot air from flat plate solar collectors at temperature level less than 80 $^{\circ}$C (Everett, 1980).

THEORETICAL ANALYSIS

The amount of heat required for water evaporation and heating the wet sand can be calculated from Eq (1):

$$Q = Lev \times [Cw1 - Cw2] + Cp \times [ts1 - ts2] \tag{1}$$

where;

Q	= amount of energy for drying one ton of sand		kJ/ton
Lev	= latent heat of evaporation of water	=2240	kJ/kg
Cw1	= water content of sand at dryer inlet	= 11	%
Cw2	= water content of sand at dryer exit	= 5	%
Cp	= specific heat of wet sand	= 2.093	kJ/kg
ts1	= inlet sand temperature	= 25	$^{\circ}$C
ts2	= exit sand temperature	= 75	''

The area of flat plate solar collector required to dry one ton of sand can be calculated from Eq (2):

2196

$$A = Q / [SI \times \eta_d]$$ (2)

where;

 A = collector area per ton sand m^2
 SI = mean yearly solar intensity in Egypt = 20 MJ/m^2.day
 η_d= overall dryer efficiency (Mohamad, 1982) = 0.25

Using the above figures the area of flat plate collector needed to dry one ton sand was calculated to be 10 m^2.

EXPERIMENTAL WORK

The first trial was carried out to test the possibility of sand drying using solar energy and to find out a suitable method of drying. Three methods with three similar sand piles were tried. The first one was the open air type, i.e. sun drying. The second one used a flat vertical reflector. The third used an aluminum frame with plastic cover and chimney for air circulation. Figure 1. shows the flat reflector used in the second method, fig. 2. shows the plastic dryer, and fig. 3. shows the sand pile used in the three methods. Tests were carried out for ten days to measure the moisture content of sand. Two samples per day were collected from each pile. The samples were collected from the vertical centre of the tested piles and at distances of 20 cm (P1) and 40 cm (P2) from the pile top as shown in fig. 3.

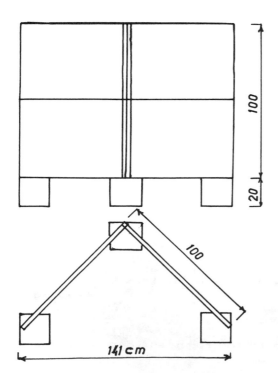

Fig. 1 Flat booster mirror

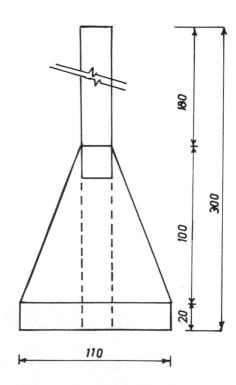

Fig. 2. Plastic dryer

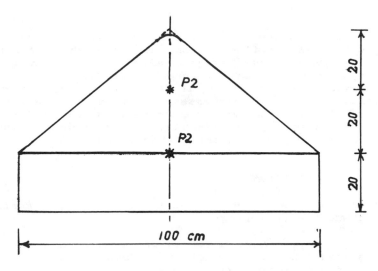

Fig. 3. Sand pile and points of samples collection.

RESULTS AND DISCUSSION.

Table 1 shows the results of measuring the moisture content of the picked samples at different points for the first two days. For all samples, the moisture content falls below 5% during these two days. These values satisfy the mixing requirements in the industry, so the remainder results are not important.

TABLE 1   The Moisture of the Picked Sampler at

Different Points at the First Two days

time		First Day		Second Day	
		9 AM	2 PM	9 AM	2 PM
Drying Method					
Sun drying	P1	8.4	3.1	2.24	0.84
	P2	8.86	4.9	2.97	1.3
With Reflector	P1	6.44	1.72	1.69	0.33
	P2	6.88	2.93	2.97	1.87
Plastic Cover	P1	10.13	2.84	1.79	0.5
	P2	11.66	4.2	2.333	0.68

The minimum moisture content was achieved at point P1 in the second method with reflectors on the second day. As shown in the table, all the moisture content values are less than 1.87% after two days, in spite of the difference in the moisture content at the beginning of the test in the three methods due to non similarity of washing processes of piles. Independent of the method of drying, the moisture content at the end of the first day is satisfactory on condition that the moisture at the beginning will not exceed prescribed value (11%)

CONCLUSIONS

From the above mentioned analysis and the experimental work, one can conclude the following;

1. Solar energy can be utilized for drying sand to the suitable moisture content for the glass industry where a considerable amount of fuel can be saved.

2. The time period of one day, or maximum two days, is sufficient to reach the drying level in all   drying methods used.

3. Design modifications and experimental efforts are needed to reach the suitable dryer configurations and economics.

ACKNOWLEDGEMENT

Thanks are offered to engineer Anwar Ezzat for his true co-operation during this work.

REFERENCES

Everett, D. H. (1980). Principles of drying and evaporation. *Sunworld. J.*, *Vol. 4, No. 6*, 182-185.

Salah, M. K., I. K. Abdou. (1983). Energy consumption in glass industry and glass products in Egypt. *First National Conf. for Glass Industry in Egypt*. Cairo, Egypt.

Mohamad, M. A., S. H. Soliman, I. A. Sakr. (1982). Comparison of performance of different configurations of plastic solar dryers. *Fourth Int. Conf. for Mechanical Power Eng*. Cairo, Egypt.

NEW ABSORBER GEOMETRIES FOR WAVELENGTH AND ANGULAR SOLAR SELECTIVE ABSORBER COVER COMBINATIONS

M.Lazarov, T.Eisenhammer, R.Sizmann

Sektion Physik, Ludwig-Maximilians-Universität München,
Amalienstr. 54, D 8000 München 40, Germany

ABSTRACT

We present numerical calculations of directional absorptances and hemispherical emittances for different solar selective absorber materials and heat mirrors in various combinations and geometries. The combinations are designed for non concentrating collectors to provide process heat in the temperature range beyond 250°C using both direct and diffuse solar irradiance. We optimized the yearly efficiency with respect to the film thicknesses of absorber and heat mirror materials and the fraction of covered aperture area. We show that under certain conditions a combination of a flat heat mirror and a convex hemispherical absorber can show a larger efficiency than a flat combination or the absorber alone.

KEYWORDS

selective absorber; heat mirror; directional emittance; directional absorptance; solar collector; non-concentrating collector; process heat

INTRODUCTION

High efficiency production of process heat with temperatures beyond 250 °C using non-concentrating solar collectors requires a pronounced selectivity of absorber and cover material. As demonstrated earlier (Lazarov, 1990), for attaining high temperatures low emittance is more important than high absorptance. To reduce emittance we combine a selective absorber with a heat mirror. Here we report on numerical calculations of absorptance and emittance of a convex hemispherical absorber in combination with a flat heat mirror which partially covers the absorber aperture (see Fig. 1 for definition of symbols). We compare the performance of this spherical geometry to flat combinations and absorbers for a few absorber materials in combination with a TiO_2-Ag-TiO_2 heat mirror.

Absorber materials with low emittance such as TiN_xO_y (ZhiQiang, 1984) and a-C:H, but also Black-Cobalt (for optical constants see references in (Eisenhammer, 1990)), which show low emittance at small thickness, were examined. Given the spectral index of refraction $\tilde{n}(\lambda)$, we calculated the directional spectral reflectance of these films on copper by Airy-summation (Knittl, 1976) for various film thicknesses. $\tilde{n}(\lambda)$ of the copper substrate was corrected for high temperatures, whereas the $\tilde{n}(\lambda)$ of the absorbing films remained uncorrected. The spectral directional transmittance and reflectance of TiO_2-Ag-TiO_2 heat mirrors was calculated by Airy-summation. The thicknesses of TiO_2 were 35nm. Calculations were performed with bulk $\tilde{n}(\lambda)$ of silver ("bulk") and with $\tilde{n}(\lambda)$ corrected for thin films applying a Drude modell ("Drude") analoguous to (Szczyrbowski, 1987).

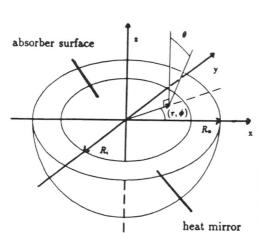

absorber surface

Fig. 1: Geometry of the hemispherical absorber. The heat mirror covers the hemisphere only to the radius $R_i \leq R_a := 1$. In the following R_i is normalized to R_a.

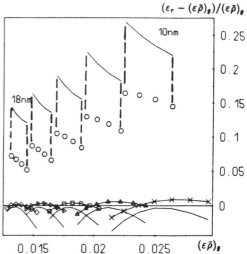

Fig. 2. $(\varepsilon_r - (\varepsilon\bar{\rho})_g)/(\varepsilon\bar{\rho})_g$ of the hemispherical ($\triangle$) and flat combination (signs+lines) versus $(\varepsilon\bar{\rho})_g$ of the flat combination. Lines show theoretical results, see text. Thicknesses of the absorber (TiN_xO_y): 30-90nm for flat and 40-70nm for hemispherical combinations. Heat mirror is TiO_2-Ag-TiO_2 with "bulk" silver thicknesses of 10-18nm.

The optical constants are assumed to be isotropic; hence, in the calculations for absorptance and emittance there is no azimuthal ϕ-dependency. For flat combinations the directional absorptance $(\alpha\tau)(\theta)$ was calculated according to

$$(\alpha\tau)(\theta, T) = \frac{1}{\int_0^\infty AM1.5(\lambda)d\lambda} \cdot \int_0^\infty \frac{AM1.5(\lambda) \cdot \alpha(\theta,\lambda,T) \cdot \tau(\theta,\lambda,T) \, d\lambda}{\alpha(\theta,\lambda,T) + \tau(\theta,\lambda,T) - \alpha(\theta,\lambda,T) \cdot \tau(\theta,\lambda,T)}, \quad (1)$$

where $\alpha(\theta,\lambda,T)$ is the absorptance of the absorber, $\tau(\theta,\lambda,T)$ the transmittance of the heat mirror and $AM1.5(\lambda)$ denotes the air mass 1.5 solar spectrum. The resulting hemispherical emittance ε_r was calculated by

$$\varepsilon_r = \left\langle \frac{\varepsilon(\theta,\lambda,T) \cdot \bar{\varrho}(\theta,\lambda,T)}{\varepsilon(\theta,\lambda,T) + \bar{\varrho}(\theta,\lambda,T) - \varepsilon(\theta,\lambda,T) \cdot \bar{\varrho}(\theta,\lambda,T)} \right\rangle, \quad (2)$$

with

$$\langle x(\theta,\lambda,T) \rangle := \int_0^\infty \int_0^{2\pi} d\phi d\lambda \frac{2\pi L_\lambda(\lambda,T)}{\sigma T^4} \sin(\theta)\cos(\theta) \; x(\theta,\lambda,T), \quad (3)$$

where $L_\lambda(\lambda,T)$ is Planck's spectral black body radiance at the absorber temperature T; σ is the Stefan-Boltzmann Constant, $\varepsilon(\theta,\lambda,T)$ the emittance of the absorber and $\bar{\varrho} = 1 - \rho$, where ρ is the reflectance of the heat mirror.

In case of the hemispherical geometry $(\alpha\tau)_K(\theta)$ and ε_r were calculated with Monte-Carlo integration and ray tracing algorithms. Further details are given in (Eisenhammer, 1990). Differing from earlier work, the calculations presented here consider not only the θ-dependency, but also the spectral dependencies of $\varepsilon(\theta,\lambda,T)$, $\bar{\varrho}(\theta,\lambda,T)$, $\alpha(\theta,\lambda,T)$ and $\tau(\theta,\lambda,T)$. Specular reflectivity was assumed. The error in $(\alpha\tau)_K$ and ε_r due to calculational errors is $\leq \pm 0.5\%$.

NUMERICAL RESULTS

Emittance

If properly designed, a combination of a selective absorber with a heat mirror can reduce the emittance of an absorber or a heat mirror (in combination with a black body) substantially. To achieve best performance, angular and wavelength dependencies have to be considered (Lazarov, 1989).

To demonstrate the influence of angular and wavelength dependence on ε_r of a flat combination (Eq. 2), we expand $\varepsilon(\theta, \lambda, T)$ and $\bar{\varrho}(\theta, \lambda, T)$ around their mean values (according to Eq. 3):

$$
\begin{aligned}
\varepsilon(\theta, \lambda, T) &:= <\varepsilon> + \; f(\theta, \lambda, T) \quad \text{with} \quad <f>= 0 \\
\bar{\varrho}(\theta, \lambda, T) &:= <\bar{\rho}> + \; g(\theta, \lambda, T) \quad \text{with} \quad <g>= 0 \; .
\end{aligned}
\tag{4}
$$

In a second order approximation, assuming small $<\varepsilon>$, $<\bar{\rho}>$, f and g, we find for ε_r :

$$
\varepsilon_r \cong (\varepsilon\bar{\rho})_g \left\{ 1 + \frac{1}{N^2} \cdot \left(2 <fg> - \frac{<\bar{\rho}><f^2>}{<\varepsilon>} - \frac{<\varepsilon><g^2>}{<\bar{\rho}>} \right) \right\} \; ,
\tag{5}
$$

where $N = \; <\varepsilon> + <\bar{\rho}> - <\varepsilon><\bar{\rho}>$ and $(\varepsilon\bar{\rho})_g$ is the resulting emittance of a combination of grey materials, i.e. constant ε and $\bar{\rho}$:

$$
(\varepsilon\bar{\rho})_g = \frac{<\varepsilon><\bar{\rho}>}{<\varepsilon> + <\bar{\rho}> - <\varepsilon><\bar{\rho}>} \; .
\tag{6}
$$

According to Eq. 6 the main influence on ε_r is given by the averages $<\varepsilon>$ and $<\bar{\rho}>$. In the second order approximation, Eq. 5, ε_r is decreased by the autocorrelations $<f^2>$ and $<g^2>$ and can be increased or decreased in dependence of the sign of the correlation $<fg>$. As an example we demonstrate the effect of f and g on ε_r for a TiN_xO_y-Cu absorber combined with silver sandwiches in Fig. 2. This graph shows $(\varepsilon_r - (\varepsilon\bar{\rho})_g)/(\varepsilon\bar{\rho})_g$ versus $(\varepsilon\bar{\rho})_g$. The points are calculated from Eq. 2 and the lines are from Eq. 5 for the same material combinations. Points below zero indicate a proper tuning of wavelength and angular dependencies of ε and $\bar{\rho}$.

Changing the absorber geometry results in a different θ-dependency of $\varepsilon(\theta, \lambda, T)$ and so in different correlations $<fg>$ and $<f^2>$. Unfortunately, every non-flat geometry increases the emitting absorber area, thus increasing ε_r . There is always some reabsorption present and the increase in ε_r is not proportional to the increase of the absorber area. In the case of a hemispherical absorber the area is increased by a factor 2, but the increase of ε_r (without heat mirror) is 1.7 to 1.9 for the present materials and film thicknesses. To compare hemispherical and flat geometry the results for hemispherical TiN_xO_y-absorbers totally covered with silver sandwiches are plotted in Fig. 2. The emittance of the hemispherical combination is 5 to 15% higher than for the corresponding flat combination. This result can be compared with ε_r of a flat combination assuming that the change in geometry results only in a rise of $<\varepsilon(\theta, \lambda, T)>$ in Eq. 6 due to the increased absorber area. The corresponding curves, calculated with Eq. 5, lie well above the result of the hemispherical combinations (points). The good performance is due to the better tuning of the angular dependencies of $\varepsilon(\theta, \lambda, T)$ and $\bar{\varrho}(\theta, \lambda, T)$. The absolute values of the emittances lie above those of a flat combination, and the only advantage of a hemispherical geometry would be the better absorptance $(\alpha\tau)_K$ resulting from the increase in absorbing area. Not covering the aperture fully with a heat mirror increases $(\alpha\tau)_K$. Unfortunately, also ε_r is increased, see Fig. 3. Analoguous results as demonstrated for a TiN_xO_y-absorber are found with a-C:H and Black-Cobalt.

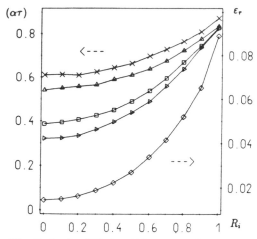
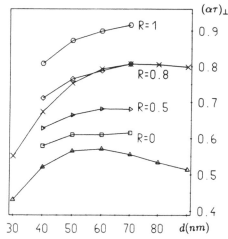

Fig. 3: $(\alpha\tau)_K(\Theta)$ for different angles and ε_r versus R_i of a hemispherical combination. The absorber is TiN_xO_y (50nm), heat mirror is TiO_2-Ag-TiO_2 with "bulk" silver (16nm). Angles are 0°, 54°, 72°, 81°.

Fig. 4: $(\alpha\tau)_\perp$ versus absorber thickness. Results for TiN_xO_y-absorber ($\times$), flat combination ($\triangle$) and hemispherical combinations. The heat mirror is "bulk" silver with a thickness of 16nm. For the hemispherical combination the plotted results are for R_i =0, 0.5, 0.8, 1.

Absorptance

Figure 3 shows some typical results for $(\alpha\tau)_K$ and ε_r as a function of R_i. Both increase with increasing R_i, corresponding to a smaller fraction of the aperture covered by the heat mirror. Figure 4 shows normal absorptance $(\alpha\tau)_\perp$ of the system with a TiN_xO_y-absorber as a function of the absorber thickness and different R_i. For comparison results for the corresponding flat absorber and flat combination are shown. $(\alpha\tau)_\perp$ of the flat absorber shows a weak maximum at a thickness of about 70nm. The reduction in $(\alpha\tau)_\perp$ at larger thicknesses is due to an interference maximum of reflection around $\lambda = 600nm$. The flat combination shows a stronger reduction of $(\alpha\tau)_\perp$ at thicknesses beyond 60nm, since the absorbers maximum of reflectance lies at the weavelength of the heat mirror's maximum transmission. $(\alpha\tau)_\perp$ of the hemispherical combination and the absorber are equal at $R_i \approx 0.8$. The maximal absorptance of the hemispherical absorber without heat mirror is essentially increased relative to the flat absorber. Also for a totally covered hemispherical absorber $(\alpha\tau)_\perp$ is larger than for the flat combination, the thickness dependence being smaller than for the flat combination. Here $(\alpha\tau)_\perp$ is increased by 7.5% to 11%. Similar results were obtained for a-C:H-absorber, especially with a large minimum of absorptance around an absorber thickness of 130nm in the cases of flat absorber and of the flat combination.

Yearly Efficiency

To compare the performance of combinations of different absorber materials and heat mirrors a yearly efficiency was calculated according to the following definition:

$$\bar{\eta} = \frac{\int [(\alpha\tau)(\theta) \cdot \cos\theta \cdot E(\theta,t) - \varepsilon_r \cdot \sigma \cdot (T^4 - T_a^4)]^+ \, dt}{\int \cos\theta \cdot E(\theta,t) dt}, \tag{7}$$

where T is the absorber temperature (300°C), T_a the ambient temperature (12°C) and $E(\theta,t)$

Table 1: $<\bar{\rho}>$, $(\alpha\tau)_\perp$ and $\bar{\eta}$ (all in percent, $T = 300\ °C$) for TiO$_2$-Ag-TiO$_2$ heat mirrors (in combination with a black body), TiO$_2$ thickness 35nm, at different thicknesses for "bulk" silver and "Drude" silver.

d (nm)	"bulk" silver			"Drude" silver		
	$<\bar{\rho}>$	$(\alpha\tau)_\perp$	$\bar{\eta}$	$<\bar{\rho}>$	$(\alpha\tau)_\perp$	$\bar{\eta}$
10	4.4	73	18	14	68	0
12	3.3	70	24	-	-	-
14	2.7	67	27	7.3	64	3
16	2.2	63	28	-	-	-
18	1.8	59	29	4.7	57	8
22	-	-	-	3.4	49	10

Table 2: ε_r, $(\alpha\tau)_\perp$ and $\bar{\eta}$ (all in percent, $T = 300°C$) for optimized systems; absorber thicknesses d in nm. The optimal R_i is 0, corresponding to a fully covered absorber (except Black-Cobalt, "Drude" silver: $R_i = 0.2$).

Material	hemisphere				flat combination				flat absorber			
	d	ε_r	$(\alpha\tau)_\perp$	$\bar{\eta}$	d	ε_r	$(\alpha\tau)_\perp$	$\bar{\eta}$	d	ε_r	$(\alpha\tau)_\perp$	$\bar{\eta}$
"bulk" silver (16nm):												
a-C:H	100	1.5	58	31	300	1.5	54	28	70	3.4	62	18
TiN$_x$O$_y$	50	1.7	61	32	50	1.5	57	30	50	4.9	75	17
Black-Cobalt	140	1.7	61	32	170	1.7	59	29	140	6.6	77	10
"Drude" silver (18nm):												
a-C:H	70	2.5	53	18	70	2.0	46	17	70	3.4	62	18
TiN$_x$O$_y$	50	2.9	56	17	50	2.4	52	18	50	4.9	75	17
TiN$_x$O$_y$(h)	50	3.2	56	15	50	2.8	52	15	50	6.4	75	10
Black-Cobalt	140	3.1	57	17	140	2.8	53	16	140	6.6	77	10

the time dependent radiant flux density with the angle of incidence $\theta(t)$ referring to the collector normal. [ ]$^+$ denotes that only positive values are taken into account. In this equation only losses due to thermal radiation are considered, as we compare only optical properties. Other losses, such as convection or thermal conduction, are excluded. For the irradiance E hourly values on a horizontal surface for Augsburg (Blümel, 1986) were used. For each system an optimization of the film thicknesses was performed. The best results are summarized in Table 1 and Table 2.

DISCUSSION

Highest values of $\bar{\eta}$ are achieved for bulk silver heat mirrors due to their small $<\bar{\rho}>$. The differences between hemispherical and flat combinations are due to larger $(\alpha\tau)$ of the hemisphere. However, the use of bulk data for the optical constants of the thin silver films produces too optimistic results.

Drude corrected silver produced more realistic data; the heat mirrors match well with experimental data (Gläser, 1980). As shown in Table 2, ε_r of the hemispherical combination is increased by up to 20% in comparison to the flat combination. The increase in $(\alpha\tau)$ is not sufficient for a better performance. Hemispherical and flat combinations and flat absorbers show nearly the same $\bar{\eta}$.

Considering calorimetric measurements on TiN_xO_y (Brunotte, 1990), the ε_r of the absorber is probably too small at elevated temperatures. To match these experimental results, we increased ε_r by 0.015 ("$TiN_xO_y(h)$"). The results for this absorber in combination with a "Drude" silver heat mirror are presented in Table 2. We consider this system as the most realistic case. ε_r of the hemispherical combination is only increased by 14% relative to the flat combination, the tuning of absorber and heat mirror is good. Both combinations show equal performance, while the flat absorber is worse than the combinations due to larger ε_r. The same conclusions can be drawn for Black-Cobalt.

For a-C:H with a "Drude" silver heat mirror we calculated $\bar{\eta}$ for a lower absorber temperature T of 250 °C , while ε_r was assumed to be unaffected. $\bar{\eta}$ is increased to about 26% for the flat absorber and the hemispherical combination, while for flat combinations $\bar{\eta}$ is not larger than 23% due to the small $(\alpha\tau)$. In this case of lower temperature, where the emittance is less important, the optimal R_i is about 0.7, corresponding to larger ε_r and $(\alpha\tau)_K$.

In the calculated yearly efficiencies only losses due to radiation are taken into account; the radiant flux density is referred to a horizontal surface. For the materials examined a combination of a selective absorber and a heat mirror can supply process heat at 300 °C with an $\bar{\eta}$ of 15-17%. A better optical quality of the absorber would not increase $\bar{\eta}$ substantially. The greatest potential lies in optical improvements of the heat mirror: a thermal efficiency $\bar{\eta}$ of 30-32% could be accessable.

ACKNOWLEDGMENT

We are grateful to Bundesministerium für Forschung und Technologie, Bonn, Federal Republic of Germany, for funding and support.

REFERENCES

Blümel, K., E. Hollan, A. Jahn, M. Kähler and R. Peter (1986). *Technical Report T 86-051, Bundesministerium für Forschung und Technologie.*

Brunotte, A., F. Liebrecht, M. Lazarov and R. Sizmann (1990). *Proceedings of the 7.Intern. Solar Forum 90, Frankfurt*, 2, 963–967.

Eisenhammer, M.Lazarov and R.Sizmann (1990). *Proceedings of the 7.Intern. Solar Forum 90, Frankfurt*, 2, 974–979.

Gläser, H.-J. (1980). *Glass Technology*, 21, 254–261.

Knittl, Z. (1976). *Optics of Thin Films.* John Wiley & Sons.

Lazarov, M. and R.Sizmann (1989). *Proceedings of the ISES Solar World Congress 1989 Kobe*, 2, 1323–1327.

Lazarov M., T.Eisenhammer and R.Sizmann (1990). *Proceedings of the 7.Intern. Solar Forum 90, Frankfurt*, 2, 968–973.

Szczyrbowski, J., A. Dietrich and K. Hartig (1987). *Solar Energy Materials*, 16, 103–111.

ZhiQiang, Y. and G.L.Harding (1984). *Thin Solid Films*, 120, 81–108.

DEVELOPMENT OF A SOLAR DRIER-GREENHOUSE PROTOTYPE FOR AGRICULTURAL PRODUCE

R. Corvalán   R. Román   I. Napoleoni

Department of Mechanical Engineering
University of Chile
Casilla 2777 Santiago – Chile

ABSTRACT

In this paper we report on two aspects of a research project on solar drying: the development of a cost–effective solar drier for fruits and vegetables with a loading capacity between 1000 and 2000 kg; and the results of a drying model that can be used to predict drying rates and times for a wide variety of products under field conditions. The model developed is of a semi–empirical nature and allows prediction of total drying times with an error of less than 10% in comparison with experimental results in the field.

The experiments conducted in the prototype show that the quality of the products is equal or better to that obtained in commercial drying. Economic analysis indicate that for this type of drier, the drying cost (including capital cost and maintenance) is about 60 % of the energy cost in a conventional thermal drier.

KEYWORDS

Solar energy; solar drying; drying models; radiative–convective solar drying process; fruit and vegetable drying.

INTRODUCTION

The project has been carried out in two stages. In the first one a complete analysis of the constant and decreasing rate drying periods was made in a laboratory drying chamber using different products and external conditions. The second one was the design, construction, and field testing of two solar dryer prototypes of the radiative–convective type. This design has the advantage that it combines the function of solar collector and drying chamber in one unit, resulting in lower costs. One, of 100 kg loading capacity, is suitable for small scale production, and the other one of 400 kg capacity is a scale model of a drier suitable for small scale industrial production. In this particular case, one can also use the drier as a greenhouse during periods when drying is not possible.

The interest in the development of a good drying model is multiple. In effect, we wanted a simple tool that would allow us to predict drying times and productivity for a drier under a wide variety of field conditions. The model had to be simple so as to easily apply it to a wide variety of products, and sufficiently accurate to predict average drying times and rates to within 10% of reality.

DRYING MODEL

There have been many drying models presented in the literature (Bertin, 1978; Sharaff–Eldeen, 1979) but usually they can be applied successfully only to a single kind of product, or else one has to know many properties for the product which are difficult to determine. The drying model presented here is an improvement of an earlier one (Alvarado, Román,

1986), which gave satisfactory results for the constant rate drying period but had important errors in the decreasing rate period.

The model is of a semi–empirical nature and is a product of the following observations and hypothesis:

- The drying rate in a solar drier, $\dot{X}$, is one or two orders of magnitudes less than in a conventional thermal drier. In a solar drying process, both product and surrounding air characteristics influence drying rate. In conventional drying, however, drying times depend mainly on the mass transfer rate inside the product.

- For most fruit types, where initial moisture content, X_i, is about 4 or 5 times the final moisture content, and for typical solar drying rates, constant and decreasing drying rate periods can normally be detected.

- In the first drying period, the drying rate, $\dot{X}_i$, depends on the exchange surface; a global mass transfer coefficient; solar energy absorbed by the product and on the water vapor pressure difference between surrounding air and product surface.

- In the decreasing rate drying period, drying rate, $\dot{X}_d$, decreases exponentially, and is a function of the drying rate when the product has critical moisture content and the difference between instantaneous and critical moisture contents.

The following basic expressions for drying rate, in both pure convective and radiative–convective processes, were found and confirmed with laboratory experiences:

$$\dot{X}_i = K \cdot A \cdot (p_{vs} - p_{va}) + C_1(c) \cdot I \tag{1}$$
$$\dot{X}_d(X) = \dot{X}_{cri} \cdot exp[B \cdot (X - X_{cri})] + C_2(c, X) \cdot I \tag{2}$$

Equation (1) is an extension of a classical relation in *Kneule (1964)* and predicts drying rate for the first period and (2) predicts it for the decreasing rate period. In the first one we see that drying rate $\dot{X}_i$ is a function of two terms; the first one takes into account the exchange area A, K (which is a function of product geometry and airspeed) and the difference in water vapor pressure between the product surface and airstream. The second term takes into account the *increase* in drying rate if solar radiation falls directly on the product being dried. Simply stated this increase is proportional to solar intensity on the product I, and an empirical constant $C_1(c)$, which is a function of product geometry and airspeed. Equation (2) states that for the decreasing rate period drying rate, $\dot{X}_d(X)$ is a function of moisture content, drying rate at critical moisture content $\dot{X}_{cri}$, radiation intensity on the product and two empirical constants which can be determined experimentally.

Laboratory and field experiences confirm the validity of equations (1) and (2) under typical solar drying operational conditions: differences between ambient and drying air temperatures about 25 C to 45°C; air speed under 1 m/s and solar energy intensity up to 1000 W/m².

EXPERIMENTAL WORK

Laboratory Experiences

The laboratory drying chamber designed and constructed (Fig. 1), permits drying tests under controlled conditions, with the following external parameters range: Drying air velocity from 0 to 3 m/s; drying air temperature from room temperature to 70°C; air humidity from ambient air humidity to saturation; radiation from 0 to 1000 W/m².

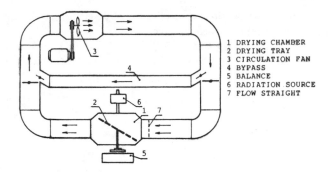

Fig. 1. Laboratory drying chamber

The drying chamber was instrumented with type T thermocouples for wet and dry temperature measurement, an anemometer and a precision balance for continuous weight control. Several drying test were conducted with grapes, apples, apricots and carrots, in order to determine the constants K; C_1, $C_2(c, X)$ and B, in equations (1) and (2).

Field Experiences

Field experiences on two solar driers were conducted. The first one was a unit of 100 kg fresh product capacity (*TEKHNE* drier, Fig. 2). We used this unit to determine which pretreatments were best for the fruit to be dried and also to validate in the field the drying model already presented. This unit is 6 m long, 1.2 m wide, 0.30 m high and permits the loading of five trays with up to 20 kg each of fresh product. The cover is U.V. resistant polyethylene and it has natural air circulation using a 4 m high chimney.

The second unit was one of 400 kg of fresh product capacity. This corresponds to a scale representation of a semi–industrial unit of a 1000 to 2000 kg capacity drier (Fig. 3). This unit is 10 m long, 6 m wide and has an elliptical cross–section 2 m high at center. Air circulation is through a 5.5 m high chimney with 0.9 m diameter. The structure is made of steel tubing and can be easily disassembled.

Both the *TEKHNE* drier and industrial prototype were equipped with copper–constantan thermocouples, anemometers and an Kipp CM–11 pyranometers. The parameters under measurement were collected using portable data–loggers. The large unit has been used as greenhouse in winter and as a drier in summer. Drying tests were conducted in both units using grapes, apricots, prunes and apples.

The main experimental results in both driers were:

- Drying times varied from three days (apricot halves and apples) to ten days (grapes). Drying time was highly sensitive to fruit pretreatment. For the large unit drying times were 20 to 30% longer than for the *TEKHNE* drier. These results were for daily insolations between 15 to 20 MJ $\cdot$ m^{-2}.
- Product quality was equal or superior to that typical of commercial driers. There was a marked improvement in taste and aroma retention compared to commercially dried products.
- Experimental data and modelling indicate that it is possible to use these units as driers at least five months per year.
- The *TEKHNE* drier had an overall efficiency of about 40% as a solar air heater, while the other unit this dropped to around 20%. This results imply that average air temperatures inside the drier were 15 to 25°C over ambient temperature during drying hours.

- Results indicate that the capacity of the 400 kg unit can be increased to 600 kg without detrimental effects in performance.

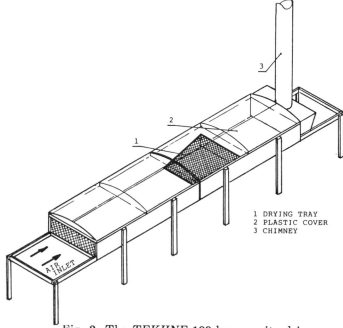

1 DRYING TRAY
2 PLASTIC COVER
3 CHIMNEY

Fig. 2. The *TEKHNE* 100 kg capacity drier

A preliminary economic analysis was made using the following hypothesis:

- Period of use: 4 months per year with 5 batches per month (these vary from 8 batches/month for apricots or apples to 3 per month for grapes).
- Average fresh load: 400 kg. Average dried product weight 80 kg.
- Initial investment US$ 700, annual interest rate 11.5%.
- Replacement of the plastic cover and maintenance, US$ 200 every three years.
- Lifetime of the drier: 9 years.

The above hypothesis give us a drying cost of US$ 0.10/kg of dried product. As a comparison, fossil fuel drying of grapes requires 0.4 kg of Diesel fuel per kg of raisins, this has a direct cost of US$ 0.17. This means that the drier is economical just considering fuel savings. The economic outlook is better if one considers a much lower investment and the possibility of using it as a greenhouse when not in use as a drier.

VALIDATION OF THE DRYING MODEL

In order to validate the drying model proposed above, the following experimental methodology was applied:

- In the laboratory drying chamber conduct two drying experiments: the first one of the pure convective type and the second one of the radiative–convective type.
- After drying the product, final moisture content should be determined using standard methodology.

From both curves one can determine K, X_{cri}, and the empirical constants C_1, B, $C_2(c, X)$.

- Determination of the scale factor between the laboratory drying chamber and the full scale drier by the analysis of the global mass transfer constant K.

- Computation of the predicted drying rates for field drying process and comparison with observed drying rates in field prototypes.
- From the drying rates, computation of weight versus time in the field drier.

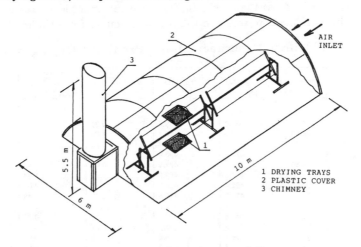

AIR INLET

5.5 m

10 m

6 m

1 DRYING TRAYS
2 PLASTIC COVER
3 CHIMNEY

Fig. 3. The greenhouse type drier

Determination of the Scale Factor in the Model

The global mass transfer coefficient K determined in the laboratory chamber is usually different than that for the field drier. This is due to differences in geometry, product density and air speed over the product. K is given by:

$$K = \frac{\beta_v}{R_v \cdot T} \cdot \frac{60 \times 10^6}{W} \tag{3}$$

and β_v is:

$$\beta_v = 0.514 \cdot Re^{-0.4526} \cdot \frac{c}{Sc^{2/3}} \tag{4}$$

for a porous bed we have:

$$Re = 4/6 \cdot \frac{1}{1 - \epsilon} \cdot \frac{c \cdot D}{\varrho \cdot \nu} \tag{5}$$

If one calls K' the global mass transfer coefficient for the field unit, and assuming the same air temperature and air speed, then we have:

$$\frac{K}{K'} = \left[\frac{1 - \epsilon}{1 - \epsilon'}\right]^{0.4526} \cdot \frac{W'}{W} \tag{6}$$

in our case, and for grapes:

$$\frac{K}{K'} = 5.87$$

further analysis shows that this same "scale factor" should be applied to $C_1(c)$ and $C_2(c, X)$. Another important difference between the laboratory chamber and the field drier is that in the first one radiation on the product is at constant intensity while for the other solar radiation varies during the day. To take this into account when one computes drying rates for a certain time interval, total *insolation* (MJ) during that interval is considered to calculate average *intensity* during the period.

This method was applied to several drying curves. Table 1 shows both experimental and modelled drying rates and weights in the case of grapes. One can see that even though drying rates have errors of up to 40%, drying *times* and the evolution of weight can be predicted with errors of less than 10%. In Fig. 4 we can graphically see the drying curves for this case, in both the real data and the predicted curve from the model.

TABLE 1 Experimental and Predicted Drying Rates

TIME [Hours]	WEIGHT		DRYING	RATES
	EXPERIMENTAL [g]	PREDICTED [g]	EXPERIMENTAL [g/Kg min]	PREDICTED [g/Kg min]
0.00	553.8	553.8		
3.00	539.0	539.2	0.85	1.23
6.50	500.0	513.1	1.69	1.72
8.50	489.3	506.8	1.24	1.22
26.00	479.4	496.9		
28.50	470.2	477.5	0.64	1.40
30.50	447.8	451.2	1.93	1.94
32.50	432.4	429.7	1.33	2.03
34.50	423.8	420.5	0.75	1.39
48.50	415.7	412.3		
50.50	413.6	408.6	0.18	0.63
54.00	401.2	393.2	0.61	0.94
56.00	388.7	386.0	1.08	0.88
58.00	381.4	381.2	0.63	0.80
75.00	374.9	374.7		
76.50	366.0	366.7	1.03	0.89
122.00	283.8	284.5		
124.00	280.2	279.6	0.31	0.57
126.50	264.6	272.4	1.08	0.55
130.00	257.1	265.5	0.37	0.52
147.00	251.3	259.6		
151.00	233.2	251.4	0.78	0.47

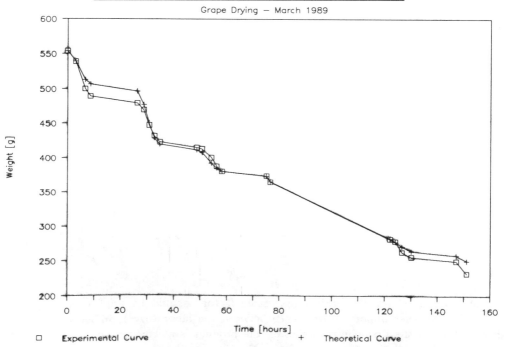

Fig. 4. Experimental and predicted drying curves

ACKNOWLEDGEMENTS

We express our acknowledgement to *FONDECYT (Fondo Nacional de Ciencia y Tecnología)* which financed most of our work through Project $N°$ 88–0573, *TEKHNE*, an Appropriate Technology Center, that built and helped us test the 100 kg prototype and the *CYTED–D* program , which through international cooperation gave essential help and direction to our work.

NOMENCLATURE

A: drying surface, m².

B: empirical coefficient.

c: air speed, m/s.

D: product diameter, m.

I: solar energy intensity on the product, W/m².

K: global mass transfer coefficient, s/m.

ϵ: product porosity.

ν: air dynamic viscosity, N · s/m².

ϱ: air density, kg/m³.

β_v: vapor diffusion coefficient, m/s.

p_{va}: water vapor pressure in drying air, Pa.

p_{vs}: water vapor pressure on the product surface, Pa.

R_v: water vapor gas constant, J/(kg K).

Re: Reynolds number.

Sc: Schmidt number, ν/D

T: air stream temperature, K.

W: product dry weight, g.

X: Moisture content, dry basis, at instant t, kg_{H_2O}/kg. $\dot{X}$ is drying rate.

X_{cri}: critical moisture content. This corresponds to the moisture content when drying rate starts to decrease.

REFERENCES

Alvarado, S. and Román, R. (1986). Solar dehydration in a radiative–convective drier. *Latin American Journal of Heat and Mass Transfer*, **10**, 53–73.

Bertin, P. et Combarnous (1978). Détermination des paramétres de séchage de certains fruits compacts. Thèse, Université de Bordeaux.

Kneule, F. (1964). Le Phénomène du Séchage a Conditions Constantes. In "Le Séchage", Eyrolles, Paris. 119 p.

Sharaff-Eldeen, Y.I. and Haidell, J.L. (1979). Falling rate drying of fully exposed biological materials. In 1979 Winter Meeting of ASAE, New Orleans, U.S.A.

-Boiling Water Collectors-

Low Cost Tubular Collectors without Convective Heat Losses

V. Heinzel, J. Holzinger, S. Petersen

University of Karlsruhe
Institute for Reactor Development
Dr. V. Heinzel
P.O. Box 6980
D-7500 Karlsruhe
Tel. 07247/822996

Abstract

Tubular collectors with, as usual in vacuum tubes, flat absorber platines have to be tracked with the sun if they are combined with parabolic trough concentrators. This renders the collector suspension complex and costly. The effect of other absorber geometries were investigated by heat loss measurements. As vacuum tubes often are not avaiable in developing countries, heat losses of glass tubes with low internal pressure are studied.

The investigations are based on a low prize, vacuum-tight adhesive glass/metal joint and a hand pump, with which under pressure down to about 1 h Pa (1m bar) are achievable. It is found that this pressure is sufficient to prevent convection in glass tubes with 100 mm diameter completely. In case of the collectors for process steam production developed at our our institute, the efficiency consequently increases from 38 % to 45 % (insolation 800 W/^{2} ,absorber temperature 100 °C) with a flat absorber platine. (The vacuum tubes would reach 52 %).

Introduction

Direct evaporating solar collectors, to provide saturated steam in the temperature range from 70°C up to 150°C for industrial appplication, were operated successfully in an open air test /1/.
The collector had a tubular receiver in the focus of a 4 times concentrating parabolic trough. As usual in vacuum tubes, flat plate absorbers, to which the boiler tubes are attached, were used. Those absorber plates keep the heat capacity of the collector and heat losses small, but require tracking of the absorber together with the concentrator. This renders the collector suspension complex and costly.

Vacuum tubes were converted for internal boiling and operated at high efficiency. However they are expensive and not available in developing countries. Therefore the development was concentrated on receivers with non evacuated glass tubes surrounding, in a first step, a flat plate absorber.

Investigations, described in this paper aim at economic features which are
system simplification: tackled with an absorber shape which makes absorber tracking unnecessary

and/or
efficiency improvement: achieved by heat loss reduction.

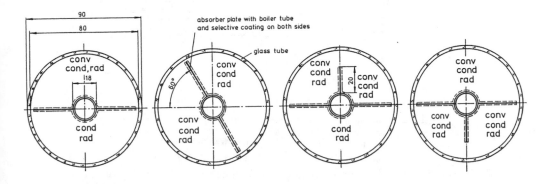

fig. 1: cross sections of tubular collectors used for heat loss measurements
geometry 1b corresponds to 1a but tilted
geometries 2a and 2b aim at larger acceptance angle

The heat losses of tubular collectors are mainly governed by the heat transfer between the absorber platine and the glass tube. This heat transfer is composed by thermal radiation exchange, heat conduction and conductivity. In case of the geometries considered here (Geometry 1a in figure 1) with air at normal pressure in the glass tube, experiments have shown that each of the heat transfer types contributes about 1/3 to the overall heat exchange at an absorber temperature of about 100°C and with a selective absorber surface (see figure 3).
Convection and conduction are eliminated in vacuum tubes, which have an internal gas pressure of less than 10^{-3} Pa. The goal of this work was to demonstrate with heat loss measurements, that the convective heat losses can be eliminated with under pressures achievable with glass/metall adhesive joints to tighten the glass tube against a metall flange and hand pump at low cost. Both were developed as precondition for this project.

The geometries, considered, are shown with the cross section in figure 1. Type 1a represents the standard absorber plate with the boiler tube for the heat removal by steam and the surrounding glass tube. The plate has to face the sun and, therefore, is usually tilted (1b). At normal air pressure inside the glass tube this creates an addition natural convection below the plate, not occuring in case 1a.

"No absorber tracking" is achieved with the T-shaped absorbers of type 2. However the fin has the higher heat loss potential.

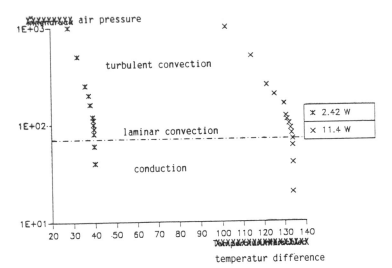

fig. 2. temperature differences absorber plate/glass tube depending on the air
        pressure inside the glass tube (geometry 1a) for heater power 2.42 W
        and 11.4 W related to a collector section of 150 mm length. length

Heat loss measurements

The heat transfer coefficient α for free convection results from dimensionless $Nu = \alpha l/\lambda$.
This Nusselt-number is derived from empirical functions depending on the dimensionless
numbers $Ra = Gr\, Pr$ (Rayleigh-number)

where $Gr = g\,\beta l^3\, \Delta T/ \nu^2$ (Grashof-number)

and for an ideal gas $Gr = g\,l^3\, p^2\, \Delta T/ (\eta^2 R^2 T^3)$

and $Pr = \eta\, cp/\lambda$ (Prandl-number).

(g = gravity, ß = air expansion coefficient, l = characteristic length, ΔT = temperature
difference between absorber plate and glass tube, η = dynamic viscosity, ν = kinematic
viscosity, R = gas constant, T = medium absolute air temperature, λ = air heat
conductivity, c_p = specific heat).
In /2/ it was found for a duct between two horizontal plates -the lower one heated - that
at Ra < 5830 the flow pattern changes from turbulent to laminar till Ra= 1708. Below
that number the fluid between the two plates became stagnant, so that heat conduction
took place , only.
For a geometry like 1a (see fuigure 1) the heat transfer functions for the convective case
are described in /3/.

The influence of the air pressure inside the glass tube on the convective heat loss can be
seen from the Grashof-number. However, the transition from the situation with and
without convection was not given in /3/ nor can be taken over from the geometry in /3/
without examination.

Therefore, an experiment was conducted with 1,5 m long sections of the collector tubes
with the cross sections from figure 1. The glass tubes were glue joint to metal flanges, to
which the boiler tubes were attached with the same silicon type glue. A electric heated
rod was inserted into the boiler tube.
Thermocouples were attached on the absorber plate and the glass tube on peripheral and
axial positions.

At constant heater power and constant temperature at all thermocouples, the differences of the mean temperatures at the absorber plate and the glass tube were used to derive the heat transfer coefficient.

Before each heat loss measurement the air pressure within the glass tube was pumped down to the desired value and kept constant during the measurement. In order to get the pure radiative heat transfer, the pressure was reduced to 10^{-3} Pa, at which neither convection nor conduction takes place. Conductive or convective heat losses resulted from the total heat losses at higher pressures reduced by the radiative heat losses at comparable temperature conditions. Heat transfer above and below the absorber platine were treated separately.

In a first step the heat transfer function and the transition pressures for geometry 1a were determined. Both coincided well with the experimental result in /2/ and /3/.

The temperature difference absorber/glass tube at constant heater power, respectively the reciprocal heat loss coefficient, is shown in figure 2 as function of the air pressure in the glass tube for two power levels (related to 150 mm collector tube length). The temperature of the absorber increases with decreasing pressure to an unsteadiness at about Ra = 6000 and then to a steadystate value below Ra = 1700 corresponding an internal air pressure of about 70 hPa. Derived from the peripheral temperature distribution of the glass tube, it can be stated that below 70 hPa only conductive heat transfer takes place.

For Geometry 2a and 2b the same transition pressure at 70 hPa was measured. In case of 1b, the absorber plate with tilted $\theta = 60°$ the transition was less well defined, as the convection below the plate caused some uncertainty when forming the mean temperature. However the deviation below 70 hPa were small. Further investigation will be conducted in order to show wether there exists a third type of convection arround the plate.

Therewith the pressure achievable with the hand pump -about 1 hPa- is sufficient to eliminate the convective heat transfer with all geometries investigated. The leak tests with the adhesive joints showed that the air pressure in the collector tube would increase afew hPa per year, or respectively remain below this value for a longer operational period.

Collector efficiency

The measured heat loss values were used to calculate the collector efficiency. Figures, 3 and 4 demonstrate comparisons of non evacuated, low pressure and high vacuum tubes for the different geometries of the absorber tubes.The assumption for the collector are 4 times concentration, insulation 800 W/m², direct insulation share 80 %, concentrator reflectivity 0,85, receiver shading on the concentrator 30%, glass tube transmission 90%, absorber coefficient 90%, absorber thermal emissivity 9%.

Both figures show that at normal air pressure in the glass tube the convection below the tiltes plate impairs the collector efficiency. Below 70 h Pa this effect vanishes, as convection is eliminated.

Under pressure in case of geometry 1a will increase the power output of the collector by about 40% (Δ T = 100°C).

The transition from geometry 1a to 2a or 2b impairs at subpressure the collector efficiency by 11 %. Future investigation will have to show that power loss will be balanced by minor losses due to simplier and shorter steam and feed water line connections.

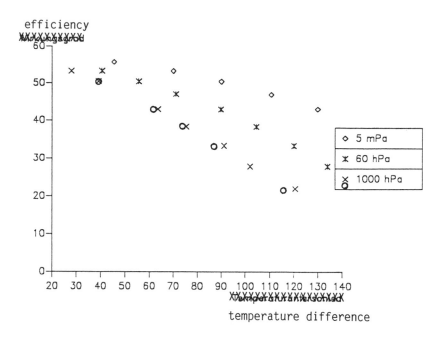

fig. 3. collector efficiency for geometry 1a/b at different air pressure within the glass tube

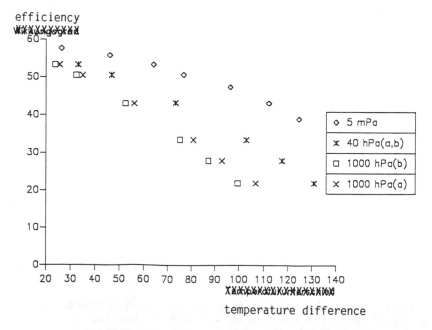

fig. 4. collector efficiency for geometry 2a/b at different air pressure within the glass tube

NINE YEARS EXPERIENCE WITH SOLAR THERMAL PUMPS IN INDONESIA

B. Bandarsyah[*], N. Suharta[*], D. Schneller[**], H. Nötzold[***]

[*]BPP Teknologi, [**]Dornier, [***]TÜV Rheinland

ABSTRACT

The REI-Project (Renewable Energies Indonesia) started in July 1985 on the basis of an already performed joint 5 years cooperation on scientific research and technological development on Solar Energy Utilization between the Federal Republic of Germany and the Republic of Indonesia.
Its targets are the testing and qualification of new technical systems under tropical conditions, technology transfer to enable Indonesian manufacturers to produce the hardware in the country and the elaboration of studies related to renewable energy utilization.
In the year 1982 the first Solar Thermal Pump STP was put into operation. Due to the promising results a field test with 5 additional STP followed on the island of Lombok in 1987.
In 1990 a technology transfer program to enable Indonesian manufacturers to produce STP in Indonesia started.
The program contains the manufacturing of all together 15 STP 4 of which will be tested as a 0-series for half a year before the field test will begin.

KEYWORDS

Solar Thermal Pumps, STP, Technology Transfer Program, Field Testing of STP.

EXPERIENCE

In the frame of the scientific and technical cooperation between the Republic of Indonesia and the Federal Republic of Germany a program on Renewable Energy Utilization has been performing since 1979.
As one of the Solar Thermal activities a first Solar Thermal Pump, STP, was commissioned in 1982 at the laboratory test field of BPPT, the Indonesian Agency for Assessment and Application of Technology, near Jakarta.
The objective of this project was to investigate the reliability and effectivity of such a system in Indonesia with the aim of drinking water supply in remote areas without electric grid by means of renewable energy sources.

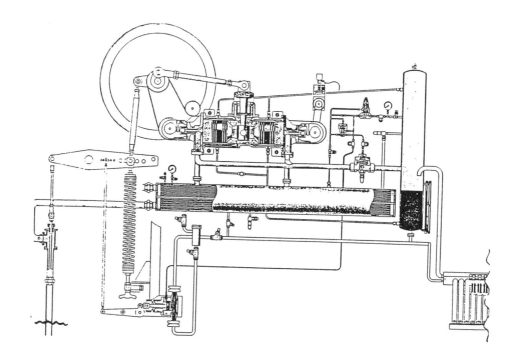

Fig. 1 . Scheme of the STP

The system consists of 4 functional subsystems :

- solar collector with evaporator
- working medium cycle
- engine with transmission
- water pump.

The double acting piston engine of the pump converts the solar thermal energy via a low temperature Ranking cycle into mechanical power is transfered to the fly-wheel and from there by a metal rod to the piston pump in the well.
The prototype STP is until now working without major problems, so that due to the promising results of the tests, a field testing followed. In 1987, 5 pump systems were installed and commissioned on Lombok island for drinking water supply of several villages. Since that time, the pumps are in operation.
The pumps are equipped with data acquisition units system, so that an extended documentation about pump efficiency, solar data, the amounts of pumped water and days of operation per year are obtained.
A typical sequence of graphs (Insolation and Hydraulic output) over a month is shown in figure 2.

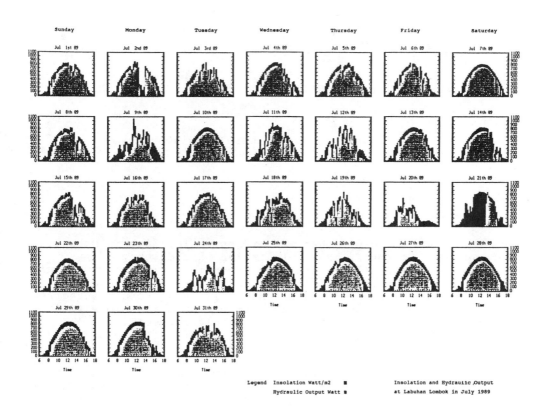

Legend Insolation Watt/m2   ■   Insolation and Hydraulic Output
        Hydraulic Output Watt ■   at Labuhan Lombok in July 1989

Fig.   2 . Insolation and Hydraulic output of STP
            for a typical month.

Regular inspections provided additional experience about the pump
operation, maintenance requirements and component improvements.

The yields and efficiencies of the pumps were so encouraging that
a technology transfer program to Indonesian manufacturers has
being started in 1990.

This project contains the manufacturing of 4 prototypes of Solar
Thermal Pumps with a half year acceptance test.   After its
completion a field test program with these and 11 additional
Indonesian made systems is envisaged.

        SUMMARY

Since 1982 test, assessment and application of STP are being
done.   The prototype, and 5 additional STP on the island of Lombok,
showed such promising results that a technology transfer program
with the aim to manufacture a O-series in Indonesia has been
initiated, aiming to supply   remote villages with drinking water
and wide spreading   this kind of technology in Indonesia.

SOLAR ASSISTED CROP DRYING
A   COMPETITIVE ALTERNATIVE ALSO IN COOL HUMID CLIMATES

Gustav Tengesdal

Department of Agricultural Engineering
Agricultural University of Norway, N-1432 AAS-NLH, Norway

ABSTRACT

Solar assisted crop drying has proved its reliability in Norway. It
is competitive in cost with simple ambient air dryers, and
competitive in reliability with more expensive oil fuelled hot air
dryers. Several hundred installations have been made since the early
1980's. The key to the success has been simple structure integrated
solar collecting systems, which the farmer can build himself with
ordinary building materials, at a relatively low cost.
We usually use corrugated steel or aluminium roofing sheets as
absorber without any cover. Efficiencies are in the range of 0.3 to
0.7, and temperature rise from 2 to 6°C is normal depending on airflow rates and many other variables. A temperature rise of just a
few degrees makes a significant contribution to the drying process.
With an average relative humidity of 70 to 80%, the ambient air has
very little or no drying potential at all. We have experienced that
as a general rule of the thumb, drying time with solar assisted
systems is reduced to approximately the half of what is normal with
ambient air drying of grains and hay. As additional benefits the
farmer also gets a better product of higher quality, reduced danger
of fungus growth and reduced dependence on good drying weather.

KEY WORDS

Crop drying in cool humid climates, grain drying, hay drying, solar
assisted crop drying, bare plate air collectors

INTRODUCTION

The annual Norwegian production of grains is about 1 to 1.5 million
tons, dominated by barley, oats and wheat. Normal moisture content
during harvest is 20 to 30% (wet base), but may, under difficult
harvesting conditions, exceed 30%. Approximately two thirds of the
production is dried and stored on the farms. Out of this, 80% is
dried with unheated ambient air. This way of drying will, with
normal to good conditions, give a satisfying result. If, however,
moisture content at harvest is high and weather conditions are
difficult during the drying time, a reduction in the grain quality
will sometimes be the result. To avoid this risk of a quality
reduction , many farmers have installed oil fuelled heaters to
assist the drying process. Even if the total economy in these oil

fuelled systems can be doubtful, we see that farmers do appreciate them.
A different way of reducing the risk associated with ambient air drying, is to include a solar collector in the drying system.
With an investment that will usually be about the same as for a conventional oil based heating system, a solar collecting system can give comparable security for a good drying result.
It must be mentioned here that the direct use of the hot exhaust air from oil burners is not allowed in Norway. A heat exchanger and a separate exhaust outlet is required, thus reducing efficiency and increasing costs. LPG burners can be used without heat exchangers.

The conservation of grass to feed cattle, sheep and goats during our 6-8 months of winter, is one of the most time-consuming jobs for the farmers during the summer. The traditional labour intensive methods of natural drying in the fields were more or less abandoned during the fifties and sixties. Even if barn hay drying was introduced during the same decades, it never came to play any dominant role in our grass conservation. The reason was probably that silage conservation was introduced and strongly supported by our agricultural authorities in the same period. The argument that 'you can harvest and conserve your crop even when it is raining' is also hard to come about in rainy Norway. Today three quarters of the conserved grass is silage, and the last quarter is hay.
In neighbouring Sweden, however, the relationship is nearly just the opposite. We acknowledge today that the benefits associated with silage compared to hay have been overestimated, and that the problems with silage have turned out to be difficult to solve properly.
These problems include soil structure problems because of transport with heavy vehicles, runoff of silage juice to nearby creeks and rivers resulting a.o. in fish death, and a feed quality problem that might cause health problems in herds depending dominantly on silage. All these factors have increased the interest in hay as an 'environmentally sound alternative'.

WHY SOLAR ASSISTED DRYING ?

The equilibrium moisture content of an agricultural product is a function of the relative humidity of the ambient air. In air with a relative humidity of approximately 70% the equilibrium moisture content of grains and hay will usually be about 15% (wet base). That will also give safe storage with our temperatures.
The average relative humidity during the summer months will, however, often be close to 80% in the coastal regions of Norway, as it is in most of the regions surrounding the North Sea. That means that the ambient air has a very low drying potential. With the exception of a few hours in the early afternoon, the average day has hardly any drying potential when the moisture content in the product is under 20% (w.b.). There are, however, very few average days. We have more or less sunny and more or less rainy days. Approximately 15 days in a month will have some rain, and the remaining will be somewhat sunny.
The daily average insolation ranges from 15 to 20MJ/m^2.
The interior parts of the country have a drier climate. With an average relative humidity of approximately 70% , the natural conditions for drying with ambient air are significantly much better than on the coast. This difference is illustrated in Fig. 1. and 2. While the daily temperature amplitude on the coast is about 8 to 10°C, it will be 12 to 14°C in the interior parts.

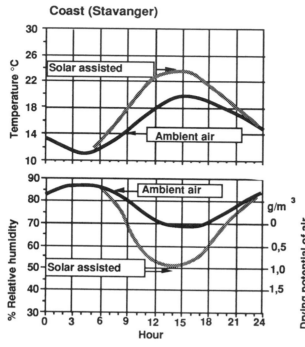

Fig. 1.
Average day variation of air temperature and relative humidity on the coast in the summer. The grey line shows how this variation can be with a solar collecting system. By amplifying the normal day variation the drying potential of the air is increased.

The drying potential of the air is here estimated in grams of water that every cubic metre of air can carry out of the drying mass in the final drying stage. Note that without a solar system hardly any drying will take place.

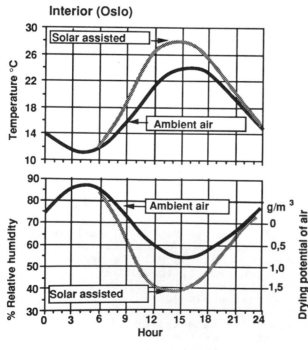

Fig. 2.
Average day variation of air temperature and relative humidity in the interior parts of Norway in the summer. The grey line shows how this variation can be with a solar collecting system.

The drying potential of the air is here estimated in grams of water that every cubic metre of air can carry out of the drying mass in the final drying stage. Note how the solar collecting system increases the drying capacity of the air.

That means that the ambient air will have a significant drying potential for 10-12 hours every day. A solar collecting system will however boost the daily temperature amplitude and thus increase the capacity of the dryer. For drying of grains in late August or September, conditions can be less favourable, and a solar collecting system much more necessary.

The final stage in the drying process requires a drier air than the first stages do. If the weather is moist, the blower can run for days and weeks without being able to dry out the final moisture from the product.These conditions can produce mould growth in the mass and thus reduce the quality of the product. With a solar assisted system it is possible to do the first stages of the drying process somewhat faster, and the final stage significantly much faster than with ambient air.
The fact that the running time of the blower is reduced perhaps to 50% of what would be normal with ambient air drying is of little significance compared to the increased security of a good drying result. Most installations have been made in farm buildings in the range of 200 to 800m^2 ground floor area. For hay dryers we usually use the entire roof, also the north facing one as solar collector, because in June-July, when the hay is harvested, most of the radiation is on the roof surfaces.
For grain dryers we will not use a north facing roof surface, but rather the south facing roof and wall as absorbing surfaces.

PERFORMANCE OF BARE PLATE AIR COLLECTORS FOR CROP DRYING

In the literature the performance of bare plate solar collectors is well documented. I will limit this short review to what has been published in Norway and some of our neighbouring countries.
In Sweden, Gustafsson and Ekstrøm (1980) published recommendations for solar collectors for crop drying. They described bare plate and covered absorber types. A recent report, Pahlman (1988), indicates that the bare plate type absorber is likely to be the most interesting alternative, because of its low price and expected longer life span. The difference in performance between the two types is of little significance according to Pahlman. A well designed bare plate system will perform even better than a not so well designed system with a covered absorber.
From Scotland, Spencer and Graham (1987) report of experiences with solar assisted drying of grains and hay. They have used both bare plate and covered absorber in their experiments. So far the economy appears to be marginal if just savings on the use of conventional energy is considered. If additional benefits such as better product quality and the 'insurance' value are also considered, solar assisted drying appears to be more promising.
Switzerland is probably the country in Europe that has done the most on solar assisted hay drying. Keller,Kyburz and Kölliker (1988) made a number of experiments on the performance of collectors with bare plate and covered absorbers. Their results have been very useful also for our conditions. The significant difference in pressure drop that was observed between airflow across and along the corrugation is of vital importance in collector design.
The first solar collecting system for crop drying in Norway was built in 1981 according to Jeksrud, Lilleng and Bjerke (1986). During the following ten years several hundred installations have been made. The use of the solar assisted dryer for several purposes can improve the total economy. Græe (1988) reported on the use of a

'universal dryer' with an integrated bare plate solar collecting
system for grains, hay, seeds, and forest products like wood chips.
Tengesdal, Græe and Skjevdal (1990) showed that the capital costs
are dominating in solar assisted hay drying. Energy savings with the
solar system are of little significance in the total cost picture.
Andersen, Espe and Herbjørnsen (1991) have investigated how well a
number of bare plate solar collecting systems are performing. They
have looked at barns on the shore of the Barent Sea at a latitude of
70° north, at the North Sea shore, and at altitudes up to 1000 m
above sea level. The average efficiency they found was approximately
0.5. They have also showed that solar assisted hay drying is very
well possible in regions where ambient air drying is far too risky.
Based on literature and a number of measurements we have found that
the performance we can expect from a bare plate solar collector is
indicated in Fig.3. A number of variables will influence   the
performance in addition to the airflowrate. Local climate and wind
is important as well as the sky temperature for radiation exchange.
The collector design is also very important. Air speed, channel
height and length, and air leakage rate will all influence the
performance.

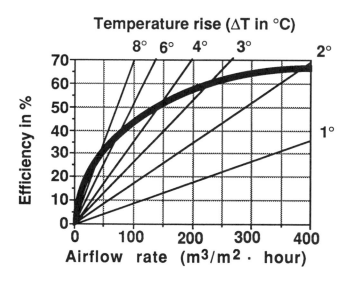

Fig.3. Approximate maximum efficiency and corresponding temperature
       rise of a bare plate solar collector with dark painted
       corrugated steel or aluminum sheets as absorber, at radiation
       intensity of 400 W/m^2, with wind up to 1 m/s and pressure drop
       up to 150 Pa.

A SIMPLE STRUCTURE INTEGRATED SOLAR COLLECTOR

The key to minimize costs is to make use of the structural
components of the roof also to build the passage through which the
heat transfer fluid air can flow. If we have a roof with high
rafters (150-200mm) spaced up to 1.2 m, the passage between the
rafters will probably be the most convenient place to let the air

flow. With frames or trusses with 3-5 m spacing the horizontal
purlins or beams will be so high that they will constitute an ideal
passage for the air. There is little reason to insulate the back
plate of the collector because the temperature inside is usually
very close to the collector temperature. Accordingly we use a cheap
($1/m^2) plastic foil to make the back wall in the collector.
A plastic foil or some other sheating is always used under a
corrugated metal sheet roof to avoid condensation and dripping.
That means there are hardly any extra costs associated with the
construction of a solar collector when it is done at the same time
as the roof is built or renewed. The only extra cost will be the
ducting from the roof and/or the wall to the blower. That is one of
the major reasons why many farmers in Norway will make the necessary
preparations for a solar collecting system when they build new or
renew the roof of an old building, even if they haven't got a dryer
yet. We also see that farmers who are considering to install oil
burners to assist drying often change their mind and build a solar
collecting system under their roof if that is possible. The capital
cost will be of similar size for the two systems, while the running
costs are clearly in favour of the solar alternative.

REFERENCES

Andersen, S.L., I. Espe and O. Herbjørnsen (1991).Solenergi
    låvetørkemålinger, nord,syd,øst,vest Norge. Rapport 91-02.
    Department of Physics, University of Oslo.
    P.O.Box 1048, 0316 Oslo 3, NORWAY.
Græe, T (1988). Universal dryer with building solar collector system
    for forest and agricultural products. IBT-artikkel 49/88.
    Department of Agricultural Engineering,
    Agricultural University of Norway, 1432 AAS-NLH, NORWAY
Gustafsson, G.,and N. Ekstrøm (1980). Solfångare för torkning.
    Aktuellt från lantbruksuniversitetet 282,
    Sveriges lantbruksuniversitet, 750 07 UPPSALA, SWEDEN.
Jeksrud, W. K.,H. Lilleng and A. Bjerke (1986). Utilization of solar
    radiation on roofs for crop drying. Report No. 112.
    Department of Building Technology in Agriculture.
    Agricultural University of Norway, 1432 AAS-NLH, NORWAY.
Keller, J.,V. Kyburz and A. Kölliker (1988). Untersuchungen an
    Luftkollektoren zu Heiz- und Trocknungszwecken.
    Paul Scherer Institut, Würelingen und Villingen,
    CH-5232 Villingen PSI, SWITZERLAND.
Pahlman, T. (1988). Field test of solar collectors for drying of
    grain and hay. Swedish University of Agricultural Sciences.
    Department of Farm Buildings, Box 624, 22006 LUND, SWEDEN
Spencer, H. B.,and R. Graham (1987). Developments in solar grain
    drying in Scotland. The Agricultural Engineer, Vol 42, No 1.
    West End Road, Silsoe, Bedford MK45 4DU, UNITED KINGDOM.
Tengesdal, G., T. Græe and T. Skjevdal (1990). Høytørker og
    solfangersystem. Småskrift 13/90.
    SFFL, Moerveien 12, 1430 AAS, NORWAY.

SOLAR DRYING - A SURVEY OF DIFFERENT TECHNOLOGIES AND THEIR INFLUENCE ON PRODUCT QUALITY

Müller, K.* and Reuss, M.**

* Fraunhofer Institute of Food Technology and Packing, Schragenhofstr. 35, 8000 Munich 50, F.R. Germany

** Institute of Agricultural Engineering, Technical University of Munich, Voettinger Str. 36, 8050 Freising, F.R. Germany

ABSTRACT

Worldwide increasing population results in an increasing demand for food and efficient conservation methods. Among others, traditional drying is and will be the most important one. A simple type of solar drying is direct exposure to the sun, which gives reasonable quality for some products with only few losses. For many other products the introduction of shading is already an improvement. Several small scale solar dryers with natural or forced convection for farm applications were developed all over the world, giving an improvement in quality. Even big plants like industrial conveyor dryers were equipped with solar collectors to provide the heat required for the process.

In a food drying process, besides economic aspects like energy saving strategies, it is particularly important to try to attain a high quality final product. The quality of food is determined by different criteria. Examples are technical, chemical, physical, biological, hygienic and physiological properties of the products. Changes in quality determine, besides the later sensory properties, colour, taste, aroma, texture and also the later hygienic status. Different quality changes in low moisture food such as dried vegetables are a function of the equilibrium moisture content. In the low moisture region mainly oxidation reactions lead to rancidity of fats and bleaching of carotenes. Quality losses caused by oxidation reactions occur mainly during storage. In the middle moisture region the non enzymatic browning or Maillard-reaction becomes dominant in food containing reducing sugars and amino acids or proteins. Due to its high temperature coefficient the Maillard-reaction is often the dominant deterioration reaction at high storage temperatures and in thermal processes. Changes caused by enzymatic activity and microorganisms appear in the higher moisture region. They influence the sensory and the hygienic properties of the products mainly during storage.

KEYWORDS

Solar drying, low temperature drying, product quality, Maillard-reaction

INTRODUCTION

Worldwide increasing population requires not only increasing
agricultural production but also reduction of postharvest food
losses by applying more efficient conservation methods. A major
part of food is produced in developing countries and therefore
reduction of losses should affect the economies of these countries. Drying in the open sun as the simpliest conservation
method is accompanied by losses of 20-50 % and sometimes more.
By applying more sophisticated drying techniques these losses
can be reduced considerably. All these require heat in an optimum manner to yield high quality products. Among others, solar
energy may be the most promising   one to meet the demands of
drying processes with respect to seasonal availability and location. Many developing countries show favorable conditions for
application of solar drying. Many products can be dried at low
temperatures (< 50 °C), which could easily be provided with solar
energy with natural or forced convection, others require medium
(50-100 °C) or high temperature (> 100 °C) drying. Additionally,
the temperature level of the process is responsable for the quality of the product. Other drying methods like freeze-drying, os-
movac or desiccant drying etc. are not considered in this paper.

DESCRIPTION OF TYPICAL SOLAR DRYERS

The most simple but also effective system is drying in the open
sun. Damages by direct radiation, wind precipitation, insects
etc. may occur. Open shading and/or transparent glazing can provide protection from direct sun and precipitation. Several systems like this were developed and reviewed in literature. They
work either with natural or forced convection with air flow over
or through the product layer. Fig.1 shows typical large and small
scale solar dryers working with natural convection.

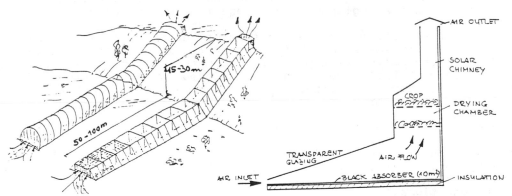

Fig. 1.   Natural convection large (left) and small
(right) scale solar dryers.

Besides a large number of small scale, specially developed solar
dryers, also industrial plants with commercial tray or belt dryers
were equipped with solar systems of several hundred square meters
of air heating collectors. In these plants usually ambient air is
preheated, the required temperature is reached by a conventional
backup system. All the plants are operated with forced convection.

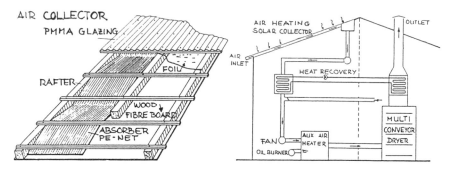

Fig. 2.   Scheme of a solar drying plant for herbs
           530 m^2 air collector, multi conveyor dryer

QUALITY CHANGES DURING THE DRYING PROCESS

The properties of dried vegetable food are influenced by chemical, physical, enzymatic and microbiological reactions and processes. The quality changes during the drying process are of a
complex nature, as several of these processes take place simultaneously. The extent and the reaction velocity are mainly influenced by the drying parameters, the moisture content and the
belonging water activity (defined by the sorption isotherm), the
product compounds, the enzymatic activity and the microorganisms.
Important chemical deterioration reactions are interactions of
the ingredients, oxidation, decomposition, cross linkage, hydro-
lysis and caramelizing of the sugars. All chemical reactions contribute to the change in the original quality properties of a dry
product: colour, taste, smell, nutritional value and shelf life.
Due to its high temperature coefficient the Maillard-reaction often becomes the dominant deterioration reaction in drying processes and at high storage temperatures. Maximum reaction velocity
is located in the region of middle equilibrium moisture contents.

Changes in the structure and the form of the products are caused
by physical processes (shrinking, crystallisation) which mainly
influence the rehydration properties and the cooking time. The
most important influence on these quality parameters is related
to the pretreatment (blanching conditions) and the temperature
treatment of the products during the drying process. A longer
blanching time will lead to a shorter cooking time and a careful
temperature treatment during the drying will lead to better rehydration properties. Contrary to careful dried products, hard
dried products (high temperature, high air velocity, low rel. humidity) will show insufficient rehydration properties due to the
'case hardening' of the product surface which affects the diffusion conditions.

Changes caused by enzymatic activity and microorganisms appear in
the higher moisture region. They influence the sensory and the
hygienic properties of the products mainly during storage. This
occurs especially if the equilibrium moisture content in the
dried food is not low enough, and if enzymes and microorganisms
have not been completely inhibited by the pretreatment and the
heating process.

RESULTS

The Maillard-reaction causes a strong change in colour (browning)
which is accompanied by changes in taste and aroma, with shorter
shelf life and with losses in the nutritional value (e.g. loss of
essential amino acids).

The influence of different drying parameters like air temperature,
air velocity, relative humidity, form and loading density of the
products on the formation of Maillard products was found out in
drying experiments. Amadori compounds, which are built in the early stage of the Maillard-reaction by the reaction of reducing sugars and amino acids, were used as indicator compounds for the
extent of the quality changes. The results of our experiments allowed to derive formulas with which the extent of the deterioration (calculated as the concentration of Amadori compounds) can
be determined at any stage of the drying process if the product
moisture, the product temperature and e.g. the air velocity is
known. Our results are available for all drying processes in consideration of the special drying curves and the temperature
treatment of the products. This means that for solar drying, too,
our results can be used to determine the deterioration by the
Maillard-reaction. As one of the most important results the activation energy and the frequency factor shall be shown as a
function of the product moisture content.

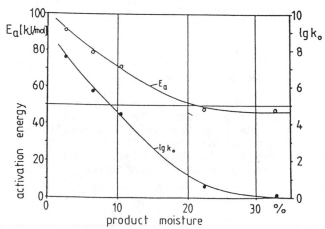

Fig. 3. Activation energy and frequency factor as
functions of the moisture content.

If the product moisture and the temperature are known the actual
deterioration degree can be calculated during the drying process.
A decrease in temperature of 10 K will lead to slower reaction
velocities of a factor of 2 to 4.5 depending on the region of
product moisture during the drying. Decreasing the temperature is
the most effective possibility to keep the deterioration small.
Besides the temperature influence on the deterioration, the influence of the air velocity could be determined. The formation of
Amadori compounds increases with increasing air velocity at constant air temperatures.

The extent of deterioration can finally be calculated using the
following equations:

$$1/C' = k*t + 1/C_0' \qquad (1)$$
$$\lg k = -E_a/2,3RT + \lg k_0 \qquad (2)$$
$$C = k^0 * v_L + C_0^0 \qquad (3)$$
$$C = 100 - C' \qquad (4)$$

C = concentration of Amadori compounds (Mol%);
C'= decrease of the reaction extent (Mol%),
　　(C_0'=100Mol%, C_0^0= 0 Mol%);
k, k^0 = rate constants (1/s); k_0 = frequency factor (1/s);
E_a= activation energy (J/mol); R = gas constant (J/molK);
T = absolute Temperature (K); v_L= air velocity (m/s);
t = drying time (s)

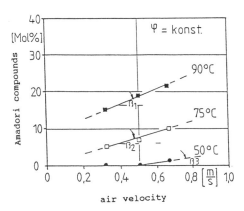

Fig. 4. Influence of the air velocity on the formation
of Amadori compounds.

The calculation of the deterioration in solar dried carrots with
the aid of a drying curve given from Bryan shows that there is no
quality loss caused by Maillard-reaction. The solar drying process needs 19 hours, and the maximum difference between the maximum air temperature was 43 °C. The maximum difference between the
analytical determination of Amadori compounds and the calculated
values is about 30 %.

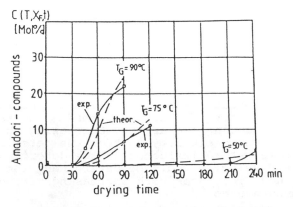

Fig. 5:　Calculated and experimental values of Amadori
compounds in carrots at different drying temperatures

In comparison to the quality obtained by solar drying the extent
of deterioration in dried carrots derived from conventional drying processes (60-100 °C) can reach about 40-60 % of the maximum
deterioration at the end of the process.

With regard to avoiding the Maillard reaction,   solar drying at
temperatures below 50 °C shows some advantages in comparison with
conventional drying processes. Another criterion to assess the
product quality and the quality of the process is the vitamin
retention in the dried products. Shakya, Moledina and Flink reported about the influence of different process parameters in a
solar drying process on the content of ascorbic acid in potatoes.
In a solar drying process of about 300 minutes, an air velocity
of 1.5 m/s and a loading density of 15 kg/m², the inlet air temperature was varied between 60-70 °C. The ascorbic acid retention
was about 70-84 % (determined by titration method) at an initial
moisture content of 340-345 g water/100g solids and moisture contents of the dried product of 11-20 g water/100g solids. Different air velocities and different loading densities showed no
significant influence on the ascorbic acid retention. In comparison to these results, Augustin reports of an ascorbic acid retention in conventional dried products of 38 to 45 % and Karel of
only 25-34.5 %.

Other important quality parameters are the rehydration properties
of the dried products. The rehydration index and the velocity of
rehydration increase with decreasing temperature treatment of the
products in careful drying processes. These results are found in
many drying experiments with carrots, champignons, bell pepper,
chives and blueberries. Pinaga compared the rehydration properties of green beans which are derived from different convective
drying processes.

- Drying with ambient air at 22-23 °C and 40-50 % rel. humidity.
- Drying in a 2 steps, with ambient air and finished with solar.
- Conventional drying at about 65 °C.

After three hours of rehydration, the rehydration index of the
carefully dried products (22-23 °C; 40-50 % R.H.) and the products
derived from the solar drying process reached about 0.8-0.9, while
the conventionally dried products reached a rehydration index of
0.6 and, finally, after 5 hours, only 0.7 as a maximum.

CONCLUSIONS

The experiments and results about the quality properties of dried
products reported from different authors show that the solar drying or drying processes using solar energy are leading to respec-
table product qualities. Due to the careful drying conditions, the
quality changes influenced by chemical and physical reactions
like changes in colour, taste, smell, shelf life, nutritional
value and case hardening are kept small.

REFERENCES

Müller K., Bauer W.: Optimierung der Trocknung pflanzl. Lebens-
mittel mit Leitsubstanzen. Teil 1: ZFL 1, 88; Teil 2: ZFL 40, 89

FULLY DEVELOPED LAMINAR FLOW AND HEAT TRANSFER IN THE PASSAGES OF V-CORRUGATED SOLAR AIR HEATER

Dai, Hui and Li, Zongnan

Guangzhou Institute of Energy Conversion,
Chinese Academy of Sciences
P.O. Box 1254 Guangzhou, China   510070

ABSTRACT

This paper is focused on the study of the fully developed laminar flow and heat transfer in passages of V-corrugated solar air heater. Nusselt numbers and friction factors for the flow are obtained from numerical solution, boundary elements method, in a range of different corrugation angle and dimension ratio H/Hv. And the effect of geometric parameters on the flow and heat transfer was shown clearly by the results. Then, the approximate formulas of fRe and Nu varying with these geometric parameters are given by regression. The results from the paper provide the part of the theoretic basis, fully developed laminar flow and heat transfer, for the optimum design of V-corrugated solar air heater.

KEYWORDS

Solar air heater; V-corrugated passage; geometric parameter; fluid flow and heat transfer; boundary elements method

INTRODUCTION

Solar air heaters have been commonly used in engineering applications, for example, air conditioning, agricultural and industrial products drying, etc. Their extensive uses await the development of economically viable solar air heaters. Till now, many kinds of solar air heaters have come out since the development of the first one. However, a number of studies (Cole-Appel, 1977; Hollands, 1979; Joudi, 1986) have pointed out that a simple well-designed V-corrugated absorber plate can offer a lot of advantages over a flat absorber plate. Firstly, it increases the heat transfer area and the heat transfer coefficient. Secondly, it suppresses the free convection. Thirdly, the corrugation will increase the effective absorbtance of the plate by "cavity" effects.

For solar air heater as commodities, optimum design is very important. For flat plate heater, the optimum design has been studied in detail. But for the corrugated plate heater, it has not been done. It is mainly due to the fact that flow and heat transfer become complicated when flat absorber is replaced by V-corrugated plate. In the past papers, many theoretical and experimental studies on corrugated air heater were restricted to the duct with the corrugation angle θ =60° and the clearance H=0. And the heat transfer coefficient between the V-corrugated plate and the heated air is simply obtained by α v=α flat/sin(θ /2). It is in the reference (Scheider 1978) that the heat transfer in triangular ducts with different apex angle from 10 degrees to 170 degrees was discussed by the finite elements method. But it is still unknown how the geometric parameters of corrugated ducts (the corrugation angle θ and the height of the corrugated absorber Hv) effect on fluid flow and heat transfer when there is a clearance H between the absorber and the base plate. It is also unknown whether the gap can enhance the heat transfer in ducts. So the problem of heat transfer process occurring in ducts is of particular importance to analysis and design of an optimum solar air heater.

In systems which employ air as the heat transfer fluid, it is common that the flow through the heat transfer passage is hydrodynamically laminar. This paper is aimed at the heat transfer for steady, fully developed laminar flow in V-corrugated passages with different geometric parameters.

MATHEMATICAL ANALYSIS

The configuration of V-corrugated solar air heater is illustrated in Fig.1. For the symmetry and periodicity of corrugated passages, the region ABCDA, shown in Fig.2, is selected for analysis purpose Its boundary conditions are the uniform heating at the corrugated plate and the adiabatic condition at the base plate. The temperature of the corrugated plate is uniform in any cross-section.

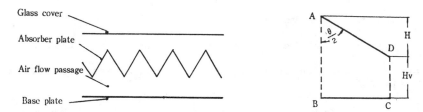

Fig.1. V-corrugated solar heater   Fig.2. The typical part of duct

The governing equations for the momentum transport and energy transport are given as follows

$$\frac{\partial^2 w}{\partial X^2} + \frac{\partial^2 w}{\partial y} = \frac{1}{\mu} \frac{dp}{dz} \tag{1}$$

$$\frac{K}{\rho Cp}\left(\frac{\partial^2 T}{\partial x^2} + \frac{\partial^2 T}{\partial y^2}\right) = w \frac{\partial T}{\partial Z} \tag{2}$$

In the equation (2), heat balance over a unit length gives

$$\partial T / \partial z = \theta / (\rho A \dot{w} Cp) \tag{3}$$

Where Q is the heat transfer rate per unit axial length on corrugated plate. A is the flow cross-sectional area, A=(Hv+2H)Hvtg(θ /2)/2. Then equation (2) becomes

$$\frac{\partial^2 T}{\partial x^2} + \frac{\partial^2 T}{\partial y^2} = \frac{w}{\dot{w}} \frac{Q}{KA} \tag{4}$$

Introduce the dimensionless quantities

$$X = x/De, \quad Y = y/De, \quad A' = A/De^2,$$

$$U = \frac{w}{[(-dp/dz)De^2/\mu]}, \quad \dot{U} = \frac{\dot{w}}{[(-dp/dz)De^2/\mu]},$$

$$\phi = (T-Tw)/(Q/K), \quad \phi b = (Tb-Tw)/(Q/K) \tag{5}$$

The equivalent hydraulic diameter De is written as

$$De = 2(H+2Hv)/[1+CSC(\theta/2)] \tag{6}$$

Then the equations (1), (4) are nondimensionalized as

$$\frac{\partial^2 U}{\partial X^2} + \frac{\partial^2 U}{\partial Y^2} + 1 = 0$$

$$\frac{\partial^2 \phi}{\partial X^2} + \frac{\partial^2 \phi}{\partial Y^2} = \frac{U}{\dot{U}} \cdot \frac{1}{A'} \tag{7}$$

the boundary conditions are
i) $\partial U/\partial x = 0$, $\partial \phi/\partial x = 0$ along lines of symmetry
ii) $U = 0$, $\partial \phi/\partial x = 0$ on the adiabatic base wall
iii) $U = 0$, $\phi = 0$ on the corrugated plate

The friction factor f is given by

$$f = (-dp/dz)\, De/(1/2\, \rho\, \dot{w}^2) \tag{8}$$

Defining $Re = wDe/\nu$ and introducing the dimensionless quantities, equation (8) yields

$$f \cdot Re = 2/\dot{U} \tag{9}$$

where the dimensionless mean velocity $\dot{U}$ is obtained by

$$\dot{U} = \int_A U dA/A \tag{10}$$

The heat transfer coefficient, α, based on DA, is defined as

$$Q\triangle Z = \alpha\,(Tw-Tb)\triangle Z\, Hv\, Sec(\theta/2)$$

$$\alpha = Qcos(\theta/2)/[Hv(T_w-T_b)] \tag{11}$$

The average Nusselt number is given by

$$Nu = \frac{\alpha\, De}{K} = \frac{Q'}{K(T_w-T_b)} \frac{De\, cos(\theta/2)}{Hv} \tag{12}$$

Introducing the dimensionless quantities, then equation(12) becomes

$$Nu = -De\, cos(\theta/2)/(Hv\, \phi_b) \tag{13}$$

where the dimensionless bulk temperature φb can be obtained by

$$\phi b = \int \int_A U \phi \, dA / (AU) \qquad (14)$$

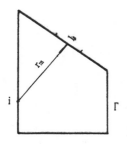 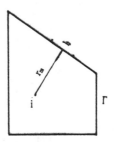

$$1/2Ti+B+ \int_r q^*Td\Gamma = \int_r T^*qd\Gamma \qquad\qquad Ti+B+ \int_r q^*Td\Gamma = \int_r T^*qd\Gamma$$

Fig.3. Boundary elements method

As shown in Fig.2. the flow in the duct is more complex than in flat plate duct. But a newly developed numerical method, boundary elements method, as shown in Fig.3., can be used to deal with this problem effectively. The character of this method is that the solution of the field is turned into the solution of the boundary, and the dimensions of problems can be lowered. Comparing with finite elements method and finite difference method, it can reduce the spot point number and the storage of computer obviously. And it has higher calculation accuracy. In the program, 70 boundary elements and 20x40 internal points are given for calculation.

DISCUSSION

In order to explore the difference of flow in ducts with different θ and H/Hv, the velocity distributions are illustrated in Fig.4. It can be found that fluid velocity is retarded in the apex corner area, and the smaller the apex angle, the larger the retarding region. It is called "corner effect". Furthermore, especially to small corrugation angle, whereas the gap H/Hv becomes larger, the maximum velocity gradually moves from corrugation area to clearance area. And, eventually, the flow becomes similar to that in parallel plate duct.

The variation of the product of fRe and Nusselt number with the half-corrugation-angle is shown in Fig.5 and Fig.6 for H/Hv=0, 0.2, 0.4, 0.6. In the studied range,it is shown that for a given dimension ratio, the product of fRe and the Nusselt number do not fully increase with an increase in corrugation angle. They exhibit a maximum for a given H/Hv and such a maximum value occurs at 20° < θ/2<55° for 0⩽H/Hv⩽0.6. This phenomena can be explained as follows.

For friction factor f, $f=(-dp/dz)De/(1/2 \rho w^2)$, shearing stress $\tau \propto dp/dzDe$ and $\tau =-Udw/dn$, it can be deduced $f \propto dw/dnDe^2$ when Re is fixed. In comparison of Fig.5(b), (c) and (d), for a given dimension ratio, the velocity gradient dw/dn increases with an increase in corrugation angle,and so does the temperature gradient. This results in increasing in friction factor and heat transfer.

2236

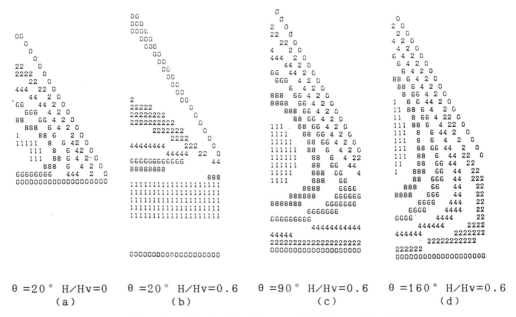

θ =20° H/Hv=0   θ =20° H/Hv=0.6   θ =90° H/Hv=0.6   θ =160° H/Hv=0.6
     (a)   (b)   (c)   (d)

Fig.4. Distribution of U/Umax field

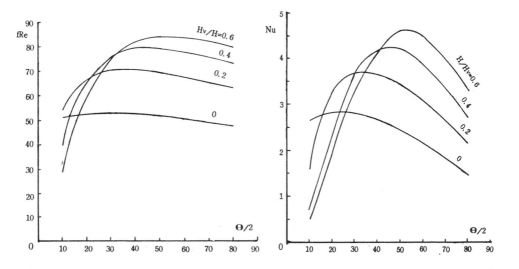

Fig.5. Variation of fRe with θ/2   Fig.6. Variation of Nu with θ/2

But De decreases with θ increasing. Then, as θ further increases, the fRe and Nu decrease with an increase in θ.

In addition, it can be found that when θ is at larger angle, for example, θ/2>45°, fRe and Nu increase with H/Hv for a given corrugation angle, and when at a smaller angle, θ/2<30°, fRe and Nu increase first, then decrease as H/Hv increasing. The explanation is that, De increases with H/Hv when an angle is given, and f increases with De^2. For small angle, the velocity gradients become much lower in the corner region for larger H/Hv, as compared Fig.5 (a) with Fig.5(b), and so do temperature gradients. Consequently the fluid acts as insulator between thinner fluid core and the wall. For low values of H/Hv, the increase in De surpasses the "fluid insulation effect", and fRe and Nu increase with H/Hv. However, as H/Hv is further increased, the effect of the fluid insulation dominates and fRe and Nu decrease with an increase in H/Hv. But for large corrugation angle, the effect of the insulation is not as much as that in small angle, shown in Fig. 5(a) and 5(b), so fRe and Nu increase monotonically with an increase in H/Hv.

Here, it should be pointed out that the geometric parameters which the maximum Nu lies at can not be regarded as the optimum values. The optimum selection must consider comprehensive effects of parameters on heat transfer, for example, on the natural convection and the radiation convection between the glass cover and the corrugated absorber, and on the forced convection in passages.

CONCLUSION

1. For the solar air heater with V-corrugated absorber, the corrugation angle and dimension ratio H/Hv are very important to optimum design, because they determine the equivalent hydraulic diameter and the fluid insulation effect area; moreover, affect the distribution of velocity and heat transfer in ducts.

2. The larger the given dimension ratio, the bigger the corrugation angle at which the maximum of fRe and Nu occur.

3. The approximate formulas fRe and Nu for $20° \leqslant \theta/2 \leqslant 80°$, H/Hv= 0, 0.4 and 0.6 are obtained by two-dimension regression as follows,

For H/Hv=0
$$f_{Re} = 50.2521 + 0.1788X - 0.0027X^2$$
$$Nu = 2.4756 + 0.0258X - 0.0005X^2$$
$$r_{Re} = 0.9847,$$
$r_{Nu} = 0.9942$, where r is the correlation coefficent, X=H/Hv.

For H/Hv = 0.4
$$f_{Re} = 28.1972 + 1.9830X - 0.0018X^2$$
$$Nu = -1.0918 + 0.2095X - 0.0021X^2$$
$$r_{Re} = 0.9394, \quad r_{Nu} = 0.9862$$

For H/Hv = 0.6
$$f_{Re} = 10.3896 + 2.6974X - 0.0234X^2$$
$$Nu = -1.6309 + 0.2258X - 0.0021X^2$$
$$r_{Re} = 0.9740, \quad r_{Nu} = 0.9973$$

REFERENCES

Cole-Appel, B.F., and G.O.G. Lof, L.E. Shaw (1977). Proc. of Annual Meeting, 2-6
Hollands, K.G.T., and E.C.Shewen (1979). Sun II 254-248
Joudy, K.A., and Mohammde (1986).Energ. Convers.Mgmt. vol.26, No.2
Schneider, G.E., and B.L. Ledain (1978). AIAA Terrestrial Energy Systems Conference. J4-6
Guo Kuanliang (1988), Numerical Heat Transfer, P.R. China

DOUBLE PASS SOLAR CONCENTRATING COLLECTOR

Y. Tripanagnostopoulos and P. Yianoulis

Department of Physics, University of Patras,
Patras 26110, Greece

ABSTRACT

A new type of collector, with low concentration ratio and non evacuated double receiver, is designed and tested. The collector consists mainly of a stationary curved mirror and two absorbing surfaces separated by insulating material. This sandwich type absorber is placed properly, relative to the mirror, so that one of the absorbing surfaces faces the sun and comprises the first passage for the working fluid, heated directly by the incoming solar radiation. The other absorbing surface forms the second passage, which is placed in an inverted manner, and receives only reflected solar radiation, but suppresses sufficiently the convective thermal losses. The experimental results for a constructed prototype model, show that the double pass solar concentrating collector is efficient and proper for applications of fluid heating in a wide temperature range.

KEYWORDS

Solar collector, stationary concentrating collector, CPC, double absorber, low concentration ratio, non evacuated receiver.

INTRODUCTION

The stationary concentrating solar collectors with low concentration ratio ($C \leq 2$), without evacuated receiver, are of particular interest in the temperature range of about 60-100°C, because of their satisfactory efficiency and the relatively low cost. The use of a CPC type mirror (Winston, 1974, Rabl, 1980), combined with flat or cylindrical absorber, is the most characteristic type of collector in this category. The radiative thermal losses from the absorber to the ambient can be reduced by the use of a selective surface and the conductive losses by the inverse position of the absorber (Rabl, 1976, Kienzlen, 1988).

The interposition of the mirror increases the optical losses and results in lower efficiency than the corresponding flat plate collectors for operation in the low temperature region. This is important, especially for the reverse flat plate collectors, which receive solar radiation only by reflection, having the upper side of their absorber well insulated.

We give an alternative collector design compared to that of

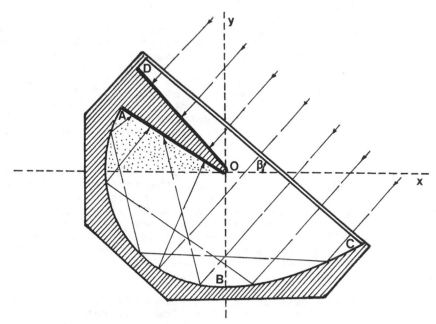

Fig. 1. Collector cross section design.

previous collectors, which combines the advantages of flat plate
and inverse absorber concentrating collectors in a single unit.
For the proposed collector we use a double pass absorber, in
series or in parallel, depending on the input fluid temperature
and the efficiency of the process for each passage. The separation
of the absorber for operation in two temperature levels may be
applicable to different absorber geometries, for liquid or air
heating, improving the efficiency of concentrating collectors,
both in low and intermediate temperatures.

COLLECTOR DESCRIPTION

The cross section of the proposed double pass concentrating
collector (DPC) design is shown in Fig. 1. The collector consists
of the curved mirror (ABC), the upper absorber [OD], the inverse
absorber [OA], the transparent cover [DC], the thermal insulation
between the two absorbers and the external thermal insulation of
the unit. The collector is closed at the two sides by flat mirrors
and is placed in an E-W direction, fixed during the year. The
solar radiation is transmitted through the transparent cover and
one part falls directly on the absorber [OD], while the rest is
reflected from the mirror on the inverse absorber [OA].

The mirror (ABC) consists of the circular section (AB), with
radius R, and the parabolic (BC) with focal lenght f=R, for
maximum solar energy collection by the stationary collector. The
collector slope and the direction of the parabola axis depend on
the site of installation. In any case, the parabola axis must
have a greater slope than the direction of solar rays at the
summer solstice, for this site, at solar noon.

Having this geometry, the DPC unit can be considered to consist of two conjugated collectors: one flat, (FPC), and one concentrating, (IPC), corresponding to the two absorbing surfaces [OD] and [OA]. These two partial collectors exhibit different optical and thermal behaviour. The proper fluid circulation can lead to the optimum operation of the unit. The thermal losses of FPC correspond to those of a flat plate collector and are mainly radiative and convective losses. Its optical efficiency can be high because of the collection of both direct and diffuse solar radiation without the interposition of a mirror. On the contrary the thermal losses of IPC are restricted because of the inverse position of the absorber, limiting the convective losses, while its optical losses are high due to the interposition of the mirror and the absorption of only reflected radiation. The suppression of convective losses from IPC is due to the trapped hot air in the space shown with dots in Fig. 1. For the restriction of radiative losses from both absorbers it is essential to use selective absorbing surface. The thermal insulation between the two absorbers is used to reduce the thermal flow from the lower hot to the upper warm surface. Finally good external insulation is used in order to reduce the thermal counductive losses from the internal space of the unit to the ambient.

The working fluid can pass through the absorbers in series or in parallel. For the first mode the low temperature fluid first passes through the upper absorber where it is preheated and consequently through the lower absorber where it is heated to a higher temperature. This process is most efficient when the flow is such that the fluid passes from one absorber to the other near a temperature T_c, for which the efficiency of the two absorbers coincides. In this way we operate the upper absorber at lower temperatures for which its efficiency is high and the lower at higher temperatures for the same reason. For the second mode of operation we use the two absorbers independently, with input fluid temperature $T_i < T_c$ for FPC and $T_i > T_c$ for IPC in order that the process efficiency is optimized.

The thermal insulation between the two absorbers is neccessary because of the different operation temperatures of FPC and IPC. Usually the second absorber is at higher temperature and there is thermal flow upwards. However, this flow does not represent thermal losses since it is captured by the FPC. At the same time the FPC loses energy only to the cover, and its efficiency is slightly improved. In a large installation of collectors the flow rate can be controlled automatically for optimum efficiency depending on the solar radiation, ambient temperature, and the input-output temperatures of the absorbers.

We have constructed and tested a model of double pass concentrating collector and we have determined the efficiency curves for the collectors, which correspond to each passage. An experimental assessment for the operation of the unit as a whole has also been made. The axis of the parabola has been taken vertical and the collector slope with the horizontal plane: $\theta \approx 40°$. We have used an aluminized mylar surface ($\rho = 0.85$), absorber of fin type with tube covered with selective Maxsorb foil ($a = 0.92$, $\varepsilon = 0.11$), and a 4 mm width transparent glass cover. The length of both absorbers is 1.00 m and the width, 0.16 m and 0.15 m, for FPC and IPC, respectively. The dimensions of the glass cover are

1.00 m x 0.35 m giving a total aperture area of $A_a=0.35$ m^2. We have used glasswool insulation which gave an average value to the sandwich of the asborbers $[K/D]_a=1.6$ W m^{-2} K^{-1} , and to the external insulation of the mirror envelope $[K/D]_m=0.7$ W m^{-2} K^{-1}. Regarding the geometry of the collector unit DPC, the corresponding aperture area for FPC and IPC is estimated to be $0.46A_a$ and $0.54A_a$ respectively therefore, the concentration ratio of the collector IPC is about 1.27.

The constructed unit is rather small, but it could be combined with others in the same enclosure. The length and the width of the absorbers could be increased, but the total volume of the unit should always be taken into account for the particular application. The absorber form can vary according to the type of working fluid used. If one uses air then the absorber form should be an air duct.

EXPERIMENTAL RESULTS

The experimental results, which are presented here, are referred to the operation of the two absorbers at relatively small temperature differences. This is essential for small thermal interaction between the absorbers. We have chosen clear sky conditions and water as the working fluid, which was heated from the ambient temperature to about 95°C.

The determination of the limits for efficient operation of the collectors FPC and IPC allows for the optimization of the double pass collector. The efficiency n is expressed as a function of $\Delta T/T$, where $\Delta T=(T_i+T_o)/2-T_a$ and I the solar radiation intensity on the aperture of the collector. T_i and T_o are the inlet and outlet temperatures of the water and T_a the ambient temperature. In Fig.2 we show the efficiency curves for the collectors FPC and IPC which are intersected at the point $\Delta T/I=0.045$ °C W^{-1} m^2. The efficiency curve, which passes between the previous two,+ corresponds to the efficiency of the whole collector unit (DPC) when the operation mode is in series, with T_i the inlet temperature at the upper pass and T_o the outlet temperature of the water from the inverted pass. From the efficiency curves we see that the collector FPC has higher efficiencly than IPC for low values of $\Delta T/I$, up to the point: $\Delta T/I=0.045$ °C W^{-1} m^2, because of the higher optical efficiency , while the collector IPC is better for higher values of $\Delta T/I$, because of lower thermal losses, although its optical efficiecny is lower.

The experimental results show that the Double Pass Concentrating Collector has satisfactory efficiency for a wide temperature range, while the combined operation of one flat and one concentrating collector, in a single unit, is more economical than the use of two separate units. Comparative tests with other types of non evacuated concentrating collectors have shown that, the DPC collector unit is more efficient than a symmetric CPC collector with flat absorber and similar concentrating ratio (curve A, Fig. 2) and that the collector IPC gives also an efficiency curve better than that of a CPC collector with concentrating ratio C=3 (curve B, Fig. 2). The proposed design of DPC type collectors with different dimensions and absorber form is straightforward. In this way, we can also realize a larger

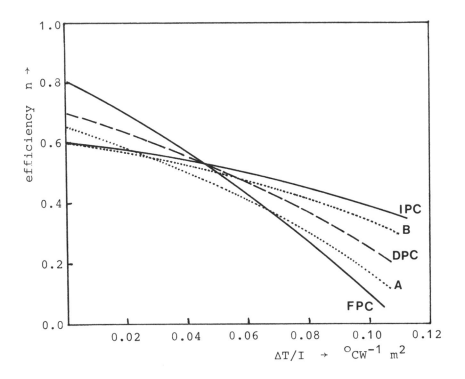

Fig. 2. Efficiency curves of the tested collectors.

collector by the use of two or more basic units in a single collector, the construction of collectors appropriate for air heating, the independent operation of the two passages, etc. For the case of large temperature differences, in order to avoid heat transfer from the lower to the upper passage one should use thicker thermal insulation between the two absorbers, in relation to the size of the absorbers and the mirror, so that the basic geometry of the unit remains unchanged.

REFERENCES

Winston , R., (1974). Solar Energy, 16, 89-95.
Rabl, A., J. O'Callagher, and R. Winston (1980). Solar Energy, 25, 335-351.
Rabl, A. (1976). Solar Energy, 18, 93-111.
Kienzlen, V., J.M. Gordon, and J.F. Kreider (1988). J. Solar Energy Eng., 110, 23-30.

MATHEMATICAL MODEL OF AN ALL-GLASS EVACUATED
TUBULAR SOLAR COLLECTOR

C.A. Estrada Gasca, G. Alvarez-Garcia[*] and R.E. Cabanillas[**]

LABORATORIO DE ENERGIA SOLAR IIM-UNAM.
AP 34, 62580 Temixco, Morelos, México

[*]Centro Nacional de Investigación y Desarrollo Tecnológico DGIT-SEP
AP 4-224, 62431 Cuernavaca Morelos, México

[**]Escuela de Ciencias Químicas, Universidad de Sonora.
AP 106, 83000 Hermosillo, Sonora, México.

ABSTRACT

A mathematical model of the thermal behavior of an all-glass evacuated tubular solar collector with the absorber film in contact with the working fluid is presented. The model is solved numerically and plots of temperature distributions for solar collector components are shown. Also, efficiency distributions as functions of parameters such as mass flow rate and heat transfer convective coefficient of the evacuated zone are included. This array has the benefit of avoiding the degassing of the film on exposure to vacuum. The numerical simulation shows that when the pressure is $\simeq 10^{-4}$ Pa, the collector efficiency has high values for a wide range of operating temperatures.

KEYWORDS: Evacuated Solar Collectors, All-Glass Tubular Collectors, Mathematical Modeling, Numerical Simulation, Thermal Behavior.

INTRODUCTION

Evacuated solar collectors appeared during the sixties, with the purpose of reaching higher temperatures and better efficiencies than those obtained in flat plate collectors. These collectors consisted of an external tubular glass cover and an internal metal tube with a solar absorber coating. The space between the cover and the receiver was evacuated to a pressure of $\sim 10^{-4}$ Pa. At this pressure the heat losses to the ambient air are substantially reduced. This first design of evacuated tubes was studied during the sixties and seventies [1,2]. However, the metal-glass sealing at the endings, insuring hermetical seal for the vacuum, still remains a costly technology.

A design of the evacuated tubes that deals with the solution of the sealing problem appeared at the end of the seventies: the all-glass evacuated tube solar collector [3,4,5]. This design also presented the innovation of magnetron sputtered selective absorber coating on the cylindrical glass surface. Despite the fact that the vacuum seal problem has been solved in this kind of collectors , the problem of degasification of the absorber coating lead to degradation of the vacuum and hence of the thermal efficiency of the collector during the operation.

Recently, a new idea has been proposed in this respect which deals with an evacuated tube in which the selective film is not in contact with the vacuum zone but is in the internal wall of the central tube, applied by chemical deposition technique [6]. We may call these collectors as *Evacuated Tubes with an Internal Absorber Film* (ETIAF) In the literature review, there is no reference to a mathematical model of this kind of collectors. To solve the problem completely we have to deal with the thermal problem in 3-dimensions, cylindrical geometry, transient state, composite media, with heat transfer by conduction, convection and radiation. In this paper, we present a mathematical model that simulates the thermal behavior of the ETIAF. In a first approximation, we will consider a steady state, one-dimensional composite media thermal problem.

DEFINITION OF THE PROBLEM

Figure 1 shows a cross section of the ETIAF. It has three essential elements: the glass cover 1, the interior glass tube 2 and the solar radiation absorber film deposited in direct contact with the working fluid in the internal glass wall. Solar radiation passes through both glasses to be absorbed by the film. The film heats up and this heat is transmitted by convection to the working fluid. The losses of heat to the ambient air will go against the direction followed by the solar radiation.

3. MATHEMATICAL MODEL

Consider the following differential control volumes shown in figure 2. Each control volume is an annular region that has a length dx. The x-axis indicates the direction of the fluid inside the tube.

Applying the first law of thermodynamics to each one of the differential control volumes and assuming steady state with constant thermophysical properties, we obtain the following equations:

Glass 1

$$\alpha_g \, r_1 q_i + r_2 hb \left(T_2 - T_1 \right) + \frac{r_2 \sigma}{\varepsilon_{12}} \left(T_2^4 - T_1^4 \right) = r_1 ha \left(T_1 - T_2 \right) + r_1 \sigma \, \varepsilon_1 \left(T_1^4 - T_a^4 \right) \quad (1)$$

Glass 2

$$\tau_1 \, \alpha_g \, r_2 q_i + \frac{k_g \left(T_3 - T_2 \right)}{\ln \left(1 + \dfrac{e}{r_2} \right)} = r_2 hb \left(T_2 - T_1 \right) + \frac{r_2 \sigma}{\varepsilon_{12}} \left(T_2^4 - T_1^4 \right) \quad (2)$$

Absorber Film

$$\tau_2 \, \tau_1 \, \alpha_f \, r_2 q_i = \frac{k_v \left(T_3 - T_2 \right)}{\ln \left(1 + \dfrac{e}{r_2} \right)} + r_2 hc \left(T_3 - T_4 \right) \quad (3)$$

Fluid

$$\dot{m} \, Cp \, \frac{dT_4}{dx} = r_2 hc \left(T_3 - T_4 \right) \quad (4)$$

where T_1, T_2, T_3 y T_4, are functions of x, $e = es/2$ and

$$\varepsilon_{12} = \frac{1}{\varepsilon_2} + \frac{1 - \varepsilon_1}{\varepsilon_1} \left(\frac{r_2}{r_1} \right).$$

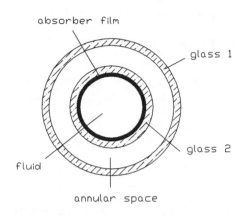

Fig. 1. Cross sectional view of a ETIFA.

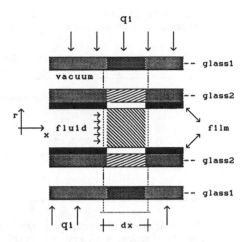

Fig. 2. Differential control volumes.

Arranging and simplifying, equations (1-4) becomes

$$A_1 T_1^4 + B_1 T_1 + C_1 T_2^4 + D_1 T_2 + G_1 = 0 \qquad (5)$$

$$A_2 T_1^4 + B_2 T_1 + C_2 T_2^4 + D_2 T_3 + E_2 T_3 + G_2 = 0 \qquad (6)$$

$$D_3 T_2 + E_3 T_3 + F_3 T_4 + G_3 = 0 \qquad (7)$$

$$\frac{dT_4}{dx} = H_4 (T_3 - T_4) \ , \quad T_4(0) = T_4^o \qquad (8)$$

where $A_i, B_i \ldots H_i$, with $i = 1 .. 4$, are constants. Those constants involve the energy transport coefficients and the properties of the materials. The equations (5-8) are algebraic equations (2 nonlinear and one linear), the another one is a first order ordinary differential equation. Those four equations are coupled.

The equations are solved numerically. The algorithm is described as follows: The axial coordinate is approximated as $x = j\Delta x$, $j = 0, 1, 2 .. N$. Then $T_1(x) = T_1(j\Delta x) = T_1^j$, similarly for T_2^j, T_3^j y T_4^j. For the initial boundary condition $j = 0$ at $x = 0$, $T_4(0) = T_4^o$, with this value, and using the iterative numerical method [7] to solve the system of equations (5-7) the values for T_3^o, T_2^o y T_1^o are found. Now applying the Euler's method to equation (8) T_4^{j+1} is obtained as:

$$T_4^{j+1} = T_4^j + \Delta x \ G_4 (T_3^j - T_4^j). \qquad (9)$$

With the values of T_3^o, T_4^o and equation (9), T_4^1 is obtained. Repeating the same procedure T_4^2, T_4^3, T_4^4 and so forth, are obtained.

RESULTS AND DISCUSSIONS

The solution of the mathematical model was carried out by considering the thermophysical properties of water and engine oil unused [8], the optical properties of borosilicate glass [9] and film deposited on the glass reported in [10]. These properties are shown in table 1. The convective heat transfer coefficients hc and ha were calculated with expressions reported in [4].

TABLE 1. Typical Values of the Parameters Used.

GLASS	WATER	Engine Oil	AIR	GEOMETRY
k_g = 1.125	k = 0.613	0.145	V_a = 1.5	Lx = 1.300
α_g = 0.05	C_p= 4186.8	1909.0	ρ_a = 1.158	r_1 = 0.030
ρ_g = 0.05	ρ = 997.424	884.1	μ_a = 1.87E-06	r_2 = 0.025
τ_g = 0.9	μ = 8.55E-04	4860.E-04	k_a = .0266	es = 0.002
ε_g = 0.85				
	Absorber Film: α_f = 0.80		ρ_f = 0.15	

The temperature variation for the elements of analysis along the x-axis is shown in figure 4. Here the variables are the dimensionless X = x/Lx and the dimensionless temperature θ = $(T-T_a)/(T_{max}-T_a)$. Also we observe in figure 4 that T_1 is almost constant for all X but T_2, T_3 and T_4 have significant variations but with the same slope. The temperatures T_2 and T_3 correspond to glass 2 and the absorber film. They have practically the same value; we did not find significant differences for other conditions of the fluid.

In figure 4 we plot the parameter f against the thermal efficiency. The parameter f could represent an *operation temperature* of the collector [11] and is defined as f = $(T_e-T_a)/q_1$. The efficiency has been defined as the ratio of the rate of the useful heat gained by the fluid and the incident solar power. The figure shows five curves for different heat transfer coefficient values; h_b = 6.0 W/m^2 K that represents the atmospheric pressure and extreme convection conditions, up to h_b = 0.026 W/m^2 K that represent the vacuum 10^{-4} Pa. At the latter pressure all the losses by convection are eliminated, leaving only minor losses due to conduction and radiation. We can observe in this figure that for high values of h_b the efficiency decreases rapidly with the increase of f, while with low values of h_b, the efficiency is almost constant with f over a wide range. Also we observe that at high values of h_b, the efficiency curve approximates to a straight line which indicates the influence of convection in the heat transfer process.

Figure 5 shows the efficiency curves for different mass flow rates. The shape of the curves is essentially the same. It is observed that at higher flows the efficiency increases, as is expected. In figure 6 the values used in figure 5 have been plotted, but the operation temperature f was parameterized. From this figure, a critical mass flow rate value can be defined, beyond which the efficiency does not increase significantly.

Results similar to those in figures 3-6 were obtained using engine oil. In figure 7, efficiency is plotted against f for the two liquids (water and oil) and the two values of the mass flow rate. It is interesting to note that the efficiency curves of water are always higher than those of oil. This is physically reasonable because the thermal conductivity and the heat capacity of water are greater than oil. Figure 8 shows a plot of efficiency vs f for different values of q_1. Here the efficiency decreases as q_1 increases as was expected.

CONCLUSIONS.

From the above results we conclude that the mathematical model has a behavior consistent with what is expected to happen physically. In particular we can conclude the following:

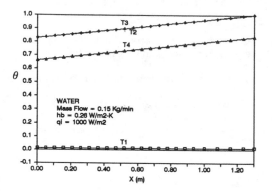

Fig. 3. Temperature distribution in
the collector.

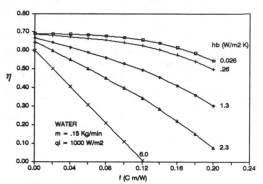

Fig. 4. Efficiency vs f for different
values of hb.

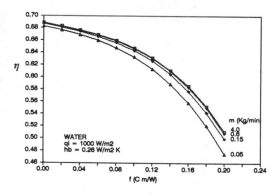

Fig. 5. Efficiency vs f for different
values of mass flow rate.

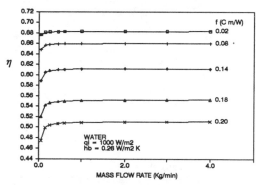

Fig. 6. Efficiency behavior vs mass
flow rate at different f's.

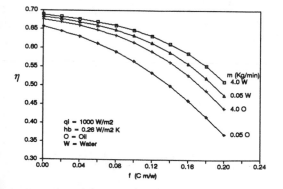

Fig. 7. Efficiency vs f for two
different values of mass flow
rate for water and oil.

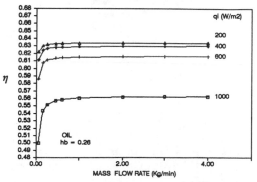

Fig. 8. Efficiency vs f for different
values of qi.

1. If $h_b = 0.026$ w/m^2 K (vacuum $\cong 10^{-4}$ Pa), then the thermal efficiencies of the collector have high values of ~ 70 % for a wide range of operation temperatures.
2. There is no significant difference between the glass temperature T_2 and the absorber film temperature T_3; hence, we suggest dealing with the system glass 2 and the absorber film as one material.
3. To consider the temperature of the glass cover 1 as constant with X is a reasonable assumption.
4. A critical value of the $\dot{m}$ (0.6 kg/min) can be defined beyond which the efficiency does not increase significantly for higher $\dot{m}$.
5. From the thermal efficiency point of view, the use of water is always better than the use of oils.

NOMENCLATURE

Cp	Specific heat, W/Kg °C.	τ	Transmittance for solar radiation.
es	Thickness of the glass tubes, m.		
h	Convective heat transfer coefficient, W/m^2 °C.	ρ	Reflectance; density, kg/m^3.
		μ	Viscosity, kg/m s.
k	Thermal conductivity, W/m K.		
L	Total length m.		SUBINDEX
$\dot{m}$	Mass flow rate, Kg/min.		
q1	Solar flux, w/m^2	a	Ambient
r	Radio, m.	b	Space between glasses
T	Temperature, °C, K.	c	Film-fluid
V	Velocity, m/s.	e	Collector entrance
		f	Film
		g	Glasses
	GREEKS	i,j	Index
		r	r-direction
α	Absorptance.	x	x-direction
Δx	Length of the element in the analysis, m.	1	glass 1
		2	glass 2
σ	Stefan-Boltzman constant.	3	film
ε	Thermal emittance.	4	working fluid

REFERENCES

1. Speyer, E. (1964). Solar Energy Collection with Evacuated Tubes. ASME, J. of Eng. for Power, Paper No. 64-WA/Sol-2.

2. Ortobasi, U. and W.M. Buehl (1975). Analysis and Performance of an Evacuated Tubular Collector. Presented at the ISES Conference, Los Angeles.

3. Analysis and Experimental Test of a High-Performance Evacuated Tubular Collector (1978). U.S. Department of Energy DOE/NASA CR-150874.

4. Evaluation of the Evacuated Solar Annular Receivers Used at the Mid temperature Solar System Test Facility (1979). Sandia Laboratories SAND78-0983.

5. Harding, G.L. Absorptance and Emittance of Metal Carbide Selective Surfaces Sputter Deposited onto Glass Tubes (1980). J. of Solar Energy Materials 2 pp 469-481.

6. Nair, P.K. and M.T.S.Nair, Chemically Deposited SnS-Cu$_x$S Thin Films with High Solar Absortance: New Approach to All-glass Tubular Solar Collectors (1991). J. Phys. D:Appl. Phys. 24 (1991) 83-87. UK.

7. Scheid, F. (1968). Numerical Analysis. Shaum's Series, McGraw-Hill Book Co.

8. Incropera, F.P. (1981).Fundamentals of Heat Transfer.John Wiley & Sons,NY.

9. Bansal, N.P. and R.H. (1986). Doremus, HANDBOOK OF GLASS PROPERTIES. Academic Press, Inc..

10.García V.M. and Nair M.T.S. and Nair P.K., Optical properties of chemically deposited PbS-Cu$_x$S and Bi$_2$S$_3$-Cu$_x$S coatings with reference to solar control and solar absorber applications (1991). J of Solar Energy Materials (in revision).

11.Duffie J.A. y W.A. Beckman , Solar Engineering of Thermal Processes (1980). John Wiley & Sons, New York.

2.27 Desalination

DESIGN STUDY OF A NOVEL SOLAR OR WASTE HEAT POWERED DESALINATION SYSTEM

John Hanneken, Manager of Engineering[1]
Dial Manufacturing, Inc.
Phoenix, Arizona

and

Warren Rice, Professor Emeritus
Department of Mechanical and Aerospace Engineering
Arizona State University
Tempe, Arizona

ABSTRACT

A relatively simple vacuum distillation system is described which uses a very effective means of removing air and condensing water vapor simultaneously. The system can use any source of energy including low grade sources such as solar heat and nuclear waste heat. It consists of three related components: a heat collection device, a thermosyphon loop in which partial evaporation occurs in or near an upper chamber, and a hydraulic air compressor which removes air and also provides condensation of the water vapor. It is important to note that the solar collector can be a simple horizontal flat plate design that operates at atmospheric pressure. The ways in which analytical modeling of each of the components, which have appeared in earlier literature, can be combined to yield an analytical model of the distillation system are indicated. Computer-implemented calculations were made to produce an example of the design of a distillation system of the subject type and the results of a brief economic study are presented. It is shown that the system compares favorably with RO and ED systems with regard to the performance factor that can be achieved, and that it is economically approximately comparable to small RO systems.

INTRODUCTION

A problem that all types of vacuum distillation systems have in common is that the feedwater contains air in solution, which travels to the condenser with the water vapor. Since the air is non-condensable, it must be removed continuously by some kind of air ejector having an energy input requirement. In addition to air entering with the feedwater, air leaks into the system, since the water in the system is at a pressure less than atmospheric pressure, and this air must also be removed.

There exists a relatively simple vacuum distillation system using a unique and very effective means of removing air and condensing water vapor simultaneously. This system was patented by Younger (1981). A nuclear waste heat application was considered by Saari (1978). All of the systems have in common the equivalent of the three components shown in Fig. 1: a heat supply that may be waste or solar heat, a thermosyphon loop in which partial evaporation occurs in or near an upper chamber, and a hydraulic air compressor which removes air and also provides condensation of the water vapor. An analytical model of such a distillation system, to provide a means of sizing and matching the components and to enable optimization of the system performance, is apparently not available in the literature.

The solar collector or waste heat collector operates at atmospheric pressure and therefore can be an inexpensive component. In the thermosyphon, the water circulates without mechanical pumping. Near the top of the thermosyphon partial evaporation of water occurs with the remaining water returning gravitationally to the heat collector. Water vapor and air produced in the chamber at the

---

[1] Formerly, Graduate Student, Department of Mechanical and Aerospace Engineering, Arizona State University, Tempe, AZ

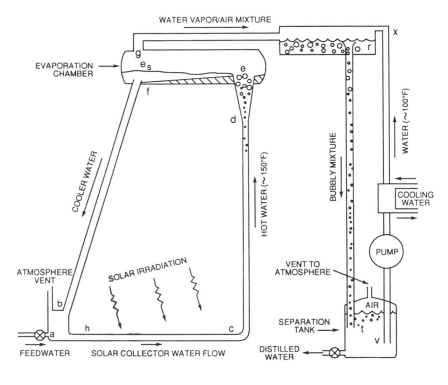

WATER VAPOR/AIR MIXTURE

EVAPORATION
CHAMBER

COOLER WATER

HOT WATER (~150°F)

BUBBLY MIXTURE

WATER (~100°F)

COOLING
WATER

PUMP

ATMOSPHERE
VENT

SOLAR IRRADIATION

VENT TO
ATMOSPHERE

SEPARATION
TANK

AIR

FEEDWATER   SOLAR COLLECTOR WATER FLOW

DISTILLED
WATER

SOLAR COLLECTOR AND THEMOSYPHON   HYDRAULIC AIR COMPRESSOR

Fig. 1   Schematic diagram of the novel solar powered desalination system.

top of the thermosyphon pass to the top chamber of the hydraulic air compressor (HAC) and are induced into the flow in the HAC downpipe in the form of bubbles. The vapor bubbles quickly condense and the air bubbles are carried downward to the separation tank, which is at atmospheric pressure, where the air is exhausted to the atmosphere. The water in the HAC must be pumped and thus a conventional water pump with an attendant power requirement must be provided. The useful product is distilled water, taken continuously from the water in the HAC, at the same mass flow rate as that of the vapor leaving the thermosyphon. The required vertical height of the pipes and tanks composing the system is approximately 11 meters and the pressure in the evaporation chamber at the top of the thermosyphon is preferably (and achievably) less than 1/20th of an atmosphere. The water circulating in the HAC must be continuously cooled by some environmentally available cooling sink. Each of the three components of the system have been extensively treated individually in the literature. The designs of solar collectors and of waste heat collectors have been exhaustively treated and the treatment given by Duffie and Beckman (1980) is an example of the information available. The hydraulic air compressor has been analytically modeled in a general way by Rice (1976). The thermosyphon is somewhat different in both purpose and operation from those treated in the literature and will be considered here in more detail than the other two system components.

ANALYTICAL MODELING OF THE SOLAR COLLECTOR

Arbitrarily, a solar source of heat was considered rather than a waste heat source. The analytical modeling of solar collectors is well understood and documented. Therefore, only the geometries and materials chosen and the modeling idealizations will be indicated here. The solar collector was assumed to be a horizontal rectangle in plan view. As indicated in Fig. 1, feedwater is supplied continuously at one edge and the thermosyphon return water reenters the collector near the feedwater entrance location. The leg of the thermosyphon up which heated water flows, leaves at the other end of the collector. Since the collector operates at atmospheric pressure, it is assumed to have an insulated bottom and to be glazed at the upper surface. An absorbing plate is provided at the bottom surface of the water flow. The analytical modeling accounted for incident radiation, convection from the outer surface of the glass to the atmosphere, radiation losses from the outer surface of the glass, conduction through the glass, absorption by the water and by the absorbing plate, and conduction through the insulation to the earth. A nodal point system was established and algebraic finite difference equations were written for each node, representing the conservation of energy. For nodal points in the flowing water, energy transported by the water was accounted for. The large number of resulting non-linear equations was then solved simultaneously for the temperature field, using an iterative method implemented in a computer program. The program was written so as to incorporate the incident solar energy as a function of date and time at a chosen geographical location.

As an example of the results, Fig. 2 shows the water temperature exiting the solar collector, to the thermosyphon, as a function of the water mass flow rate through the collector, and with the time of day as a parameter. The date is assumed to be June 21 and the location to be Phoenix, Arizona. These results are used later herein as part of the design information for an example water distillation system.

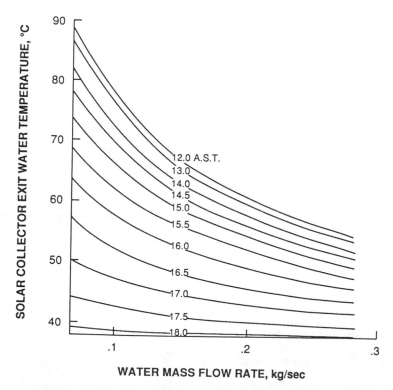

Fig. 2   An example of the calculated exit water temperature from the solar collector as a function of water flow rate and time of day.

ANALYTICAL MODELING OF THE HYDRAULIC AIR COMPRESSOR (HAC)

Ordinarily, an HAC is used to provide compressed air at its separation chamber. In this application, however, the separation chamber is vented to the atmosphere and the chamber at the top of the HAC, indicated in Fig. 1, is held at a pressure slightly less than that in the evaporation chamber at the top of the thermosyphon. Since Rice (1976) gives details of the analytical modeling of a generic HAC, the modeling is not reproduced here. Calculations based on the modeling were implemented in a computer program. All fluid flow losses are accounted for and the two-phase bubbly flow in the downflow pipe is modeled in fairly realistic detail. Using the program, parameter studies were made in which the pressure in the upper chamber was varied in a range of interest from .035 atm to .102 atm. For a specified downflow pipe diameter, the calculations resulted in values for the flow properties throughout the HAC, the required vertical distances, the air flow rate, and the required pump power, for optimum performance of the HAC. Example results are used later herein as part of the design information for an example water distillation system.

ANALYTICAL MODELING OF THE THERMOSYPHON

The flow in the thermosyphon was treated as six flow regions or sections as indicated in Fig. 1. The thermosyphon was considered to be in steady flow. Each flow region was considered by enclosing it in a fixed control volume for which work was defined as positive outward. The flow in each region was considered to be one-dimensional with properties changing only in the direction of flow. The flow in the solar collector is, of course, a part of the thermosyphon hydraulically but was earlier treated separately as an energy collector. All other parts of the thermosyphon were considered to exchange no heat with the surroundings. Each flow region is considered separately and the resulting equations were solved simultaneously using a computer program.

Region $b \rightarrow h$, $h \rightarrow c$, $c \rightarrow d$, $a \rightarrow h$ and $f \rightarrow b$

In these regions the flow is practically incompressible and the appropriate modeling equations are

$$p_b = p_h + \rho_b \, g \, (Z_h - Z_b) \tag{1}$$

$$p_a = p_h + \rho_b \, g \, (Z_h - Z_a) \tag{2}$$

$$p_c = p_h - \rho_c \, (1 + K_c) \frac{V_c^2}{2} + \rho_c g \, (Z_h - Z_c) \tag{3}$$

$$p_c = p_d + \rho_c \, (Z_d - Z_c) \left(g + \frac{f_{cd} V_c^2}{2 D_{cd}} \right) \tag{4}$$

$$p_b = p_f + \rho_f \, g \, (Z_f - Z_b) + \rho_f \, f_{bf} \, V_f^2 \, L_{bf}/(2 d_f) \tag{5}$$

Region $d \rightarrow e \rightarrow e_s \rightarrow f$

In this region the pipe is considered to be tapered with increasing cross-sectional area in the direction of flow. Application of the conservation of mass and energy, with the use of appropriate thermodynamic relations resulted in

The conservation of mass between locations d and e is expressed as

$$Z_e - Z_d = (h_d - h_e + (V_d^2 - V_e^2)/2)/g \tag{6}$$

$$x_{es} = x_e + \frac{V_e^2}{2(h_{ge} + h_{fe})} \tag{7}$$

$$p_f = p_e - \rho_f V_f^2/2 - \rho_f K_f V_f^2/2 + g\rho_f (Z_e - Z_f) \tag{8}$$

where $p_e = p_{es}$ and $Z_e = Z_{es}$ have been used and where K_f is the entrance loss coefficient. It is noted that $T_f = T_{es} = T_e$.

Solution Procedure

In a computer program, it is efficient and practical to solve for the unknowns serially through the zones starting with zone b → h. With thermodynamic data and viscosity data available for lookup at local conditions in the program, and with values of T_c and p_e and p_b known, the developed equations can be arrayed with p_c, V_c, (Z_d-Z_c), x_e, $(Z_e - Z_d)$ and p_f as unknowns and the algebraic solution can proceed. Of course, the pipe sizes throughout the thermosyphon must be prescribed in advance. The air in solution at location c is assumed to come out of solution in region $d \rightarrow e_s$ and therefore leaves the evaporation chamber with the saturated vapor at pressure $p_g = p_e$ at location g.

Example of the Design of a Desalination System

The analytical models for the solar collector, thermosyphon and hydraulic air compressor were used together to design (in the sense of specifying all pipe sizes, fluid flow rates, etc.) a desalination system to provide 50 USgal/day (.1893 m³/day) of distilled water on the "best" day of the year, June 21, at the location of Phoenix, Arizona. The design is not optimized and many possible design variables are fixed at arbitrary but reasonable values. The coupling between the three components is such that extensive iteration was necessary to determine the size of solar collector needed to achieve the chosen "best" day performance. With the assumption of a standard day on that date, iteration showed that the performance would be produced by a collector that provides 409 kg/hr of water flow rate to the thermosyphon, at 2.50 PM A.S.T. The resulting solar collector proved to be approximately 6.1m in length, 5.5m in width, using a .076m depth of water. The details of producing these results are rather routine but very lengthy and are not given here.

As part of the results, it was determined that $x_e > 0$ only after 6.25 A.S.T. and before 17.75 A.S.T; otherwise T_c was too small for water to circulate in the thermosyphon. It also was determined that pipes with nominal diameters $d_c = d_f = 1.0$ inch (Schedule 80, A.S.A. Standards), and vertical distances $(Z_f - Z_b) = 10.0$m and $(Z_d - Z_c) = 9.14$m correspond with the specified design parameters. The water at location c was assumed saturated with air. This corresponds for the design with an air flow rate to the HAC of .000817 kg/hr. The vapor flow rate is 29.41 kg/hr. These are maximums, for the highest water production rate.

For the HAC, pipes of 1.0 inch nominal diameter (Schedule 80, A.S.A. Standards), and $(Z_r - Z_t)$ = 11.95m and $(Z_x - Z_v) = 12.56$m. The HAC would circulate water at the rate of 15 USgpm (.0057 m³/min) and the pump would require an input power of 75 W. Exhaustive further details concerning additional assumptions not given here and the explicit means by which the required component matching and iteration were carried out are given in a thesis by Hanneken (1986).

Having determined a design to produce 50 USgal/day (.1892m³/day) per day of distilled water on June 21, standard annual insolation information was used, with the analytical models of the system components, to produce the performance at other times of the year, and to produce an integrated value for the distilled water produced in one year. That value proved to be, closely, 10950 USgal/year (41.45m³/year) or an average value of approximately 30 USgal/day (.1135m³/day). The time intervals during each day for which water could be produced over the year were also calculated. The very small power required to drive the pump for the HAC, about 75 watts, suggests that it might be powered by a photovoltaic array. With appropriate assumptions and calculations, it was determined that an area of approximately 1.21m² would be required, tilted up at an angle of 13.4 degrees. The performance factor over one year for the distillation system was determined to be approximately 5.673kJ/kg. The range of performance factors for other systems, including distillation (MSF), freezing, reverse osmosis and electrodialysis appears to be from approximately 10 to 2500 (Lior, 1981). In comparison, the system studied herein is efficient and attractive from an energy utilization point of view.

ECONOMIC CONSIDERATIONS AND CONCLUSION

An analysis was conducted in a manner similar to that recommended in references (Duffie and Beckman, 1980; Gerofi and Fenton, 1984). Initial capital investment, operation and maintenance costs and certain other system costs were taken into account. Details are presented in the thesis Hanneken (1986) and only a summary of results will be given herein. The results depend on a long list of choices of materials and construction methods, which were attempted to be made realistically and conservatively. Operation and maintainance costs were neglected. The solar collector glass and the photovoltaic panels would require cleaning much more often but at very small cost. A lifetime of 20 years was assumed with a zero salvage value. Time-value-of-money type calculations were applied with appropriate estimates of inflation rates, interest rates and energy costs over the lifetime of the system.

For the system described, built as a one-of a kind experimental system, the resulting unit cost of the water produced was determined to be 5.43 ¢US/USgal. It was further determined that; if produced in quantity, the unit cost cost could be reduced to approximately 3.0 ¢US/USgal. Economic data for small RO plants are very scarce but Pleass, Hicks and Mitcheson (1990) recently provided cost information for a system which produced water at a rate of approximately 2000 USgal/day. It was a conventional RO system and the unit cost was 2.72 ¢/gal. If considered over a 20 year period into the future, the unit cost would become about the same as estimated for the novel desalination system described herein.

NOMENCLATURE

A	cross-sectional area	p	pressure
f	Darcy fluid friction factor	T	temperature
g	acceleration due to gravity	v_f	specific volume of saturated liquid
h_f	enthalpy of saturated liquid	v_g	specific volume of saturated vapor
h_g	enthalpy of saturated vapor	V	velocity
K	hydraulic loss coefficient	x	quality of liquid/vapor mixture
L	length	Z	vertical distance
m	mass flow rate	ρ	density

REFERENCES

Duffie, John A. and William A Beckman (1980), **Solar Engineering of Thermal Processes**. John Wiley and Sons, New York.

Gerofi, J.P. and G.G. Fenton (1984). A simple, more accurate criterion for economic evaluation of desalination projects. *Desalination*, **49**, pp. 255-270.

Glueckstern, P.(1982). Comparative energy requirements and economics of desalting processes based on current and advanced technology. *Desalination*, **40**, pp. 63-74.

Hanneken, John B. (1986) A Feasibility Study of a Novel Solar Powered Desalination System. MS Thesis, Mechanical and Aerospace Department, Arizona State University, Tempe, Arizona.

Lior, N. (1981). Principles of Desalination, pp. 1-11. *Solar Water Desalination-Proceedings of the Second SOLERAS Workshop*, Denver, Colorado.

Pleass, C.M., D.C. Hicks and G.R. Mitcheson (1990). Wave-Driven R.O. harvests untapped energy, *Water Conditioning and Purification*, pp. 41-44.

Rice, W. (1976). Performance of hydraulic gas compressors. *ASME Transactions, Journal of Fluids Engineering*, **98, No. 4.**, pp. 645-653.

Saari, Riato (1978). Desalination by Very Low-Temperature Nuclear Heat, *Nuclear Technology*, **38**, Mid-Apr. 1978, pp. 209-214.

Younger, Philip G.(1981). Liquid Treating and Distillation Apparatus. *U.S. Patent* No. 4, 269, 664.

MULTI-EFFECT AMBIENT PRESSURE DESALINATION
WITH FREE CIRCULATION OF AIR

T. Baumgartner, D. Jung, F. Kössinger and R. Sizmann

Sektion Physik, Ludwig-Maximilians-Universität München,
Amalienstr. 54, D 8000 München 40, FRG

ABSTRACT

A two-stage distillation desalination unit operated with low temperature process heat of 70 - 85 °C has been built. Heat of evaporation is recovered in an ambient pressure, free convection driven counter-current process which amounts to a gained-output-ratio GOR up to 6 (only about 115 kWh thermal process heat is necessary to produce 1 m^3 of fresh water from saline water, compared to the 690 kWh/m^3 heat of evaporation of water).
Outstanding features are also a simple construction, easy maintenance and auxiliary energy limited for pumping the saline water through the (plastic) heat exchanger tubing to the top of the about-250 cm high desalination tower.

KEYWORDS

Thermal Desalination, Natural Convection, Gained-Output-Ratio GOR

INTRODUCTION

The simplest installations for solar driven thermal desalination are solar stills. They are working successfully worldwide since more than 100 years (Harding, 1883; Delyannis, 1981; Malik, 1982). The fresh water (distillate) production rate of (even the most effective) stills is rather poor (3 to 5 kg/m^{2}d (Cooper, 1973; Tanaka, 1983)), because no heat of evaporation is recovered. This implies that at least the specific heat of evaporation of about 690 kWh is necessary to gain 1 m^3 of fresh water.

To reduce the specific costs of thermal desalination units it is necessary to lower the specific process heat input by recovering heat of evaporation without a critical increase of the technical expenditure.

Based on previous research work (Heschl, 1987; Baumgartner, 1990) a two-stage multi-effect distillation desalination unit has been developed. It operates by employing a closed air loop at ambient pressure which carries by natural (free) thermal convection the water vapor from the evaporator to the condenser area, where heat of condensation is partially recovered for repeated evaporation. A gained-output-ratio GOR of 6 is reached (only 115 kWh of thermal process heat is required to produce 1 m^3 of fresh water). This is comparable to conventional 8-stage Multi-Stage-Flash plants. The upper operation temperature is 85 °C.

Process heat of that temperature level is available from solar collectors, solar ponds, geothermal heat or industrial waste heat.

MULTI-EFFECT DESALINATION PROCESS

In Fig. 1 the vertical cross-cut illustrates the principle of the present two-stage unit.

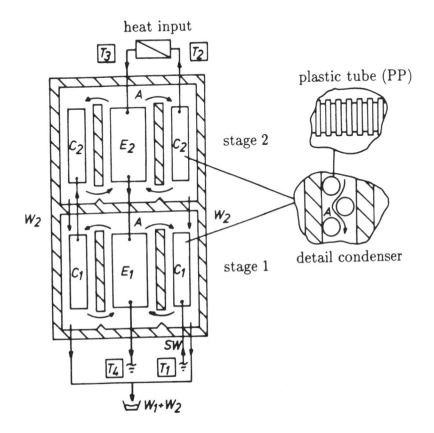

Fig. 1: Cross-cut of a two-stage unit. E_i: evaporators; C_i: condensers; W_i: water production rates; SW: saline water; A: air circulation; T_1: condenser saline water inlet temperature; T_2: condenser saline water outlet temperature; T_3: evaporator saline water inlet temperature; T_4: brine temperature

Stage 1 and 2 are identical units. The casing is made of a waterproof heat insulation material. The overall height is 250 cm. The base area is 1 x 0.7 m². The two main elements in each stage are the evaporators E_i (heat and mass exchanger surface: 30 m² in every stage) and the condensers C_i (exchanger surface: 17 m² in every stage). The evaporators E_i are built of vertical parallel sheets of jute fabric. They are suspended from the top of each stage, where the hot saline water is distributed to wetten the sheets uniformly.
The hot saline water flows with temperature T_3 from the top of the sheets in E_2 and leaves E_1 as waste water (brine) with temperature T_4. This induces an ascending natural convection of air in each stage. The air becomes saturated with water vapor and leaves the evaporators at almost the same temperature as the saline water enters the evaporators (e.g. T_3 in stage 2).

By evaporation the saline water gradually drops in temperature, at the bottom of stage 1 its temperature T_4 is close to temperature T_1, at which the saline water enters the condenser in stage 1.

The condensers C_i are coils of plastic tubing wrapped around an inner casing. Air can freely pass along the coils. The used material is polypropylene (PP) with an annular, fluting structure to increase surface area (see detail picture). At the bottom of each stage the cold saline water enters C_i, counter-current to the downstream of the hot, water vapor saturated air. Condensation occurs on the outer surface of the plastic coil (which provides an excellent heat transfer to the water inside the coil). The heat of condensation is carried upwards with the saline water, which attains at the top of stage 2 a temperature T_2. The temperature increase $T_2 \rightarrow T_3$ is provided by the heat input

$$\dot{Q}_{input} = \dot{m}_{SW} c (T_3 - T_2) \tag{1}$$

with $\dot{m}_{SW}$ the mass flow rate of the saline water and c its specific heat capacity.

Product water mass flow $\dot{m}_W$ is collected at the bottom of stage 1; it is available at a temperature close to the inlet temperature T_1 of the saline water.

The counter-current flow of air and saline water in each stage guarantees recycling of heat of evaporation (condensation). Because of the remaining temperature difference T_4-T_1 the saline water carries away a waste heat

$$\dot{Q}_{waste} = (\dot{m}_{SW} - \dot{m}_W) c (T_4 - T_1). \tag{2}$$

In an ideal desalination process the difference $\dot{Q}_{input}$-$\dot{Q}_{waste}$ would amount to the (integral) heat of separation of a flow of pure water $\dot{m}_W$ from a flow of saline water $\dot{m}_{SW}$.

RESULTS

Fig. 2 shows the gained-output-ratio GOR as a function of the saline water load $\dot{m}_{SW}$ at various temperatures T_3. GOR, which is a measure of the multi-effect efficiency, is defined as

$$GOR = \dot{m}_W l / \dot{Q}_{input} \tag{3}$$

with l the specific heat of evaporation of water (about 2390 kJ/kg). The saline water inlet temperature T_1 is 25 °C.

Fig. 3 shows the dependence of the fresh water production rate $\dot{m}_W$ on saline water load $\dot{m}_{SW}$ at various temperatures T_3. With a saline water load of 80 kg/h and a process heat input of 85 °C (in the evaporator E_2), we find a fresh water production of about 6 kg/h and a GOR of 4.1.

Fig. 2 and 3 exhibit that a compromise is necessary between a maximum GOR and a maximum fresh water production. A saline water load of 60 to 70 kg/h and a temperature T_3 of 85 °C seems to be a good compromise for a steady-state-operation of the present desalination unit. A GOR of about 5 and a fresh water production of about 5 kg/h is the result.

The salinity of the brine output increases by only 4%. Recycling of the brine can be of economic advantage, particularly when brackish water is used as the primary input. In this case, since

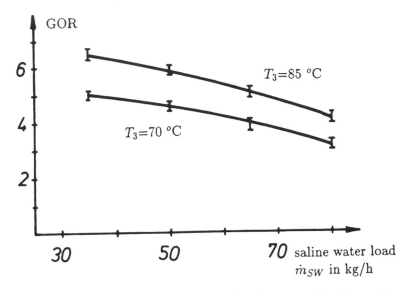

Fig. 2: Gained-output-ratio GOR as a function of the saline water load $\dot{m}_{SW}$ at temperatures $T_3 = 70$ and 85 °C

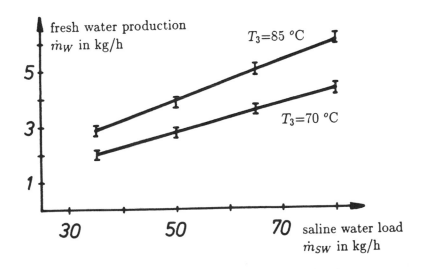

Fig. 3: Fresh water production rate $\dot{m}_W$ as a function of the saline water load $\dot{m}_{SW}$ at temperatures $T_3 = 70$ and 85 °C

temperature T_1 is not allowed to rise, a cooling device is required which levels T_4 down to T_1. The desalination unit operates as a thermodynamic cycle. The entropy entering with the high temperature heat and the additional entropy which is produced by irreversibilities has to be discharged via low-temperature waste heat.

Maintenance of the desalination unit is found to be easy. Any deposition on the evaporator textile surfaces can be knocked off after drying. Depositions inside the polypropylene tubes can be removed by beating the piping and a simultaneous fast flow of water through the tubes. Auxiliary (electrical) energy is required for saline water pumping through the condenser tubing to the top of the desalination tower (there are no fans to circulate the working medium air). It amounts to be less than 1 kWh (depending on the efficiency of the pumping system) for 1 m^3 of fresh water.

CONCLUSIONS

- The ambient pressure free convection circulation of water vapor saturated air system is a simple but highly efficient construction for production of potable water from saline water of almost any salinity.

- Its gained-output-ratio is about 5 and, thereby, equivalent to standard 8-stage MSF devices.

- The materials used for construction are corrosion resistive jute cloth for the evaporator part and polypropylene tubing for the condenser part. The construction is simple, which guarantees low cost construction and easy maintenance.

- The process is driven by process heat of maximum 85 °C; auxiliary power is only needed for water pumping.

REFERENCES

Baumgartner, T., O. Heschl, D. Jung, and R. Sizmann (1990). Meerwasserentsalzung mit Niedertemperaturwärme. In *Tagungsband 7. Internationales Sonnenforum*, pages 811–816, Frankfurt.

Cooper, P.I. (1973). The Maximum Efficiency of Single Effect Solar Stills. *Solar Energy*, 15:205–217.

Delyannis, A. and E. Delyannis (1981). Solar Distillation in Greece. In *Proceedings of the Solar Desalination Workshop*, pages 131–142, Denver.

Harding, J. (1883). Apparatus for Solar Desalination. *Proceedings of the Institute of Civil Engineers*, 73:284 ff.

Heschl, O. and R. Sizmann (1987). Solar Sea Water Desalination with a High Efficiency Multi Effect Solar Still. In *Proceedings of the ISES Solar World Congress*, pages 2814–2818, Hamburg.

Malik, M.A.S., G.N. Tiwani, A. Kumar, and M.S. Sodha (1982). *Solar Distillation*. Pergamon Press Inc.

Tanaka, K. (1983). Improvement of the Performance of the Tilted Wick Type Solar Stills. In *Proceedings of the Solar World Congress*, page 251 ff., Perth.

ACKNOWLEDGEMENTS

The authors are grateful to Dipl.- Phys. O.Heschl for consulting and the Bundesministerium für Forschung und Technologie Bonn, FRG, for financial support.

EFFECTS OF PEBBLES AND WICK ON THE PERFORMANCE OF A SHALLOW BASIN SOLAR STILL

M.A.C. CHENDO* AND S.U. EGARIEWE**

*DEPT. OF PHYSICS, UNIVERSITY OF LAGOS

AKOKA - LAGOS, NIGERIA.

**DEPT. OF PHYSICS, BENDEL STATE UNIVERITY EKPOMA,NIGERIA.

ABSTRACT

A comparative performance study on the effects of pebbles, charcoal and wick on single sloped shallow basin solar still is presented.

The results show that distillation rates of the participating stills between sun rise and sun set increased in the following order:- charcoal, wick, control still and pebble. Compared to the daily solar irradiance, the maximum irradiance point occurred hours before that of the respective stills in the order above.

Comparison of the respective stills' daily productivity shows that the pebble lined still has the greatest yield with its maximum yield ocaufting four hours after sunset (due to its better heat storage capability). For more productivity, a combinaiton of charcoal and pebble or dye and pebble may be further investigated.

KEY WORDS:

SOLAR DISTILLATION, SHALLOW BASIN STILL CHARCOAL BED, PEBBLE BED AND WICK TYPE PERFORMANCE.

INTRODUCTION.

Shortage or non existence of fresh potable drinking water is a major problem in most developing countries like Nigeria. Most people in these areas live in poor villages,shanty towns and semi urban areas where house hold wastes, agricultural wastes, community refuse and sewage are disposed directly into water bodies used fordrinking, bathing laundry, food preparation etc. The result of these unsanitary practices is heavy organic or fecal pollution with a prevalence of water borne diseases such as cholera, typhoid fever, bacillary dysentry etc culminating in deaths decline in agricultural activities etc. Most conventional methods of purifying water are energy intensive hence most developing nations are not able to meet the demands of fresh potable water adequately.

As a consequence of this limitation, other methods of water purification have to be addressed by the inhabitants. One such method is by solar distillation. Solar distillaiton has been in practice for a longtime and during this period varieties of solar distillation plants have been designed and fabricated. Solar distillaiton has an important role to play in the improvement of quality of life especially for those in rural areas of developing nations like Nigeria.

Although many experimental and theoretical contributions on the performance of basin type stills etc have been established, not much has been reported in literature on performance characteristics of skills with some additives like charcoal, dyes, pebbles/rocks etc. Hence, this paper will address the experimental study on wick-type stills, charcoal bed and pebble bed lined solar stills. In addition, a comparative analysis of the results obtained will be presented. Work on their theoretical analysis is in progress and will be presented in due course.

THEORETICAL CONSIDERATIONS.

The basic heat and mass transfer relations for a single basin still etc have been extensively discussed by Malik et al (1) and other researchers. These relations are summerized as follows:-
(1) Convection heat transfer from impure water to the cover, given by:-

$$q_{cw} = 0.884 \left[Tw - Tgi + \left\{ \frac{(Pw - Pg)(Tw - 273)}{268.9 \times 10^3 - Pw} \right\} \right] (Tw - Tgi)$$

$$= hcw \ (Tw - Tgi). \qquad (1)$$

where Tw and Tgi are temperatures of saline water and glass cover respectively, Pw and Pg are partial vapour pressures at temperatures Tw and Tgi and hcw is the convective heat transfer coefficient.

(ii) The radiative heat transfer from water to glass cover:-

$$q_{rw} = \frac{(Tw^4 - Tg^4)}{\frac{1}{Ew} + \frac{1}{Eg} - 1} \qquad (2)$$

(iii) Evaporation heat transfer (q_{cw}) from water surface to glass cover is :-

$$q_{ew} = 16.273 \times 10^3 \ hcw \ (Pw - Pg) \qquad (3)$$

(iv) Mass transfer rate;

$$Me = \frac{qew}{L} \qquad (4)$$

L is the latent heat of vaporization of water.

(v)   The Heat balance on the contents of the still (q_{aw}) is

$$q_{aw} = q_{cw} + q_{ew} + q_{rw} + q_c \qquad (5)$$

(vi)   Radiation and convection heat transfer between the glass cover and the environment q_{ra} and q_{ca} respectively are:-

$$q_{ra} = Eg\sigma\left[(Tg + 273)^4 - (Tsky + 273)^4\right]$$

$$q_{ca} = hca \ (\ Tg - Ta) \qquad (6)$$

Tsky and Ta are sky and ambient temperatures, respectively.

Equations 1 - 6, together with some of the recent refinements put forward for the prediction of the steady state performance of a single basin still by Clark (2), was used in computing the energy fractions within the still.

EXPERIMENTAL PROCEDURE.

Three identical single-roofed basin type solar stills each with an effective area of 0.42 m² and glass cover 3.0 mm thick were constructed. The basins are lined with a Poluvinyle chloride (PVC) material and the experimental additives.

A shematic diagram of the typical still is shown in fig. 1. Saw dust was used as the insulating material between the basin and wooden chambers. The stills were oriented in the east - west direction and charged periodically.

Two sets of experiments were conducted for a comparative study of the performance of the constructed stills:-

    (i) Charcoal and pebble lined basins still
    and
    (ii) Pebble lined and wick - type basin still .

In addition to the above, a control experiment with an identical conventional single-roofed basin solar still was monitored.

The stills' productivities, solar irradiance measurments and other variables were carried out at the Univerity of Lagos (Nigeria) Science Complex (Latitude 6.5°N).

RESULTS AND DISCUSSION.

Figure 2. shows the productivities of both the charcoal and pebble lined stills for a typical day, while fig. 3 shows the water/basin temperature variations for the pebble, wick and control stills.

The productivity data for a typical day for the wick-type and
pebbles is summarized in Table 1. The effect of the atmospheric
conditions on the productivities of the charcoal and pebble
line lag between the maximum points of hourly yield and global
insolation for the respective stills. The pebble lined profile
show that its maximum porductivity occurred about four hours
after sun set and production continued throughout the night.

This singular. productivity pattern of the pebble lined still
is not unexpected, since the pebbles have a much ligher thermal
capacity than either the wick or charcoal. Thus, it uses the
stored or cumulated energy during the day for distillation.
Another factor that is also responsible for its night time productivity is the fall in ambient temperature which consequently
increases the rate of condensation because of the lowered temp.
of the glass cover. Wind effect is also a contributory factor.

The higher day time yield for both the wick and charcoal may
be attributed to :-
   (i)   the increase in the evaporation area by the wick's
           porosity.
   (ii)   Both charcoal and wick exhibit capillary actions,
           and hence maintaining a wetted surface.
   (iii)   Charcoal is reasonably "black" to solar radiation,
           hence it enhances its absorption.
   (iv)   The rough surface nature of the charcoal pieces scatters
           rather than reflects incident solar radiation, thereby
           reducing the reflected losses.
   (v)   Charcoal increases the effective area of the still
           and also enhances the utilization of diffuse irradiance
           by the still more than the conventional still without
any additive . Chendo and Schmitter (3)

Examination of the various temperature profiles indicates that at
the early hours of the day their differences are negligible.

Table 1. Summary of a typical day'sdata for wick and Pebbles

Still	Day time yield (l/m^2)	Overnight yield (l/m^2)	Daily yield (l/m^2)
Control	0.466	0.350	0.816
Wick	0.769	0.280	1.049
Pebbles	0.536	1.189	1.725

Table 2. Summary of daily data for various weather conditions for charcoal and pebbles

Day	Daily Insolation (W/m^2)	Daily yield for charcoal (L/m^2)	Daily yield for pebbles (L/m^2	Weather conditi
1	135,8	0.240	0.440	Overcas
2	303.5	0.448	0.531	Fairly Cloudy
3	375.0	0.555	0.836	Variabl
4	782.2	0.995	1.040	Clear

CONCLUSION

From the experimental results obtained Pebble lined solar stills have a significantly higher daily productivity than others tested.

A chemical analysis of the yield from the participating stills is necessary   to ascertain if the additives have any effect on the quality of the distillate.

For a round-the-clock maximization of the productivity, a comb. of pebbles and charcoal has to be studied. Also a thorough theoretical analysis of the physics of the system has to be undertaken. The authors are currently looking into these two areas.

The overall efficiency of the stills studied can be improved by the use of good quality sealants and improved insulation material.

REFERENCES:

1. MALIK M.A.S., Tiwari, G.N, Kumar A., and Sodha. M.S. (1982) Solar distillation Pergamon Press, Oxford.

2. Clark. J.A. (1990) The Steady State Performance of a Solar Still, Solar Energy Vol. 44, pp43 - 49.

3. Chendo. M.A.C. and Schmitter E.F. (1982) Observed Perspex Covered Solar Still, Nigeria Journal of Solar Energy. Vol. 2. pp 33 - 46.

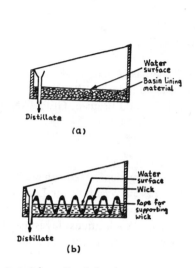

Fig·1: Schematic of the shallow-basin still.

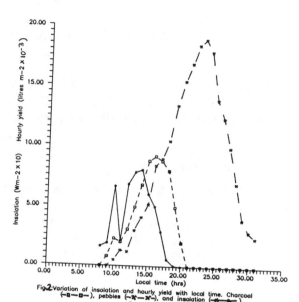

Fig.2:Variation of insolation and hourly yield with local time. Charcoal (—□—□—), pebbles (—✗—✗—), and insolation (—✱—✱—).

Fig.3:Variation of basin temperature with time for the control (solid line), pebbles (—▲--▲----▲), wick (—■—■—■) and the ambient temperature (✱——✱——✱).

SOLAR THERMAL DESALINATION PROJECT AT THE PLATAFORMA SOLAR DE ALMERIA

Eduardo Zarza*, J. I. Ajona**, J. León*
Dr. K. Genthner***, A. Gregorzewski***,

*Plataforma Solar de Almeria, Apartado 22, E-04200 Tabernas (Almeria) SPAIN
**CIEMAT-IER, Avda. Complutense 22, Edif. 42, 28040 Madrid, SPAIN
***Universität Bremen, Technische Thermodynamik, Pf. 330440, 2800 Bremen, GERMANY

ABSTRACT

Two research institutions, CIEMAT (Centro de Investigaciones Energéticas, Medioambientales y Tecnológicas), a Spanish, and the German DLR (Deutsche Forschungsanstalt für Luft- und Raumfahrt e.V.) are studying solar desalination of brackish and seawater in a Solar Thermal Desalination (STD) Project, part of the Industrial Solar Energy Applications Investigation Program at the Plataforma Solar de Almeria (PSA), a solar energy research center in Southern Spain.

The technical feasibility and high reliability of this industrial application of solar energy were proven during Phase 1 of this project. During Phase 2, which is presently underway, the solar desalination system is being optimized and improved to make it more competitive with conventional desalination systems. The addition of an absorption heat pump in June, 1991, will make the system twice as efficient.

KEYWORDS

Solar desalination systems, multi-effect distillation, absorption heat pump, parabolic trough collectors, industrial process heat

STD PROJECT OBJECTIVES

Project objectives are:

- study the technical and financial feasibility of this industrial application of solar energy
- optimize the solar desalination system by introducing and evaluating those improvements which could make it more reliable and competitive.

Tasks required to achieve these objectives were scheduled in two phases as follows:

Phase 1

- Implement and study a solar desalination system which integrates a conventional M.E.D. plant and the solar collector facilities at the Plataforma Solar de Almeria.
- Analyze desalination plant performance when run by a solar system under diverse operating conditions using this
  experimental data to develop desalination plant and solar collector field subsystem

models.
- Study possible efficiency and cost effective improvements in the solar desalination system.

Phase 2

- Implement the most cost effective improvements, according to Phase 1 results, into the solar desalination system.
- Analyze improved desalination system behaviour under diverse operating conditions and compare with theoretical phase 1 predictions.
- Develop solar desalination system (solar collector field, thermal energy storage system and M.E.D. desalination plant) simulation design software for applications with defined requirements (weather conditions, nominal production, storage medium, etc.).
- Reduce the cost of the solar system by optimizing solar components (collector field and thermal storage system), thus reducing the specific cost of the solar energy source, and increasing competitiveness.

DESALINATION SYSTEM DESCRIPTION

The solar desalination system implemented during Phase 1 of the STD project is composed of:

- A 14 cell or stage Multi-Effect Distillation (M.E.D.) plant
- A thermocline thermal storage oil tank
- A one-axis tracking parabolic trough solar collector field

These subsystems are interconnected as shown in Fig. 1. A synthetic oil heat transfer fluid (Santotherm 55) is heated by the solar collectors, where the solar energy is thus converted into the sensible thermal energy of the oil, and stored in the thermal storage tank. Further description of the M.E.D. plant, thermal storage tank and solar collector field follows.

DESALINATION PLANT

The STD Project "Sol-14" desalination plant, manufactured by the French company, ENTROPIE, S.A., is a Multi-Effect Distillation (MED) plant consisting of 14 vertically stacked cells or effects, in which horizontal tube bundles are sprayed with seawater at around 70°C for evaporation. At this temperature scale formation is limited, so that seawater need not be pretreated with acid. The only feedwater treatment is the injection of a scale inhibitor to avoid frequent acid cleanings. The Sol-14 plant design seems well adapted to small units (up to a few hundred m³/day output) and is highly efficient. (See process diagram in Fig. 1.).

In the fourteen cells of the plant evaporator body, temperatures and pressures decrease successively from cell (1) to (14). Vapour is introduced at 70°C into the first cell tube bundle and condenses as it is sprayed by feedwater. The latent heat released evaporates part of this feedwater at 67°C, 0.28 bar. This steam goes to cell (2), where it is also condensed in a tube bundle sprayed with feedwater, part of which enters the second cell and evaporates at the lower temperature/pressure of 64°C/0.24 bar. The same condensation/evaporation process is repeated in cells (3) to (14), the water condensed in each cell passes through a U-shape tube to the next cell, until the vapour produced in cell (14) at 33°C/0.05 bar is condensed in a final condenser cooled by seawater. Part of this seawater is used as feedwater for the MED process, being preheated from condenser outlet to cell (1) inlet in 13 preheaters. From cell 1, the seawater passes on from one cell to another by gravity before being extracted from cell (14) by the brine pump.

Cell tube bundles are made up of straight tubes connected at both ends to tube plates. Cell vacuum is maintained by a hydroejector system, consisting of two ejectors driven by seawater at 3 bar pressure from an electric pump operating in a closed tank cycle. The low pressure boiler is a heat exchanger which generates low pressure steam under vacuum (at approx. 70°C) using hot oil from the thermal storage tank.

Table 1. Technical specifications of the Sol-14 desalination plant

Nominal output	3m3/h
Heat source energy consumption	190Kw (aprox.)
Performance Ratio (Kg distillate/2300 KJ heat input)	10 (aprox.)
Output Salinity	50 ppm TDS
Seawater flow: at 10°C at 25°C	8 m3/h 20 m3/h
Feedwater flow	8 m3/h
Brine reject	5 m3/h
Number of cells	14

Solar Collector Field

The collector field consists of one-axis tracking East-West oriented parabolic trough collectors manufactured by ACUREX (USA), model 3001. The reflectors are thin glass (0.6-0.88 mm) with silvered backside, glued on flexible steel sheets (manufacutred by GLAVERBEL, Belgium). Other technical specifications of the collectors are:

Table 2. Collector specifications

Rim angle	90°
Geometrical concentration ratio	18.2
Absorber	Selective black chrome-coated steel tube receiver
Peak efficiency at noon vs. Operating temp. oil	180°C: 58% 220°C: 55% 290°C: 50%

Thermal Storage System

The desalination plant is connected to a thermal storage system which consists of a single thermocline vessel. The hot oil, acting simultaneously as a heat transfer fluid and heat storage medium, enters the tank, which is inertized by nitrogen, through an upper manifold. The thermal utilization factor (efficiency) is 92%.

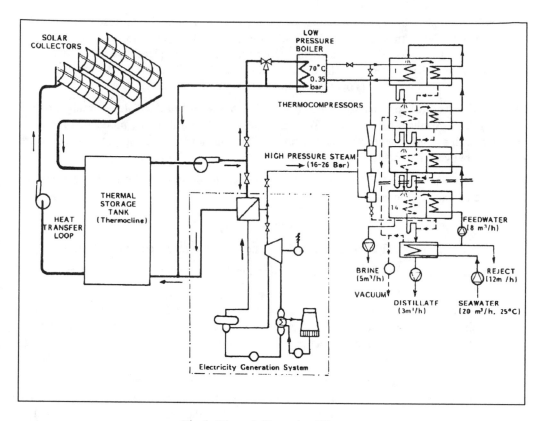

Fig.1, Phase 1 Operating Diagram

PHASE 1 RESULTS

This solar desalination system, tested and evaluated from 1988 through 1990, showed high operational reliability. Integration of the conventional MED plant into a solar system has not shown any major problems. The operating and maintenance requirements are quite similar to those of a conventional MED plant. Only small mechanical repairs have been made for the last two years.

The MED plant has shown low thermal inertia. It takes only 22 minutes to reach nominal steady distillate flow from the moment the oil starts entering the low pressure boiler. When the plant operation is changed over to different loads, it adapts to new conditions in less than 20 minutes. The low thermal inertia and low overnight vacuum loss make quick morning startup possible. It takes 35 minutes to reach the nominal production of distillate.

The plant performance ratio (P.R.) depends on the operating mode, of which there are two:

Boiler mode: The hot oil is sent from the storage tank to the low pressure boiler where vapour is produced at 70°C. This vapour enters the cell 1 evaporator tube bundle where it condenses, providing the multi-effect distillation process with the required thermal energy. (See Fig. 1). When the plant is operated in boiler mode, the P.R. is between $9.3 < P.R. < 10.7$, depending on how dirty the evaporator tube bundle surfaces are.

Thermocompressor mode: A small fraction of the high pressure (16-26 bar) steam produced for the PSA Electricity Generation System turbine and generator, is introduced into the two desalination plant thermocompressors, where it is mixed with the low pressure steam (0.05 bar) from cell 14, thus delivering 70°C steam into the cell 1 evaporator tube bundle where it is condensed, producing the thermal energy required by the plant. (See Fig. 1). This condensate is then mixed with the distillate in cell 2. The P.R. is within the range of $12 <$ P.R. < 14 when the plant is operated in the thermocompressor mode.

The specific electrical consumption varies from 3.2 to 5 Kwh/m³ of distillate, depending on the operating load (distillate production). Test results also show that distillate production increases with increasing final condenser temperatures, when operating under a given temperature difference between first cell and end condenser.

Phase 1 data were used by the University of Bremen to develop a Sol-14 Plant computer modelling and simulation code.

PHASE 2

The Phase 1 study of possible solar desalination system improvements concluded that:

-    The desalination process itself could be improved, thus achieving greater efficiency and lower specific costs.

-    The importance of the cost of the solar system (collectors and storage system) in the total specific cost of distillate water production, is very high. Therefore, any optimization of the solar system will have a direct repercussion on the final specific cost of the distillate produced by the system.

The desalination process may be made more efficient by recovering the approx. 110 KW thermal energy currently wasted by rejecting part of the seawater used to cool the final condenser, where the temperature increases to 35°C. For the recovery of this waste heat, a double-effect absorption heat-pump was considered the best possibility. The heat pump, coupled to the first cell and final condenser of the desalination plant, will receive 90 KW, 180°C input thermal power from the solar field and recover 110 KW from the final condenser, thus delivering 200 KW thermal power at 65°C and 0.25 bar to the first cell of the desalination plant (See Fig. 2).

Another improvement will be the replacement of the present electric pump/hydroejector vacuum system by steam ejectors and a new boiler, reducing the electrical consumption by more than 50%. Fig. 2 shows the operating diagram of the improved solar desalination system which will be implemented in June, 1991.

A new low cost Spanish parabolic trough collector was designed simultaneously with Phase 1 development. The first collector prototype was finished in December, 1990, and is currently being tested at the PSA facilities.

All the above mentioned improvements could reduce the specific cost of distillate produced by the optimized solar desalination system to about $3/m³ of distillate, assuming a lifetime of 15 years and an annual interest rate of 15%. Final Phase 2 report is due for December, 1992.

CONCLUSIONS

The coupling of an MED desalination plant to a solar system composed of a parabolic trough collector field and a thermocline storage tank, has been proven feasible. A specific heat consumption of about 30 Kwh/m³ distillate and a specific cost of $3/m³ distillate seem to be achievable by adding an absorption heat pump to the MED plant. A new low-cost parabolic trough collector has been designed to work at temperatures of about 250°C. An installation cost of 25,000 Pts./m2 aperture area could be possible with some improvements in this prototype.

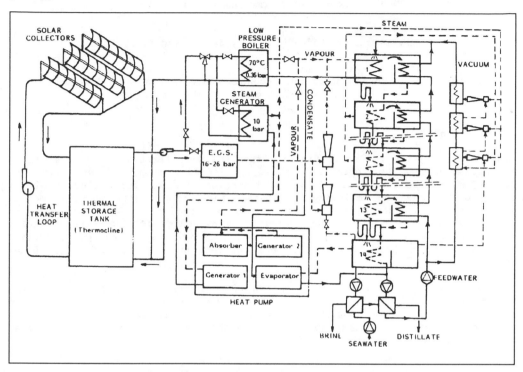

Fig. 2. Improved System Operating Diagram

ACKNOWLEDGEMENTS

The authors would like to express their gratitude to the PSA maintenance and instrumentation teams for their help in the installation and operation of the system. Thanks are also due to our CIEMAT colleagues for their technical advice and help, as well as to the University of Munich for its work on the design of the heat pump unit. We also want to thank ENTROPIE, S.A. for its continuous help and support during the development of this project. And last, but not least, the authors would also like to thank Mrs. D. Fuldauer and Mrs. C. Montesinos, who assisted in the preparation of the manuscript.

REFERENCES

Hömig, H.E. (1978). *Seawater and Seawater Distillation*. Vulkan-Verlag, Essen.

Lior, N. (1981). *Principles of Desalination*. Solar Desalination Workshop, Denver.

A Novel Solar Sterilization Water and Distillation System:
Experiment and Thermodynamic Analysis

C. Schwarzer, M. Wiedenfeld, N.K. Bansal and K.-H. Ertl

Energy and Environmental Group, Fachhochschule Aachen, Jülich Campus
Jülich, Federal Republic of Germany

ABSTRACT

A novel solar system for water Sterilization and distillation to procure drinking water has been developed, fabricated, monitored and thermodynamically analyzed. The unit is essentially a modified version of flat plat collector with a
heat recovery unit and employing transparent insulation as the glazing. Measurements performed under German summer conditions showed that for days with clear
sky, it was easy to obtain fluid temperatures exceeding 100 °C. Experimentally
sofar the V/Q-value (volume of the distillate / incident solar energy) increased
from 0.20 .1/MJ to 0.36 1/MJ with a heat recovery unit for a 1.7 m^2 collector
area.

A thermodynamic analysis of the system shows that the expected distillate from
the system depends very sensitively on the steam fraction in it. With improved
heat recovery unit it is possible to decrease the steam fraction and achieve a
much higher distillate output up to a theoretical maximum of 66 1/m^2 day and an
incident solar energy of about 21 MJ.

KEYWORDS

Solar sterilization; distillation; transparent insulation; heat recovery unit,
coefficient of performance

INTRODUCTION

Solar distillation has been one of the most investigated technology in the field
of solar thermal applications because the United Nations declared the decade
1980 - 1990 as the need for drinking water in developing countries. A large number of simple solar stills have been developed today. A description of these
units is contained in Malik et al (1985) (2). One of the drawbacks of all these
units is their lower efficiency and hence a lower distillate output per m^2 solar
collection area besides maintenance and the cost. Recently there have been
advance developments (3, 4); these units however require electrical power and
are therefore not suitable for regions where electricity is not available.
Besides it is not ensured in all these developments that water gets boiling and
the distilled water does not contain any microorganism. Bansal et al (1) developed a system in which water boiling was ensured. The efficiency of this system
was however restricted up to 50 %. By simple modifying a flat plat collector and
incorporating a waste heat recovery system, it has been possible to get a much
higher distillate simultaneously ensuring no microorganism in the distilled
water. A detailed thermodynamic analysis of the system has also been carried
out. The description of the system, results of measurements and the achievable
performance based on the theory have been presented in this paper.

SYSTEM DESIGN

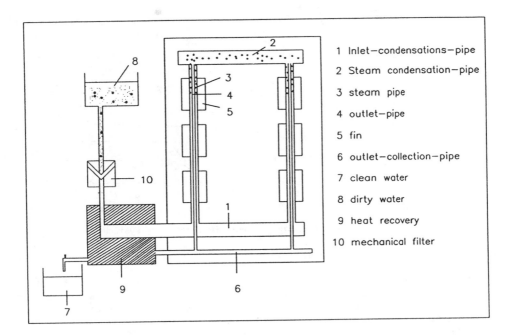

Fig. 1: Overview of the System

The system shown in Fig. 1 is a modified version of a flat plate collector with double coaxial riser tubes and a heat recovery unit at the inlet for recovering the latent heat of evaporation and the sensible heat of the condensed water. The glazing system in the collector consists of two 5 mm thick glass separated by a distance of 10 cm to accommodate translucent polycarbonate as honeycomb material, the so called transparent insulation(5). The coaxial tube first itself functions as a heat recovery unit and then an additional heat exchanger allows the recovery of the remaining waste heat, which is utilized to preheat the inlet water into the collector.

The water level in the riser tube -3- is controlled by the level of the unsterilized water in the container connected to the inlet of the collector. The incident solar energy is absorbed by the absorber and transferred to water in the riser. Water gets heated, vaporizes and the water vapors rise to the collecting channel -2-. The vapors could pass through the inner coaxial tube -4- and a part gets condensed; the heat of condensation gets partly transferred to water in the inner tube and the rising water flow in opposite directions thus allowing an effective heat exchange, which is used by the heat recovery unit -9- at the inlet of the collector.

EXPERIMENT

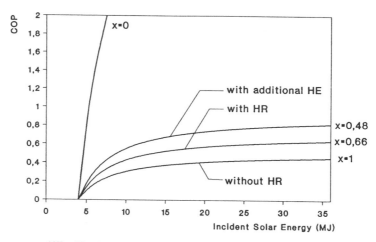

HE = Heat exchanger   HR = Heat recovery system

Fig. 2:   Theoretical curves of destillate output for different systems

The experiments on the unit were performed during summer days at Jülich in Germany (latitude 51°55'N). Fig. 2 shows the obtained distillate output for different incident solar energy. It is to be noted that the collector system itself requires an energy of 4 MJ for heating before evaporation starts. The temperature of water in the collector was recorded to be 120°C.

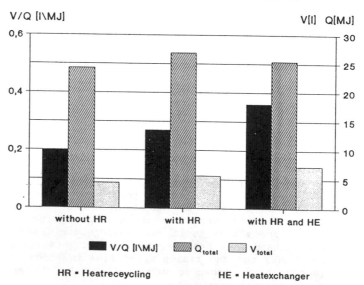

HR = Heatreceycling   HE = Heatexchanger

Fig. 3: Comparisation of solar purification system with and
without heat exchanger and heat recovery systems

Fig. 3 shows the values of the incident solar energy for three cases
1. Only collector unit allowing the water vapor to condense near the top of the collector
2. Collector unit without heat recovery at the collector inlet allowing to assess the influence of coaxial tubes
3. Collector with coaxial tubes and heat recovery unit at the inlet

The results of the experiments are summarized in Fig. 3. It is observed that the distillate output increased from an initial value of 0.20 1/MJ without any heat recovery unit to 0.36 1/MJ with the heat recovery system of the designed unit.

THERMODYNAMIC ANALYSIS

The performance of the water sterilizer can be estimated in terms of the actual quantity of the distillate and by making a comparison of this quantity with maximum theoretical achievable purified water. Naming this quantity as the coefficient of performance (COP) mathematically defined as

$$COP = \frac{m_c}{m_{th}} \tag{1}$$

with

$$m_c = \frac{\text{useful energy}}{\text{spec. energy to get 1 kg destillate}}$$

$$= \frac{Q_{useful}}{q_{dest}}$$

and

$$m_{th} = \frac{\text{Incident solar energy}}{\text{spec. theoretical energy}}$$

$$= \frac{Q_{sol}}{q_{th}} \tag{2}$$

$q_{collector}$ and $q_{theoretical}$ are given by the following expressions

$$q_{collector} = c_w \Delta T + xr \tag{3}$$

$$q_{theoretical} = c_w \Delta T + r \tag{4}$$

Useful energy from the collector that is available for water purification

$$Q_{useful} = \int A(\eta_0 R(t) - U_L(T_c(t) - T_a(t)))dt - Q_0 \tag{5}$$

The incident solar energy on the collector area is the integrated solar insolation over the entire period of system's use

$$Q_{solar} = \int_0^t A R(t)dt \tag{6}$$

Defined by Eq. (1), the COP therefor can be written as

$$COP = \frac{Q_{useful}}{Q_{sol}} * \frac{q_{th}}{q_c}$$

$$= \frac{\int_0^t A(\eta_0 \, R(t) - U_L(T_c(t) - T_a(t)))dt - Q_0}{A \int_0^t R(t)dt} * \frac{c_w\Delta T + r}{c_w\Delta T + x \, r} \quad (7)$$

Without any heat recovery system COP is simply the system's efficiency defined as

COP(without heat recovery unit)

$$= \eta_s = A \int_0^t \frac{(\eta_0 \, R(t) - U_L(T_c(t) - T_a(t))dt - Q_0}{A \int_0^t R(t)dt} \quad (8)$$

The COP therefor with heat recovery unit can be written as

$$COP = \eta_s * \frac{c_w\Delta T + r}{c_w\Delta T + xr} \quad (9)$$

It is clear that the case x=1 corresponds to a system without heat recovery and the case x=0 corresponds to the maximum possible efficiency. The actual results will lie between these two values.

RESULTS AND DISCUSSION

Eq. (9) clearly shows that the quantity of distillate depends very sensitively on the steam fraction x in it. Theoretically expected COP and the distillate output for different values of steam fractions are shown in Fig. 2 and 3 respectively. The detailed experimental results with the theoretical expectancy curve are shown in Fig. 4.

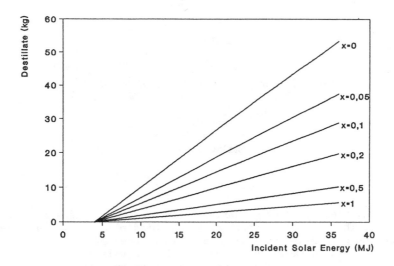

Fig. 4: Destillate dependency on solar energy and the steam fraction

The results clearly show that one should try to improve the heat recovery system as much as possible rather concentrating on further improving the collector's efficiency.

Nomenclature:

q_{th}	=	$c_w * \Delta T + r$	(J/kg)
q_e	=	$c_w * \Delta T + x*r$	(J/kg)
Q_0	=	Threshold energy at which evaporation takes place	(J)
Q_{sol}	=	Incident solar energy	(J)
R	=	Average intensity of solar radiation	(W/m²)
r	=	Latent heat of vaporization	(J/kg)
x	=	Steam fraction	(-)
c_w	=	spec. heat capacity of water	(kJ/kg K)
U_L	=	Collector's heat loss coefficient	(W/m²*K)
T_c	=	Average collector's temperature	(K)
T_a	=	Average ambient temperature	(K)
η_0	=	Optical efficiency	(-)
η_s	=	System's efficiency without heat recovery unit	(-)

REFERENCES

1. Bansal N.K. (1986) Misra Anil, şolar sterilizer, Solar Energy
2. Hesel O. (1987) Desalination with Renewable Energy,
   Ph. D. thesis, Technical University München
3. Malik M.A.S. et al (1985), Solar Distillation, Pergamon Press,
   New York
4. Delfnis A., Solar Distillation, Greece (1986)
5. Platzer W., Solar transmission und heat transfer effects in transluzent
   thermal-isolation materials, Dissertation, institut of physics,
   Albert-Ludwig-Universität Freiburg (1988)

FEASIBILITY OF SEA WATER DESALINATION USING WIND
ENERGY RESOURCES IN THE MIDDLE EAST

Essam Mitwally
Executive Director
African Regional Centre for Solar Energy

ABSTRACT

Economic development of many countries is impeded by the lack of fresh and safe water supply. The problem is critical in the Middle East in areas where only sea and/or brackish waters are the only available sources of water. Large desalination schemes are already deployed, but serving relatively large communities in places like Saudi Arabia, Kuwait and Libya. Smaller communities, on the other hand, still suffer due to water shortage. Small to medium size desalination plants are already on the world market, but it is the lack of adequate and reliable energy supply which is at the core of the problem. In areas where enough wind energy is available, wind-powered systems can solve this problem. This study addresses two sizes: one scheme providing daily output of 3000 ton of fresh water and another stand-alone desalination system producing 100 - 150 ton per day. Using the wind power resources available in the Middle East, the two schemes are shown to be economically feasible under various financial conditions.

KEYWORDS

Desalination; seawater; brackish; Middle East; wind; wind-farm; stand-alone.

INTRODUCTION

The development of many areas in the world, particularly in the Middle East, is impeded by the lack of adequate water supply for drinking, agricultural, industrial and touristic purposes. Seawater and brackish waters are available but need to be desalinated prior to the final usage. In some areas water having up to 1800 ppm of salt is being used with adverse results on the people´s health, livestock or agriculture. Some States have supplied potable water to their larger settlements using complex-fueled desalination plants, for example, Saudi Arabia, Kuwait and Libya. However, the used technologies are not readily adaptable to the needs of small remote communities at reasonable economic scales, let alone their dependence on depletable fossil fuel resources and the need for skilled operation and continued maintenance. Small to medium-scale desalination technologies for seawater and for brackish water have advanced rapidly during the last decade, and their outputs are in the ranges needed for small to medium size communities of 1000 to 50,000 people, depending on the use of water. Systems producing around 100 to 3000 ton per day of fresh water will not be uncommon. The final decision regarding using small, medium or large-scale desalination for any of the above purposes depends in the end on the economic feasibility of utilization taking into account the cost of importing the required water to the same sites.

This study focuses on communities which are either in-land running agricultural, industrial or mining settlements, or on the coast engaged in fishing or tourist activities as Summer or Winter resorts.

Desalination Market

Detailed studies of the desalination market (Fluor, 1978; Gerofi, 1982) divided the world into two areas – the U.S., and "overseas". The demand outside the U.S. is dominated by the Middle East, and by Saudi Arabia in particular. Saudi Arabia had expanded its desalination capacity extremely rapidly during the last decade, with reported annual expenditures of $1.5 billion. The Gulf crisis of 1990/91 had imposed more conditions on the construction of large scale desalination plants to replace the damaged plants in Kuwait and the demand in Libya requires doubling of the present desalination capacity by the year 2000. Therefore, it is estimated that the installed capacity in the Middle East will grow by 10 to 15% per annum till the year 2000.

In the Red Sea and Arabian Gulf , the waters are extremely saline (over 44,000 ppm), and in the Mediterranean Sea this reaches 37,000 ppm. Also, the ground-water is quite saline in some places, and its salinity rises if there are excessive withdrawals. The latter phenomenon had been observed in Bahrain, Tripoli in Libya and Matruh in Egypt.

The cost of importing water to the sites concerned can also be an obstacle. For example, at Abu Ghossoun on the Red Sea in Egypt, this cost reached $20 per ton (Mitwally, 1986) representing mostly transportation cost. Even the water quality had been reported to be inadequate depending on the cleanliness of the transportation tanks which might had been transporting oil on a previous errand.

Desalination Technologies

The main technologies employed for desalination are classified under:

Evaporation. The main source here is thermal energy, and the specific energy needed per ton of fresh water is high particularly for seawater desalination. The advantage of this old technology is its flexibility regarding the source of thermal energy. Desalination could take place under pressure or vacuum. Vacuum vapor compression was determined to be the most appropriate method considering specific energy consumption; however, its operational and maintenance requirements make them less competitive with reverse osmosis systems at this time.

Membrane technologies. Two types of technologies employing membranes are popular. Reverse osmosis (RO) is the more popular one, where the source of energy is mechanical power, although the pump is usually driven by an electric motor. The specific energy consumption could be reduced by over 60% using energy recovery systems. For high salinity feedwater, pretreatment is usually necessary, and the membranes utilized need to be changed at some regular intervals. DC or AC motors can be used. The other type is "electrodialysis" which is most suited for low salinity feedwater like brackish water. The source of energy is DC power, and the modified systems of "electrodialysis reversal" (EDR) cut down on the maintenance of electrodes. As in RO systems, membranes have to be changed regularly but at longer intervals.

Other technologies also exist, like freezing, but are neither popular nor economical.

SOURCES OF ENERGY

Remote areas, where small scale desalination systems are needed, suffer from shortages of reliable sources of fuel. Moreover , in the oil-importing States, using conventional energy sources would be a real burden on their economies. Fortunately, in the States needing desalination the most, solar energy is abundant, and coastal areas are blessed with moderate to high wind regimes. It should be noted that if good wind potential exists at one place, even with high insolation, then it would be cheaper to use wind rather than solar technology (Mitwally,1986). In addition, when dealing with water production, the excess energy produced at peak wind velocities can be used for more water production or storage, and there will be no need for the expensive means of energy storage.

Wind Energy Potential

The World map on wind energy distribution (WMO, 1981) shows that coastal areas in the Arabian Gulf have wind potential in the range of 5 - 6 m/sec average wind speed producing some 160 - 275 Watt/sq.m at a height of 10 m. At the hub heights of 30 to 50 m wind generators will produce even more power of 300 to 550 Watt/sq.m. Muntasser (1984) and El Osta (1990) showed that some of the coastal areas in Libya have wind potential of up to 8 m/sec average wind speed producing around 1100 Watt/sq.m at a height of 10 m. At 50 m hub height these reaches 1600 Watt/sq.m. Therefore, there is no question about the presence of suitable wind regimes in the Middle East. A study by the United States Department of Energy (US DOE) concluded that electricity generation using wind turbines at Ras Ghareb in Egypt, where the wind conditions are similar to those at Derna in Libya, would be economical, on which basis a 4-turbine wind farm was constructed. The project is being evaluated at present to assess its technical performance and economical feasibility.

The Wind Energy Configurations

The large desalination scheme of 3000 ton/day will require daily consumption of energy up to 40,000 kwhr which includes around 5% extra for parasitic powers here and there. The available energy can only be considered once wind machines are selected and after matching their performance characteristics with the available wind regimes. This has been done for available 100 Kw, 200 Kw and 250 Kw machines. When an order is placed with suppliers, they will also match the performance of their machines with the supplied wind data, but very little difference will be expected. On this basis, this study recommends the following wind power generators: either 16, 20 or 40 wind turbines, each rated at 250, 200 or 100 Kw, respectively, operating in a wind regime of 6 m/sec annual average speed measured at a height of 10 m. Such machines exist with the following design characteristics and cost data as given in Table 1 (Mikhail, 1981).

The small-scale desalination stand-alone scheme of 100 - 150 ton/day is a straightforward corollary of the above. Around 1500 Kwhr will be needed every day for the operation. This can be easily obtained from one 200 Kw machine or two 100 Kw machines.

ECONOMIC CONSIDERATIONS

The total cost of water from a desalination plant must include the capital and the energy equipment charges, maintenance and operation charges and cost of supplies. The latter cost for an RO plant includes an amount for membrane replacement which is periodic, but not annual, and depends greatly on the pretreatment of the feed water, its salinity and periodic back washing of the membranes. Membrane replacements could be as long as five years with proper operation and maintenance.

Capital Charges

The desalination plant. There are numerous estimates of desalination plant capital costs which are updated periodically. This study focuses on desalination plants using RO with energy recovery because these had been judged to be the most suitable for seawater desalination with reduced energy consumption. For the sizes of interest in this study, i.e. 3000 ton/day and 100 - 150 ton/day, the total installation cost varies between $1200 and $1500/ton/day, respectively. Table 2 gives RO cost data excluding energy costs.

Energy Equipment Cost

It is obvious that for the large desalination scheme of 3000 ton/day, the choice of 16 wind machines rated at 250 Kw with towers will cost less ($4,400,000) as compared to 20 wind machines rated at 200 Kw with towers ($6,000,000). In addition to that cost, the electric network with 10% diesel backup will cost about $1,600,000, bringing the total energy equipment cost to $6,000,000.

For the small-scale desalination scheme of 100 ton/day, the selection of one 200 Kw wind machine ($300,000) will be cheaper than two 100 Kw machines ($350,000). Due to the small difference in costs, and for the sake of reliability of operation, the second choice of two 100 Kw machines is highly recommended. In

TABLE 1. Wind Machine Design Characteristics and Cost Data

	250 Kw	200 Kw	100 Kw
Rated power at Wind Speed	13.4 m/s	10 m/s	11.1 m/s
Configuration			
Blades	2	2	3
Axis	Horizontal	Horizontal	Horizontal
Orient	Down-wind	Up-wind	Down-wind
Cut in speed	4 m/s	4.2 m/s	3.5 m/s
Cut out speed	26.8 m/s	17.8 m/s	26.8 m/s
Weight in kg	2750	2200	1800
WTGS cost	$300,000	$250,000	$160,000
Rated rpm	94	40	50
Blade material	Fiberglass	Aluminum	Aluminum
Speed control	Mechanical pitching	Mechanical pitching	Stalling
Alternator/gen type	Induction generator	Synchronous	Induction 1800 rpm
Tower	tubular	tubular	tubular/lattice
Cost of tower	$50,000	$50,000	$50,000/$15,000
Total Cost	$350,000	$300,000	$210,000/$175,000

this case there shall be no need for diesel backup except for parasitic loads during no wind conditions. The energy available, however, would be enough to produce 150 ton/day.

Table 2. Seawater Desalination Costs for RO Systems with 30% Recovery

Production	100-150 ton/day	3000 ton/day
Capital costs		
Plant	$150,000	$3,600,000
Installation	$30,000	$750,000
Intake and Outfall	$10,000	$300,000
Subtotal	$190,000	$4,650,000
Engineering/management	$9,000	$250,000
Contingency	$10,000	$500,000
Total	$209,000	$5,400,000

Annual Costs

The annual costs are based on 15 year plant life, excluding membranes. Table 3 gives the annualized costs for the selected options. It is assumed that the small-scale desalination system will be purchased without recourse to financing, and thus there will be no capital charges. For the large-scale desalination system it is assumed that $3,000,000 would be available, and the remaining $8,400,000 will be financed. However, for the sake of completeness, financing of the total required capital of $11,400,000 will also be considered.

For the large-scale desalination scheme, and with the above arrangement, the cost of producing the water would be around $1.5 /ton. If the water is sold at price of $6/ton, the payback period will be around 31 months. For lower priced water, the payback period would, of course, be longer. For example, if the water is sold at $5/ton, the payback period would be 40 months. In addition to the base case, calculations were performed for a few cases, and the results are shown graphically in Fig. 2 which gives the payback period in months.

Table 3. Annual Costs for the Two Desalination Schemes

	100/150 t/d	3000 t/d (partial financing)	3000 t/d (total financing)
Capital charges	-----------	$1,104,380	$1,498,801
Operation costs	$6,000	$30,000	$30,000
Administration	$4,000	$12,000	$12,000
Chemicals	$6,000/$9,000	$90,000	$90,000
Maintenance	$3,000	$18,000	$18,000
Membranes	17,000/$25,000	$250,000	$250,000
Total	36,000/$47,000	$1,504,380	$1,798,800

At the water selling price of $6/ton, all of the cases give, more or less, the same result. Since a $5/ton selling price would be more competitive, then it may be worthwhile to look into the case of total financing. If the capital cost for the large-scale desalination scheme cannot be raised all at once, due may be to some skepticism on the part of the investors, then the project can be implemented in phases. This fits beautifully with the wind-powered scheme, because there shall be at least 16 wind machines of 250 KW rated capacity each. These machines will be interconnected to form a local electric grid, or a "wind-farm". The desalination plant may also be modular with parallel connected units.

For the small-scale desalination scheme and with the above arrangement, the water produced will be 150 ton/day for a total capital investment of $909,000. At a selling price of $5/ton, the payback period will be 54 months. If the water is sold to touristic fishing boats at the present rate of $40/ton, then the investment is recouped in less than half a year only. Obviously, the true situation will be between these two extremes. Either case, the economic feasibility has been proven.

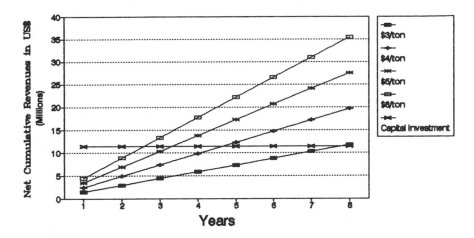

Fig. 1. Revenues variation with water selling price.

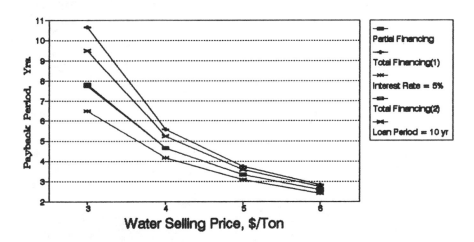

Fig. 2. Payback period vs. water selling price.

REFERENCES

Fluor Engineers and Constructors Inc.(1978) Desalting Plans and Progress and Evaluation of the State of the Art and Future Development Requirements. 2nd Edition. NTIS # PB259 272.

Gerofi, John P.(1982). The Technical and Economic Feasibility of Solar Desalination. Tech. Report. Dept of Chemical Engineering. University of Sydney. N.S.W. Australia.

Mitwally, Essam.(1986). Economics of Solar/Diesel Desalination on the Red Sea. The Energy and Water Magazine of Egypt. Issue VI.

WMO (1981). Meteorological Aspects of the Utilization of Wind As An Energy Source. Technical Note No. 175 (Also WMO-No.575).

Mikhail, Amir S.(1981). Wind Power for Developing Nations. SERI/TR,762-966.

Muntasser, M. A.(1984). Promising Role for Wind Energy in Libya. The Libyan J. of Energy. Vol. 1. Issue No.1.

EL Osta, Wedad and Fahima Al Shakshouki (1990). Wind Energy Applications: The Libyan Experience. Workshop on "Women and NRSE". Tripoli. Libya.

2.28 Solar Ponds

MICRO-FILTRATION TECHNIQUES FOR EFFECTIVE USE OF BITTERNS IN SALT GRADIENT SOLAR PONDS

W.W.S. Charters; R.W.G. Macdonald; B.P. Marett; D.R. Kaye

University of Melbourne, Parkville, Victoria 3052, Australia

Abstract

If solar ponds are sited close to sea water based salt works, as was the case for the Laverton solar ponds in Australia, bitterns is available as a low cost waste product from the salt production process. This sea water concentrate residue is also preferable for use in solar ponds because it has a higher density than saturated sodium chloride solutions and, in addition, includes a substantial percentage of magnesium chloride.

In many inland Australian locations it is possible to find underground bitterns under salt lakes and salt pans providing, therefore, an ideal source of salt and water for solar pond construction.

With the Laverton bitterns ponds a substantial difficulty experienced over several years involved algal contamination leading to solution coloration and reduced pond transparency.[1]

Chemical cleaning techniques were extensively tested and ultimately discarded as the stratified nature of the ponds was not conducive to even distribution of chemicals. In practice the high ionic concentrations of the bitterns pose further difficulties for chemical treatment. After extensive laboratory and field tests of various filtration methods ultra filtration, using micropore filter elements, has been shown to be both practical and cost effective in bitterns clarification.

Key Words

Bitterns, solar ponds, algal control, flocculation, microfiltration.

Bitterns and Water Clarity

At the Laverton site the bitterns used to build the salinity gradient is vivid pink in colour due to the presence of carrotene. This coloration is caused by the alga Dunaliella salina and other minor unidentified algae. Some properties of this Laverton bitterns are detailed in Appendix 1(i) and Appendix 1(ii) Table 1.

A major reason for using bitterns rather than normal sodium chloride solution is that it provides a high density solution enabling maintenance of a stable density gradient in the solar pond. Low cost and wide availability are additional reasons for use of this seawater concentrate residue but little operational experience has been obtained with bitterns solar ponds. It is evident, however, from the limited published material available that treatment is required to improve the optical transparency of the bitterns solution before use to ensure adequate thermal performance of the solar pond in practice.[2]

Chemical Cleaning Techniques

Many chemical techniques exist to clean natural waters but none of these are directly applicable to bitterns.[3] Extensive chemical tests were conducted during the period of the Laverton project 1980 to 1988. A range of chemical flocculation methods was tried and discarded on the basis of insufficient optical transparency achieved and the high cost of the materials involved. It was definitively concluded from these tests that standard chemical water treatments are inadequate for bitterns clarification. Only very high concentrations of some chemicals such as hydrogen peroxide (1-2%) and alum (200-300 ppm) provided near to acceptable levels of water clarity and this was at an unacceptable and uneconomic cost - see Figure 1.

All of the optical clarity measurements obtained during the water clarification programme were obtained using an UV-Visible wavelength spectrophotometer with distilled water used in the reference cell. The absence of any peaks on the spectrophotometer traces confirmed the supposition that the lack of transparency of the bitterns solutions was due to the presence of particulate matter.

Filtration Techniques

Following on from the failure of the chemical treatments it was decided to undertake filtration experiments. Ultra filtration is required as, although the algae can grow to 20 microns in diameter, a large fraction of the algal population is below the filtration capacity of common sand and diatomaeceous earth filter (15-20 microns). Various multiple bed filtration systems were tested but without success and following investigation of filtration systems with effective filtration diameters of several microns close attention was focussed on ultra filtration techniques.

Laboratory ultra filtration Tests

After testing and discarding the use of fibrous rope filters of 10 micron and 1 micron nominal ratings because of the residual colour of the filtrate the filter type finally chosen consisted of fibrous glass straw filters encased in a canister and with a filtration range of 0.2 micron absolute. The filter canister had an effective area of $2.7m^2$ ($26ft^2$).

Water is passed through the straws and the tangential pressure across the glass walls leads to the transport of filtered water across the membrane. The concentrate is then returned to the prefiltered bitterns storage tank. Material which collects on the glass walls is sheared off by the flow, by reversing the flow direction, or by back flushing across the glass wall. Chemical flushing of any built-up material can also be usefully employed as a cleaning technique if required.

This filtration system was found to be very effective in removing all coloured material and in producing crystal clear water using a 0.5 micron filter element. No other clarification technique trialled had provided similar clarity - see Figure 2.

At lower wavelengths the optical transparancy of the filtrate sample was sensitive to the pumping pressure with the filtrate obtained at higher pressure showing a decrease in transparency.

The filtrate flow rate, as indicated by the laboratory tests, gradually decreases with time. From the data obtained the indication is that filter operation may exceed 20 minutes before any back flushing becomes necessary.

Ultra Filtration Field Tests

A single filter canister of $2.7m^2$ ($26ft^2$) surface area was first installed at the field test site. As the filtrate flow was low the influence of operating pressure and bitterns concentration on flow rate were investigated. The pressure and back pressure of the filter were varied within the range recommended in the operational manual and it was discovered that the flow rate is relatively insensitive to pressure.

The inlet bitterns concentration was next varied within the density range 1.18-1.32 tonne/m^3- the lower value being the minimum possible for setting up a stable salinity gradient. The field test data illustrate a slow increase in filter flow rate with decreasing salt concentration (measured at 18°C).

The tests with the single filtration canister were then stepped up to larger scale field trials to enable a full investigation of this system. To operate such a filtration system in a commercial solar pond, several canisters would be required and these would be operated automatically to maximise the effective filtration rate of the system. Three additional filter canisters were purchased and installed in-line with solenoid valves and electronic control gear to allow reversible flow through the filters. The control gear, which was built in-house, comprises a timing circuit which opens two valves and closes two others - see Figure 3. After each twenty minute period the flow direction was reversed. The filtrate was then passed into a 10m diameter, 1m deep inspection tank. It was then passed, under the influence of gravity, into the solar pond via an injection diffuser set at the appropriate height.

Several days of continual filter operation were without any apparent build-up of algal or particulate matter on the glass walls. This indicated that these filters and the filtration system were suitable for field operations - see Figure 4. After this period of operation, a malfunction in the sandprefilter led to the filter operation being postponed. Other delays such as the loss of the three phase power to the pump due to an earth leakage problem, prevented continuance of the filtration program.

It can be concluded however that the actual glass fibre filters would have sufficient life to be economic in the cleaning of solar pond bitterns. The first filter canister purchased has been operated over a period of more than 250 hours without a significant increase in the pressure drop across the fibres and without any other major signs of deterioration in performance. The systems filtrate production rate of 10 litres per minute is slow for filling solar ponds the size of the Laverton solar ponds each of 1000m^2 surface area. However, if operated for twenty four hours a day, filtration times of forty to sixty days would provide enough clean water to fill the solar pond. Since bitterns filtration is a once-off requirement, such time scales would normally prove to be economically feasible. A large number of canisters would decrease the period of filtration but add significantly to the overall filtration costs.

Once clean filtrate were available, the filtration system could routinely be used to maintain water clarity. By the addition of acid to the clean water, algal growth in the bottom convective or the gradient zones of the pond can be inhibited. However algae will re-establish themselves in the surface convective zone where the temperatures and salinities are most favourable for their growth. The use of standard swimming pool or ship board chlorinators to regularly chlorinate this water could help to reduce algal growth.

It has been observed at Laverton that the algae will tend to penetrate into the upper most centimetres of the gradient zone where the temperature fluctuations are less, and conditions are possibly more suitable for algal growth. Under favourable conditions, a layer of algae may form in the statified zone as well. In these operational situations, the water is stratified and is therefore not available continuously or routinely for extraction and chlorination. In this event it is possible to temporarily extract a thin layer of algal continimated water and to move it to a storage area where it can be treated by the filtration system. This water once clean can be returned to its position in the pond. For large ponds such a filtration practice would reduce the overall pond salt requirements and with clearer water, higher thermal efficiencies would result.[4]

Conclusions

Standard chemical treatments have been found to be ineffectual in controlling the water clarity at the Laverton bitterns solar ponds in Australia. All chemical techniques trialled proved to be uneconomic in terms of the high level of chemical concentrations required, which also impose major chemical waste disposal problems and high transport costs.

A major success of this project has been the design and application of a suitable and cost effective filtration system for bitterns treatment. This system is based on commercially available glass fibre filters designed for ultra filtration.

Economic viability of any filtration system will of course vary from site to site. However, as the system is light and compact transportation costs are not likely to be excessive even in very remote locations such as in Central Australia. This is in sharp contrast to the substantial costs incurred with the use of chemical treatment systems.

Appendix 1[5]

Biological and Chemical Testing of Laverton Bitterns

(i) **Particulate Matter in Water Samples**
Microscopic examination of water samples collected April 5, 1986, resulted in the following laboratory analysis:

1) **Bitterns storage pond.** A large amount of background particulate material, particles predominantly 5 μm diameter, although some particles larger than 5 μm. These particles were unidentifiable, and it was not possible to perform a particle count.

Also present were flagellated algae, cell length 10-20 μm, at least two species including *Dunaliella.* Algal count: 7300 cells/ml.

2) **Bore water sample.** Lesser amount of background particulates than sample 1, similar size but tending to be consolidated into clumps rather than dispersed through the sample. Algal count: *Cymbella* -like diatoms: 400 cells/ml.

3) **Channel water sample.** Less background material than both samples 1 and 2. Algal count: Diatoms only (*Synedra, Gyrosigma,*) 400 cells/ml.

4) **Lined pond sample, 1.5 metre height.** Sample contained a large amount of small particulates, predominantly <5 μm diameter, some larger. No healthy algae were detected, although some particles which could not be positively identified may have been dead and distorted algae or flagellated protozoans.

(ii) **Description of the Laverton Bitterns**
The chemical composition of the bitterns, bore water and channel water are given in the table below;-

TABLE 1

EXAMINATION	SAMPLE 1 BORE	SAMPLE 2 CHANNEL	SAMPLE 3 BITTERNS
Total Organic Carbon	4	8	130
Suspended Solids	23	19	1,500
Nitrate as Nitrogen	0.09	0.05	0.01
Organic Nitrogen	1.5	1.7	16
Sulphate as SO_4	5,600	1,900	46,000
Total Phosphorus	0.07	0.18	1.3
Chloride	44,000	15,000	190,000
Bicarbonate Alkalinity	320	170	1,100
Carbonate Alkalinity	<1	<1	430
Total Filtrable Residue(105^{10}C)	74,000	27,000	340,000
Iron	0.80	5.2	1.0
Sodium	18,000	6,500	51,000
Calcium	900	400	230
Magnesium	4,000	1,100	36,000
Potassium	700	320	12,000

Typical compositions of bore, channel and bitterns waters at Laverton, Australia.

1 WWS Charters; RWG Macdonald
Development of Solar Saline Ponds in Australia
NERDDP EG85/444 Australian Government Publishing Services
Canberra, Australia 1985

2 BP Marett; WWS Charters; RWG Macdonald; DR Kaye
Recent Developments at the Laverton Solar Ponds
ISES Solar World Congress
Hamburg, Germany 1987

3 RWG Macdonald; WWS Charters; BP Marett; DR Kaye
Project Report - Solar Pond Project
Victorian Solar Energy Council
Melbourne, Australia 1988

4 The Office of Saline Water
Saline Water Conversion Engineering Data Book
US Department of the Interior
Washington DC, USA 1971

5 B.P. Marett
Hydrodynamic Stability in Diffusive Interface
Ph.D. Thesis University of Melbourne
Melbourne, Australia 1988

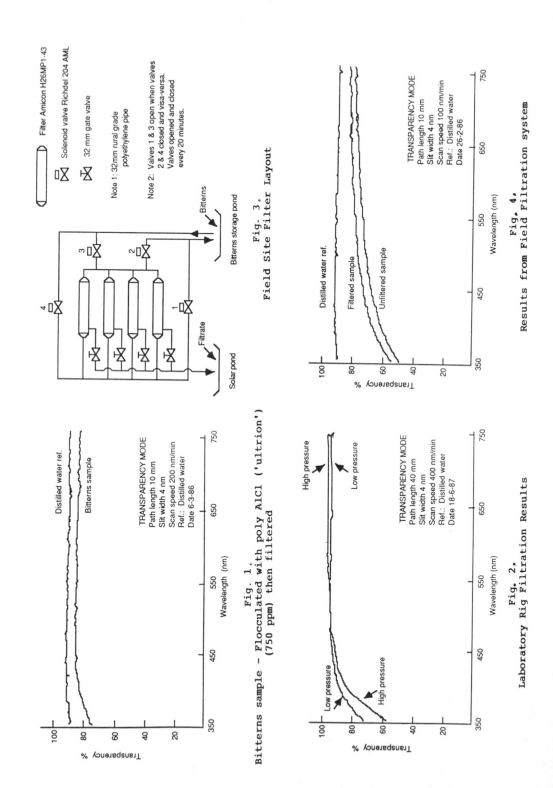

Fig. 1,
Bitterns sample – Flocculated with poly AlCl ('ultrion')
(750 ppm) then filtered

Fig. 2,
Laboratory Rig Filtration Results

Fig. 3,
Field Site Filter Layout

Fig. 4,
Results from Field Filtration system

EXPERIMENTAL STUDY OF MINI-SATURATED SOLAR PONDS UNDER NATURAL CONDITIONS

Li Shensheng   and   Lü Zengan

Physics Department, Beijing Teachers' College
Beijing, P.R.China

ABSTRACT

Two mini-solar ponds using potassium alum and sodium carbonate respectively were constructed and operated under natural conditions to achieve fully-saturated ponds A mini-pond of the same size with sodium chloride was set up simultaneously for comparison. Experimental results showed that the concentration gradient could be adjusted and modified satisfactorily by simply adding salt crystals at the proper depths in the ponds. A new concept of "winter-saturated pond" was presented.

KEYWORDS

Solar pond; sodium carbonate; potassium alum; self-generation mechanism; winter-saturated pond.

INTRODUCTION

In a conventional solar pond with sodium chloride, the salt gradient zone is not stable due to salt diffusion from lower to upper layers, thus continuous operation of sodium chloride pond requires proper adjustment and maintenance constantly. But these maintenance processes are very inconvenient and rather expensive (Nielsen, 1977, 1979; Tabor, 1981). Thus, some new concepts of more stable ponds have been proposed, such as membrane stratified pond (Hull, 1980), viscosity stabilized pond (Shaffer, 1986) and gel pond (Wilkins and Lee, 1987). The fully-saturated pond (Li and Li, 1989) is also one of these new concepts.

The fully-saturated pond is built up with one kind of salt whose solubility is a strongly increasing function of temperature, and the concentrations of salt solution in the pond are maintained at or near the solubility limits by applying enough amounts of salt at all depths. Thus salt diffusion from lower to upper layers will cause supersaturation and crystalization at upper layers, and the crystals falling down to lower layers will dissolve and achieve saturation again. Therefore, the fully-saturated pond has the ability to self-maintain its salt gradient profile. There is no need to adjust the salt gradient, theoretically. Practically, it is a desirable advantage over the conventional, especially large-scale, sodium chloride pond. It is necessary to point out that the perfect steady state of fully-saturated pond described above is only an ideal model. In practice, the actually operated pond can only be a quasi-saturated pond. That means most layers of the gradient zone are just close to the saturation states, but have supersaturation or undersaturation to some extent. This is due to the diurnal and seasonal variations of insolation and other environmental causes. Moreover, keeping the concentration of salt solution at all depths in saturated states would need a large amount of salt and result in the precipitation of salt crystals and form a deposition on the bottom of the pond.

The concept of saturated pond was first proposed in the middle of seventies (Nielsen and Rabl, 1975), and had been developed quickly and widely during recent years (Edsen, 1979; Ochs and Bradley, 1979; Mangussi, Saravia and Lessino, 1980; Kooi, 1981; Vitner, Reisfield and Sarig, 1984; Vitner and Sarig, 1986; Hull, 1986). But most of these studies were only within the range of theoretical analyses or laboratory experiments by heating the simulating system at the bottom with electricity which are quite different from the practical heating processes by solar irradiation. Therefore, the actual performance of saturated pond is still unknown. In this paper, the construction and operation under natural conditions of three mini-saturated ponds of area about 2.5 m^2 with sodium carbonate, potassium alum and sodium chloride respectively are presented and the experimental results are discussed.

EXPERIMENTAL RESULTS

Material Selection

Materials suitable to construct saturated ponds must satisfy following four conditions simultaneously, (1) the solubility of the salt should increase rapidly with temperature within the temperature range designed; (2) the solubility of the salt should be small at the freezing point in order to save the salt consumed; (3) the aqueous solution of the salt should be transparent, chemically stable, nonpoisonous and noncorrosive; and (4) the salt resource should be abundantly available and cheaply obtainable.

There are many kinds of salt which can satisfy the first three conditions, but some of them are strong oxidizers or even poisons, and some are rather expensive. In this experiment, only $Na_2CO_3 \cdot H_2O$ and $KAl(SO_4)_2 \cdot 12H_2O$ were chosen to construct two mini-saturated ponds and a conventional $NaCl$ pond was built up for comparison.

Experimental Measurements

The main physical quantities to be measured in this experiment were the temperature, density and concentration of the solutions in three ponds.

The temperature profiles were obtained by lowering a semi-conductor thermometer slowly and progressively in the ponds, and readings were taken at intervals of 0.05 m from top to bottom. The measurements were made three times a day and the atmospheric temperature was measured at the same time.

The density profiles were obtained by means of mass-volume ratios, and the samples of solution were taken at the same places where temperatures were measured. The measurements were made once a week.

The concentration profiles were obtained from the concentration-refractive index curves (Ye, Song and Li, 1987), and the refractive indexes of the salt solution were determined by Abbé refractometer.

Experimental Results

The experiment began from June 26, 1989. When the temperature gradients in three ponds were established successively, different salt columns were put into the seperated ponds once and again until the designed amount of salts were used up.

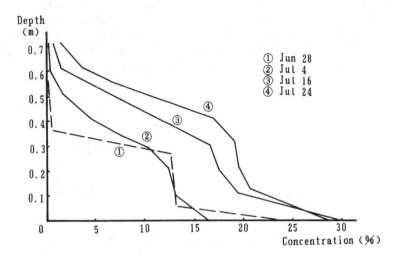

Fig.1   Development of concentration gradient
of the sodium carbonate pond.

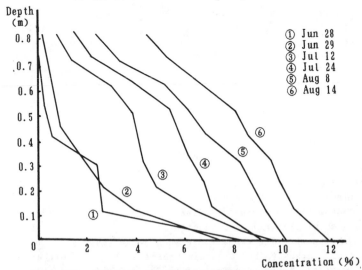

Fig.2   Development of concentration gradient
of the potassium alum pond.

This method of building up a saturated pond proved practical and available. The results of developing the concentration gradient of sodium carbonate and potassium alum ponds are shown in Fig.1 and Fig.2 respectively.

During the experiment, some green micro-organisms appeared in the potassium alum pond on July 9, 1989 which grew rapidly and affected seriously the transparency of the pond. For controlling the growth of such green micro-organisms, $CuSO_4 \cdot 5H_2O$ was added into the pond more than once but had little effect, so bleaching powder was used instead. As a result, the transparency of the pond was still worse than the other two. It was the main embarrassment of the potassium alum pond.

In autumn, salt crystals appeared in the potassium alum pond on September 21, 1989 along with the decreasing of atmospheric temperature and a convective zone arised at the bottom. Till October 18, 1989 the pond was fully convective and the concentration gradient was absolutely destroyed. However, during the following observations, the self-generation mechanism of concentration gradient had been found twice for the first time under natural conditions. The developing process is shown in Fig.3.

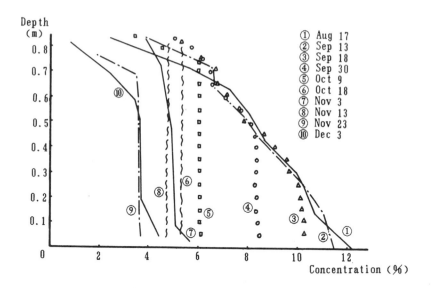

Fig.3 Self-generation mechanism of concentration
gradient of the potassium alum pond.

The experiment lasted for five and half months. The temperatures at the bottom of three ponds along with the atmospheric temperature are shown in Fig.4. On the one hand, the maximum temperature of sodium carbonate pond reached 65°C in summer, which exceeded about 18°C than sodium chloride pond; and on the other hand, it was not fully saturated when its temperature much higher than 34°C, as the solubility curve of sodium carbonate has a turning point at 34°C. Furthermore, its heat capacity is several times than that of sodium chloride pond (Elwell, Short and Badger, 1977).

However, the sodium carbonate pond was fully saturated in winter time. Although there were a lot of salt crystals precipitated on the bottom, it was still very stable and the concentration gradient maintained nicely, in spite of the reflectivity of the bottom being increased significantly. This result was in distinct contradiction with the conclusion of theoretical analysis (Kooi.,1981).

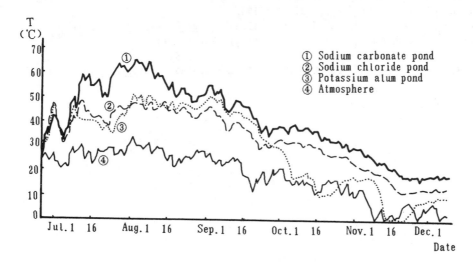

Fig.4 Temperature variations at the bottom of
three ponds and that of atmosphere

CONCLUSION

A concept of "winter-saturated pond" can be developed from above discussion. In this model, the pond is not saturated in summer time, but can be saturated in winter time. In this kind of solar pond, the amount of salt for building up a saturated pond can be much saved, and the possible problem of crystalization can also be avoided. Furthermore, its concentration gradient can be self-maintained automatically. Thus, it is more practicable and available than the fully-saturated pond.

REFERENCES

Edesess, M., D. Benson, J. Henderson, and T.S. Jayadev (1979). Proc.14 Intersociety Energy Conversion Engineering Conf., Boston, 56-61.
Elwell, D.L., T.H. Short, and P.C. Badger (1977). Proc. 1977 Congress of ASES, Orlando, 16.29-16.33.
Hull, J.R. (1980). Solar Energy, 25, 317-325.
Hull, J.R. (1986). Solar Energy, 36, 551-558.
Kooi, C.F. (1981). Solar Energy, 26, 113-120.
Li, Q.W., and S.S. Li (1989). Proc. 1989 Congress of ISES, Kobe, Vol. 2, 1428-1432.
Mangussi,J., L.Saravia and G.Lessino (1980). Solar Energy, 25, 455-457.
Nielsen, C.E., A. Rabl, J. Watson, and P. Weiler (1977). Solar Energy, 19, 763-766.

2302

Nielsen, C.E. (1979). *Proc. 1979 Congress of ISES*, Atlanta, SUN II, 1010-1014.

Nielsen, C.E., and A. Rabl (1975). *Solar Energy*, 17, 1-12.

Ochs, T.L., and J.O. Bradley (1979). *Proc. 1979 Congress of ISES*, Atlanta, SUN II, 1026-1028.

Shaffer, L.H. (1986). *US Patent* 438992.

Tabor, H. (1981). *Solar Energy*, 27, 181-194.

Vitner, A., R. Reisfield, and S. Sarig (1984). *Solar Energy*, 32, 671-675.

Vitner. A., and S. Sarig (1986). *Solar Energy*, 36, 133-140.

Wilkins, E., and T. Lee (1987). *Solar Energy*, 39, 33-51.

Ye, X.M., A.G. Song, and S.S. Li (1987). *Acta Energiae Solaris Sinica*, 8, 95-97.

PERIODIC SOLAR POND FOR A RURAL COMMUNITY

BY

J.Srinivasan
Department of Mechanical Engineering
Indian Institute of Science
Bangalore, India 560 012

ABSTRACT

The construction of a periodic solar pond to meet the hot water needs of a rural community is discussed. The solar pond is designed to operate in a periodic mode because the hot water needs for bathing is primarily in winter. The details of construction of a 400 square metre solar pond (for a rural island community on the west coast of India) are provided. The evolution of temperature and density profiles in the pond is highlighted.

KEY WORDS:

Solar energy; Periodic Solar pond; Process heat; Rural community

INTRODUCTION

The primary energy needs of a rural community are for cooking and bathing. The fuel used for this purpose is mainly firewood. Masur is an island community located on the west coast of India in the estuary of the river Aganashini ($14°28'N$, $74°23'E$). The village has a population density of 949 per square kilometer. The tree cover in the minor forest in this island was reduced drastically in the last twenty years. The firewood used for heating water for bathing in this village is estimated to be 50 tonnes per year. Professor Madhav Gadgil, Convenor, Centre for Ecological Sciences, Indian Institute of Science, Bangalore suggested that a solar pond could be built at Masur to supply hot water for bathing and thus reduce the dependence on firewood.

The hot water used (for bathing) in this community has a strong seasonal variation. The demand for hot water is primarily during the winter season and almost non-existent during the summer. We decided, therefore, to operate the solar pond in a periodic mode. This approach was suggested by Professor Carl Nielson at the second conference on `Progress in Solar ponds' at Rome in March 1990. A periodic solar pond is used both as a solar collector and an evaporating pond. Thus the cost and complexity of salt recycling is avoided. This is particularly important for small solar ponds in rural areas. Solar ponds will be successful in rural areas if they are operated as close to a natural solar pond as possible. The climate in the west coast of India is ideally suited for the operation of a periodic solar pond. This region receives more than 3000 mm of rainfall during the

period June to September and hardly any rainfall during October
to May (see figure 1). During the period January to May, when the
demand for hot water is low, the solar pond can be used as an
evaporation pond.

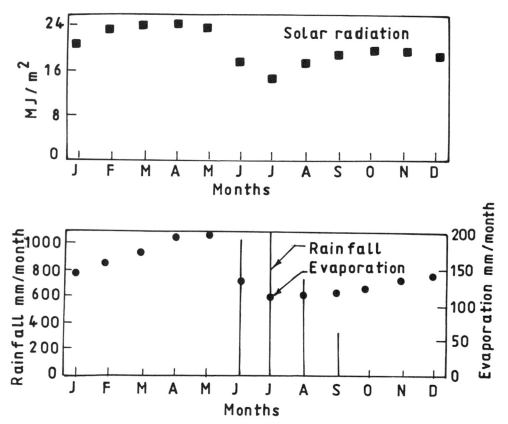

Fig.1.Climatological Variation of Solar Radiation,
Rainfall and Evaporation near 14°N,73°E.

POND CONSTRUCTION

The site chosen for the construction of the solar pond was near
the high school in Masur (Figure 2). The site permitted the
construction of a pond 16m by 16m at the bottom and 26m x 26m at
the top with a side slope of 1:2 and a depth of 2.5m. The
excavation took a longer time than envisaged because of the
presence of large boulders. Two layers of low density
polyethylene (LDPE) of thickness 0.25mm were heat sealed in the
factory and brought as a single piece (34m by 34m) and laid. A
third LDPE liner was used as a sacrificial liner on the sides of
the pond. 150 tonnes of salt were dumped into the pond during
the last week of May. Heavy monsoon rainfall began soon after
the salt was dumped and the pond was filled to the brim within
one month. Hence we could not establish the gradient zone and

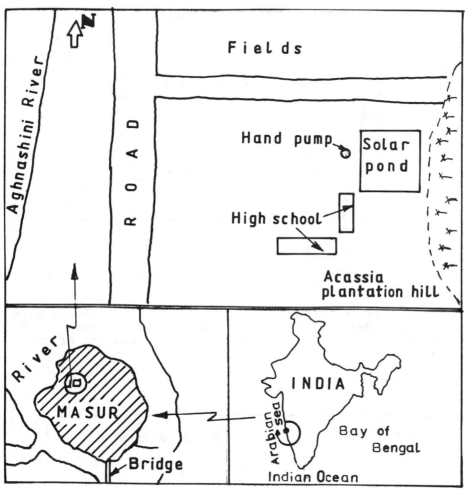

Fig.2. Location of Periodic Solar pond at Masur Island in India.

allowed the gradient zone to evolve naturally. The variation of
of salinity with depth in the pond on 1st October 90 and 31st
January '91 is shown in Figure 3. The density of the samples from
various depths was measured using hydrometers and salinity
calculated from the density and temperature of the sample.We find
that the thickness of the gradient zone has increased from 50 cm
on 1st October 90 to 55 cm on 31st January 91. The salinity in
the storage zone has remained constant at 28 % during the period
on account of undissolved salt in the bottom of the pond. The
amount of salt dissolved can be obtained from the salinity
profile,this was 40.6 tonnes on 1st October 90 and 57.3 tonnes on
31st January 91.Hence we can conclude that the undissolved salt

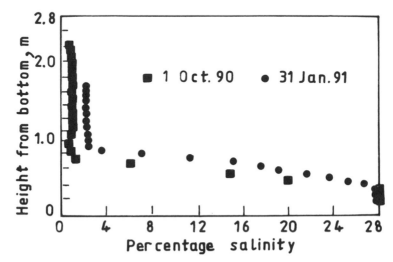

Fig.3. Salinity variation from 1st October 90 to 31st January 91

at the bottom was 92.7 tonnes on 31st January 91. The temperature
profile on 1st October 90 and 31st January 91 is shown in Figure
4. We find that there has been a substantial heating of the lower

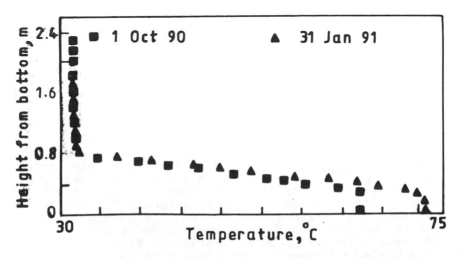

Fig.4. Temperature profiles on 1st October 90 and 31st January 91

region during this period. The evolution of the temperature of the storage zone from 31st August 90 to 31st January 91 is shown in Figure 5. We find that there has been a linear increase of the temperature in the storage zone in the last 8 weeks. The linear increase in temperature has been on account of constant solar radiation during this period and almost a constant gradient zone thickness. The decrease in the level of water in the pond on account of evaporation is shown in Figure 6. The decrease in water level in the last 10 weeks is consistent with the evaporation rate of 180 mm/month obtained from climatology (Figure 1). Heat extraction from the pond will be performed using immersed copper coils (diameter 13 mm and length 15 m). This technique has worked successfully in the Bangalore solar pond (see Srinivasan (1990) for details).

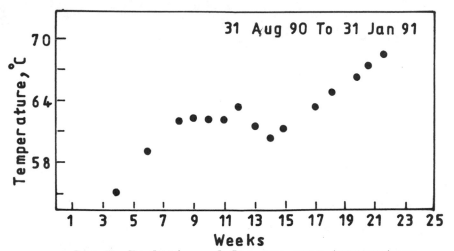

Fig.5. Evolution of Storage zone temperature

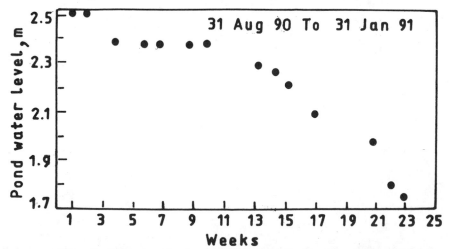

Fig.6. Decrease in Water level due to Evaporation.

Acknowledgement

This project was supported by The Department of Non-Conventional Energy Sources,New Delhi, Centre for Ecological Sciences,Indian Institute of Science,Bangalore and Karnataka State Council for Science and Technology,Bangalore. I thank Mr. Somasekhara and Mr.Harikantra of the Centre for Ecological Sciences at Kumta for their assistance in implementing the project.I thank Mr.P.N.Vedagiri, Mr.Narayanswamy and Mr.Ganesh Kumar for their help during the construction and monitoring of the pond.

References

Srinivasan,J.(1990) The performance of a Small Solar pond in the tropics.Solar Energy,45,221-230.

SOLAR PONDS FOR GRAIN DRYING:
A FEASIBILITY STUDY FOR SEED CORN DRYING

G.L. Cler, P. Hartstirn*, & T.A. Newell*

U.S. Army Construction Engineering Research Lab, Champaign, IL
*Department of Mechanical & Industrial Engineering
University of Illinois at Urbana-Champaign

ABSTRACT

A salt gradient solar pond has been proposed as a means of reducing fossil fuel consumption by a large hybrid seed corn production facility. This paper examines the existing drying system and evaluates the economic benefits and thermal performance of solar ponds for supplying a portion of the energy requirements. The facility design has been analyzed considering the possibility of a solar pond meeting the load of one, two, or all three dryers. The proposed solar pond configurations have been optimized to maximize the savings-to-investment ratio based on current economic conditions and system cost estimates. The optimal solar pond configurations all tended to have a solar fraction of about 25% and discounted payback periods ranged from 15 to 16 years.

BACKGROUND

Salt gradient solar ponds (SGSP) have been used for many different applications where low temperature thermal energy is required. These include water desalination, space heating, electricity production, mineral processing, swimming pool heating, and many others. Crop drying is an area that consumes large quantities of energy and may be an excellent application for solar ponds.

The annual energy consumed for Illinois corn drying is about 15,000,000 GJ [1]. Much of this low temperature heat is supplied by natural gas and LPG. Many attempts have been made to replace these fossil fuels in this application with solar energy [2]. These projects have met various degrees of success. The typical drawback to using solar energy in this application is that grain drying occurs in the fall over a relatively short time period. If this period happens to be cloudy, very little useful solar energy is collected. SGSP, because of their seasonal collection and storage capabilities, have the potential to overcome this problem associated with flat plate collectors. While SGSP have a lower overall efficiency than flat plate collectors (FPC), their lower cost per unit of area can more than make up for this difference, resulting in a lower cost per unit of energy harvested [3].

The facility analyzed in the work is located in east central Illinois and belongs to a large hybrid seed corn company. The corn at this facility is dried while on the cob by heating air from ambient temperature to between 35 - 43°C and forcing this low relative humidity air through the dryers. The facility operates three dryers simultaneously during the drying season, each with 4 fans supplying air at approximately 90 m³/s. This air is heated as it comes in direct contact with the LPG fired burner within the dryer. Typical fuel consumption for the drying process is approximately 2200 GJ/day. The drying season depends on the moisture content and quantity of corn to be dried and varies from four to eight weeks. In a typical short season, all drying would be completed during the month of September. A long season would start about one week earlier and extend about three weeks into October. A "long season" was used in this evaluation. It was decided to evaluate three separate solar pond configurations. Each proposed system would be optimized to meet the needs of one, two, or all three driers.

SGSP optimizations in this study were performed using PONDFEAS [4,5], a PC based solar pond feasibility and design tool. PONDFEAS allows for variable thermal and economic parameters and can optimize the SGSP configuration to maximize the project savings-to-investment ratio (SIR) or life cycle savings (LCS). PONDFEAS is not a research tool, but rather a framework for analyzing SGSP with respect to potential processes.

ANALYSIS

Input parameters for the PONDFEAS simulations were obtained from a variety of sources, the facility manager being the primary source of information for dryer physical characteristics and operating strategies [6]. SGSP construction costs were based on past experience gained during the construction of the University of Illinois solar pond and current cost estimates [7]. Site specific weather information is contained in a PONDFEAS data file along with U.S. Department of Energy annual fuel escalation rates. Input parameters used are listed in Table 1. The nonconvective zone (NCZ) was set equal to 1.25 meters and not changed for each simulation. Determination of this value was a result of many simulations and was found to be very near the optimal NCZ depth in all cases.

Table 1
Input parameters for PONDFEAS simulations

Pond		Drying Process	
UCZ depth	0.5 m	Dryer final temperature	38°C
NCZ depth	1.25 m	Dryer inlet temperature	ambient
LCZ depth, optimal	1.1 - 1.4 m	Air flow/dryer	360 m³/s
ground temp	10°C	Conventional fuel	LPG
		Combustion Efficiency	100%
Operating Schedule			
August	last week	**System Component Costs**	
September	entire month	Excavation	$2/m³
October	first 3 weeks	Liner, installed, for pond	
dryers operate continuously		and evaporation surface	$9/m²
		Salt, delivered	$0.025/kg
Economic Data		Wave control devices	$1/m²
Study life	25 years	Controls	$2500
Energy cost	$5.21/GJ	Security fence around pond	$20/meter
Discount rate	7%	Land cost	$5000/ha
Fuel Escalation rate	annual DOE rates	Project administration	10% of cost
	average: 5.0%	Average annual M&R cost	1.2% of cost
Down Payment	25%	Heat extraction equipment	
Mortgage term	15 years	for optimal case	
Loan interest rate	11%	1 dryer	$88000
Depreciation	12 years	2 dryers	$14100
Renewable energy tax credit	10%	3 dryers	$210000

In order to reuse the salt transported from the LCZ to the surface of the SGSP, a brine recycling system similar to the one used at the UIUC solar pond was included in the cost estimates [7]. The evaporation area required for brine control in this geographic region is approximately one half the surface area of the SGSP and it should be located on the south facing berm of the pond. This surface is initially used as an unloading ramp for the trucks delivering salt and is then lined with a similar material as is used to line the pond.

Three scenarios were evaluated: one, two, and all three dryers. The loads in each case were calculated based on heating the required volume of air from ambient temperature to the final desired temperature, in this case 38°C. The thermal energy required by all dryers are equal so the total load is three times that of an individual dryer. The total load was then compared to the average annual LPG consumption and found to be in good agreement. The thermal load for each scenario is listed below in Table 2.

Table 2.
Thermal energy required (GJ)

# Dryers	1	2	3
August	3528	7057	10585
September	17109	34217	51326
October	16951	33903	50854
Total	37588	75177	112765

DISCUSSION

The optimal pond configuration for each load is summarized in Table 3.

Table 3.
Summary of optimization results

# Dryers	1	2	3
pond area (m²)	14000	25000	39000
SIR	2.08	2.27	2.35
LCS ($1000)	181	357	569
DPB (years)	16.2	15.3	14.8
solar fraction	0.29	0.25	0.26
SGSP cost ($1000)	670	1122	1686
SGSP cost ($/m²)	48	45	43

In Figure 1, pond areas are plotted with savings to investment ratio for each of the three loads. It can be seen that the economies of scale are still important, increasing the SIR by about 13% from the small pond to the largest.

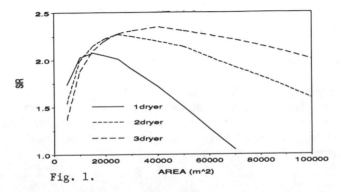

Fig. 1.

It can be seen that the SIR curves are relatively flat over a fairly large range of pond areas (+/- 10%). This suggests that, within reasonable limits, the economic benefits of owning a pond area are not critically linked to the optimal area. Figure 2 shows the SIR (a), DPB (b), LCS (c) and SF (d) versus pond area and optimal pond depth for the entire dryer load. The discounted payback over a very large range of pond areas, is extremely flat. Variations of +/- 50% of the optimal surface area result in less than a one year variation in DPB. The LCS steadily increase up to a surface area of about 70,000 m^2 and then begins to flattens out. Solar fraction, as one would expect, steadily increases with pond area.

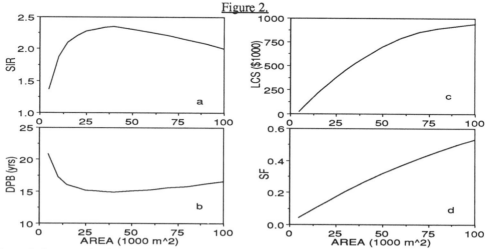

Figure 2.

Figure 3 shows a pie chart indicating the estimated percentage of the total system cost for each of the ponds major construction activities and components for the optimal pond designed for the system consisting of all three dryers.

Figure 3.

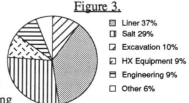

Liner 37%
Salt 29%
Excavation 10%
HX Equipment 9%
Engineering 9%
Other 6%

The "OTHER" pie slice collectively displays the balance of the SGSP construction costs. One readily can see that two thirds of the project cost are for the liner and salt. The cost of both items can vary widely. Depending on site specific conditions, salt can be nearly free. The other extreme would be a location requiring the salt be purchased and trucked a long distance. For the case in this study, the liner is the largest single item. Approximately one third of the liner cost is due to the evaporation surface area requirements for this region. Price estimates presented here are based on using a high density polyethylene liner. This material is strong and durable but somewhat expensive when compared to clay lined ponds. For SGSP to make a significant energy impact in agricultural areas such as this, liner integrity must be assured. An artificial liner would be required for this purpose.

Table 4 compares the results of a +/- 10% variation in the cost of the installed liner and delivered cost of salt relative to that used for this evaluation. The initial estimated liner cost for this project was $9.00/$m^2$ and the delivered salt was $0.025/kg. Simulations with these variations were performed with the optimal pond configuration as determined using the actual cost estimates. As one would expect, reducing the initial cost of one or more components improves the economic performance of the SGSP. A 10% reduction in liner or salt cost each represent less than 4% of the total system cost but they can represent a significant reduction in the discounted payback period for the project on the order of 25%.

It is also interesting to compare the energy supplied by the solar pond to the actual energy incident on the pond area during the drying season. As noted earlier, one advantage of SGSP over typical FPC is the long term energy storage capability. Table 5 shows the energy supplied by the pond, the incident solar radiation during the drying season and annually, and the pond system efficiency.

Table 4.

Cost Variations from Initial Estimates

		-10%	0%	+10%
liner	SIR	2.56	2.35	2.16
	LCS(1000)	$630	$569	$508
	DPB(yrs)	10.8	14.8	15.8
salt	SIR	2.51	2.35	2.20
	LCS(1000)	$616	$569	$523
	DPB(yrs)	11.3	14.8	15.6
liner	SIR	2.73	2.35	2.02
& salt	LCS(1000)	$677	$569	$462
	DPB(yrs)	9.5	14.8	16.5

Table 5.

# Dryers	1	2	3
Energy delivered from SGSP (GJ)	11000	19200	29400
Incident solar energy w/o* (GJ)	12900	23100	36000
Efficiency w/o	0.85	0.83	0.82
Incident solar energy annual (GJ)	75500	134900	210400
Efficiency annual	0.15	0.14	0.14

*note: w/o = while operating (8 week drying season)

The annual efficiencies calculated here, 14 - 15%, would not be uncommon for a SGSP in this application. While this value is low compared to FPC, system costs are often ten times more expensive per collection area. Also, as the processing season is extended the solar pond performance tends to improves. Particularly if processing can begin earlier. The long term storage capabilities of solar ponds can be seen in Table 6, a summary of the weekly energy extracted from the SGSP optimized for all three dryers. It can be seen that during the early part of the drying season the energy extracted from the pond is significantly greater than the incident solar radiation available during that time period.

Table 6.

SGSP Energy Extracted (GJ) vs. Incident Solar Radiation (GJ)

Week	SGSP Energy	Solar Radiation	ratio
1	10585	5560	1.90
2	5045	5270	.96
3	3243	4980	.65
4	2577	4670	.55
5	2259	4350	.52
6	2035	4030	.50
7	1861	3710	.50
8	1771	3390	.52

During the last week of processing, the ratio of energy extraction vs. incident solar radiation increases slightly. This is due to the decrease in incident solar energy and average ambient temperature. As the ambient temperature decreases from week to week, the minimum useful pond temperature, which is the ambient temperature + 5°C in this study, also decreases.

CONCLUSIONS

Several conclusions can be drawn from this work. First, solar ponds, because of their seasonal storage capabilities, can supply large quanties of low temperature thermal energy for grain drying activities during a relatively short time period. While the discounted payback period is relatively long, on the order of 15 years, a reasonable return on one's investment can be expected. Table 3 summarizes the optimal pond configuration and cost for each of the three cases considered. Optimal pond size need not be exactly matched to the load. Pond areas of +/- 10% resulted in SIR reductions of less than 2%. For the three cases considered here, the larger the load, and optimal pond area, the greater the SIR. The optimal pond configuration for all three dryers shows an increase of about 13% in SIR over the case with one dryer. Therefore, economies of scale are still important. In this particular application it would be very easy to construct SGSP in modules to meet the load of individual dryers. While some of the economic benefits of a single large pond are reduced, the risk of thermal inversion and loss of all thermal energy is also reduced. Maximum pond temperatures are not expected to exceed 80°C. Therefore with reasonable gradient zone maintenance, this should not be a major problem.

The installed cost of the liner is the single most expensive element of the SGSP considered here, accounting for over 1/3 of the entire project budget. The delivered salt cost followed and accounted for over 1/4 of the budget. Depending on local conditions and requirements, these costs could be significantly reduced. A reduction by 1/3 of the cost of each of these items reduces the DPB to about 5 years. This payback period is probably the longest that would generally acceptable by industry. Many areas have better transportation access (river or train) and a considerable reduction in salt costs are possible. Less expensive lining techniques have been used in some SGSP but may not be acceptable for highly productive agricultural areas. Therefore, cost reductions for liner systems will only come about through material cost reductions and automated techniques for liner installation.

Salt gradient solar ponds are currently competitive with fossil fuels for many applications including grain drying and should be considered as a viable alternative. With the current renewed interest in environmental issues, SGSP are ready to play a major role in supplying energy for our agricultural processing needs.

ACKNOWLEDGEMENTS

The authors wish to thank Mr. Mike Gumina and the staff at Pioneer Hybrid International in St. Joseph, IL for their help and interest in this project.

REFERENCES

"Survey of Grain Drying Practices in Illinois," American Society of Agricultural Engineers Meeting, 1978.

Guinn, G.R., "Process Drying of Soybeans Using Heat from Solar Energy," Proceedings of the Solar Industrial Process Heat Symposium, 1978, p.63.

Healey, Henry and Gerald L. Cler, An Investigation of Privately Financed Renewable Energy Projects for Army Installations, Technical Report E-90/12 (USACERL, September 1990).

PONDFEAS User's Manual, Version FY90, U.S. Army Construction Engineering Research Laboratory, 1990.

Cler, Gerald L., C. Jantzen and T. Newell, "A Microcomputer-Based Solar Pond Design Tool," Progress In Solar Ponds, 2nd International Conference, Rome, Italy, March 1990.

Hartstirn, P., "Feasibility Study: Salt Gradient Solar Pond for Pioneer Hi-Bred International at St. Joseph, IL," Senior Project, Department of Mechanical and Industrial Engineering, University of Illinois, April 1990.

Newell, T.A., R.G. Cowie, J.M. Upper, M.K. Smith, and G.L. Cler, "Construction and Operation Activities at the University of Illinois Salt Gradient Solar Pond," Solar Energy, Vol. 45, No. 4, pp. 231-239.

CONDUCTIVITY PROBE FOR MEASURING SMALL VARIATIONS IN THE SALINITY GRADIENT OF A SOLAR POND

A. Joyce , P. Raposo, * M. Collares Pereira
Departamento de Energias Renováveis do L.N.E.T.I.
Azinhaga dos Lameiros à Estrada do Paço do Lumiar 1699 LISBOA CODEX - PORTUGAL.
* Centro para a Conservação de Energia.
Estrada de Alfragide, Praceta 1, AMADORA - PORTUGAL

ABSTRACT

Operating a salt gradient solar pond demands a precise but yet easy and inexpensive control of the salinity gradient and its boundaries. Traditionaly the method used is to extract samples at different heights and measure salt content either using a Mohr Westphall scale [1] or conductivity meters usually of two types: platinum electrodes type and electrodeless type [2].

An idea of Prof. Prof. Carl Nielsen, led to the development of a conductivity probe, using cheap electrodes, which can be used directly in the pond avoiding the time consuming process of taking samples to be analysed outside. This probe has a small spatial resolution (5 mm) that provides good information on pond stability.

The present work describes the construction and calibration of the probe and compares the measurements made in calibrated samples with those obtained with a small model utilizing a network of resistances simulating the conductivity between the electrodes. Suggestions are made to construct a set of these probes closely spaced and instaled inside a pond to have continuous readings of the salinity profile.

KEYWORDS

SOLAR POND, SALT GRADIENT, ELECTRIC CONDUCTIVITY PROBE.

PROBE DESCRIPTION

The conductivity probe has four small gold plated electrodes placed at the corners of a square 5mm side. A fifth electrode in the center of the square connects all shielded cables to the pond. A sine wave oscilator of 0.5 V amplitude and 10 KHz frequency is applied to two of the electrodes placed in one of the vertical sides of the square. The applied voltage is sensed through the solution in the other two electrodes and after being amplified 100 times, is measured by a voltmeter.
The electrodes used are pins used in PC computer sockets, mounted in small holes on a PVC rod.

Around the electrodes there's a 10 mm diameter plastic cylinder that confines the solution. The cylinder is provided with small holes to allow for solution circulation. Figure 1 shows a diagram of the probe and its electric connections.

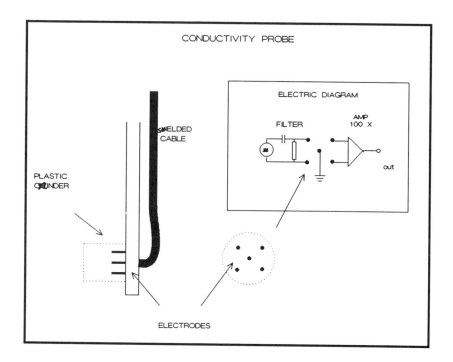

Fig. 1 - Probe description.

Conductivity of solutions becomes higher with increasing concentration and temperature, and so the impedance between the injection electrodes is smaller (signal short circuted). The signal measured in the sensing electrodes becomes also smaller. Probe signal is as a consequence higher for small concentrations.

The dimensions of the probe are such that it permits a spatial resolution of 5 mm in the salinity measurement inside the pond.

PROBE CALIBRATION

To calibrate the probe 6 solutions of salt (Sodium Clorhide) in distilled water were prepared with concentrations in weight percent equal to 4.8 %, 9.1 %, 13.0% , 16.7% , 20.0% and 23.1%.

To measure the dependence on temperature the solutions were heated in a thermostatic bath with 0.1 °C resolution from 10 °C to 60 °C in 5 °C jumps.

The temperature of the solution was measured by a thermistor placed along the probe. For each solution and temperature the signal coming out from the probe was measured.

Figure 2 shows the results obtained. Dashed lines represent an exponential fit to the obtained data.

Figure 2 also shows the high sensitivity of the probe to small values of concentration which makes it particularly useful in measuring the upper boundary of the gradient zone.

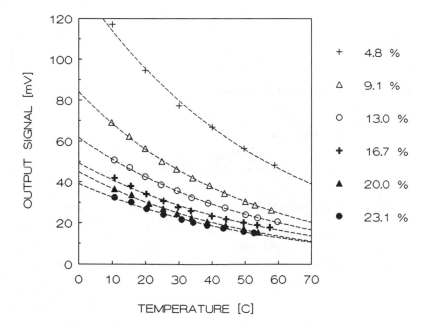

Fig. 2. Values obtained during calibration.

To obtain a relation between the output voltage and concentration, we used the equation developed by Fynn et al [3] , to calculate electric conductivity for each concentration and temperature. The results obtained, and presented in Fig. 3, show that a fit to the experimental values, of the type $y = a\,x^b$, is very good and constitutes a convenient way to obtain the concentration values.

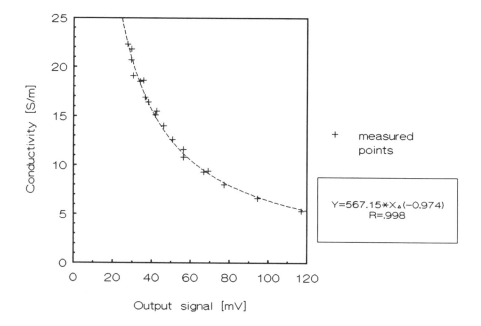

Fig. 3. Conversion between probe signal and conductivity.

A MODEL OF PROBE BEHAVIOUR

To analyse the electric behaviour of the probe, a small model consisting of a network of resistances was developed which permits the simulation of different electrodes configuration and the influence of cables with different lengths.The model considers the solution between electrodes to be a resistance whose value depends on solution conductivity , distance between electrodes and equivalent solution cross section. Therefore its value can be calculated using the well known relation:

$$r = \rho \; \frac{l}{s}$$

with: r - resistance between electrodes [Ω].

                ρ - solution resistivity [Ω.m]

                l - equivalent lenght [m]

                s - equivalent cross section [m^2]

Figure 4 represents the equivalent circuit of the probe.

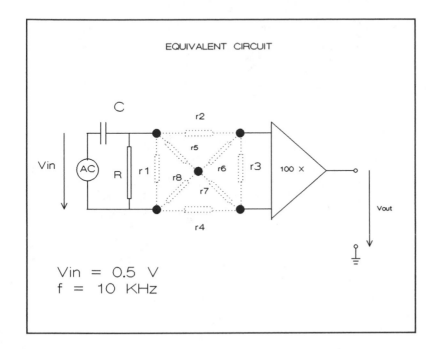

Fig. 4 - Equivalent circuit of probe behaviour.

Equating currents and voltages in the network we can obtain:

$$Z\,I_1 + R\,(\,I_1 - I_2\,) = V_{in}$$
$$R\,(\,I_2 - I_1\,) + r1\,(\,I_2 - I_3\,) = 0$$
$$r1\,(\,I_3 - I_2\,) + r5\,(\,I_3 - I_4\,) + r8\,(\,I_3 - I_6\,) = 0$$
$$r2\,I_4 + r6\,(\,I_4 - I_5\,) + r5\,(\,I_4 - I_3\,) = 0$$
$$r3\,I_5 + r7\,(\,I_5 - I_6\,) + r6\,(\,I_5 - I_4\,) = 0$$
$$r4\,I_6 + r8\,(\,I_6 - I_3\,) + r7\,(\,I_6 - I_5\,) = 0$$
$$V_{out} = 100\ r3\,I_5$$

Where $Z = 1/\omega C$ and $\omega = 2\pi f$. The model can simulate the presence of long cables between the probe and the measuring device as is generally the case in pond measurements. Figure 5 represents simulated values together with values measured during calibration.

ELECTRODE DEGRADATION

To analyse possible degradation of the electrodes, we have compared new electrodes with electrodes submitted to conditions harder than normal (2 V and 5 KHz). Comparison has been made, using a

2320

scanning electron microscope with semi quantitative chemical analyses.

Degradation of the used electrode was detected by the lower value of the peak corresponding to gold and higher value of the copper peak due to the polarization suffered by the probe. Nevertheless the electrodes are very cheap (0.5 USD) and their replacement is a simple matter.

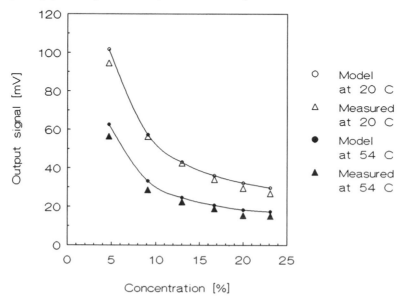

Fig. 5 - Simulated values and measured points.

CONCLUSIONS

Using cheap electrodes and electronics a device was developed   permitting good resolution for small values of concentration. Utilizing the same principle, several probes can be closely spaced and connected to the same sine wave oscillator, providing continuous readings of the salinity profile.

ACKNOWLEDGEMENT

The authors which to thank Maria Bárbara and Beatriz who assisted in the laboratory measurements.

REFERENCES

[1] – Joyce, A. (1984). Construção e estudo de um Lago Solar de gradiente de salinidade. Master Thesis.

[2] – Collares Pereira, M. and Joyce, A. (1987). Lago Solar de Pégões: funcionamento com extracção de energia. III Congresso Ibérico de Energias Renováveis.

[3] – Fynn, R. P. and Short, T. H. (1983). Solar Ponds - A basic manual. The Ohio State University Ohio Agricultural Research and Development Center.

WIND EROSION MODELING OF SOLAR POND GRADIENT ZONES

R.N. Abdel-Messih[*] and T.A. Newell[**]

[*] Energy and Automotive Dept., Faculty of Engineering, Ain Shams University, Cairo, EGYPT
[**] Mechanical and Industrial Engineering, University of Illinois at Urbana-Champaign, USA.

ABSTRACT

A mathematical model is described for prediction of transient and steady state velocity profiles in confined and unconfined fluid layers. As an application of the model to solar ponds, a simulation of the wind effect on core velocity and upper layer depth, during periods where wind speed changes, is examined. This effect can be calculated either analytically or numerically.

The time and velocity dependence of the erosion rate are studied and a mathematical relation is developed. A comparison between the erosion effect, due to wind speed, and that due to the average wind speed, in the same period of time, is presented. It was found that the effect of the average velocity on erosion is smaller than that caused by a fluctuating wind speed.

INTRODUCTION

Various models have been formulated in an attempt to predict the response of the upper mixed layer in solar ponds or oceans to changes in meteorological inputs and boundary conditions. In recent years, the use of salt gradient solar ponds for collecting and storing energy has received strong attention. The change of depth of the upper layer of the solar pond, which is present due to various processes such as wind mixing and double diffusive convection, should be minimized as much as possible to increase overall solar pond efficiency (Kooi [1]). Erosion of the non-convective zone is thus a major problem of solar pond operation (Nielsen [2]). Among the suggested causes for rapid erosion is wind mixing (Tabor [3], Hull [4] and Meyer [5]). Experimental results showed that the increase in potential energy of a stratified fluid due to turbulent entrainment across the density interface is a constant fraction of the input of turbulent kinetic energy (TKE) (Linden [6], Atkinson [7], Rohr [8], Kato [9] and Siegfried [10]). Adopting the experimental relation found by Kato [9] and using a turbulent boundary layer criterion by Stern [11], a mathematical model is proposed to predict the velocity gradient, surface velocity as well as the evolution of the depth of the upper convective zone.

Comparisons between the proposed model and experimental results of Kato [9] and Kantha [12] support the validity of the model. A unique feature of the model is the ability to predict transient changes of mixed layer erosion rates as wind speed changes. This feature will be discussed with respect to simulation of erosion in a solar pond's upper mixed layer.

The present model differs from previous models due to the inclusion of transient changes of the "core" (upper convecting zone) velocity. That is, the acceleration/deceleration of the upper convecting zone is accounted for as the wind speed changes. As will be discussed, fluctuating winds, such as typical diurnal variations in wind speed, can lead to significantly different erosion rates than a time averaged wind speed.

MODEL FORMULATION AND VALIDATION

A model has been formulated that deals with the homogeneous upper fluid layer confined between the water's free surface and the parallel interface below that free surface. Some reasonable assumptions were made to formulate the model, they are: 1)- Linear velocity distribution is considered in laminar sublayer region. 2)- Uniform velocity distribution in the fluid core of the upper convective zone, and in case of circulation as in small containers the return velocity is of the same magnitude as the upper core velocity and opposite in direction. 3)- A critical Reynolds number is maintained across boundary layers.

Figure 1 shows an element of the upper layer of fluid in a container where all the velocity profiles are drawn. The element dimensions are represented by l,w and h for its length, width and height, respectively. The following terms describe all the energy quantities in the fluid element.
i- Energy communicated to water surface by the wind through the surface shear stress $\{\tau_s (u_s - u_c) lw\}$. ii- Kinetic energy in the core, $\{(1/2)\rho u_c^2 lwh\}$. iii- Dissipated energy in the two adjacent middle boundary layers, $\{2(\tau_m u_c lw)\}$. iv- the dissipated energy in the boundary layer just before the interface which is equal to $\{\tau_b u_c lw\}$.
Equating the energy communicated by the wind to all other energy forms leads to:

$$\{\tau_s(u_s-u_c)lw\} = \{2(\tau_m u_c lw)\} + \{\tau_b u_c lw\} + d\{(1/2)\rho u_c^2 lwh\}/dt \tag{1}$$

From the boundary layers we can develop the following relations:
$\tau = \mu\,(\partial u /\partial y) = \mu u_r / \zeta$ and $R_{ec} = \rho\, u_r\, \zeta /\mu$, which leads to: $\tau = \rho u_r^2 / R_{ec}$. The boundary thickness resulting from an applied shear stress is calculated based on a criterion postulated by Stern [11]. Stern's hypothesis assumes a natural balance is maintained between dissipation and momentum processes across the boundary layer. A critical Reynolds number characterizes the balance and allows the boundary layer thickness to be calculated in a convenient manner. Critical Reynolds number for pipe flow [11] and falling films [13] has been found to be approximately 100. Replacing the former relations in eqn. (1), and after some arrangements, it becomes:

$$R_{ec}\{d(\rho u_c^2 h)/dt\} = 2\rho u_c^3\{(\alpha-1)^3-3\} \tag{2}$$

where $\alpha = u_s /u_c$. Equating the left hand side of eqn. (2) to zero, the steady state solution is found or "$\alpha = 2.44225$". As the shear stress in the water side, τ_s is equal to the shear stress in the air side, τ_a then: $\mu_w (u_s-u_c) / \zeta_w = \mu_a (u_a-u_s) / \zeta_a$. But, $R_{ec} = \rho_w (u_s-u_c) \zeta_w /\mu_w = \rho_a (u_a-u_s) \zeta_a /\mu_a$, therefore: $\alpha = \{(u_a/u_c) +(\rho_w /\rho_a)^{1/2}\}/\{1+(\rho_w /\rho_a)^{1/2}\}$. This leads to "$(u_a/u_c) = 45.928$" or "$(u_c/u_a) = 0.02178$". In case of an unconfined fluid layer, where end wall effects disappear, i.e. there is no circulation, then the middle mixing zone or the middle boundary layer is dropped from eqn. (1), therefore eqn. (2) becomes:

$$R_{ec}\{d(\rho u_c^2 h) / dt\} = 2\rho u_c^3\{(\alpha-1)^3 -1\} \tag{3}$$

which is the same as eqn. (2) except replacing 1 instead of 3 in the right hand side. The corresponding steady state solution of eqn. (3) leads to the ratio, "$(u_a/u_c) = 32.1511$", or "$(u_c/u_a) = 0.0311$". When an interface exists, ρ and h are time dependent. Therefore eqn. (2), after some arrangement will be ;

$$du_c / dt = \{u_c^2/ R_{ec}\, h\}\{(\alpha-1)^3 -3\} - [dh /dt] / 2h - [d\rho /dt] / 2\rho \tag{4}$$

It is known that the increment in depth is equal to the entrainment velocity multiplied by the time increment or $dh=u_e dt$. From experimental results obtained by Kato et al [9], the relation between the increase in the potential energy and the turbulent kinetic energy can be written as:

$$\{gu_e\Delta\rho_{st}\, h / 2\} = C\{(1/2)\rho_i\, u_*^3\} \tag{5}$$

where $\Delta\rho_{st}$ is the density step, ρ_i is the initial upper layer water density, $u_* = (\tau_s/\rho)^{1/2}$ is the friction velocity and C is a constant equal to 2.5 for the case of no end walls or fluid system without recirculating flow.

Before going to the solution of eqn. (6), the density step and the increase of fluid density in the upper layer would be calculated as follows:

i) Case of two homogeneous layers of different density: If ρ_i and h_i are the initial upper layer density and depth respectively and $\Delta\rho_{si}$ is the initial density step and if h_e is the increase of the upper layer depth due to erosion during time t, then $\delta\rho_i$, the increase in the upper layer fluid density, and $\Delta\rho_{st}$, the new density step at any time t, are given by the formulae: $\delta\rho_i = \Delta\rho_{si}h_e/h$ and $\Delta\rho_{st} = \Delta\rho_{si}h_i/h$, where $h = h_i + h_e$. It must be noted that $\delta\rho_i$ increases while $\Delta\rho_{st}$ decreases with increasing h_e or h. If there is solid bottom this means that there is no erosion which is identical to infinite value of a density step.

ii) Case when the upper layer is homogeneous and the lower layer underneath the interface has linear density gradient, $\partial\rho / \partial y = \Gamma$: The definitions are the same as in the previous case except that $\Delta\rho_{si}$ is equal to zero, and $\delta\rho_i$ and $\Delta\rho_{st}$ are given by: $\delta\rho_i = \Gamma h_e^2/2h$ and $\Delta\rho_{st} = \Gamma h_e(2h_i + h_e)/2h$. In this case both $\delta\rho_i$ and $\Delta\rho_{st}$ increase with increasing h_e or h.

Considering a linear density gradient, the solution of eqn. (4) , for confined and unconfined fluid layers will be discussed. The experimental data found by Kato et al [9] were used to calculate the critical Reynolds number, R_{ec}. A value of "$R_{ec}=120$" was found to be reasonable. If the fluid layer is initially at rest, the solution shows that the core velocity increases to its steady state value depending on the value of the wind speed. This is an acceleration phase. A deceleration phase occurs as the wind speed decreases while the core velocity is higher than its steady state value corresponding to this wind speed. A comparison between the model and experimental results [9,12] is given elsewhere [14].

In either confined or unconfined fluid layers initially at rest, the necessary time to attain a steady state depends on the wind speed and the initial upper layer depth. Using dimensionless plotting, Fig. 2 shows the solution of the model when the value of the density gradient is equal to 100 kg/m^4. It is obvious that the time needed to reach a quasi-steady state, which is of about one to two hours, increases with decreasing wind speed and/or increasing the initial depth and vice versa. The effects of changing the density gradient on the transient evolution of the core velocity with time is found to be insignificant. But in general the core velocity increases as the density gradient increases for the same wind speed. This can be explained in the following manner; a stronger density gradient results in a smaller mixed zone due to a smaller erosion which contains more kinetic energy. Fig.3 shows the effect of the density gradient on the erosion rate for a wind speed of 5 m/s.

APPLICATION TO SOLAR PONDS

Actual wind speed: In this section, an application of the mathematical model to solar ponds is made to show the effects of the wind during periods in which wind speed changes. Since the erosion rate, u_e is inversely proportional to the depth, and the depth increment during a time "Δt" is $\Delta h = u_e(\Delta t)$, therefore; $\Delta h \propto \Delta t / h$. For the same wind speed and the same time period, two depth increments are different for two initial different depths h_a and h_b, and they are related as; $\Delta h_b = \Delta h_a (h_a / h_b)$. ne depth, h, after this period will be: $h = h_b + \Delta h_b$. Depth, h_n at the end of that period "n" and depth increment, Δh_n during that period are generalized to be:

$$\Delta h_n = (h_{ne} - h_{nb}) (h_i / h_{n-1})$$ (6)

$$h_n = h_{n-1} + \Delta h_n$$ (7)

where subscripts "nb" and "ne" are the numbers referring to beginning and end of the period "n" on the plot and h_i and h_{n-1} are the initial depth used to plot the curve and the actual depth calculated in the previous step respectively. The core velocity and depth calculated by this method at the end of the time history of table [1] were found to be 0.027 m/s, 1.0679 m, respectively.

TABLE 1   Time History of Wind Speed

No.	1	2	3	4	5	6	7
t [s]	1000	200	100	200	1000	200	900
u_a[m/s]	1	3	5	3	1	3	1

Fig. 4.a shows the core velocity and the depth calculated by the numerical method, at any instant, using the same time history of table [1]. The previous analytical solution gives approximately the same results. In the same figure the variation of core velocity with time due to average wind speed is also plotted. Also the depth variation with time due to this average velocity is shown in Fig. 4.b. It is found that the effect of the actual wind speed variations is always greater than that of the average value. The model results show that the increase in depth, " $\Delta h = (h - h_i)$ " can be written on the following form:

$$\Delta h = [f(u_a)]^{1/2} \ (t^{1/2}) \tag{8}$$

where $f(u_a)$ is a function of the third degree of the wind speed. From eqn. (8) it is obvious that wind speed influence on the erosion rate is very much higher than that of the period of time. From eqn. (5) the entrainment velocity, u_e is rewritten as "$u_e = k \ u_*^3 / h$", therefore the erosion during a time "dt" can be written as: $dh = \{k \ u_*^3 / h\} \ dt$, where "k" is equal to " $C \rho_i / g \ \Delta \rho_{st}$". The density step, $\Delta \rho_{st}$ varies very slowly with time therefore "k" is considered to be a constant. As the friction velocity, "$[u_* = (\tau_s / \rho)^{1/2}]$ ", and $\tau_s = \rho \ (u_s - u_c)^2 / R_{ec}$, then "$u_* \propto u_a$" therefore, after integration a similar equation to eqn. (8) is found as:

$$h = (k)^{1/2} \ \{ u_a^{3/2} \} \ \{ t^{1/2} \} \tag{9}$$

Average wind speed: During a period of time "t", the value and the direction of wind speed change. If direction changes are neglected, the effect of wind speed on solar ponds, during that period of time, is compared with that due to the average wind speed during the same time period. Suppose that the actual wind speed values, during intervals of time $t_1, t_2, ..., t_i, ..., t_n$ are respectively, $u_{a1}, u_{a2}, ..., u_{ai}, ..., u_{an}$. The average wind speed, u_{av} can be written as:

$$u_{av} = u_{a1} \ \phi_1 + u_{a2} \ \phi_2 + \cdots + u_{a \ i} \ \phi_i + \cdots + u_{an} \ \phi_n \tag{10}$$

where, $\phi_i = t_i / t$ and $t = \Sigma \ (t_i)$, and $\Sigma \ \phi_i = 1$. From eqn. (9), the total erosion due to different wind speeds can be written as follow:

$$h_i + \Delta h = \{ u_{a1}^{3/2} \ \phi_1^{1/2} + u_{a2}^{3/2} \ \phi_2^{1/2} + \cdots + u_{ai}^{3/2} \ \phi_i^{1/2} + \cdots + u_{an}^{3/2} \ \phi_n^{1/2} \} \{k \ t \}^{1/2} \tag{11}$$

The erosion due to the average wind speed, during the total period of time is also given by:

$$h_i + \Delta h_v = \{ (u_{a1} \ \phi_1 + u_{a2} \ \phi_2 + \cdots + u_{ai} \ \phi_i + \cdots + u_{an} \ \phi_n)^{3/2} \} \{k \ t \}^{1/2} \tag{12}$$

By inspection, the value between brackets in eqn. (12) is smaller than that of eqn. (11) for all values greater than or equal to zero of both u_{ai}, ϕ_i. This shows that the estimation of the erosion effect of solar ponds, due to the average wind speed, can be incorrect, unless the wind speed variation is very small.

CONCLUSION

The model developed in this work allows calculation of gradient zone erosion during transient conditions to be predicted. The acceleration and deceleration of the upper convective zone of a solar pond generally requires a time period of an hour to reach a quasi-steady flow.

Another important area examined by the model is the effect effect of fluctuating wind velocities. The results show that a time average wind speed will underpredict the erosion of a time varying (e.g., diurnal variation) wind speed.

Several areas of investigation are needed in order to more fully understand the effects of wind. Measurement of pond velocities relative to wind speed in operating solar ponds is needed in order to validate proposed models. Although the present model has been validated with experimental data that represents two limiting conditions (density gradients versus sharp step interfaces), the effects of wave formation and wave suppression devices need to be examined.

NOMENCLATURES

C: constant in equation (7).

g: gravitational acceleration, $[m/s^2]$.

k: constant

r: radius, $[m]$.

t: time, $[s]$.

w: fluid system wide, $[m]$.

y: vertical axis

$\Delta\rho$: density step, $[kg/m^3]$.

ρ: density, $[kg/m^3]$.

μ: dynamic viscosity, $[N.s/m^2]$.

τ: shear stress, $[N/m^2]$.

f: frequency $[1/s]$

h: height or depth, $[m]$.

l: fluid system length, $[m]$.

R_{ec}: critical Reynolds number.

u: velocity or speed, $[m/s]$.

x: horizontal axis.

α: velocity ratio, u_s/u_c.

$\delta\rho$: density increase, $[kg/m^3]$.

Γ: density gradient, $(\partial\rho/\partial y)$, $[kg/m^4]$.

v: kinematic viscosity, $[m^2/s]$.

ζ: boundary layer thickness, $[m]$.

subscripts:

a: air.

e: entrainment or end.

i: initial or number, takes values from 1 to n.

m: middle.

r: relative.

v: average

b: bottom or beginning.

c: core or actual

si: initial step.

st: step at time "t".

s: surface or screen.

w: water.

REFERENCES

1. C. F. Kooi, The steady state solar ponds. J. Solar Energy 23, 37-45 (1979).
2. C. E. Nielsen, Conditions for absolute stability of salt gradient solar ponds. Proc. ISES Cong. 2, 1177, (New Delhi, Jan. 16-20, 1978).
3. H. Z. Tabor and B. Doron, The Beith Ha' Arava 5MW(e) solar pond power plant (sppp). Progress Report (1989).
4. J. R. Hull, Wind induced instability in salt gradient solar ponds. ASES, conf. proceedings 3.1, 371-375 (1980).
5. K. A. Meyer, A numerical model to describe the layer behavior in salt-gradient solar ponds. J. Solar Energy Engineering, 105, 341-347 (1983).
6. P. F. Crapper and P. F. Linden, The structure of turbulent density interface. J. Fluid Mech. 65.1, 45-63 (1974).
7. J. F. Atkinson and D. R. F. Harleman, A wind-mixed layer model for solar ponds. J. Solar Energy Engineering 31.3, 243-259 (1983).
8. J. J. Rohr, Mixing efficiency in stably-stratified decaying turbulent. Geophys. Astrophys. Fluid Dynamics, 29, 221-236 (1984).
9. H. Kato and O. M. Phillips, On the penetration of a turbulent layer into a stratified fluid. J. Fluid Mech. 31, 643-655 (1969).
10. B. Siegfried and D. R. F. Harleman, Effect of wind-induced mixing on the seasonal thermocline in lakes and reservoirs. Second International Symposium On Stratified Flows, The Norwegian Institute Of Technology Trondheim, Norway, 24.-27. June, (1980).
11. M. E. Stern, Inequalities and variational principles in double diffusive turbulence. J. Fluid Mech. vol. 114, pp 105-121 (1982).
12. L. H. Kantha, O. M. Phillips and S. R. Azad, On turbulent entrainment at a stable density interface. J. Fluid Mech. 79, 753-768 (1977).
13. T. A. Newell, K. Y. Wang and R. J. Copeland, Film flow characteristics for direct absorption solar receiver surfaces. ASME Winter Annual Meeting, Miami, Fl., Nov. 1985.
14. R. N. Abdel-Messih and T. A. Newell, Transient wind shear effects on solar ponds gradient zones. Report, Solar Energy Program, Dept. of Mech. & Industrial Eng., University of Illinois at Urbana-Champaign, January, 1991.

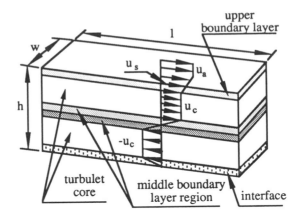

Fig. 1. Velocity distribution in an element
inside a confined fluid layer

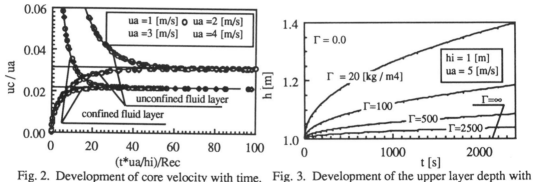

Fig. 2. Development of core velocity with time.

Fig. 3. Development of the upper layer depth with
time for various values of the density gradient

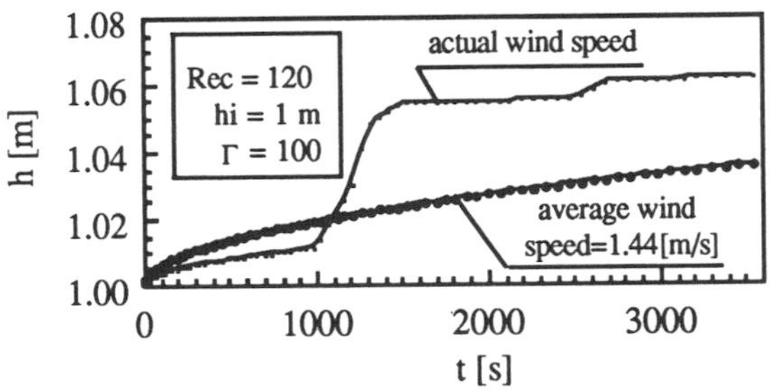

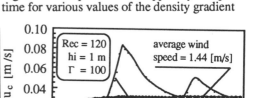

Fig. 4.b. Variation of the depth with
time for actual and average wind speed.

Fig. 4.a. Variation of core velocity with
time for actual and average wind speed.

2.29 Desalination and Solar Ponds Posters

SOLAR HEATED MEMBRANE DISTILLATION

G.L.Morrison[+], Sudjito[+], A.G. Fane[*] and P.Hogan[*]

[+] School of Mechanical and Manufacturing Engineering,
[*]School of Chemical Engineering and Industrial Chemistry
University of New South Wales. Sydney Australia

ABSTRACT

The possibility of using membrane distillation for solar powered desalination is described. A simulation model of a membrane distillation system has been developed, and combined with the TRNSYS solar simulation program to evaluate the process of a solar powered membrane distillation system. A preliminary design for a 50 litres/day pilot plant is presented.

KEYWORDS   Distillation, Membrane, Solar, Thermal

INTRODUCTION

Solar powered distillation plants have been developed on the principles of solar stills, solar multiple effect distillation (SMED), solar multistage flash distillation (SMSF), solar photovoltaic powered reverse osmosis (SPRO), and solar photovoltaic powered electrodyalisis (SPED). Basin type solar stills received a considerable amount of research and development between 1950 and early 1970's. This system is suitable for small capacity plants such as a household or a small community in rural/arid zones, but it is not economically viable for large capacity plants due to its low efficiency. The other four systems compete to gain acceptance in applications, which are mostly in the Middle East. In contrast to the solar still, these four systems are usually applicable for large capacities only. A costing of SMSF and SPRO plants in Australia, showed that the product cost increases dramatically for plant capacity less than 300 m³/day. A study for a SPRO plant with a capacity of 50 m³/day concluded that such a system was not economically viable unless the cost of photovoltaic panels decreased to about 5 percent of the cost at that time (about A\$ 20/watt).

Membrane distillation (MD), a new process in membrane technology, has some features that make it suitable for solar powered distillation plants. The process can operate at temperatures of 50 to 90°C and it is possible to recover latent heat in the product stream. The process also operates at atmospheric pressure and hence the associated components are not expensive. The application of MD for water desalination has been studied by a number of workers [1-3]. A preliminary cost calculation for small capacity MD plants [3] showed production costs marginally higher than conventional RO plants with the same capacity. However, the MD plant has practical advantages for application in rural/remote areas where electricity is not available. Compared to solar stills, MD systems offer better heat utilization due to heat recovery and better system efficiency. This paper presents a study of a small solar powered MD plant to produce potable water from brackish water in rural/arid zones.

SOLAR POWERED MEMBRANE DISTILLATION (SPMD) PROCESS

Membrane distillation is a thermally driven membrane process in which a hydrophobic microporous membrane separates hot and cold streams of water. The hydrophobic nature of the membrane prevents the passage of liquid through the pores whilst allowing the passage of water vapour. The temperature difference produces a vapour pressure gradient which causes water vapour to pass through the membrane and condense on the colder surface. The result is a distillate of very high purity which, unlike conventional distillation, does not suffer from entrainment of species which are non-volatile.

A flow diagram of the SPMD process is shown in Fig. 1. A hot and cold stream are contacted counter-currently in the MD module, at which point mass and energy are transferred from the feed to the permeate stream. The module contains hollow fibre membranes tightly packed in a shell in order to enhance the heat transfer characteristics of the shell side (7). The streams are recontacted in the main heat recovery heat exchanger (HX1) where energy is transferred back from the permeate to the feed. To provide process continuity the feed and permeate must be returned to their original temperatures before being recycled to the MD module. The feed stream is reheated using energy from a solar collector, which in this case is isolated from the feed circuit. Water is taken from the feed tank heated by the collector and returned to the tank, allowing the flow conditions through the collector to be set independently of those chosen for the MD circuit. The tank enables the system to store solar collector output when the collector output temperature is to low for efficient membreane operation.

The primary energy sink is a second heat exchanger HX2 that is cooled by accumulated product or cold excess feed. If cold feed is used in HX2 the make up water is taken from the outlet of HX2 as shown in Fig. 1. As the process continues the accumulating product water is removed from the permeate stream. The feed stream is depleted of water by the distillation, and becomes more concentrated in the retained impurities. This build-up must be limited by a controlled bleed stream.

MD Module Simulation Model

Many membrane configurations have been investigated for MD modules [1-3]. In this study a hollow fibre module was used as shown in Fig. 2. A simulation model of the MD process has been developed, based on the theory of Schofield [7]. The MD model has been integrated with the TRNSYS solar package to produce a full system simulation model. The membrane model computes mass transfer, temperature and pressure gradients in the fibre using an iterative solution procedure along the feed stream direction.

Membrane Characteristics

The simulation model has been used to provide design information with respect to the MD module, by predicting the system performance for a range of operating conditions. Figure 3 shows the benefits of operating at high feed temperatures and high liquid flow rates. This information is based on a pilot module which has a $0.17 m^2$ membrane area consisting of 1100 fibres (0.17m length, 0.3mm i.d. and 0.6mm o.d.) with a pore size of 0.22 micron and 70% porosity. The non-linear increase in flux with increasing temperature reflects the exponential increase in the vapour pressure which provides the driving force. The effect of a higher liquid flowrate is to increase the heat transfer coefficient, and thus reduce the effect of temperature polarization. This means that the temperatures at the membrane surface more closely approximate that of the bulk streams, and thus the trans-membrane temperature difference is greater. This produces a greater driving force, and consequently enhances the flux. Accompanying the increases in flux is a decrease in the heat loss factor. This represents the fraction of the total heat transferred across the membrane that does not contribute to the flux. This occurs primarily by conduction through the membrane structure, and through the air and water vapour in the pores. Heat transfer by conduction increases approximately linearly with temperature gradient, unlike the vapour pressure driving force and thus the mass flux which increases exonentially. This means that although more heat is lost by conduction at higher temperature difference, it is a lower proportion of the total heat transfer.

Figure 4 shows the effect of increasing the membrane area with constant stream inlet temperatures and flowrates. The increase in area in this case is achieved by increasing the length of the fibres. The overall effect of larger membrane area is that a greater amount of heat is transferred from feed to permeate. As the approach temperature difference becomes smaller the driving force across the membrane is reduced, and the flux drops. For the conditions shown in Fig. 4 the drop in flux is offset by the increase in area and so the production rate (kg/hr) is observed to climb. As the area becomes larger the rate of increase in production rate diminishes until the point where a further increase in area has negligible effect.

One of the main advantages of the MD process is its ability to recover the latent heat of vaporisation. The proportion of heat transferred during distillation that can be recovered depends primarily on the approach temperatures of the liquid streams. Large membrane areas and lower flowrates both lead to increased contact time in the module, which gives rise to closer approach temperatures and thus more recoverable heat. However one of the inherent trade-offs in MD, is that whilst higher flowrates and smaller membrane areas yield higher flux, the possibility of heat recovery is reduced. The conditions of choice then become a function of the application.

SYSTEM DESIGN SENSITIVITY

The pilot plant (Fig. 1) has been modelled by combining the membrane module simulation program with the TRNSYS solar simulation program [8]. The operating conditions, module specifications, and solar collector type were kept constant during the sensitivity study. The module specifications chosen were as for Fig. 4.

Heat exchanger size, defined by the product of the heat transfer coefficient and area (UA), was varied from 25 to 550 W/k. for HX1. The performance measure chosen was the productivity, defined as the kg water produced per unit of external energy provided (kg of permeate per MJ energy from the solar collector). Figure 5 indicates that increasing the size of the main heat exchanger reduces the need for external energy, as would be expected. This trend is shown for two sizes of the cooling heat exchanger, HX2

The capacity of the plant is affected by the efficiency of the collector in converting the incident radiation into sensible heat. Figure 6 shows the system performance for a flat plate and evacuated tube collector over a range of feed temperatures. The performance of a system with a flat plate collector drops off above 70°C due to lower collector output at high temperatures. In contrast a system utilizing evacuated tubes increases in efficiency past 100°C.

The storage tank for solar thermal input is included to avoid inefficient operation of the module at times when the collector outlet temperature is low. Large storage volumes reduce the mass transfer, because the average feed delivery temperature decreases as the storage volume is increased. A small storage volume allows energy from the early morning and during transient periods to be collected and stored over periods of 30 to 60 minutes without significantly reducing the feed delivery temperature during high irradiation periods.

PROTOTYPE PERFORMANCE

Figure 7 shows the simulated variation in production rate over an eight hour day-time operation, using radiation data for Sydney in summer. This information is based on a module with the same number and type of fibres as described previously. Comparative data for the actual pilot plant under the same conditions are also graphed (with energy input from a heating element). The chosen operating conditions were for an average feed temperature of 70°C and flowrates of 30kg/hr. The simulation gives a reasonably accurate picture of the production rate trend, although the figure shows that it typically predicts 10% higher than determined experimentally. This may be due to the heat losses or to imperfect flow distribution in the membrane module.

CONCLUSIONS

A distillation process using solar thermal energy to power a membrane distillation system has been shown to be feasible. Heat recovery is the most critical factor governing the cost of solar powered membrane distillation plants as the solar collectors are likely to be the major cost item. High heat recovery can be achieved either by using low mass flow rates of the feed and permeate, or using long membrane fibres. Evacuated tube collectors are more suitable for membrane distillation than flat plate collectors, since the membrane works best at feed temperatures of 60 to 90°C for a non-pressurized system.

The capital cost of the unit is very sensitive to the extent to which heat is recovered, especially above a heat recovery factor of 0.8. The minimum capital cost requires 60 to 80% heat recovery. To obtain this level of energy conservation the process conditions that must be used are generally those that result in fairly low MD fluxes. These are low flowrates and large heat tranfer areas, which lower the driving force.

For a domestic sized plant of 50kg/day the optimum configuration appears to be a solar collector area of around $3m^2$, a membrane area of $1.8m^2$ and a total heat exchange area of 0.7 m^2. The capital cost for this unit is conservatively estimated at $3500 (Aust.).

ACKNOWLEDGEMENTS The authors would like to thank Memtec Ltd, for material support and advice, and NERDDC for financial support.

REFERENCES

Anderson S.I. et al., "Design And Field Tests Of A New Membrane Distillation Desalination", Desalination Vol.56 (1985) pp.345-354.

Sarti G.C., "Low Energy Cost Desalination Processes Using Hydrophobic Membranes", Desalination Vol.56 (1985) pp.277-286.

Schofield R.W., Fane A.G., And Fell C.J.D., "The Efficient Use Of Energy In Membrane Distillation", Desalination Vol.64 (1987) pp.231-143.

Schofield,R.W. P.A.Hogan, A.G.Fane and C.J.D.Fell, "Developments in Membrane Distillation", Proceedings ICOM '90, p728-730.

Schneider,K and T.S.Van Gassel, "Membrandestillation", Chem.Ing.Tech. (1984), pp514-521.

Schofield,R.W. A.G.Fane and C.J.D.Fell, "The Efficient Use of Energy in Membrane Distillation", Desalination, 64(1987) pp231-243.

Schofield R.W., "Membrane Distillation", PhD. Thesis School of Chemical Engineering and Industrial Chemistry, UNSW, 1989.

KlineS.A. et al"TRNSYS 12.2 Users Manual", University of Wisconsin Solar Energy Laboratory, 1988.

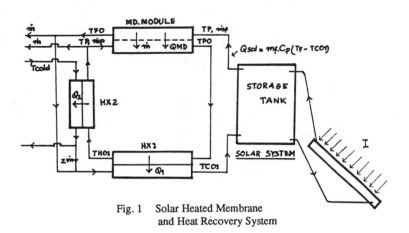

Fig. 1   Solar Heated Membrane
and Heat Recovery System

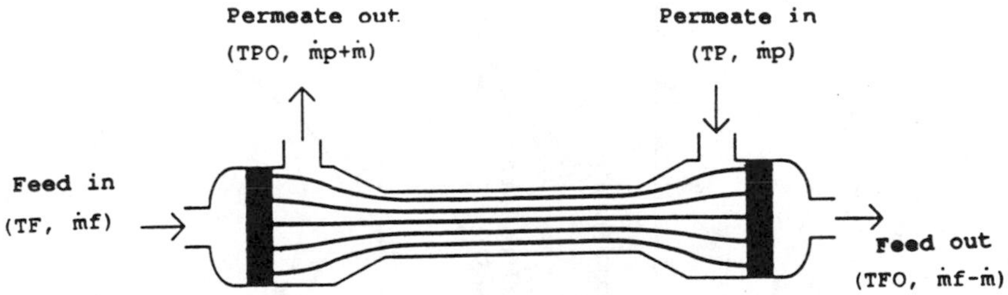

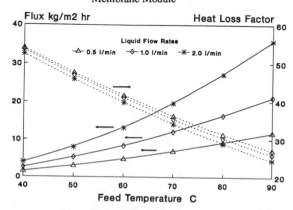

Fig. 2   Schematic of a Hollow Fibre
Membrane Module

Fig. 3.   Mass Flux and Heat Loss
versus Feed Temperature

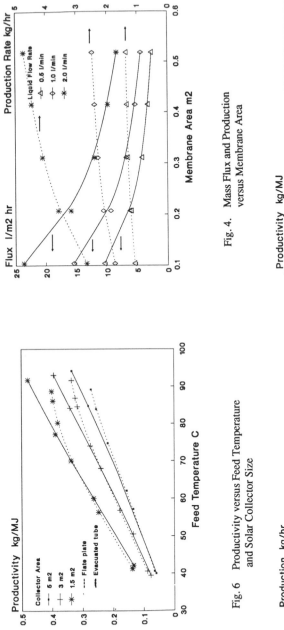

Fig. 4. Mass Flux and Production
versus Membrane Area

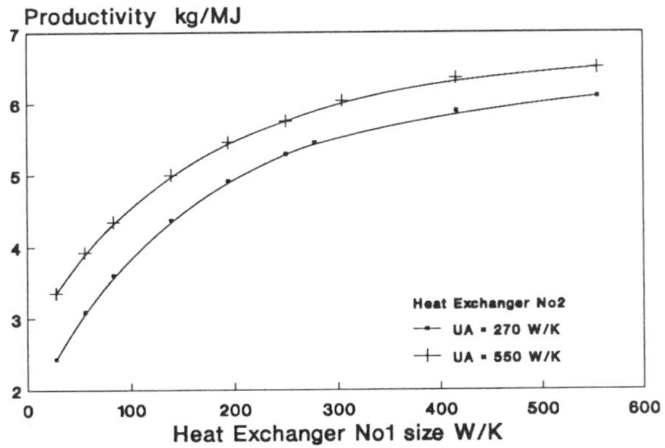

Fig. 6 Productivity versus Feed Temperature
and Solar Collector Size

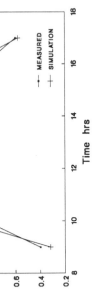

Fig. 5. Productivity versus Heat Exchanger Size

Fig. 7 Diurnal variation of production

COMPUTER MODELLING OF SOLAR POND WITH LATENT HEAT STORAGE

Li Shensheng[*], Mo Zhen[**], Lu Weide[**]

[*] Beijing Teachers' College, Beijing, P.R. China
[**] Beijing Solar Energy Institute, Beijing, P.R. China

ABSTRACT

This paper presents a mathematical model of the thermal process of solar ponds and that of phase change materials paved on the bottom of the pond. The possibility of enhancing the thermal storage capacity of solar ponds with salt hydrates is explored for the first time. Results of calculation show that it is not so efficient under technical and economical conditions now available.

KEYWORDS

Solar pond; thermal storage; phase change; latent heat; salt hydrates.

INTRODUCTION

A solar pond is a thermal collecting and storaging device which can achieve low temperature thermal storage in large quantity and long term. The key problem in its practical application is the storage capacity. In conventional solar ponds, the main storage capacity is the sensible heat of its salt solution in the lower convective layer.

Rabl and Nielsen (1975) suggested enhancing the thermal storage capacity by using latent heat of the phase change materials (PCMs). This method of thermal storage would enhance the density of storage capacity while reduce the amplitude of temperature variation during thermal storage and heat extraction. This paper presents a preliminary exploration of the thermal performance of a solar pond with PCMs, and provides information for further research in this field.

MATHEMATICAL MODEL OF SOLAR POND WITH PCMS

This paper attempts to construct a mathematical model of solar pond with PCMs on the basis of known theory of its thermal performance (Weinberger, 1964).

Suppose that solar pond with PCMs is constructed by paving the PCMs in closed container on the bottom of the pond. Salt hydrates are chosen as PCMs. The thickness of PCMs should be very thin (less than 1 cm in general) due to the separation of crystals and solution of salt hydrates.

The thermal process of phase change of PCMs on the bottom can be simplified as a one–dimensional conduction with moving boundary (Kreith 1973; Garg, Mullick and Bhargave, 1985). Assume that the heat capacity of the PCMs can be neglected and the temperature distribution in the PCMs is linear vertically, following relation can be obtained from the thermal equilibruim:

$$\frac{K_1[(T)_{z=H} - T_m]}{s_1} = \rho_1 \gamma_1 \frac{ds_1}{dt_1} \tag{1}$$

where the left side represents the heat conducted into PCMs from the bottom of the pond, while the right side represents the heat absorbed during phase change. Heat absorbed or extracted during phase change is

$$q_1 = \frac{\sqrt{2}}{2} \frac{K_1[(T_1)_{z=H} - T_m]}{h} \tau_1^{-\frac{1}{2}} \tag{2}$$

where τ_1 is the dimensionless time.

Now, on the foundation of Rubin's model (1984), different dimensonless boundary condition is applied when phase change occurs,

$$\left(k\frac{\partial\theta}{\partial z}\right)_{z=1} = \alpha BJK \sum_{j=1}^{4} \eta_j \exp\left(\frac{-a_j K}{\cos\varphi_2}\right) - \frac{\sqrt{2} K_1 H}{K_0 h}[(\theta)_{z=1} - \frac{T_m - T_0}{\Delta T}]\tau_1^{-\frac{1}{2}} \tag{3}$$

where the parameters are the same as that in Rubin's model.

Using finite difference equations, the temperature profile of solar pond with PCMs in any time interval during operation can be calculated by repeated substitution.

EXPERIMENTAL EXAMINATION OF THE MATHEMATICAL MODEL

To examine the calculated results of mathematical model of solar pond with PCMs, two mini–solar ponds of dimensions 1.57m × 1.57m × 0.8m were used and $Na_2SO_4 \cdot 10H_2O$ and $Na_2S_2O_3 \cdot 5H_2O$ were chosen as PCMs, and the experiments were achieved in summer and autumn of 1989 under natural conditions. But no satisfactory results were obtained because of the rainy weather in those seasons, so simulating experiments were undertaken instead.

Mathematical Model of Simulating Experiments Indoor

The heating process of simulating experiments indoors is different from that of practical solar pond operating under natural conditions. Although the heat source of the former is from the bottom of the simulating ponds, the phase change processes in both cases are approximately similar.

Mathematical model of simulating experiments with isothermal heating at the bottom. The surrounding and top of the simulating ponds were thermally insulated, hence the thermal process in the ponds can be described similarly to the Rubin's model, but with following dif-

ferent boundary conditions:

$$(k\frac{\partial\theta}{\partial z})_{z=1} = \frac{k_s H(T_m - T_a)}{h(1-z_1)\Delta T} - \frac{k_s H}{h(1-z_1)}(\theta)_{z=1} \tag{4}$$

$$(k\frac{\partial\theta}{\partial z})_{z=1} = \frac{H(T_{HW} - T_a)}{\left(\frac{\delta_g}{k_g} + \frac{h}{k_1}\right)\Delta T} - \frac{H(\theta)_{z=1}}{\left(\frac{\delta_g}{k_g} + \frac{h}{k_1}\right)} \tag{5}$$

where k, k_s, k_g, k_1 are relative coefficients of heat conduction, δ_g and T_{HW} are thickness and temperature at the bottom of simulating ponds respectively.

From the thermal equilibrium of phase change process a dimensionless thickness can be obtained:

$$z_1^2 - A_1 z_1 - B_1 \ln(1 - \frac{C_1}{D_1} z_1) = E_1 \tau_1 \tag{6}$$

where A_1, B_1, C_1, D_1, E_1 are constants deduced from thermal equilibrium. The temperature profiles in simulating ponds would be obtained if the initial temperature profile is given as in a conventional solar pond.

Mathematical model of simulating experiments with steady flow of heat from the bottom. The dimensionless boundary conditions of the thermal process in simulating ponds are:

$$(k\frac{\partial\theta}{\partial z})_{z=1} = \frac{k_s H(T_m - T_a)}{h(1-z_1)\Delta T} - \frac{k_s H}{h(1-z_1)}(\theta)_{z=1} \tag{7}$$

$$(k\frac{\partial\theta}{\partial z})_{z=1} = \frac{Hq}{\rho ck_0 \Delta T} \tag{8}$$

where q is the heat flux through the bottom of simulating ponds.

Similarly, z_1 can be obtained through thermal equilibrium:

$$z_1 - F_1 \ln(1 - G_1 z_1) = \tau_1 \tag{9}$$

where F_1, G_1 are constants through deduction. The temperature profiles in simulating ponds would be obtained by solving the numerical finite difference equations.

Comparision Between Values Obtained by Simulating Experiments and Calculation

The container of simulating pond is made of glass, the solution of salt hydrates is contained in a bag of thin film. PCMs are paved between the container and the bottom of the bag. The surrounding and top of the container is insulated, and heating is through the bottom of container from outside. Isothermal heating is achieved by a thermostat, and steady heat flow is achieved by electro−thermic film printed on the bottom outside of the container. $Na_2SO_4 \cdot 10H_2O$ of thickness 8mm and area $0.01m^2$ is chosen as the PCM in the experiment. The temperature of the bottom is maintained at 45℃ during isothermal heating; while the electric power of 10W and 20W are maintained respectively during experiment of steady heat flow.

The experimental and calculated results of isothermal heating are shown in Fig. 1. Phase change occurs at 32.4℃, and the maximum difference between experimental and theoretical temperatures is about 7%.

The experimental and calculated results of steady heat flow are shown in Fig. 2. The maximum difference between experimental and theoretical tempertures is about 9—10%.

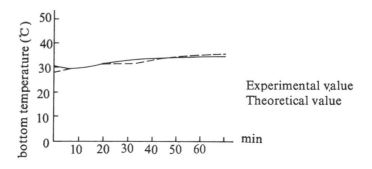

Fig. 1. Comparison between and experimental theoretical values under isothermal heating.

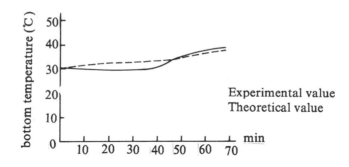

Fig. 2. Comparison between theoretical and experimental values under steady flow heating.

The temperature differences between experimental and theoretical values under two different heating processes are less than 10% It indicates that the mathematical model of phase change process described above is essentially available.

CALCULATED RESULTS OF SOLAR POND WITH PCMS AND DISCUSSIONS

In simulating the temperature variation at the bottom of solar pond with PCMs under long—term operation and without heat extraction, the temperature profile in this pond can be obtained by numerical difference method. The treatment of J factor, B value and heat transfer in upper and lower convective zones is similar to that of Rubin's model (1984). Suppose the pond locates at 40 ° N, then annual variation of atmospheric temperature of Beijing area can be given by:

$$T_a = 11.6 + 15.4\cos[\frac{2\pi(D-27)}{365.25}] \tag{10}$$

The most commonly used salt hydrate Na_2SO_4 $10H_2O$ is chosen as PCM.

Suppose the solar pond sets in operation from the Spring Equinox (March 21). When the temperature at the bottom rises to higher than that of the melting point of $Na_2SO_4 \cdot 10H_2O$, then phase change begins. Comparison of maximum temperatures between solar pond with PCMand aconventional one with depth of 1m and dimensionless number (related to solar irradtion) B = 312 is given in Fig. 3. After the beginning of phase change, the rate of melting becomes smaller, so that only one part of radiation provides the PCM to continue its melting, while other part makes the temperature at the bottom rise again, but rate of temperature rising is slower than that of the conventional pond. But after the conclusion of phase change, the temperature at the bottom of solar pond with PCM is lower than that of conventional one. Its heat loss to environment is smaller too, so its rate of temperature rise is larger, hence the temperature at the bottom would approach to that of conventional pond in a very short time. From Fig. 3 it can be seen that during the long-term operation of solar pond, the effect of PCM to the thermal storage capacity of solar pond only exists nearby the melting point.

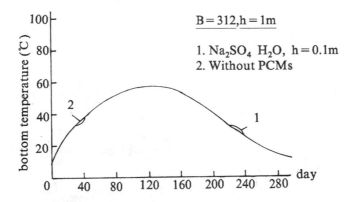

Fig. 3. Comparison of temperatures between solar ponds with
PCMs and conventional solar pond

Fig. 4 represents the temperature at the bottom of solar pond with different intensities of solar irradiation. It can be seen from the figure that the contribution of PCM to enhance the thermal storage capacity is not evident too.

In general, the values of $\rho_1\gamma_1$ of different salt hydrates are rather near, so the difference of their effects to the thermal storage capacity of solar pond is not significant. Now suppose an ideal PCM with same melting poind as that of $Na_2SO_4 \cdot 10H_2O$, but its value of $\rho_1\gamma_1$ is 100 times that of the latter. Comparison of the temperatures at the bottom of solar pond with this ideal PCM to that of $Na_2SO_4 \cdot 10H_2O$ is shown in Fig. 5. It is obviously that PCM with such a large value of $\rho_1\gamma_1$ can significantly enhances the thermal storage capacity of solar pond and reduce its heat loss to environment. Although the existence of such a PCM has not been found yet, but to search such PCMs with large values of $\rho_1\gamma_1$ and to apply them in solar ponds, would be an available way to improve the performance of solar ponds effectively.

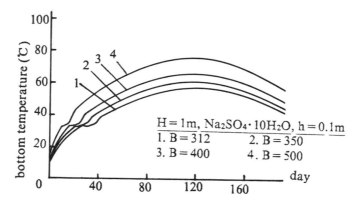

Fig. 4. Comparison of temperature between solar ponds with PCMs
under different solar irradiations

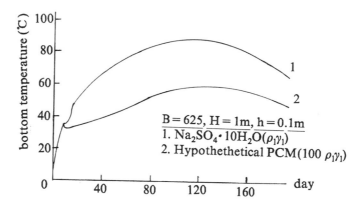

Fig. 5. Comparison of temperatures at the bottom of solar ponds
with different PCMs

REFERENCES

Garg, H. P., S. C. Mullick, and A. K. Bhargava (1985). Solar Thermal Energy Storage, D. Reidel Publishing Company, Dordrecht. p. 227–240.

Kreith, F. (1973). Priiciples of Heat Transfer, 3rd ed. Intext Educational Publishers, New York.

Rabl, A. , and C. E. Nielson (1975). Solar ponds for space heating. Solar Energy, 17,1–12.

Rubin, H., and B. A. Benedict (1984). Modeling the performance of a solar pond as a source of thermal energy. Solar Energy, 32. 771–778.

Weinberger, H. (1964). The physics of the solar pond. Solar Energy, 8, 45–56.

CAN ANY CA-CLAY BE USED AS A LINER FOR NACL SOLAR PONDS?

R.Almanza and R. Castañeda

Instituto de Ingeniería, Universidad Nacional Autónoma
de México, Ciudad Universitaria, 04510, D.F., México
FAX (5) 548-3044

ABSTRACT

A calcic nonswelling clay is being studied as a liner for solar ponds. Its chemical analysis showed the following cations: Ca^{++} 31.9 meq/100g of soil, Mg^{++} 15.96 meq/100g of soil, Na^+ 1.96 meq/100g of soil and K^+ 0.66 meq/100g of soil. The soil has a texture of 68% clay, 14% sand and 18% silt. Under hot water the permeability was of the order of 10^{-8} cm/s, and the thermal conductivity was about 1.5 W/m°C. Nonetheless, when this clay was under hot NaCl brine (50°C), the permeability started to increase until 10^{-6} cm/s or more. In order to try to understand the behavior of this clay, additional studies are in progress. An X-ray diffraction analysis, a differential thermal analysis, an electron microscopy study and tests under a permeameter are being realized. It seems that the clay under the influence of an electrolyte as the NaCl brine, flocculates and makes the voids higher and therefore more premeable.

As a conclusion, it is possible to affirm that not any Ca-clay can be used as a liner for NaCl solar ponds. To know which one is convenient for liners, it is necessary to analyze its mineral structure mainly.

KEYWORDS

Liners; solar ponds; clay liners; calcic clays; flocculation.

INTRODUCTION

The state of the art of liners for large-scale artificial ponds is available only to get permeabilities from 10^{-7} to 2 x 10^{-8} cm/s (Daniel, 1987; Montague, 1982; Ghassemi, 1985) with clays, membrane liners, or a combination of both. This technology has been developed mainly for waste disposal facilities. The cost of compact clays according to Kays (1977) is about 2 to 3 times cheaper than the one for membrane liners.

Compact clays can be an option to be used as liners for solar ponds. Two clays have been studied, a kaolinite and a bentonite by Almanza (1989, 1990). Both clays gave a permeability of the order of 10^{-7} cm/s and their thermal conductivities were less than 0.6 W/m°C, after have interacted with NaCl brine, and also due to an ion exchange. This means that one Ca^{++} ion was changed by two Na^+ ions. Before the interaction between the clay and the brine, the permeabilities and the thermal conductivities were higher, and both clays were calcics, and after such interaction they were sodics. The advantage of one clay over the other was that the cation exchange capacity and the specific surface were higher in the bentonite clay than in the kaolinite one. However, the bentonite is a swelling clay and under strong changes of temperature (10°C or 20°C in one hour) some fractures may appear.

For NaCl solar ponds, the clays must be Ca-clays because -during the ion ex-change- two Na^+ ions are interchanged for one Ca^{++} ion permitting the clay to be more impermeable with the time until all the Ca^{++} are exchanged for the Na^+.

For these reasons a third Ca-clay is being studied. This clay is a calcic non-swelling clay and the results are shown in this paper.

CA-CLAY

In a third study, a nonswelling clay is being tested.

This soil had the following texture: 68% clay, 18% silt, and 14% sand. A soil with more than 30% of clay has a very good impermeability (Grim, 1962).

This soil can be found in the State of Michoacán, Mexico. Its chemical analy-sis showed thàt it is a Ca-clay. It has Na^+ with 1.96 meq/100g of soil, Ca^{++} with 31.9 meq/100g of soil, Mg^{++} with 15.96 meq/100g of soil and K^+ with 0.66 meq/100g of soil.

The mechanical properties demonstrated that the LL was 76.6% (Fig. 1), the PL was 37.1%, and the Ip was 39%, before interacting with a NaCl brine. The soil can be classified as a CH clay of high plasticity according to Lambe (1969).

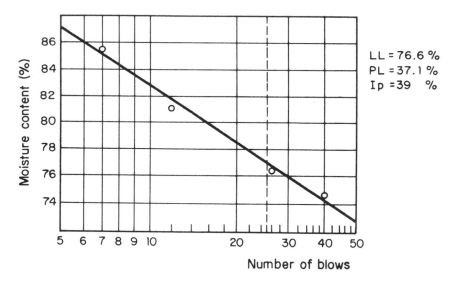

Fig. 1. Liquid limit before interaction with hot brine
(defined at the Casagrande cup at 25 blows).

In soil mechanics, the Proctor Compaction Method is a standard practice to de-termine the humidity necessary for maximum impermeability (Lambe, 1969, 1951) Fig. 2 shows this test getting such a humidity. The humidity found to produce the lowest permeability was 36%. This value was obtained by adding 5% to the maximum position of the curve (Grim, 1966). This clay was compressed in the laboratory apparatus shown in Fig.3. This apparatus has been described by Al-manza (1989), and it makes possible to measure the thermal conductivity and the vertical permeability. These values were at 50°C: 1.5 W/m°C and $\sim 10^{-8}$ cm/s, respectively.

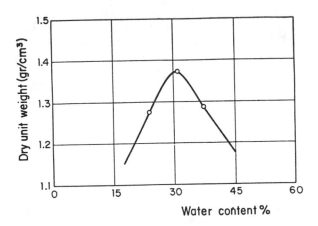

Fig.2. Proctor test for determining the optimum water content.

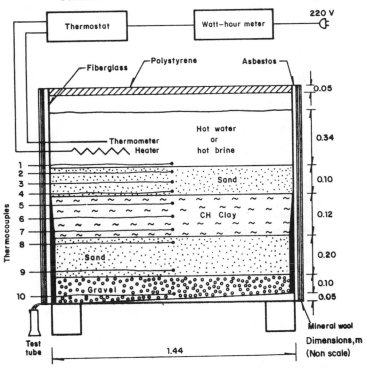

Fig.3. Laboratory apparatus for obtaining permeability, thermal conductivity and ion exchange.

After this clay was under the hot NaCl brine (50°C), the permeability started to increase until 10^{-6} cm/s or more. In order to try to understand the behavior of this clay, the following analyses are in progress:

a) X-ray diffraction analysis. Two diffractometers have been obtained, one as the soil was found in the field, and the other by identification of the principal clay minerals (<2 μ m) after the minerals were under the effects of heating (200°C and 560°C) and glycol treatment. Figures 4 and 5 show the diffractometers obtained. As can be seen in these figures it is possible to identify this clay as an illite-montmorillonite.

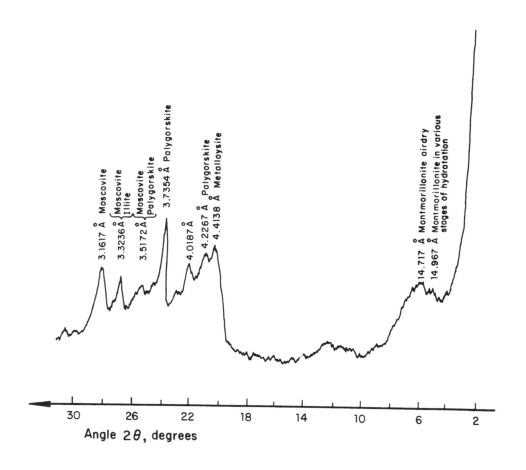

Fig. 4. Diffractometers obtained from the soil that were found in the field.

b) Differential thermal analysis. This analysis was obtained by using only clay minerals (<2 μ m). The thermogram gave additional information about the probable minerals in the soil.It showed that the clay could be illite-chlorite or illite-vermiculite.

c) Electron microscopy. A scanning electron microscope was used to try to identify the minerals. Through this technique it is possible to say that the mineral structure of the soil could be an illite-montomrilonite one.

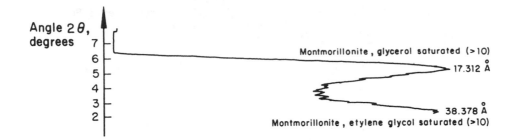

Fig. 5. Diffractometer obtained after the clay was
treated with the orientated aggregate technique.

d) Permeameter. In order to understand the mechanisms that occur in the clay
under water and NaCl brine, a permeameter was used. This device is a laboratory apparatus where can be obtained lower permeabilities than in the
field because there is a better control of the different parameteres involved in the sample forming. With distilled cold water, a permeability
of about 10^{-10} cm/s was obtained. While using cold NaCl brine the permeability was 10^{-8} cm/s, and with hot NaCl brine the permeability was 10^{-7} cm/s.

CONCLUSIONS

The main conclusion of this study is that not any Ca-clay can be used as a liner for solar ponds. It seems that a clay with mineral structure of an illite
with another mineral and un r an electrolite as the NaCl flocculates, being
the void ratio higher than with distilled water only.

The use of clays as liners is a very good option in large area solar ponds.
However, in order to know if any clay can flocculate, it is important to be
aware of its mineral structure. Mainly, the illite flocculates under the influence of an electrolite as the NaCl brine, as it is shown in this study.
Soil mechanics has demonstrated that some clays can flocculate (Mitchell, 1976)
and also can be more permeable.

REFERENCES

Almanza, R., A. Martínez and G. Segura (1989). Study of a kaolinite clay as a
liner for solar ponds. Solar Energy Vol. 42, No. 5, pp. 395-403

Almanza, R., and M. C. Lozano (1990). Mechanical and thermal tests of a bentonite clay for use as a liner for solar ponds. Solar Energy Vol. 45, No. 4,
pp. 241-245.

Daniel, D.E. (1987). Earthen liners for land disposal facilities. Geotechnical
Practice for Waste Disposal 87, R.D. Woods (ed.) ASCE, New York.

Ghassemi, M., and M. Haro (1985). Hazardous waste surface impundment technology. Journal of Environmental Engineering, ASCE, Vol. 11 No. 5, pp.602-617

Grim, R. E. (1962). Applied Clay Mineralogy, McGraw Hill, New York.

Kays, W.B. (1977). Construction of Linings for Reservoir, Tanks and Pollution
Control Facilities, Willey, New York.

Lambe, T.W. (1951). Soil Testing, Willey, New York.

Lambe, T.W., and R. V. Whitman (1969). Soil Mechanics, Willey, New York.
Mitchell, J.K. (1976). Fundamentals of Soil Behavior, John Wiley & Sons.
Montague, P. (1982). Hazardous waste landfills: some lessons fron New Jersey.
  Civil Engineering- ASCE, September, pp. 53-56

EVAPORATION FROM A FREE WATER SURFACE WITH SALT CONCENTRATION

Ernani Sartori

Laboratório de Energia Solar, Universidade Federal da Paraíba
58059 João Pessoa PB, Brazil

ABSTRACT

In the present work a free water surface with salt concentrations of 3.5%, 10% and 20% that operates during 24 hours a day is considered in theoretical simulations. The influence of salts in water on evaporation is determined through the application of the Raoult's Law. As evaporation proceeds, salt precipitation occurs and alters the evaporator bottom colour, which effect upon evaporation is also investigated. It is shown that the salt alone has a relatively small effect on evaporation, mainly for salinities under ten per cent, but the bottom reflec - tivity due to salt deposition has a determinant influence on the pond perform - ance. Other results show, as expected, that evaporation increases with the water temperature increasing but with higher salinities this increase is less than with lower ones.

KEYWORDS

Solar evaporator; salinity; bottom reflectivity, mole fraction; water tempera - ture; transient operation.

INTRODUCTION

Salt concentration in free water surfaces is especially encountered in salt works evaporators and in solar ponds, having a marked effect on the corresponding per - formances. The presence of salt in water reduces the water vapor partial pressure of the solution, to which term the evaporation is directly proportional. Consider - ing a free water surface not thermally stratified (shallow pond), besides this influence, the salt still generates a big damage to its efficiency due to the bot - tom reflectivity caused by the salt deposition. Solid salt generally has a high optical reflectivity.

Taking into account the referred system, in this paper two kinds of investigations are performed. In the first the evaporation is determined according to the sali - nity and temperature of the solution, keeping constant all other variables. In the second three distinct situations in the transient mode are simulated: a) the free water surface without salt concentration; b) with salt in water, but the bottom is still black; c) with salt in water, but due to salt precipitation the bottom is already white. An insulated solar evaporator with a water surface of 50 m X 20 m and a water layer of 20 cm is utilized for the analyses. In this sec - ond investigation, values of 24 hours of operation are given, showing the influ - ence of the salt alone and of the bottom reflectivity on evaporation. The behavior of the water temperature in this period is also presented.

THEORETICAL CONSIDERATIONS

The transient operation of free water surfaces has been mathematically modelled and discussed in Sartori (1988, 1990a). Since the present case considers the turbulent flow, we have from Sartori (1990a) the corresponding equation for the evaporative heat transfer rate which is

$$q_{ew} = 0.6799h_c(P_w - P_d)h_w/P_T \qquad (1)$$

where h_c is the average forced convective heat transfer coefficient, P_w and P_d are the partial pressures of water vapor at the water and the air dew point temperatures, respectively, P_T is the atmospheric pressure, and h_w is the latent heat of vaporization of water. Expressions and units for the above terms can be found in Sartori (1988, 1990a, 1990b).

To take into account the influence of salt on evaporation, the Raoult's Law is employed, which states that the partial pressure of the solution is equal to the product of the partial pressure of the pure solvent (at the same temperature) by the mole fraction of the solvent in the considered solution (Nehmi, 1967), or

$$P_s = P_w \cdot x_m \qquad (2)$$

This equation satisfies the condition that the partial pressure of the solution is always lower than the partial pressure of the pure solvent.

The mole fraction of the water x_m is expressed by (Cooper, 1970)

$$x_m = 1/(1 + 0.621[S/(100 - S)]) \qquad (3)$$

where the salinity S is the total mass of salt, divided by the total mass of water and times 100, and is the equivalent salinity in terms of NaCl.

Thus, equation (1) can be written as

$$q_{ew} = 0.6799h_c(P_s - P_d)h_w/P_T \qquad (4)$$

The remainder equations of the model are kept without alterations.

THEORETICAL SIMULATIONS

In this work two kinds of investigations are performed. In the first the following equation is employed to find a correlation among evaporation, water (or solution) temperature and salinity

$$m = 0.6799h_c(P_s - P_d)/P_T \qquad (5)$$

Evaporation is then determined according to the temperature and salinity of the solution, keeping constant all other variables.

The second, simulates the transient operation during 24 hours a day (using input data of July 22, 1987) of an insulated solar evaporator which has a water surface of 50 m X 20 m and has a layer depth 20 cm. One dimensional flow in the length direction is considered. Salt concentrations of 3.5% (like the average sea water con - centration), 10% and 20% are assumed. As evaporation proceeds, salt precipitation occurs and alters the evaporator bottom colour. So, the effect of the solar re - flectivity on evaporation due to solids concentration onto the bottom is also investigated. Therefore, this second analysis can be divided into three distinct situations: a) the free water surface without salt concentration; b) with salt in water, but the bottom is still black; c) with salt in water but due to salt precipitation the bottom is already white, i.e. the operation occurs when the bottom area is uniformly full of salt. For each considered salinity, cases b and c are applied.

Some other considerations for the simulations are done in the following para -
graphs. The solar absorptivity of the particle-free salt water is here considered
to be the same as for pure water (Hawlader, 1980) but when the bottom is already
white, an absorptivity value of 0.20 for the solution-bottom system is admitted,
since a reflectivity value of 0.8 is found to be reasonable (Kooi, 1981) when
solid salt covers the bottom.

For the energy balance of the evaporator the thermal inertia of the salt is also
computed which mC_p term is added to that of the original system. Although the
variation of the specific heat of the salt with salinity is relatively small, the
corresponding value is accounted in the calculations. The hourly salinity is also
calculated according to the hourly evaporation. The salinities utilized in this
investigation are within the solubility limit of the salt.

RESULTS AND DISCUSSIONS

Figure 1 shows the results obtained with the use of eqn (5). It can be seen that
the evaporation of salt water increases with the water (or solution) temperature
increasing, as is also the case for pure water. With higher salinities this in -
crease is less than with lower ones. It is also depicted from these results
that the percentage reduction in evaporation is approximately equivalent to the
salinity, i.e. a salinity of 3.5% produces an evaporation around 3.5% lower than
that for pure water.

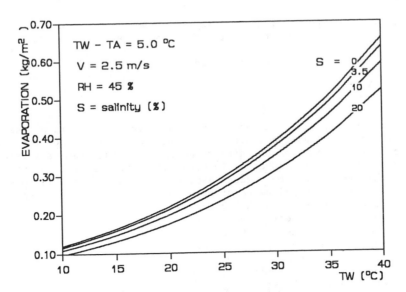

Fig. 1. Evaporation according to salinities and solution temperatures

Table 1 summarizes the results of the second kind of investigation of this work,
i.e. simulations of the transient operation of the evaporator in different conditions, as described previously.

As shown, for a whole day the salt alone reduces the evaporation in 2.4%, 6.6%
and 14.2% in relation to a zero percent salinity and respectively for the 3.5%,
10% and 20% salinities. Such reduction represents around 0.7 times the correspond
ing salinity. The difference between this factor and that found from Figure 1 can
be attributed to the several variables that interfere on the transient operation,
like the system thermal inertia, solar absorptivity, thermal insulation, atmos -

2350

Table 1

RUN	S (initial) %	ABSORP	TW_{ave} °C	TW_{max} °C	EVAP kg/m²day	S (final) %	REDUC EVAP. %
1	0	0.90	28.11	34.27	5.44	0	0
2	3.50	0.90	28.42	34.42	5.31	3.59	2.4
3	3.50	0.20	23.06	25.24	0.93	3.52	83.0
4	10.00	0.90	29.01	34.81	5.08	10.26	6.6
5	10.00	0.20	23.50	25.60	0.75	10.04	86.2
6	20.00	0.90	30.00	35.54	4.67	20.47	14.2
7	20.00	0.20	24.24	26.26	0.44	20.05	92.0

pheric conditions, etc. which are not taken into account in eqn (5). The salinity can be considered as having a relatively small effect on evaporation, mainly for percentages under ten. But that is not the case when we consider the bottom reflectivity caused by salt deposition. As can be seen, with the absorptivities values used the evaporation is reduced 83.0%, 86.2% and 92.0%, respectively, for the corresponding salinities. This is according to the conclusions of Kooi (1981) , for whom the performance of a shallow pond with a high bottom reflectivity can be disastrous.

The average water temperature as well as the maximum temperature increase with the salinity increasing, which results are in accordance with those from Cooper (1970). As expected, the final salinity after 24 hours increased due to the evaporation, showing higher increase for higher evaporation and vice-versa.

CONCLUSIONS

Evaporation from salt water increases with the water (or solution) temperature increasing, but with higher salinities the evaporation is lesser than with lower ones. As a rule of thumb the salinity value can be considered as a parameter for estimating the percentage reduction in evaporation in relation to that from pure water, but this factor can be reduced because of several variables present in the transient operation. The influence of each one of these variables on evaporation considering the salinity is another matter of investigation. The salinity alone has a relatively small effect on evaporation, mainly for salinities under ten per cent. The bottom reflectivity due to salt precipitation, however, has a determinant effect on the pond performance.

REFERENCES

Cooper, P.I. (1970). The Transient Analysis of Glass Covered Solar Stills, Ph.D. Thesis, University of Western Australia, Perth.
Hawlader, M.N.A. (1980). The influence of the extinction coefficient on the effectiveness of solar ponds. Solar Energy, 25, 461-464.
Kooi, C.F. (1981). Salt gradient solar pond with reflective bottom: application to the "saturated" pond. Solar Energy, 26, 113-120.
Nehmi, V.A. (1967). Química Geral, 5th ed. Escola Politécnica, São Paulo.
Sartori, E. (1988). A mathematical model for predicting heat and mass transfer from a free water surface. In: Advances in Solar Energy Technology (W.H. Bloss and F. Pfisterer, ed.), Vol. 4, pp. 3160-3164. Pergamon Press, GB.

Sartori, E. (1990a). Prediction of the heat and mass transfer from a free water surface in the turbulent flow case. Proceedings of the ISES Solar World Congress Kobe, Japan, pp. 2343-2347.

Sartori, E. (1990b). The thermal inertia and the conduction heat loss effects on the solar evaporator. Proceedings of the World Renewable Energy Congress, Reading, U.K.

THERMAL ANALYSIS ON A SMALL CYLINDRICAL KIER POND

W. G. Chun, H. Y. Kwak, T. K. Lee, S. H. Cho, and P. C. Auh

New and Renewable Energy Research Center
Korea Institute of Energy and Resources(KIER),
Daejeon, Korea

ABSTRACT

This paper reports the thermal performance of a small cylindrical salt-stratified solar pond ("KIER Pond") built at the Korea Institute of Energy and Resources(KIER). The pond is 4m in diameter and 3m deep with 1.3m gradient zone. The thermal storage element is 1.2m deep and is a fully mixed body of NaCl solution. Towards the end of September, temperatures well above 80°C were recorded for the storage zone with collection efficiency of 18%. In the present analysis, operational characteristics of the pond are analyzed experimentally and numerically(2-D model), extending on a wealth of fundamental information already available in this area. The primary quantities monitored during the operation of the pond are temperature, salinity, and pond transparency. Special concern is directed toward predicting and understanding data from the KIER pond for the possible implementation of solar pond technologies in Korea.

KEYWORDS

A cylindrical solar pond; numerical study; experimental investigation; performance; economics

INTRODUCTION

A country with virtually no indigenous oil resources, Korea has put enormous research efforts in developing renewable energy technologies since the late 70's. Especially, the solar resource has drawn a great deal of interest because it appears to have the greatest potential & reliability. Growing environmental concerns would further expedite its technological development as solar energy is continually renewable and pollution-free.

The concept of solar pond is one of the leading technologies in harnessing solar power on a large scale. While it's too soon to exclude any renewable energy technology to ease very large reliance on petroleum-based energy, solar pond technology seems to outclass the others in its scale. There are various research activities and utilization schemes reported worldwide in the related area of solar pond.

In Korea, however, only a handful of limited studies have been conducted in the area, and very little is tried to implement this technology for any practical utilization. Its various aspects are confined to a small but growing community. This sluggishness is partially due to the public consensus and government policy which often shifted from a need to replace fossil fuels to one of heavy reliance on them.

The purpose of this study is to investigate the performance of a salt-stratified solar pond under the typical Korean weather condition and to analyze its economic aspect for possible utilization in the future.

POND DESCRIPTION

KIER pond was originally designed to facilitate the modeling of the decanting method for thermal energy extraction under various controlled conditions, as principal advantage of the solar pond over other collector systems is its inherent capacity for long term thermal energy storage. The pond is 4m in diameter and 3m deep with 1.3m gradient zone. Filling of the pond has been carried out according to the method proposed by Zangrando. To ensure dynamic stability of the gradient zone, the actual density gradient was thicker than the value given by static criterion. The thermal storage element is 1.2m deep and is a fully mixed body of NaCl solution. The bottom and perimeter of the pond are well-insulated with 0.2m polyurethane foam. Fig. 1 is a schematic diagram of the KIER pond.

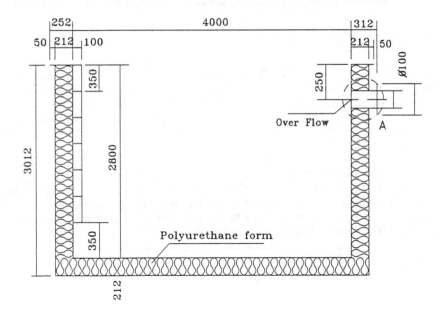

Fig. 1. Schematic cross section of the KIER pond.

MEASUREMENTS

Temperatures in the fluid column were continuously monitored with the use of thermocouples(T-type) and a data logger. Thermocouples are embedded along the center and perimeter of the fluid column at 20 cm intervals. Soil temperatures near the pond(1m from the pond wall) are also measured for the ground heat transfer calculations. Measured data are transmitted to the off-site data logger computer and hourly averaged values are calculated for analysis. The attenuation of solar radiation(the pond transparency) has been constantly checked by an underwater pyranometer installed just below the gradient zone.

NUMERICAL EXPERIMENT

The transient effect of decanting schemes within the storage zone has been studied numerically (2-D) to examine the spread of flow associated with the removal of the decanting layer. Especially, the effects of inlet and outlet jet placement are extensively investigated since the characteristics of the flow field largely depend upon the arrangement of the jets. Three two-dimensional partial differential equations (the heat transport equation, vorticity transport equation, and stream function equation) that are coupled by the nature of the problem were solved by using the ADI(Alternating Direction Implicit) method and Successive Over-Relaxation.

A simple one-dimensional energy balance model has also been developed, which allows the modeling of solar pond performance under various environmental conditions and thermal energy extraction schemes. This model is similar to the other one-dimensional models developed elsewhere. Using the one-dimensional model, simple evaluations have been performed here on the operation and economics of the solar pond for various applications, including process heating and power generation.

DISCUSSION OF RESULTS

Fig. 2 shows a predicted result of the flow pattern when injection&withdrawal ports are located one below the other along one side of the storage zone. This arrangement may be desirable for decanting schemes, for it could save the cost of piping considerably as compared with the other cases. However, it is likely to extract less thermal energy from the storage zone unless the horizontal dimension of the pond is within its maximum thermal penetration length.

Fig. 3 represents the effects of pond depth on temperature profiles in the storage zone. The rate of temperature rise in each profile is almost identical after 60 days with one another. This figure has been obtained by running our 1-D model with I_o = 114 W/m^2 (average irradiance for autumn and winter seasons). As shown, a fairly good agreement exists between the predicted and measured results.

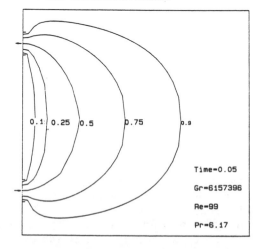

Fig. 2. Streamlines when injection and withdrawal
ports are located one below the other along
one side of the storage zone.

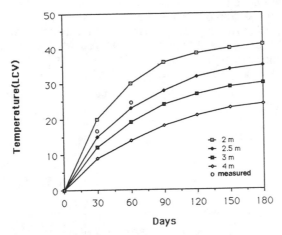

Fig. 3. Effects of pond depth on temperature
profiles (given relative to the ambient)
in the storage zone.

Fig. 4 shows salinity profiles immediately after the filling and 18 days thereafter. No appreciable differences are observed in the profiles, which demonstrates, more or less, the stability of the gradient zone resulting from the combined effect of the salinity and temperature gradients. Salinity layers remain gravitationally stable and nonconvective despite the heating of the bottom, because the salinity gradient is sufficient to maintain a stabilizing density gradient.

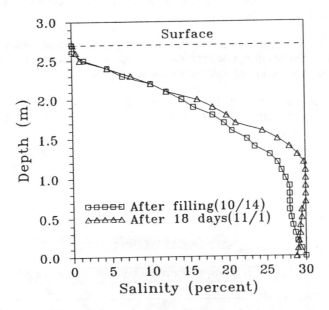

Fig. 4. Salinity profiles immediately after the filling
process and 18 days thereafter.

Data from two different days has been considered in Figs. 5 and 6. These figures show temperature profiles of the fluid column and soil as a function of pond depth, which are typical of a solar pond. As day progresses toward November, the overall pond temperature rises, whereas the soil temperature decreases. The data measured for the storage zone are somewhat higher than our 1-D model has predicted. This may be the effect of filling process, which was initiated in mid summer.

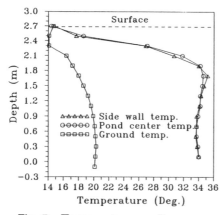

Fig. 5. Temperature profiles
(after filling:10/14/90).

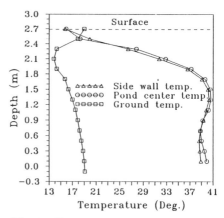

Fig. 6. Temperature profiles(18 days
after filling:11/1/90).

Solar pond could become technically feasible and economically viable for a wide range of applications in Korea till turn of the century. Especially, its application for space heating and crop drying (on a large scale) seems most feasible. By introducing a heat pump adequately, the pond could be operated over an extended period despite the reduction in available solar radiation during winter.

New experiments with solar ponds are currently being initiated at a number of research institutions around the country, and their results may help solve some of the difficulties in adapting the solar pond concept suitable to our situations.

The projections of feasible solar technologies with oil prices are given in Fig. 7. This figure reflects current market prices(1990) of various items required to implement the corresponding technology in Korea.

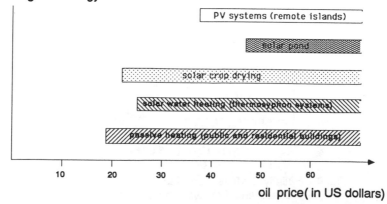

Fig. 7. Projections for feasible solar technologies in Korea.

CONCLUSIONS

Some aspects of the operation of solar pond (KIER Pond) understood in the present analysis were very informative in assessing the physical parameters involved in operating a feasible solar pond. However, in order to quantify these physical parameters more accurately and determine their interrelationship, the present study has carried out a systematic investigation to evaluate the operation and economics of the solar pond for possible industrial applications in Korea.

ACKNOWLEDGEMENTS

The authors would like to express their gratitude to the Ministry of Science and Technology for the financial support.

REFERENCES

Hirschmann, J. R. (1970). Salt flats as solar heat collectors for industrial purposes. Solar Energy, 13, 83-97.

Kooi, C. F. (1981). Salt gradient solar pond with reflective bottom: application to the "saturated" pond. Solar Energy, 26, 113-120.

Leshuk, J. P., R. J. Zaworski, D. L. Styris, and O. K. Haring (1978). Solar pond stability experiments. Solar Energy, 21, 237-244.

Nielsen, C. E. (1975). Salt-gradient solar ponds for solar energy utilization. Energy Conservation, 2, 289-292.

Rabl, A., and C. E. Nielsen (1975). Solar ponds for space heating. Solar Energy, 18, 1-12.

Tabor, H. (1963). Solar ponds. Solar Energy, 7, 189-194.

FERTILIZER SALT CHARGED SOLAR POND
INTEGRATED WITH "PROGRAMMED AGRICULTURE"

K. R. Woods* and R. R. Dedolph**

*Energy-Environment Inc., 1264 Harvest Ct.,
Naperville, IL. 60564,USA

**Gravi-Mechanics Co., 22W510 71st Street,
Naperville, IL. 60540, USA

ABSTRACT

The use of fertilizer salts has the advantage of utilizing the solar pond
in dual function applications. In this way the total cost of developing the
solar pond may be allocated proportionally to each function.

The fertilizer salt charged solar pond enhances the environment by providing
nutrients to plant life and an environmentally benign source of heat.

"Storm water management" ponds are required by law in many areas to control
the storm water run-off from developed land to a rate not exceeding the rate
prior to development. The usual requirement mandates a design for a statistical
100 year storm or a rainfall that might occur once every 100 years. The
required storm water management ponds can easily be designed as fertilizer
salt charged solar ponds that not only provide useful heat for many purposes
but also provide a labor-saving method of applying fertilizer to the
adjoining landscape.

"Borrow pits" used as a source of fill in road or building construction may
also be designed as fertilizer salt charged solar ponds. These "borrow pits"
are found in many parts of the world as a consequence of construction activity.

"Programmed Agriculture" provides nutrients and water to plants at precisely
the time in the plant growth cycle that the plant can maximize its use and
thereby produce a superior crop and yield.

KEYWORDS

Fertilizer salt solar pond; programmed agriculture; living filter; borrow pit;
storm water management.

INTRODUCTION

Salt gradient solar ponds using sodium chloride as the salt medium has a proven
record of being a low cost method of collecting and storing solar energy for a
number of low temperature applications in many diverse climates. See Nielsen
(1980) and Tabor (1981). Pollution of the soil or ground water from sodium
chloride has deterred the use of this practical and proven method of collecting
and storing solar energy from more widespread use worldwide.

The use of fertilizer salts in an integrated system of "programmed agriculture" with design flexibility to cover a broad range of applications should provide a low cost method of storing and applying fertilizer as well as collecting and storing solar energy.

Soil Conservation

A disturbing consequence of past farming techniques in developing a competitive agricultural economy is the erosion of the rich Illinois prairie soils, estimated to be over 300 mm in the last century. The construction of irrigation ponds provide a technique to store rain water and fertilizer to maximize the use of these essentials to healthy plant growth and specifically tailor these to the needs of the crop. The intelligent use of these precious prairie soils can maximize yields at minimum cost and enhance those soils as well as provide a means to place agriculture in a more competitive position in world markets without destruction of fragile eco-systems.

Living filter for land disposal of human and animal waste

The attempt to clean up our streams and prevent the further deterioration of our ground water source of drinking water is leading to a third generation of sewage treatment (tertiary treatment). The USEPA has stated that the contamination of the USA's ground water supply is "epidemic".

As in air pollution, dilution is no longer an acceptable solution to pollution.

A systems approach to our human waste disposal problem has led to land application of effluent from primary sewage treatment plants at locations as far North as Wisconsin and as far South as Georgia.

The project described provides for the land application of animal wastes utilizing a 40 Hectare field in La Salle County, IL. as the living filter. The lesson that can be learned from this project has application for land application of human waste as well.

In many areas of the USA the discharge from sewage treatment plants to the rivers and streams adds significantly to the total flow of those rivers and streams.

The additional flow is due to the "mining" of water from our deep and shallow wells at a rate faster than the natural recharge rate and the subsequent discharge of the effluent from our sewage treatment plants to the rivers and streams.

The land application of these effluent flows to provide shallow well recharge and make use of the "living filter" concept is a logical extension of the systems approach to our waste water disposal problem.

A search of the literature reveals an urgent need to assess and quantify the environmental impact and crop yield response to the land disposal of human and animal waste.

This project can also provide crop yield data as a basis for decelerating the current trend of a chemical approach to agriculture.

DESCRIPTION

The end product of this operation is compost used in the production of mushrooms. The raw material is horse manure and straw bedding transported by truck to the compost wharf. The composting is performed outdoors as an open air process.

Storm and process water run-off from the concrete compost wharf is diverted to a separator structure to remove solids. The solids are returned to the wharf. The liquids are diverted to a zero discharge holding pond of the enriched liquid material. The chemical analysis of the liquid material in the holding pond is in the following table.

Sample	Potassium	Phosphorus as PO$_4$	Nitrite	Nitrate	Kjeldahl Nitrogen	Ash	Total Solids
1)	740 mg/l	1mg/l	<0.05mg/l	8mg/l	90mg/l	0.16%	0.37%
2)	770 mg/l	3mg/l	<0.05mg/l	10mg/l	60mg/l	0.12%	0.28%
3)	700 mg/l	2.5mg/l	<0.05mg/l	10mg/l	60mg/l	0.14%	0.27%
4)	700 mg/l	2.5mg/l	<0.05mg/l	8mg/l	80mg/l	0.12%	0.25%
5)	740 mg/l	3mg/l	<0.05mg/l	15mg/l	70mg/l	0.14%	0.25%
6)	700 mg/l	3mg/l	<0.05mg/l	10mg/l	60mg/l	0.14%	0.26%

The pond serves as a storm water management and waste stabilization pond of 0.49 Ha-M (4 Ac-Ft) capacity. This enriched liquid is returned to the compost operation or is blended with irrigation water to "fertigate" an adjoining forage crop. The forage crop acts as a living filter. See Appendix A for a flow diagram of the system.

The irrigation water is stored in an adjoining pond having a capacity of 3.2 Ha-M (26 Ac-Ft). The irrigation pond straddles an intermittent stream that serves a drainage area of 299 Hectares (740 Acres).

The irrigation water pond serves as a settling basin to precipitate out the contaminants and top soil from upstream farming operations that are carried with the storm water run-off.

The source of irrigation water will be the run-off from upstream property plus pumpage from shallow wells to assure a dependable source of irrigation water during drought periods.

The irrigation pond is configured to be converted to a fertilizer salt charged solar pond. The heat from the solar pond will be used to heat and cool the buildings and office and to pasteurize the compost.

The feasibility of using fertilizer salts in a salt gradient solar pond has been reported in the literature. See Hull (1986) and Hull and co-workers (1989). We have developed proprietary methodology that will substantially decrease the salt concentrations necessary for solar ponds. Small scale tests have been performed at Argonne National Laboratory that verify the performance of fertilizer salts for solar pond use. See Hull and co-workers (1989).

The living filter for this project is a forage crop of alfalfa fertigated from the storm water and waste stabilization pond. See Stanberry (1955) for the Irrigation Practices for the Production of Alfalfa.

Initial test results of ground water samples taken from two test wells support the concept that a living filter system, when well designed, properly installed and judiciously operated is not only environmentally benign but can reverse some aspects of environmental deterioration. See the following table for the results of analysis of water samples taken from two ground water sampling wells located on the Compost Supply Co. property in La Salle County, IL.

Analyzed Constituent	Upper Well (Input)	Lower Well (Output)
Nitrate Nitrogen (ppm)	All less than 1	All less than 1
Nitrite Nitrogen (ppm)	All less than 1	All less than 1
Ammonium Nitrogen(ppm)	0.4, 0.2, 0.8	0.1, 1.0, 0.6
Mean	0.47	0.57
Phosphorous (ppm)	220, 270, 220	100, 110, 150
Mean	236*	120*
Potash (ppm)	6.7, 6.4, 6.5	3.7, 3.6, 3.6
Mean	6.5*	3.6*
Total Ash (%)	0.02, 0.06, 0.04	0.02, 0.01, 0.03
Mean	0.04	0.02

* Means differ at odds greater than 99:1

CONCLUSIONS:

1. Solar ponds using fertilizer salts have the potential of serving the dual function of providing useful heat and enhancing the environment at competitive costs.

2. Land application of human and animal waste through "fertigation" is a logical extension of the age-old environmentally benign land disposal of solid manures.

3. Water storage ponds provide the means to intercept non-point pollution sources before they enter the food chain and provide a method to recover top soils carried in the storm water run-off from intensive cultivation of upstream land.

4. Water storage ponds provide the means to manage increased storm water run-off from upstream man-made development and to control the application of this vital resource to maximize plant health and yield and enhance and stabilize ground water levels.

5. Composting utilizes a waste product to develop a food product (mushrooms).

6. This system is compatible for production of optimum yields for all crops including high protein forages.

7. The solar heated storage ponds for irrigation provides a supply of tempered water to prevent "thermal shock" to delicate plants.

8. The results of ground water testing to date indicate that the living filter system, when properly designed, properly installed and judiciously operated is not only environmentally benign, but can actually reverse some aspects of environmental deterioration.

2362

ACKNOWLEDGEMENTS

This work was privately funded. The author is indebted to the late Frederick C. Noorlag, former president of Compost Supply, Inc., for his creative vision in financing this project.

I also wish to acknowledge the assistance of John R. Hull for his expertise in solar pond technology and to acknowledge the co-operation of the local representatives of the Soil Conservation Service, the Rockford and Springfield offices of the Illinois Environmental Protection Agency, the Engineers in District #3 of the Illinois Department of Transportation and officials in LaSalle County who had jurisdiction over this project.

REFERENCES

Hull, J. R. (1986). Stability and economics of solar ponds using ammonium salts. Proceedings ASME Solar Energy Division Conference, Anaheim, CA.330-337.

Hull, J. R. (1986). Solar ponds using ammonium salts. Solar Energy, vol.36, 551-558.

Hull, J. R. and co-workers (1989). Ammonium sulfate solar pond: small scale experimental results. Solar Energy, vol. 43, 57-67.

Hull, J. R., C. E. Nielsen, P. Golding (1989). Salinity Gradient Solar Ponds, C.R.C. Press, Boca Raton, FL.

Nielsen, C.E. (1980). Nonconvective salt gradient solar ponds. Solar Energy Technology Handbook, Marcel Dekker, New York, N.Y.

Stanberry, C.O. (1955). Yearbrook of Agriculture 1955, 435-443. U.S. Government Printing Office, Washington, D.C.

Tabor, H. (1981). Solar ponds. Solar Energy, 27, 181-194.

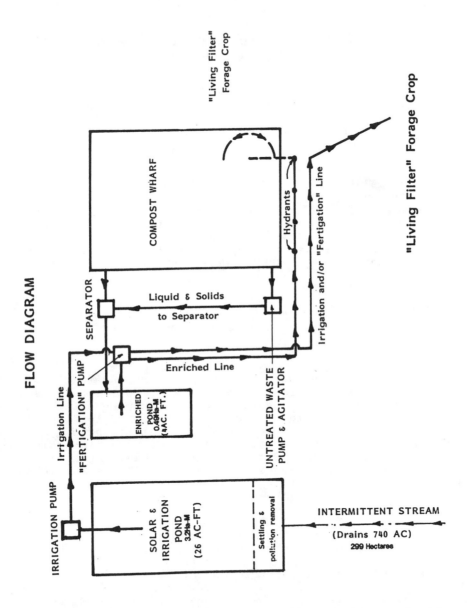

FLOW DIAGRAM

THE DEVELOPMENT OF A LOW-COST FLAT RESPONSE DETECTOR FOR USE UNDERWATER

P.T.Tsilingiris

Council of Technological Research, Commercial Bank of Greece, Athinas 14, Athens, Greece.
National Technical University of Athens, Department of Electrical Engineering, 42,28 October street, 106 82 Athens.

ABSTRACT

The design of a low cost flat response net radiometer is presented. Preliminary laboratory testing of the developed device shows that it would prove useful for fixed level or scanning radiation transmission measurements in solar ponds and in the euphotic zone of bulk masses of natural waters.

KEYWORDS

Radiation transmission measurements, submersible net radiometer, underwater transmission measurements, solar ponds.

INTRODUCTION

The lack of flat response, temperature compensated instruments for radiation transmission measurements was earlier addressed by Nielsen (1980) who suggested for the purpose either the use of standard pyranometers or calibrated selective devices. However these are expensive, they do not meet optical, spectral and thermal response requirements for use in solar ponds (Enshayan and others, 1988), while problems associated with their practical use were also reported by Wittenberg,(1981) and Almanza,(1983). Although appreciable progress in many aspects of solar pond research has been made during the last decade, there is no report yet on the development of a low cost instrument for use in the field. Appreciable research and development work has earlier been made towards low-cost instruments by Funk (1959), while Monteith (1959), optimised and Funk (1962), proposed the use of miniaturised ribbon thermopiles to improve sensitivity. The aim of the present work is to discuss briefly the theory and design of a low cost device, which was preliminary tested in the laboratory and which may be found useful for radiation transmission measurements underwater.

THE FUNDAMENTAL THEORY AND DESIGN

The incident radiation in water bodies is attenuated as a result of absorption and scattering being responsible for the develo-pment of a diffuse light field within the body of natural waters,

The energy budget of an incremental water layer between two horizontal planes 1 and 2, with $I_{1\downarrow}, I_{2\downarrow}$ and $I_{1\uparrow}, I_{2\uparrow}$, the downwelling and upwelling irradiances at the planes 1 and 2 respectively, lead to the net radiative input (N.R.I.), which is defined as

$$NRI = (I_{1\downarrow} - I_{1\uparrow}) + (I_{2\downarrow} - I_{2\uparrow}) \qquad (1)$$

Evaluation of the NRI requires the measurement of the difference between the downwelling and upwelling irradiance and therefore the use of a net radiometer detector, which should also be a rigit device for submersible use in the highly corrosive environment of hot solar pond brines. A flat response over the range of 0.3 to 1.3 µm and temperature compensation would be necessary. A thermopile is employed to detect small temperature differences developed between two exposed parallel absorbing plates of cross sectional area A, at a distance d appart, covered by flat glass windows. The thermopile is wound with a constantan wire of diameter D on a solid perspex form. Thermocouple junctions, twice as many as the number of winding turns N, are developed by copper plating up to a diameter D'. Assuming that $\Delta I = I_{1\downarrow} - I_{1\uparrow}$, $\Delta T = T_1 - T_2$ is small, and that overall (convective and radiative) heat transfer coefficients $h_{o,1}$ and $h_{o,2}$ from plates to glass covers are close enough and equal to h_o, from heat balance equations it is derived that,

$$(\Delta T / \Delta I) = (\tau\alpha) / \qquad (2)$$

$$[h_o + 2.\frac{k_p}{d}.[1 - \frac{f}{A}.(1+\epsilon)] + 2.\frac{k_f}{d}.\frac{f}{A} + 4.\frac{k_{co}}{d}.N.\frac{a}{A} + 2.\frac{k_{cu}}{d}.N.\frac{a'-a}{A}]$$

with f the cross sectional area of former across the axis of discs at distance d appart, ϵ the numerical factor which takes into account the per cent clearance of thermopile cross sectional area f to fit former into supporting plate, a' and a cross sectional areas of plated and non plated constantan wire. k is the thermal conductivity and subscripts p,f,co and cu correspond to plate, former, constantan and copper. The additive terms in the denominator of (2), correspond to convective and radiative conductance from plate to glass and to conduction conductance through perspex disc, coil form, constantan wire and copper plating layer respectively. For a typical design and parameter values, these conductances were estimated to about 8, 30, 4, 6, and 52 % of the overall plate heat loss, so sensitivity improvement is possible, by reduction of copper deposit and increase of perspex plate thickness.

Convective heat transfer coefficient is expressed as a function of Nusselt number, which according to Hollands and others (1976), is described as a function of Raileigh number in the air gap. This was found to be less than 1706 for the extreme values of plate temperature and thermophysical constants which means Nu = 1 and conduction across air gaps up to about 8 mm.

The responsivity of the device,defined as (ΔV/ΔI),is calculated from (9) as,

$$(\Delta V/\Delta I) = N.e.RF.(\Delta T/\Delta I) \qquad (3)$$

where e is the thermocouple coefficient and RF a reduction factor due to the shunting effect of copper plating, which according to Funk (1962), is given by,

$$RF = [\ 1 + [(a'/a - 1).(r/r')]^{-1}\ ]^{-1} \qquad (4)$$

The output voltage is directly proportional to the incident radiation at thermal equilibrium, which when it is disturbed due to a rapid change of ambient fluid temperature, causes the development of an output error signal which slowly decays as it approaches an equilibrium corresponding to the new temperature. Therefore, to avoid errors due to temperature changes for fixed level or scanning applications, a low thermal capacity design is desirable.
The temperature stability of the present design depends on the effects of temperature of various heat transfer conductances in (2), and may be calculated by the expression,

$$\partial(\Delta V / \Delta I)/\partial T = 0 \qquad (5)$$

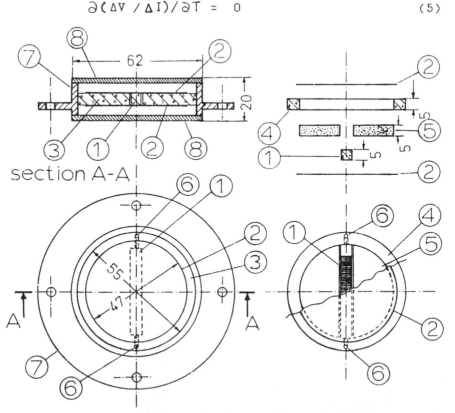

Fig.1. Plan and section of the developed design and thermopile assembly. 1 Thermopile, 2 absorber plates, 3 perspex disk, 4 perspex ring, 5 insulation disk, 6 thermopile terminals, 7 Aluminium case, 8 flat glass windows.

after introduction of the temperature dependance expressions of thermophysical properties for employed materials in (2). Taking in to account the variation of the thermal conductivity of copper, constantan, perspex and air and of the radiative loss coefficient on temperature in the range between 20 to 100 deg.C. it was found that the effect of ambient temperature on the responsivity is negligible, being typically less than 0.1 % .

PRACTICAL DESIGN AND TESTING.

The typical design consists of a thermopile (1), sandwiched between two black thin aluminium discs (2), in a thin wall aluminium watertight construction (7) and (8), as it is shown in fig.1.Thermopile is fitted inside a slot cut along the diameter of the perspex disk (3) in the opposite sides of which absorbing discs (2) are attached in thermal contact with opposite thermopile junction surfaces (1). In a second version, a perspex ring (4) supports absorbing plates (2) and thermopile (1) with the space between filled with DOW structural thermal insulation discs (5). To improve bond conductance, a high thermal conducti- vity silicon compound jelly was spread between junctions and absorbing plates while epoxy adhesives were succesfully used for up to 80 deg.C. temperatures. Glass windows of 2.5 mm,(8), are used to seal the assembly using silicone sealant.
The thermopile is a 50 to 100 turn coil wound on a perspex prism of cross sectional area of 6 to 25 sq.mm with a 0.15 mm

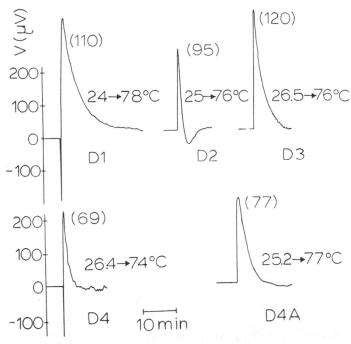

Fig.2. The response to a step change of surrounding fluid temperature for the developed prototypes D1,D2,D3,D4 and D4A. Output voltage, time scale and cold and hot bath temperatures are also shown.

constantan wire, which is copper plated so as a sequence of a series connected copper-constantan junctions in the opposite surface of the perspex prism are formed. Thermopile calibration have shown a thermocouple constant of 10-20 µV/deg.C for the developed junctions which is appreciably lower than the predictions from (4),probably due to copper plating defects. Outdoor testing have shown a typical calibration factor of 3.5 µV.sqm/W. Transient response of five prototypes to a step change of surrounding fluid temperature can be seen in fig.2, in which cold and hot bath temperatures, output voltage, time scale and the maximum equivalent energy density in w/sq.m. in parentheses, are also shown. Immersion in a higher temperature bath is responsible for a rapid temperature change of detector shell and a strong heat exchange leading to the development of an output pulse of duration proportional to the thermal mass of radiometer which gradually disappears when it assumes the bath temperature. Further laboratory testing under simulated actual conditions using a high power tungsten-halogen lamp with the radiometer submersed in a varying temperature water bath is shown in fig.3.a,b. Lines 1, 2, 3 and 4 correspond to thermopile output microvolts, surrounding average fluid temperature, heater input power and fan timing respectively. The rapid heating of bath is responsible for the development of a negative polarity output

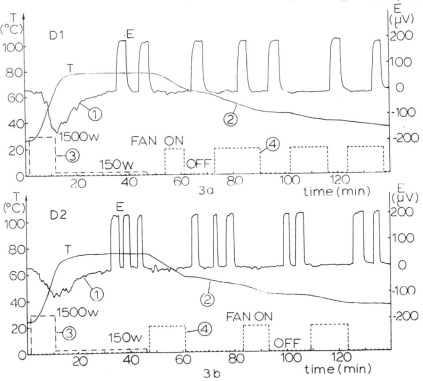

Fig.3. Laboratory testing of prototypes D1 and D2. Lines 1,2,3 and 4 correspond to output voltage,bath temperature, heating input rate and fan timing respectively.Pulse height of line 1 is proportional to the radiation transmitted at the level of the summersed radiometer.

pulse whith amplitude proportional to the thermal mass of radiometer, which is due to the development of a temperature difference owing to the flow restriction by the submerged radiometer body, responsible for turbulent heat transport from the base to the top free water surface. As soon as power input ceases, bath temperature is slowly cooling down, though acceleration of cooling rate is possible by blowing air to enhance evaporation losses as it is shown by the change of slope of lines 2. The height of the positive output pulses of lines 1 is proportional to the energy transmission at a fixed depth for a wide range of ambient fluid temperatures. An analysis of derived data shows a statistically insignificant effect of measured energy transmission on temperature, varying in the range between 20 to 80 deg.C.. An about 20 µV offset fluctuating output voltage of typical period of 1 min, ceasing at zero wind speed and at lower bath temperatures, is attributed to the heat transport to the free water bath surface and plume activity. This is also confirmed by temporary suppression of surface evaporation using a thin floating plastic film.

CONCLUSIONS

A brief analysis and details of the practical design of a low cost flat response radiometer is presented in the present paper. A brief report of results from a preliminary laboratory testing shows that the device would be useful for fixed level measurements using a number of sensors at selected levels in the pond and a reference one is mounted just underneath the pond surface. Thermal inertia is a factor of relatively minor importance since temperature variations at fixed depth are normally extremely slow and transmission is simply measured as the ratio of microvolt output of each radiometer to the microvolt output of the surface radiometer.

For scanning measurements lower thermal inertia design require the use of thin bismuth wire windings up to 0.05 mm diameter on thin wall hollow prismatic formers. Low traversing speed would be required to allow radiometer to assume the temperature of its environment which, though itshould be high enough to minimise errors owing to insolation changes. Density stratification in the gradient zone would also have a favorable effect on the operational behaviour of fixed level or scanning radiometers. However further testing is required to access optimum traversing speed for reliable routine radiation transmission scanning in solar ponds.

REFERENCES.

Almanza R. and Bryant H.C.(1983),J.Sol.En.Eng.,105,378-379.
Enshayan K.,Golding P. and Nielsen C.E. (1988), Proc.Int.Progr. in Solar Ponds Conf.,66-73,Cuernavaca, Mexico.
Funk J.P.(1959), Journ.Scient.Instr.,36,267-270.
Funk J.P.(1962), Journ.Geoph.Res.,67,2753-2760.
Hollands K.G.T.,Unny T.E.,Raithby G.D. and Konicek L. (1976), Journ.of Heat Transfer,98,189-193.
Montieth J.L.(1959), Journ. of Scient.Instr.,36,341-346.
Nielsen C.E.(1980), Chapter 11, Solar Energy Tech.Handbook, edited by Dickson W.C. and Cheremisinoff P.C.
Wittenberg L.J.and Harris M.J., J.of Solar En. Eng.,103,11-16.

MONITORING AN EXPERIMENTAL SOLAR POND FITTED WITH TRANSPARENT INSULATION

A. Javed and L. F. Jesch

Solar Energy Laboratory
University of Birmingham
Birmingham B15 2TT, UK.

ABSTRACT

The use of solar pond (1) is becoming more attractive in today's energy scene. A major advantage of solar ponds over other collectors is the ability to store large quantity of thermal energy for longer periods of time. The best documented type of pond is the salt-gradient solar pond. Several built examples are now operating successfully, but considerable difficulties still exist.

The latest developments in solar ponds are in transparently insulated solar ponds. (2) The transparent insulation is so effective that the pond's thermal behavior can match and outperform those energy collector / energy storage devices which rely on salt gradient imposed thermal stratification for top surface insulation. The sweet water pond is cheaper to build and to operate, it does not require corrosion protection and in case of accidental leak it does not pollute the soil around it.

KEYWORDS

Solar pond, Convective, Transparent Insulation, Monitoring, Experimental results

INTRODUCTION

A Solar pond of 2.4x3.6m area, 1.2m deep was constructed in Birmingham for study and demonstration purposes. It has transparent insulation on the top and it does not use salt gradient. The transparently insulated solar pond offers many operational advantages over a salt gradient solar pond.

a) By having a cover on top, the pond surface is protected against wind disturbances.

b) Elimination of cost as related to the purchase and transport of salt.

c) Elimination of salt gradient maintenance.

d) Elimination of environmental contamination hazard due to possible escape of highly concentrated salt solution.

e) Can provide storage zones more economical than salt gradient solar pond.

f) Corrosion problems caused by salt are not experienced in the heat exchangers, pipes, valves and pumps.

g) The use of fresh water may allow development of solar ponds by adding transparent insulation covers to existing lakes and reservoirs

CONCEPT

A honeycomb array can stabilize an air layer to be non convective, (3) if the honeycomb dimensions are matched to the temperature difference between the bottom and top surfaces of the layer. Such a layer will be solar transparent yet provide good thermal insulation. These characteristics of the air layer are due to reduction in convective and IR radiation losses without an appreciable increase in heat conduction due to the honeycomb walls. The effect of confining barriers on convection in fluid heated from below has been treated by several authors over the last decade. It has been shown that convection is suppressed until the Rayleigh number(Ra) exceeds a critical value known as the critical Releigh number (Rac) which depends on the aspect ratio of the elementry cell of the honeycomb structure (The aspect ratio, A is defined as the depth (L) divided by the equivalent cell diameter (D), equivalent diameter is the diameter of the circular cylinder having the same cross-sectional area as that of a square.)

Solar gains depend on the transmission of solar radiation through the honeycomb. The major portion of insolation is intercepted by the honeycomb cells and the rest falls on the top of the cell walls. The portion intercepted by the cell undergoes reflective(specular as well as diffuse),refraction and absorption through the vertical walls. The radiation falling on top of the cell walls is propogated downward through the walls due to total internal reflection and, in the process, is absorbed and scattered inside the material of the walls. In the case of a solar pond the floating honeycomb matrix therefore form a "nonconvective layer" traping the solar gain within the pond.

EXPERIMENT AND RESULTS

A experimental solar pond with an aperture of 2.4x3.6m and with a design depth of 1.2 meter was built in Birmingham, UK at 52.4 Northern Latitude. A picture of experimental solar pond is presented in Fig. 1. The pond consist of several parts: water for heat storage, the surface insulation system for reducing surface heat losses, the fibre glass lining for holding water, and a 125mm thick thermal insulation. The surface insulation system consists of honeycomb panels. The pond was constructed above the ground to avoid shadowing effect of hedges around it. Plywood was used for the construction. A 125mm thick thermal insulation was used in the side walls and bottom of the solar

Fig. 1. *View of the solar pond in Birmingham, UK*

pond. In order to shed rain water from the transparent insulation cover it has to be positioned in a sloping manner. This angle of the slope, together with the optical characteristics of the honeycomb cells will determine the maximum collectable solar energy within the limits of solar radiation available on a sloping surface. Fibre glass lining was applied on all inside surfaces which withstands the pressure. The honeycomb insulation was covered with polycarbonate film.

It is desirable that cover material should have good transmittance and good resistance to u.v. and polycarbonate has both qualities. Pillows of different sizes were made to fit in the pond. Ten K-type thermocouples each were installed vertically at two different positions in the pond, one vertical column in the northeast corner and the other southwest corner. In addition to these ten thermocouples, one thermocouple was buried at a depth of 70cm to measure the ground temperature and other one was installed on the side to measure the ambient air temperature. The mean pond temperature was calculated using the following equation:

$$Tp = [Sum(Ti + Ti+1)* dli] / (2lg)$$

where lg is the pond depth and dli is the distance between two adjacent thermocouples. To record the incident solar radiation on the pond a Kipp and Zonen CM 11 type solarimeter was installed horizontally. The Microdata 1600 microprocessor based datalogger was used for recording all thermocouples and solarimeter signals. The thermocouples are connected to an isothermal junction box and solarimeter feed information into a integrator which allows both intantaneous and time integrated values to be recorded. The time step of 6 minutes for scan gives the time resolution of the data to be 10 readings per hour.

The solar pond has operated since July 1990. It was found that the mean pond temperature is sensitive to short term variations in meteorological conditions such as daily horizontal insolation and ambient temperature. It is also sensitive to the contamination due to water vapour condensation within the surface insulation system. Therefore the insulation system (pillows) was occasionally cleaned by

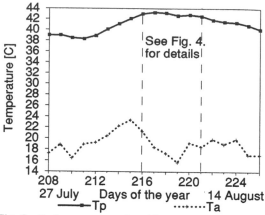

Fig. 2. *Daily average of ambient air and pond water temperature (27th July - 14 August 1990)*

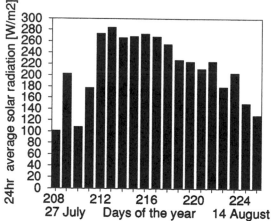

Fig. 3. *Daily average of solar radiation on horizontal surface (27th July - 14 August 1990)*

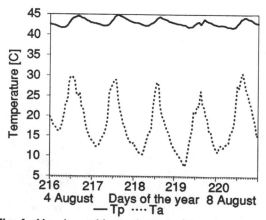

Fig. 4. *Hourly ambient air and solar pond water temperature (4th August - 8th August 1990)*

cotton gauzes by removing the pillows from the pond. Fig.2. shows the daily averages of the ambient air and pond water temperature. Fig.3. shows the daily averages of solar radiation on a horizontal surface. Fig.4. shows four days of results in more details (4 Aug - 8 Aug 1990). It can be seen from Fig.4. that the mean pond temperature is 10-20 degree higher than the daily mean ambient temperature.

REFERENCES

(1) Jesch, L.F. and Abdul-Sada, G.K.: *Solar ponds for heating green houses. Clean and Safe Energy Forever* vol 2, pp.1394-1397. Proceedings of the ISES congress in Cobe, Japan, Pergamon Press 1990

(2) Ortabasi, U., Oyksterhuis, F.H.,: *Honeycomb stabalised solar pond, Department of resources and energy remote area power supply,* workshop, pp.266-272, 1983.

(3) Hollands, K.G.T., Marshall, K.N. and Wedel, K.N.,: *An approximate equation for predicting the solar transmittance of transparent honeycombs,* Solar energy, vol 21, pp 231-236, 1978.

(4) Sharma, M.S. and Kaushika, N.D.: *Design and performance characteristics of honeycomb solar pond, Energy conversion management,* vol 27, No.1, pp. 111-116, 1987

Salt Water Desalination
Using Solar Energy

Nurreddin Sishena
Department of Physics

University of 7th of
April
Zawia - Libya

ABSTRACT

This paper presents experimental data and calculations of a modified solar basin type solar still in comparison with other existing wick and basin types stills. The increase of efficiency of this modified model is due to the increase of condensing surface and absorption rates and extra reflection plates . The increase of condensing surfaces was achieved by using transparent waved plastic cover inclined by $20^{\hat{o}}$ above the basin level . The increase of absorption rate is achieved by addition of dye material to the incoming salt water .

Keywords :

Basin type solar still , disallination , multiple v-type cover waved cover , condensor surface , dye, over-all effeciency.

INTRODUCTION

Needs for fresh water supply in costal areas for drinking at in-land areas & small farm irrigation increased in the last ten years.

Insreased consumption of ground water and low rain falls helped the increase of sea water flow towards in-land reserves . [1]

This paper describes a .small disallination system which is a modified basin type solar still to be used in residence units or a large model for buildings , taking advantages of an abundance of sea water or salty water sources (2, 3) . Two small basin type solar stills were con-structed , the dimensions as shown in the fig. (1) are one meter long and 8 m. wide and .25 m. high of well sealed galvanized steel. This basin is provided with a number of inlet and outlet water supply pipes. Flushout

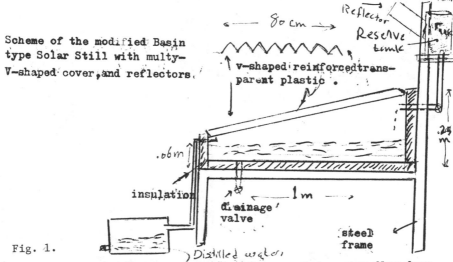

Scheme of the modified Basin
type Solar Still with multy-
V-shaped cover, and reflectors.

Fig. 1.

supply pipe was provided with control valve for a certain water flow from
a reserve tank or city networks . A transparent plastic waved   shap or
multiple-type cover was used , and compared with a flat plate cover.   The
inside of the basin was painted with a normal blank paint .While the whole
basin was insullated by a 5-cm. thick layer of polyurethan foam. [4]

Water inside the basin and outside air temperature were recorded
. [5]. Wind speed and radiation data from early works used in calculating
and comparing maximum effeciency of the system [2]. Maximum incoming radiation and absorption rates were achieved by the use of a pair of reflecting mirrors and addition of dye to one of these systems (6,7,8,).   This
design as it stands , is very cheap and can be cost effective by using
other materials. Also it has a good effeciency and very low maintenance
over a long life time. [9]

EXPERIMENTAL SETUP AND RESULTS

The basin was filled with water to about .06 meter . water wuality
was a 6000 p.p.m salt . The volume is about .098 m^3 and surface area of .8m
The experimental measurments were carried out at the city of Zawia at   sea
level and latitude 3^0 north . fig.1 . The avarage global radiation   was
nearly 5 Kwh/m^2-day ., Table 1 shows tha daily ave. global radiation   over
twelve monthes period. Measurments are taken for the two systems for comparisons . The systems were operated for ten monthes which account   for all
climatic changes .

A—The ave. monthly GLOBAL RADIATION ,and B,C ,the ave. monthly productivity of the two systems.

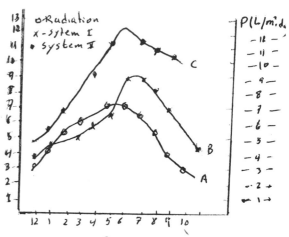

Fig. 2.

TABLE 1

Daily ave. global radiation (inkwh/m^2-day)

Month	Jan	Feb	Mar.	Apr.	May	June	July	Aug.	Sep.	Oct.	Nov.	Dec.
Kwh $\frac{}{m^2 day}$	3.0	4.1	5.2	6.1	6.7	7.3	7.3	6.6	5.6	4.0	3.1	2.1

From Dec. to Oct. and spans the temperature range from 12 to 35 C^o. The incident solar radiation was recorded on an hourly basis between 8. am. to 6 pm. in summer month and from 9 am. to 4 pm. during winter mothes . The ambient air temperature was recorded by A.T-type thermocouple . Avarage wind speed data from a previous work was also tabulated . The two systems productivities were measured also on an hourly basis . The recorded data was tabulated in table (2, 3) which represents the two systems water productivities for the ten monthes period and for a selected summer day . The whole systems were supported by a steel frames and beams to a height of about 80 cm. off the ground .

TABLE 2 Avarage monthly output of the two systems in L/m^2-day
- System I flat cover , no dye no reflecting plates .

Month	Dec.	Jan.	Feb.	Mar.	Apr.	May	June	July	Aug.	Sep.	Oct.
System I	4.2	3.9	4.1	4.6	5.0	5.9	6.7	9.0	9.0	8.6	8.1
System II	4.6	4.5	5.5	6.9	8.2	9.3	11.4	12.3	12.4	11.5	10.6

TABLE 3

Avarage daily output of the two systems in L/m^2-day (July 10th.1990)

Time	8:00	9:00	10:00	11:00	12:00	13:00	14:00	15:00	16:00	17:00	18:00
System I	1.32	1.62	2.51	3.33	4.66	6.50	7.32	8.71	9.43	11.25	12.68
System II	1.32	2.42	3.10	4.01	5.34	6.83	8.41	10.95	11.86	12.69	13.4

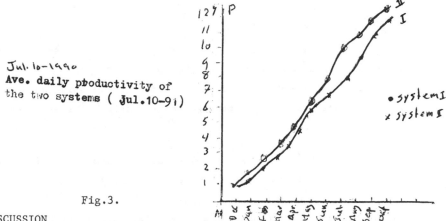

Jul. 10-1990
Ave. daily productivity of
the two systems (Jul.10-91)

Fig.3.

DISCUSSION

From tables (I, II, III) and the correspondant curves (fig.
2 , 3) we can understand the effect of the multiple V-shaped cover in
comparison with flat plate cover , this due to increase of condensing service which is proportional to heat loss by conviction of the cover service
The addition of dye (Potasium Permanganite) to salt water increases the
productivity of the system by nearly ten percent due to increase of the
radiation absorption and reduction of reflective radiation (6, 7, 8) .
We also noticed that the productivity of the two types increases noticibly
at the late hours of the day , this is due to thermal lag of the basin type
still and its thermal storage effects (9, 10, 11) .

Although ; air speed effects , the productivity of the systems
by a small factor yet a noticible increase is recorded with the increase of
globale radiation and ambient air temperature ,(See. figs.2 , 3) (12).
Finally ; all these extra additions will not affect the over-all costs by
a big portion (12 , 13) .

CONCLUSIONS

The increase of service area of the condensing cover tends to
increase the effeciency of the system . Also addition of dye materials and
reflecting plates have raised the effeciency of the system from 9 L/m^2-day
to about 13 L/m^2-day . The simple structure of the system make it easy to
use and cost effective to compete with other systems for the span of its
long life and with minimum maintanance .

ACKNOWLEDGEMENT

        The author extends his text to the national Academy of Scientific Research and the Basic Science Research Centre for their support . Also thanks are extended to Secretariat of High Education and University of Seventh of April for their encouragement and help for this research .

REFERENCES

1- Secretariat of Agriculture Annual Report (1988 , Tripoli - Libya).

2- M.S.Smeda , "Tilted Solar Radiation in North Africa , Proceeding of Energics 88, Nov.25-30 ,1988 Tripoli - Libya

3- D.Yul Kwow , "Economics Analysis of Solar Energy for Industerial Usage proceeding Energics 88 , Nov 25-30 1988 Tripoli - Libya.

4- M.M.Miselaty , and A.I. El-Twaty ; Solar Energy for Space Heating in Libya Proc. energics 88 Nov.25-30 1988 , Tripoli- Libya

5- S.Subramanyam , (1989) , Sun World , 13, No.3 , P. 84 . .

6- J.W. Eloemer . J.R. Irwin , J.A.Eibling and G.O.G. Lof (1965)A practical basin type solar still , Solar Energy , 9 , P.197

7- V.A. Akinseti , C.V. Durn (1979) , A Cheap Methods of Improving The Performance of Roof Type Solar Still , Solar Energy , 23 P.271

8- A.Hegazi (1988) , Experimental and Theoritical Study of Solar Stills Thermal Performance with Periodical Adjusted Refelectors , North Sun 1988, Solar Energy at High Latitude Cqnference ,Aug. 29-31, Borlang Sweden .

9- Chr. V.Zabeltitz , (1990) , Solar Energy for Green Houses , First Renewable Energy Conference Sep.23-28, breeding UK.

10- S. Mustafa and G. Brusewitz (1979) Direct Use of Solar Energy for Water Desallination , Solar Energy 22 P.141.

11. S. Talbert , S.Eibling and G.Lof (1970),Manual on Solar Destillation of Saline Water , U.S. Department of the interior , O.S.W. Report N. 546 .

12- R.Klasic , Muller , Chr. Von. Zabeltitz (1989) " A Comparison of Solar Desallination Systems " Gartenaunwissenschaft ,54, P.13-15.

13- D.I. Cooper (1973) Maximum Effeciency of A Single Effect Solar Still ,; Solar Energy 15, 205

COMPREHENSIVE APPLICATION OF SOLAR STILLS IN PAKISTAN

S.W. Ali

Solar Energy Research Centre (PCSIR)
B-12, Latifabad Unit-2, Hyderabad, Pakistan

ABSTRACT

There is acute shortage of drinking water in some parts of Pakistan, especially the Arid Zone and the western coastal areas. But it has high solar insolation and mean sunny day; this advantage could be well exploited for solar energy technology applications. It was this factor which promoted design and installation a 27 cubic meter solar desalination plant (Delyannis type) 20 years hence. The unit has an evaporating surface of 8870 m^2.

The unit was refurbished/recommissioned in 1988 by the Solar Energy Research Centre (SERC), after it was lying in disuse with heavy damage. This paper reveals the factors responsible for the renewed interest in the solar still technology, which suffered declining interest especially in developed countries. Two years of its operation/maintenance has revealed that the unit requires lowtech manpower and involves low recurring operational/maintenance cost, resulting in very low production cost for distilled water (ca 0.25 cents per liter).

Based on these results the SERC proposed adoption of solar stills for the problem areas in Pakistan. This has been actively taken up by the federal as well as provincial authorities.

KEYWORDS

Solar energy; drinking water; solar stills; solar passive; Pakistan; solar thermal; desalination.

INTRODUCTION

Solar distillation is one of many technologies having the potential to meet the need of fresh water in arid and semi-arid areas and especially in coastal areas with low annual rainfall. The technology has been applied only on experimental or pilot plant scale in many parts of the world, but has not met acceptance even in areas acutely suffering from non-availability of drinking water. The reasons for this situation are many and varying from case to case. A historical review of the growth of the technology has been given by Talbert and co-workers (1970). In view of the interest shown in the past three decade, a wide variety of designs has evolved comprising single effect stills (basin type), multiple effect stills, inclined stills, forced convection stills and solar concentration stills but only basin type single effect stills has met any degree of success, Table 1 lists major stills in the range of more than 1000 meter square evaporating surface, alongwith year of installation. Although there has been extensive discussion in the literature about solar stills and a number of pilot units have been installed(while others have been scattered around the world), there are not many documented cases of long-term operation. Recent reporting by

Table 1. Large Basin-type Solar Stills

Country	Location	Year Built	Basin Size m^2
Australia	Cooper Pedy*	1966	3160
Chile	has salinas*	1872	4459
Greece	Symi	1964	2687
	Aegina	1965	1490
	Patmos*	1967	8640
	Kimolos	1968	2508
	Nisiros	1969	2043
Tunisia	Mahdia	1968	1300
West Indias	Petit St. Vincent	1967	1709

* no longer operating

Twidell(1989) of dismantling of the Patmos(Greece)solar still (a Delyannis type) has incidently coincided with refurbishing of another old solar still (a modified Delyannis type) at Gawader (Pakistan). The news item reported by Twidell carries comments by the author that the important unit was broken up for no good reason other than lack of maintenance and possibly because of design flaw. Besides, the plant was abandoned between 1973 and 1988.

PLANS FOR USE OF SOLAR STILLS IN PAKISTAN.

Pakistan is an agricultural country where part of water requirement is met by the largest irrigation canal system in the world,and the remaining depends solely on rain water. In the latter case, drinking water for animals as well as human consumption is either met by encatchment of rain water or dug-wells. In arid and semi-arid areas of Baluchistan and Sindh, the majority of wells,with depth up to 100 meter,yield brackish water with TDS as 12000 ppm. Typically such areas also lack facilities of life including electric supply.

In 1986-87 the Government of Pakistan launched a large scale willage electrification programme and concurrently gave a cautious go-ahead signal for installation of a large number of solar water desalination plants(based on brackish and sea water)for areas specially remote and without electric power.

The SERC-PCSIR based at Hyderabad was entrusted to mobilize the thrust. This was accompanied by assignment to the SERC the repair and refurbishing of the 27 cubic meter per day Gawader Solar Water Desalination Plant which was lying in disuse and badly damaged for a number of years. The Gawader plant was first installed in 1972.

Two small stills with evoporating surface area of 30 m^2 were installed in Thar Desert of Sindh, followed by another two relatively larger stills of 290 m^2 each again in the same locality. Cleaning, repair and recomissioning of the worlds largest unit,with evoporating area of 8870 m^2 was also completed in mid-1988 (see Table 2 and fig 2). Design of smaller stills was similar to the design of the CSIRO (Australia). The main redeeming feature for the four smaller plants are almost total indigenisation;

Fig. 1. CSIRO Australia type Solar Still installed
at Vajuto, Thar Desert, Sindh, Pakistan.

Fig. 2. Gwader Solar Still (Modified Delyannis type)
recommision in 1988.

TABLE 2. Major Solar Stills installed in Pakistan

Location	Year Built	Basin Size m^2
Man Bai Jo-tar	1987	290
Vajuto	1990	290
Gawader	1972	8870
(Recommissioned)	1988	6200
Gawader (under construction)	1991	8870

the only item imported (but locally available) was the silicone sealant.

By the successful installation and operation of stills, backed by the experience gained in the repair work of the Gawader plant in 1988,the SERC had obtained enough confidence and expertise to express its readiness to fabricate more such plants scattered throughout Pakistan.

The provincial authorities of Sindh came up with a Rs.28 million plan for installing as many as 100 smaller and medium size stills,while the Pakistan Navy awarded a contract of Rs.7.6 million to the SERC for duplicating the modified-Delyannis Solar Still for its establishment.

SOLAR STILLS VERSUS OTHER TECHNOLOGIES.

Acceptability of solar stills in Pakistan in comparison with high technologies is based on the fact that (i) the technology has been totally indigenised, (ii) no foreign exchange component is involved, (iii) operation and maintenance by locally available semi-skilled manpower, (iv) low over-all cost of product water, (v) high plant life-expentancy and (vi) inherent modular concept enabling continuous operation even during cleaning periods or repair.

On the other hand,high tech methodology of Reverse Osmosis (RO) and Electrodoialysis involve continuous replacement of imported component and high sensitivity of the plant to departure from precscribed operation instructions. The normalised cost of water using various technologies have been calculated to vary from $ 0.80/m^3 for R.O if brackish water, $ 0.94/m^3 for Electrodialysis for brackish water to $ 6.56/m^3 for sea-water R.O. while the cost for sea-water (or brackish-water) solar still is about $ 3.25/m^3 with no foreign exchange component. These figures are valid for Pakistan. The figures for conventional flash evaporation and multi-stage flash evaporation are not given,but in energy deficient areas like Pakistan, the cost is visualized to be not favourable. Other technologies such as freezing and vapour compression have not yet been considered as possible choices although the latter has obvious advantage when run with solar thermal energy.

However, the factors that could be cited against the solar still are (i) high initial cost (ii) low yield and (iii) large evoporating surface to yield ratio.

REFERENCES

1.Talbert, S.G., Eibling, J.A. and Löf, G.O.G. (1970) Manual on Solar Dist of Saline water, Office ot Saline water, U.S. Dept of Interion, Res. and Dev.Rep. No. 546.
2. Dickinson, W.C. and Cheremisinott, P.N. (1980)
    Solar Energy Technology Handbook (part B). Marcel Dekker, Inc.pp 209.
3. Twidell, J. (1989). Sunworld. 13,26.

ENHANCEMENT OF ENERGY STORAGE IN THE GROUND BENEATH SOLAR PONDS

Ram Prasad[*] and D.P. Rao[**]

[*]Department of Chemical Engineering
H.B. Technological Institute, Kanpur-208002 (India)
[**]Department of Chemical Engineering
Indian Institute of Technology, Kanpur-208016 (India)

ABSTRACT

The salt used in a solar pond accounts for a significant fraction of the cost of pond if it is located far away from the source of salt and if seasonal energy storage is required. Some saving in the cost can be effected by making use of the energy storage in the ground beneath the pond. A method for enhancing energy storage in the ground beneath the pond is proposed. The method involves formation of trapezoidal shaped trenches at the bottom of the pond, running from one end of the pond to the opposite end. We present here feasibility studies on the enhanced energy storage in the ground. The effective energy storage in terms of equivalent depth of saturated salt solution and heat losses to the ground have been evaluated for different amplitudes of temperature of the lower-convective layer, trench geometries and depths of water table. The results show that it is feasible to achieve enhanced energy storage if the water table is large. The results also indicate that the heat loss to the ground decreases in the case of pond with trenches compared to the one with flat bottom.

KEYWORDS

Energy storage; Ground storage; solar pond.

INTRODUCTION

One of the advantages of solar ponds is that a built-in energy storage can be provided to take care of the seasonal variations of the insolation as well as energy demand. This can be achieved by providing a thick lower-convective layer for energy storage. However, the studies (Srivastava, 1980; Ozakcay and Gurgenci, 1985; Nielsen, 1988) indicate that the cost of the salt could be as high as 60-70% of the total pond cost, if the pond is located far away from the source of the salt. In principle, some saving in the cost can be effected by making use of the energy storage in the ground beneath the pond. But the studies (Akbarzadeh and Ahmadi, 1979; Zhang and Wang, 1990) indicate that this energy storage is marginal. In this paper we propose a method to enhance energy storage in the ground beneath the pond and present studies on its effectiveness.

PROPOSED METHOD

The proposed method of enhancing energy storage requires the formation of trapezoidal shaped trenches at the bottom of the pond, spanning from one end to the opposite end of the pond, as shown in Fig. 1. The salt solution in the lower-convective layer circulates either heating or cooling the 'fin-like' projection of the ground in between trenches. To aid circulation, the salt solution can be withdrawn from the bottom of the trenches and reinjected into the convective layer after heat extraction, using a circulating pump as shown in Fig. 1.

MODEL

The pond-bottom geometry is given in Fig. 1, along with the notation employed in this work. To keep mathematical complexities to a minimum in modelling the heat transfer, we have made the following assumptions:

- The temperature of the convective layer varies as a sinusoidal function of time, with a time period of one year.
- The edge losses are negligible.
- The heat-transfer resistance on the solution side is negligible.
- The movement of the ground moisture above the water table is negligible.
- The ground at the depth of water table is at a constant temperature which is equal to the yearly average ambient temperature.
- Thermal properties of the ground are constant.

With these assumptions, the heat transfer between the lower-convective layer and the ground below the pond can be modelled as given below.

Consider the lower-convective layer-ground system shown in Fig. 1. The governing heat conduction equation is:

$$\frac{\partial \theta}{\partial \tau} = \frac{\partial^2 \theta}{\partial X^2} + \frac{\partial^2 \theta}{\partial Y^2} \tag{1}$$

The initial and boundary conditions are:

$$\theta = \theta_m + \theta_0 \cos \tau \qquad \text{for } 0 \leq X \leq X_L \ , Y=0 \ , \tau > 0 \tag{2}$$

$$\theta = \theta_m + \theta_0 \cos \tau \qquad \text{for } X'_L \leq X \leq X_L^+ \ , Y=Y_d \ , \tau > 0 \tag{3}$$

$$\theta = \theta_\infty \qquad \text{for } 0 \leq X \leq X_L^+ \ , Y=Y_D \ , \tau > 0 \tag{4}$$

$$\frac{\partial \theta}{\partial X} = 0 \qquad \text{for } X=0, \ 0 < Y < Y_D \ , \tau > 0 \tag{5}$$

$$\theta = \theta_m + \theta_0 \cos \tau \qquad \text{for } X = X_L^\bullet \ , 0 \leq Y \leq Y_d \ , \tau > 0 \tag{6}$$

$$\frac{\partial \theta}{\partial X} = 0 \qquad \text{for } X = X_L^+ \ , Y_d < Y < Y_D \ , \tau > 0 \tag{7}$$

$$\theta = \theta_\infty \qquad \text{for all X and Y} \ , \tau = 0 \tag{8}$$

where

$$\theta = T/(T_m - T_\infty) \ ; \ X = x\sqrt{\omega/\alpha_g} \ ; \ Y = y\sqrt{\omega/\alpha_g} \ ; \ \tau = \omega t$$

$$\theta_m = T_m/(T_m - T_\infty) \ ; \ \theta_0 = T_0/(T_m - T_\infty) \ ; \ \theta_\infty = T_\infty/(T_m - T_\infty)$$

$$X_L = L\sqrt{\omega/\alpha_g} \quad ; \quad X'_L = (L+d \cot \beta)\sqrt{\omega/\alpha_g} \quad ; \quad X_L^+ = (L+d \cot \beta + b)\sqrt{\omega/\alpha_g}$$

$$X_L^\bullet = (L+y \cot \beta)\sqrt{\omega/\alpha_g} \quad ; \quad Y_d = d\sqrt{\omega/\alpha_g} \quad ; \quad Y_D = D\sqrt{\omega/\alpha_g} \tag{9}$$

Equation 1 is solved with the initial and boundary conditions (Eqs. 2-8) to obtain the temperature field in the ground beneath the pond. We have employed a finite-difference scheme based on the alternating-direction-implicit method (Carnahan, Luther and Wilkes, 1969) to solve the equations.

The rate of heat transfer per unit length of the trench in the z-direction, $q_b(\tau)$, from the convective layer to the ground, has been determined from the equation

$$q_b(\tau) = -k_g(T_m - T_\infty)\left[\int_0^{X_L}\frac{\partial\theta}{\partial Y}\bigg|_{(X,0,\tau)}dX + \int_0^{Y'_d}\frac{\partial\theta}{\partial X}\bigg|_{(X_L^\bullet,Y,\tau)}dY + \int_{X'_L}^{X_L^+}\frac{\partial\theta}{\partial Y}\bigg|_{(X,Y_d,\tau)}dX\right]$$

where $\tag{10}$

$$Y'_d = (d/\sin \beta)\sqrt{\omega/\alpha_g} \tag{11}$$

The rate of heat transfer per unit area of flat bottom, $q'_b(\tau)$, has been calculated using the relation

$$q'_b(\tau) = q_b(\tau)/(L+d \cot \beta + b) \tag{12}$$

The integration of Eq. 10 has been carried out using Simpson's rule.

RESULT AND DISCUSSION

Extensive numerical experiments were carried out to assess the efficacy of the proposed method of enhancement of energy storage. Considering the temperature of the convective layer varies as $T = T_m + T_0 \cos \omega t$, the temperature profiles in the ground, heat flux into and out of the pond with time, energy storage and net heat losses to the ground have been computed for various parameters. The ranges of parameters employed in this study are given in Table 1. A few significant results are presented below.

Heat Fluxes at the Bottom

Figure 2 shows the heat flux at the pond bottom with time. As expected, the heat flux into the ground is large in the beginning but decreases with time and attains near constant value within two cycles. The areas A_1 and A_2 indicated in the Fig. 2 represent the energy recovered from the ground and energy transferred to the ground, respectively, in a cycle for $T_0 = 20°C$. The difference $(A_2 - A_1)$ represents the net heat loss to the ground. The net heat losses to the ground are found to be 107.1 and 95.5 MJ/$(m^2 \cdot y)$ for flat bottom and bottom with trenches, respectively.

Equivalent Energy Storage

The energy storage in the convective layer of area A per unit depth, E_s, in a cycle is

$$E_s = 2 A \rho_s C_{PS} T_0 \tag{13}$$

The energy storage in the ground can be expressed as equivalent depth of convective layer, E, defined as

$$E = \int_0^{\tau_c} A\, q_+ d\tau \,/\, E_s \qquad (14)$$

where q_+ is the heat flux into the convective layer based on unit area of the pond and τ_c is the dimensionless cycle time. Figure 3 shows the equivalent storage for the two types of pond bottom. The energy storage increases with increase in amplitude and the water-table depth. It has been observed that the energy storage can be increased by increasing the trench depth. The energy storage greater than 2 m equivalent storage is feasible. However, increased trench depth can pose construction problems. But, the energy storage by the proposed method is infeasible if the water-table depth is less than 5 m. The equivalent energy storage is found to be maximum for trench spacing of about 1.75 m. But there is a little variation in the equivalent storage beyond 1.75 m up to 3 m. We have observed that the rectangular trenches are the most effective for energy storage. In practice, the rectangular trenches are costly to construct as they require lining to prevent the land slide. The land slide can be avoided using trapezoidal trenches with appropriate side angles. The equivalent storage increases monotonically with increase in β up to 90°. Hence, the maximum side angle that is permissible (i.e. the one that prevent the land slide and a 3 m trench spacing) may be used.

CONCLUSION

The trenches are found to be effective in enhancing energy storage in the ground. The results show that the proposed method of enhancement of energy storage is feasible if the water-table depth is large. Also the heat loss to the ground decreases marginally. The trenches not also reduce the salt requirement but offer flexibility and ease in energy extraction.

NOMENCLATURE

A   area, m^2
b   half of trench width at bottom, m
C_{ps} specific heat of salt solution, $J\ kg^{-1}\ K^{-1}$
d   depth of trench, m
D   depth of water table, m
E   equivalent depth, defined by Eq. 14, m
E_s energy storage in a cycle, given by Eq. 13, $W\ m^{-1}$
k_g thermal conductivity of ground, $W\ m^{-1}\ K^{-1}$
L   half-trench spacing of projected ground, m
q_b heat-transfer rate per unit length of trench, $W\ m^{-1}$
q_b' heat flux, given by Eq. 12, $W\ m^{-2}$
q_+ heat flux into the convective layer, $W\ m^{-2}$
t   time, s
T   temperature, ^oC

T_m mean temperature of convective layer, $^\circ C$

T_0 amplitude of temperature of convective layer, $^\circ C$

T_∞ average ambient temperature, $^\circ C$

x horizontal space variable, m

$X_L, X_L', X_L^+, X_L^*, X$ dimensionless distances, defined by Eq. 9

y vertical space variable, m

Y, Y_d, Y_D dimensionless distances, defined by Eq. 9

Y_d' dimensionless distance, defined by Eq. 11

α_g thermal diffusivity of ground, $m^2 s^{-1}$

β angle shown in Fig. 1

ρ_s density, $kg\ m^{-3}$

ω angular velocity, $rad\ s^{-1}$

τ dimensionless time, defind by Eq. 9

τ_c dimensionless cycle time

$\theta, \theta_m, \theta_0, \theta_\infty$ dimensionless temperatures, defined by Eq. 9

REFERENCES

Srivastava, A. K. (1980). Studies on solar water pumping systems. *M.Tech. Thesis,* I.I.T. Kanpur.

Ozakcay, L. and H. Gurgenci (1985). In K. G. T. Holland (Ed.), *Extended abstracts. INTERSOL-85,* p.333.

Nielsen, C. E. (1988). Salinity-gradient solar ponds. In K. W. Boer (Ed.), *Advances in Solar Energy,* Vol. 4, American Solar Energy Society, Inc. and Plenum Press, pp.445-498.

Akbarzadeh, A. and G. Ahmadi (1979). Under ground thermal storage in the operation of solar pond. *Energy,* **4**, pp. 1119-1125.

Zhang, Z. M. and Y. F. Wang (1990). A study on the thermal storage of the ground beneath solar ponds by computer simulation. *Solar Energy,* **44**, pp.243-248.

Carnahan, B., H. A. Luther and J. O. Wilkes (1969). *Applied numerical methods.* John Wiley and Sons, New York.

TABLE 1 Range of Parameters Employed

Parameter	Range
Trench geometry	
Trench depth	0 – 5 m
Trench width	0.2 m
Trench spacing	0 – 3.8 m
Angle β	0 – 90°
Depth of water table	1 – 20 m
Ground properties	
Thermal conductivity	0.5-2.5 $W\ m^{-1}k^{-1}$
Thermal diffusivity	4×10^{-7}-$5.2\times10^{-7} m^2 s^{-1}$
Other parameters	
Amplitude of temperature of convective layer	0 – 20$^\circ C$
Mean temperature of convective layer	65 – 85$^\circ C$
Yearly average ambient temperature	10 – 30$^\circ C$

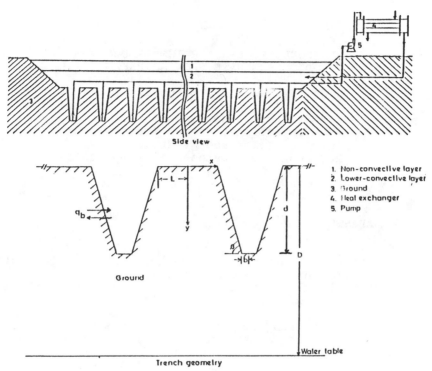

Side view

1. Non-convective layer
2. Lower-convective layer
3. Ground
4. Heat exchanger
5. Pump

Ground

Trench geometry

Fig. 1. Schematic Diagram of Solar Pond with Trapezoidal Shaped Trenches.

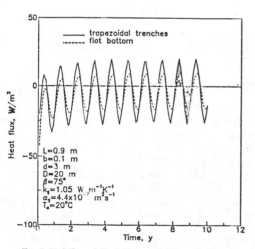

Fig. 2. Variation of Heat Flux at Bottom with Time.

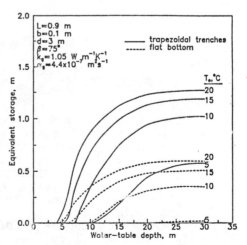

Fig. 3. Effect of Water-table Depth and Temperature Amplitude on Equivalent Storage.

SOLAR-POND PROCESS FOR THE PRODUCTION OF
SODIUM SULPHATE FROM ASTRAKANITE

Jorge M. Huacuz V., Margarita Silis C.

Electrical Research Institute
Non-Conventional Energy Sources Dept.
Solar Energy Group
P.O. Box 475, Cuernavaca, Mor. 62000
MEXICO

ABASTRACT

In this work a process to reclaim sodium sulphate from astrakanite by means of a solar pond is proposed. The process has its theoretical basis on the phase changes that the system Na_2SO_4-$MgSO_4$-H_2O undergoes at different temperatures and concentrations. A block diagram of the proposed process is presented. The process is functionally described. Results of computer simulations aimed at establishing the suitability of using solar ponds for this process are discussed.

KEYWORDS

Solar pond; astrakanite; sodium sulphate; reclaiming process.

INTRODUCTION

Solar evaporation lagoons are frequently used to concentrate natural brines to recover valuable mineral products. Upon concentration, different salts precipitate sequentially. In some processes, the concentrated brine is removed from the evaporation laggon just before precipitation of the desired product starts, and is sent to other parts of the process where precipitation takes place under more controlled conditions. At times undersired byproducts, in the form of double or triple salts, precipitate in the lagoon before the brine is removed. This normally causes loss of raw material and operational difficulties.

Astrakanite is a double salt ($Na_2SO_4 \cdot MgSO_4 \cdot 4H_2O$) which precipitates as an unwanted byproduct in solar evaporation lagoons, from brines containing both sodium sulphate and magnesium sulphate in solution. It forms a hard layer of solid material rapidly growing in thickness at the bottom of the lagoon. Eventually, the lagoon may become unoperative due to the large quantitites of astrakanite filling its space. When this happens, the lagoon has to be either abandoned and a new one built, or rehabilitated by mechanically removing the solid material. Both operations may involve large expenses plus the waste of valuable materials.

In this paper a novel solar-pond process, conceived to reclaim sodium sulphate trapped in astrakanite crystals, is proposed. This process may allow at the same time rehabilitation of the otherwise abandoned lagoons, by non-mechanical means.

THEORETICAL BASIS

The chemical system Na_2SO_4-$MgSO_4$-H_2O contains two salts with a common ion. Hence, specific equilibrium relationships are maintained among its components. These are exemplified in the hypothetical phase diagram of Fig.1. Here, point a would represent a saturated solution of $MgSO_4$ when Na_2SO_4 is not present. Line aA would represent that saturated solution in equilibrium with increasing quantities of solid $MgSO_4$, and so forth. The vertices of the diagram would represent pure components, while the inner regions of the diagram would represent chemical systems either solid, in solution or a mixture of these, with varying amounts of each component.

Such diagrams are normally used to graphically represent the processes that a chemical system may undergo. For instance, isothermal evaporation of the non-saturated solution P, would result in increased concentration along the line Pd. Further evaporation would result in precipitation of $MgSO_4$ and a

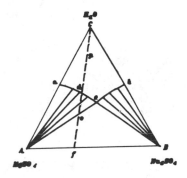

Fig. 1 "Simplified phase diagram for the chemical system: two salts with a common ion"

corresponding change in the composition of the solution. Eventualy, at point e Na_2SO_4 would start precipitating and a double salt would be obtained.

The actual structure of an equilibrium diagram is a function of temperature. Figure 2 shows equilibrium diagrams for the system Na_2SO_4-$MgSO_4$-H_2O for temperatures 25°C, 60°C, 80°C and 100°C.

These diagrams were constructed using information given in Archivald and Gale (1924). Point <u>a</u> in all these diagrams represents the composition of astrakanite. The triangular region with <u>a</u> as vertix corresponds to the region of stability for this compound (see legend to Fig.2). Thus, astrakanite can not be broken into its components by simple hydration and dehydration processes at temperatures between 25°C and 60°C, since the process will reversibly take place along line <u>a-H$_2$O</u>.

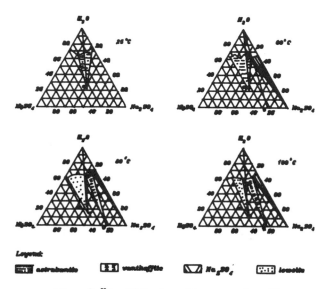

Fig. 2 "Equilibrium diagrams for the system Na$_2$SO$_4$-MgSO$_4$-H$_2$O

On the other hand, if temperature of the system is increased to 80°C, astrakanite will change phase to loweite ($_M$gSO$_4 \cdot$ Na$_2$SO$_4 \cdot$ 5/2H$_2$O). Hydrating loweite crystals at around 100°C will eventualy yield vanthoffite [(Na$_2$SO$_4$)$_3 \cdot$ MgSO$_4$], which can be hydrated at temperatures between 40°C and 60°C to yield anhydrous sodium sulphate (tenardite). This process demands energy.

RECLAIMING PROCESS

The separation process proposed to reclaim Na$_2$SO$_4$ from astrakanite will include the following steps: a) dissolving astrakanite crystals to obtain a saturated solution, and heating of such solution to temperatures between 80°C and 100°C to turn astrakanite into loweite; b) hydration of loweite in the same temperature range to turn it into vanthoffite; c) separation and dissolution of vanthoffite crystals at temperatures between 40°C and 60°C to obtain tenardite; and d) final separation of tenardite.

To provide heat required by the process, salt-gradient solar ponds will be used: first, on top of the solid astrakanite layer, a NaCl gradient will be created; heating at the bottom of this gradient by the incoming sun rays will help accomplish both the dissolution and heating operations required in step a) of the process. Hot NaCl brine from a second pond will be used to hydrate loweite. Heat from a third pond will be used to carry out step c) of the process.

Figure 3 shows a simplified block diagram of the process. Two high-temperature (~90°C) and one mid-temperature (~50°C) solar ponds are included. Sodium chloride will be used to create the gradient. It will be recovered at the end of the process and fed back to the solar ponds. MgSO₄ is also expected as byproduct. One of the high-temperature ponds will be constructed using otherwise abandoned evaporation lagoons fouled up with astrakanite. This way the lagoons may be rehabilitated at no extra cost.

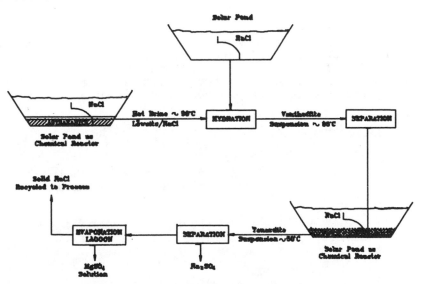

Fig.3 "Simplified block diagram of proposed reclaiming process"

PRELIMINARY THEORETICAL RESULTS

The ability of NaCl solar ponds to sustain the process was investigated by computer simulations. A modified version of the

SOLPOND computer program (Henderson and Leboeuf, 1980) was used for this purpose. Solarimetric and climatological data for a region in Mexico where the astrakanite problem is present were used as input parameters to the program. The minimum desired temperature was used as the design parameter. Thermal performance of several ponds of different sizes was then evaluated. A load of ten tons per day of astrakanite was used as the basis for calculations.

Results obtained show that the low-temperature pond can easily meet the requirements of a continuous reclaiming process. The high-temperature ponds, on the other hand, could provide enough hot brine (slightly above 90°C) between May and September only. Effects of this restriction on the economics and practicality of the process need to be further investigated.

Hydrodinamic stability at the bottom of the gradient is not expected to be of concern for the first high-temperature pond, since the bottom layer of brine would be extracted from the pond as soon as it reaches the desired temperature. However, the second high-temperature pond could pose some stability problems.

CONCLUSIONS

Several important questions need still to be solved before the process is practically implemented. For instance, information on some physical and chemical parameters of the reactions involved are still needed in order to establish engineering parameters such as residence times, controlability of the process and so forth. Laboratory work will proceed to solve such questions.

REFERENCES

Archivald and Gale (1924). The System Magnesium Sulphate-Sodium Sulphate-Water and a Method for the Separation of the Salts. J.Amer.Chem.Soc., 46b, 1760.

Henderson J. and C.M. Leboeuf (1980). "SOLPOND-A Simulation Program for Salinity Gradient Solar Ponds", Second Annual Systems Simulation and Economics Analysis Conference, SERI/TP-351-559, San Diego, California.

EXPERIMENTAL STUDY ABOUT CONVECTIVE SOLAR STILLS

J.J. Hermosillo

Unidad Académica de Tecnología Intermedia
División de Ingeniería, ITESO.
Fuego 1031, Jardines del Bosque,
Guadalajara, Jal., 44520, MEXICO.

ABSTRACT

In this paper we present an experimental study about the behavior of a convective solar still. Temperatures and rates of distillation in two different stills are studied to obtain the effect of natural internal convection. on the average, on sunny days, the convective solar still produces a volume of distilled water about 2.5% higher than a non-convective with the same dimensions.

KEY WORDS

Solar still; convection; inclined solar still; desalination.

INTRODUCTION

For many years, solar stills have been used to produce salt-free water. Conventional single basin solar stills are surprisingly simple and work reasonably well, but have efficiency limitations that have been well studied (Morse, 1975; Sodha, 1982). In many semidesert places, this is not a serious drawback, because the cost of land is low. However, there are many applications where a higher efficiency is desirable.

Various authors have presented ingenious designs of solar stills that improve the efficiency while keeping them 100% solar powered, or requiring only a small amount of extra energy. Improving the thermic efficiency of these stills without losing its technological simplicity is an interesting goal for some researchers. Sodha (1982) presents a lot of variations and designs of solar stills, developed by various authors, that perform better than the single basin solar still, producing even 75% more water under the same conditions.

One interesting design is the inclined solar still, first constructed by Tleimat and Howe in 1966 (Sodha, 1982). This solar still has several advantages over the conventional ones. First, in non-equatorial latitudes, it can be installed facing southwards in northen hemisphere (and viceversa), to improve direct solar collection. Second, the distance between evaporating and condensing surfaces is minimized, thus improving mass transport by diffusion.

Third, some optical properties for collecting solar radiation can be improved, because of the geometric design of the still.

In most of the inclined solar stills that have been developed, the ratio of condensing area to evaporating area is similar to this ratio in a single basin still. (Except perhaps in the textile type of Moustafa and Brusewitz). This is because even if the cover (condensing area) is inclined, the water in the various trays is always horizontal. However, the inclined array of the trays permits presenting a larger effective collecting area towards the sun.

A variation of the inclined solar still was presented by Alvarado and Hermosillo (1983), as shown in Fig 1. This is an inclined solar still that has a an extra area for condensation of water vapor. The position and design of this condenser at the back of the still, has several features. First, it is not exposed to sunshine. Therefore, it can be mantained at lower temperatures than the glass (or plastic) cover, favoring condensation. Second, it can be built of a metalic sheet, having a heat conducting surface that contributes to the same process. Third, this condensing surface, allocated as shown in Fig. 1, forms a duct that is capable of inducing natural convection of the humid air in the still. In this scheme, the humid air goes upwards while its heated by sunshine in the "duct" formed between trays and cover. Meanwhile, the internal air in the back duct is being cooled, tending therefore to go downwards.

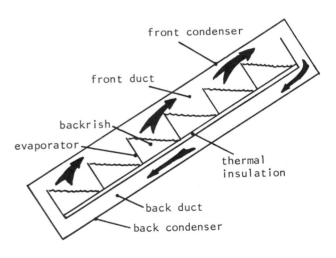

Fig. 1. Sketch of an inclined solar still with back condensing surface.

This still has not been conceived for great-scale production of distilled water, but for home scale, especially in places where space efficiency is important. However, posterior studies would show if this idea is still effective at a greater scale.

In former papers it was shown that this inclined still is capable of producing a daily volume of water about 20% higher than a single basin still of the same area (Alvarado and Hermosillo, 1985). This average was obtained from several months of operation near March equinox, in a range of sunshine conditions from cloudy (8000 kJ m^{-2} of global irradiation) to sunny (19000 kJ m^{-2}). However, the improvement of distilled water production may be attributed to either collector inclination, to larger condensing area, to better mass transfer due to internal convection of humid air in the still, or to a combination of these causes.

The aim of this paper is to show the effect of internal convection on the internal temperatures and, more important, on the daily production of distillate.

EXPERIMENTAL STUDY

Two identical solar stills were built, with the sole exception of the possibility of having internal convection. (In almost every solar still exists internal convection in a strict sense. In this paper, when we refer to internal convection we mean the transport of mass from the front duct to the back one (Fig. 1) due to natural convection, as explained before).

Both stills have identical inclined collector-evaporators with 10 horizontal trays. The slope of these units is 20°. Their dimensions are 1.15 m (length) by 0.90 m (width). This is the size of the black collector-evaporator without thermal insulation. The thickness of thermal insulation is 0.05 m. The distance between the collector-evaporator unit and both the front and the back condenser is 0.05 m. The external box is 0.10 m wider than the collector-evaporator, connecting the front duct with the back one not only at the upper and lower regions of the still, but at both sides.

The unique difference between both stills is described as follows. The first one has obstruction panels at both sides of collector-evaporator to avoid lateral movement of humid air (or diffusion), but permitting the upwards and downwards movement in the upper and lower ducts respectively. This still is supposed to have convective mass transfer and will be referred to as "convective" (CON). This is the scheme shown in Fig. 1.

The second still has obstruction panels at the top and the bottom of the still, but permits the lateral movement of humid air (supposed to be mainly due to diffusion). This is referred to as "non-convective" (NOC).

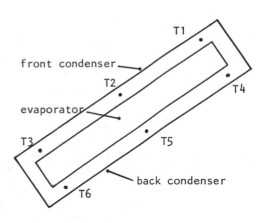

Fig. 2. Localization of thermometers inside both solar stills.

Then, both stills have exactly the same collecting area, the same orientation and inclination, the same brine capacity. Both are fed with the same amount of brine (fresh water, for this study) at the same temperature and are exposed to the same irradiance, wind, and ambient temperature.

Temperatures were measured at six points inside each still along the solar day, in the points shown in Fig 2. The mass (actually volume) of distilled water was recorded every hour.

RESULTS

Experiments have been carried out in a variety of sunshine conditions. The "typical" results reported in the following graphics were obtained averaging the results of sunny days in mid february.

Figure 3 shows temperature vs. time in both stills. The shown temperatures are T1, T3 and T6. Note that T1 is about 4°C higher in NOC still than in CON still, while T6 is almost identical in both stills.

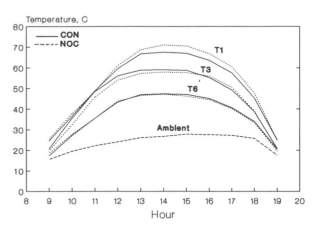

Figure 4 shows the temperature difference between both stills (CON minus NOC) in points 1, 2 and 3 of Fig, 2. Note that in front duct temperatures in CON still tend to be lower than in the NOC still, especially in the upper part. The oscillation of the difference of temperatures during the first hours of the morning is an observed phenomenon during sunny days, that can be related to the beginning of the convection from front to back ducts. Other points tend to have smaller temperature differences.

Fig. 3. Temperature vs. time in convective solar still (CON) and not convective solar still (NOC).

Figure 5 shows the temperature difference between both stills (CON minus NOC) in points 4, 5 and 6 of Fig. 2. Temperatures in back duct tend to be higher in convective still than in the non-convective one, especially in the upper part. The smallest difference of temperatures along the day is found in T6, at the bottom of the back duct.

Figure 6 shows the accumulated mass of distilled water vs. time in both stills. Three pairs of curves are shown: The lower one corresponds to distillate obtained at back condenser. The medium one, shows the distillate from the front condenser (glass). The upper one is the sum of both front and back. Note that the water mass condensed in the front condenser is practically the same amount in both stills, being about 68% of total. However, in the convective still the back condenser systematically produces more distilled water. The mean mass of water condensed in back condenser is about 5.8% higher in CON still compared to NOC still. The overall water production (fron plus back cover) is increased by about 2.5%.

Figure 7 shows the rate of distillation, in kg/h, in both stills, along with solar irradiance. The rate of distillation was obtained after adjusting cubic splines to the cumulative data. Note that the increment in rate of distillation in back condenser appears in the near noon hours, corresponding to the maximum temperature differences of Figs. 4 and 5. Those effects can be attributed to the internal convection.

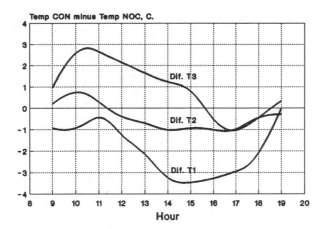

Fig. 4. Temperature differences (CON minus NOC) in front duct of the stills.

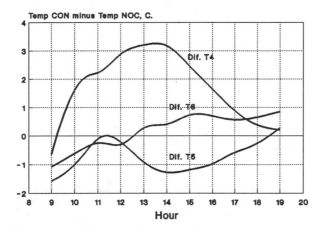

Fig. 5. Temperature differences (CON minus NOC) in back duct.

CONCLUSIONS

It has been shown that natural convection inside an inclined solar still modifies temperatures and rates of distillation, compared with a non-convective still. The percentage of distillate increment is small but systematically observed. This suggests that the still performance could be improved if forced convection existed inside the still. Further works will show these results.

2400

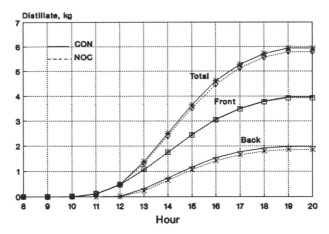

Fig. 6. Accumulated mass of distilled water along the day.

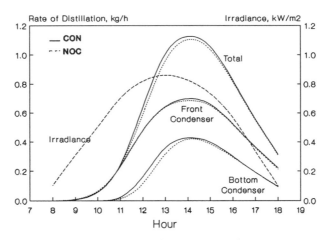

Fig. 7. Rate of distillation in both stills along the day.

REFERENCES

Alvarado, Ramón and Hermosillo, Juan Jorge. <u>Utilización de Destiladores Solares para Obtención de Agua Potable a Escala Familiar.</u> Proceedings of the VII Reunión Nacional de Energía Solar. Saltillo, México. 1983.

Alvarado, Ramón and Hermosillo, Juan Jorge. <u>Destilador Solar por Convección Natural.</u> Proceedings of the IX Reunión Nacional de Energía Solar. Mérida, México, 1985.

Morse, R.N. <u>Desalination and Drying.</u> article in Messel, H. and Butler, S.T. <u>Solar Energy.</u> Pergamon Press, England, 1976.

Sodha, M.S. et al. <u>Solar Distillation.</u> Pergamon Press, London, 1982.

2.30 Solar Drying I

Heat Pumps for Agricultural & Industrial Drying Processes

W.W.S. Charters; S. Theerakulpisut

University of Melbourne, Parkville, Victoria 3052, Australia

Abstract

Many agricultural and industrial products require controlled drying as part of the overall production process. In practice much of the commercial drying plant is not energy efficient for various reasons, one of which is that commercially available driers are not generally equipped with heat recovery facilities. With heat pumps, however, it is possible to recover both sensible and latent heat and to use to best advantage the high grade energy input to the heat pump compressor.

Because of the inherent system complexity it is important to formulate accurate simulation models of the heat pump and of the drying process and to combine them into a single system model for drier performance evaluation. It is equally important to establish the validity of any such system simulation models by comparison with actual test results of product drying.

In this paper a brief overview is given of the development of an electrically driven vapour compression heat pump drying system at the University of Melbourne, the formulation of system models based on a fixed bed thin layer drying model, and the validation of the desk top computer simulation against a series of drying tests using rough rice as the product to be dried.

Testing of various system configurations has confirmed that the best performance in terms of specific moisture extraction rate (kg/kWh) is not necessarily commensurate with choice of the configuration which optimises the heat pump coefficient of performance.

Key Words

Heat pumps grain drying, computer modelling, experimental validation.

Introduction

Drying is a common process in a number of industries. Review of the literature reveals that commercial dryers are highly inefficient due to various factors, one of which is that commercial dryers are not generally equipped with heat recovery facilities. With heat pumps, however, it is possible to recover both sensible and latent heat from the dryer exhaust, hence energy loss can be substantially reduced. Furthermore, heat pumps always deliver more heat than the work input to the compressor. These two salient features render heat pump drying a premium alternative for efficient use and conservation of energy in drying industries.

A brief coverage of the development of a heat pump model is given in this paper. The model is combined with a fixed-bed drying model to form five heat pump dryer configurations for system comparison by simulation. Theoretical and experimental results used in model verification are also presented.

Heat Pump Model

The main component models which form the heat pump model are condenser, evaporator, thermostatic expansion valve and compressor models. The governing equations for these component models are rather complex and numerous and the solution methods are lengthy. For this reason, only a brief overview is given of the development of the models without presenting the equations. Interested readers are referred to the Ph.D. thesis by Theerakulpisut[1](1990) for further details.

Fixed-Bed Drying Model

The development of modern computing systems together with more efficient, accurate and stable computational techniques has enabled research workers to implement more accurate grain drying models of the partial differential equation type. After surveying most of the PDE models available in the literature, the fixed-bed drying model selected for use with the heat pump model in this investigation was based on the model by Bakker-Arkema et al[2] (1974) which is also known as the Michigan State University (MSU) model. This model was formulated by considering energy and mass balances of air and grains within a differential volume located at an arbitrary location in the stationary bed. The model consists of four partial differential equations which must be solved simultaneously for four unknowns, namely grain moisture content, grain temperature, air humidity ratio and air temperature, at any depth of the grain bed.

Heat Pump Drying Models

The heat pump model can be combined with the fixed-bed drying model in a number of configurations which represent different drying systems. Figure 1 shows all of the preferred arrangements chosen for this investigation. It should be noted that the arrangement of the heat pump components is basically the same for all configurations, with different arrangements for air flow through the rig. Configurations 2 and 4 were eventually chosen as the preferred arrangements during the course of this study.

Heat Pump Dryer Rig

The main components of the heat pump dryer rig include evaporator, condenser, thermostatic expansion valve, compressor, motor, fans and associated air and refrigerant duct work. The overall view of the experimental rig is shown in Figure 2. The experimental rig was equipped with measuring equipment to determine refrigerant and air states, power inputs, and grain condition. The parameters measured include refrigerant temperature and pressure, refrigerant mass flow rate, air temperature, air relative humidity, grain moisture content and grain temperature.

Validation of the Heat Pump Model

In order to determine the accuracy of the heat pump model, the model was simulated separately without the fixed-bed model by using the operating conditions of the actual drying experiments. Corresponding to a given set of operating conditions, the heat pump model was used to predict the compressor suction and discharge pressures, conditions of air at the exit of the evaporator and condenser, heating capacity, refrigerant mass flow rate, compressor power input, and heat pump coefficient of performance (COP). These predicted quantities were then compared with the corresponding experimental values and found to be in good agreement throughout.

Simulation of Heat Pump Dryers

For the purpose of comparing the five heat pump dryer configurations proposed in this study, the five configurations were simulated using the same operating conditions which were defined by the compressor speed, level of suction superheat, ambient air condition, initial grain condition, and air flow rates through the evaporator and condenser. Plots of SMER of the five heat pump dryer configurations are given in Figure 3.

In the previous discussion it has been assumed that only 25% of the evaporator air flow is recirculated back into the system in Configuration 4. The other 75% of the flow is discharged to the environment. The effect of varying the amount of air recirculation from the evaporator is shown in Figure 4.

Plots of the COP's for Configuration 4 corresponding to the conditions of Figure 3 show that the heat pump COP increases with the increase of air recirculation from the evaporator. A higher COP does not necessarily correspond to a better SMER for the heat pump dryer. This confirms the finding of Geeraert[3] (1976) and Baines[4] (1986).

Validation of the Heat Pump Dryer Model

For all the ten experimental runs conducted, it was found that the heat pump dryer model accurately predicted the refrigerant mass flow rate, compressor suction and discharge pressures, heat pump heating capacity and compressor power input. The model was however found to be less accurate in predicting grain moisture content, grain temperature, air relative humidity and air temperature within the grain bed. Detailed presentation and discussion on this subject may be found in Theerakulpisut (1990).

Discussions and conclusions

An electrically driven vapour compression heat pump model has been developed by combining accurate simulation models of the heat pump and of the drying process into a single system model for drier performance evaluation. The model was used to compare the performance of five heat pump dryer configurations and it was found that Configuration 2 which is a once-through open system is the most energy-efficient.

Testing of various system configurations has confirmed not only that the model is accurate but also that the best performance in terms of specific moisture extraction rate is not necessarily commensurate with choice of the configuration which optimizes the heat pump coefficient of performance.

References

1.   Theerakulpisut, S. Modelling heat pump grain drying system. Ph.D. Thesis, University of Melbourne, Australia, 1990

2.   Bakker-Arkema, F.W., Lerew, L.E., De Boer, S.F., Roth, M.G. Grain dryer simulation. Res. Rep. 224. Agr. Eng. Dept., Michigan State University 1974.

3.   Geeraert, B. Air drying by heat pumps with special reference to timber drying. Heat pumps and their contribution to energy conservation. Camatini, E., and Kester, T. (eds.). NATO advanced study institute series, 1976.

4.   Baines, P.G. A comparative analysis of heat pump driers. Ph.D. Thesis, University of Otago, New Zealand, 1986.

Acknowledgements

This work was funded by the University of Melbourne and the Australian Electrical Research Board. The funding from these two sources is gratefully acknowledged.

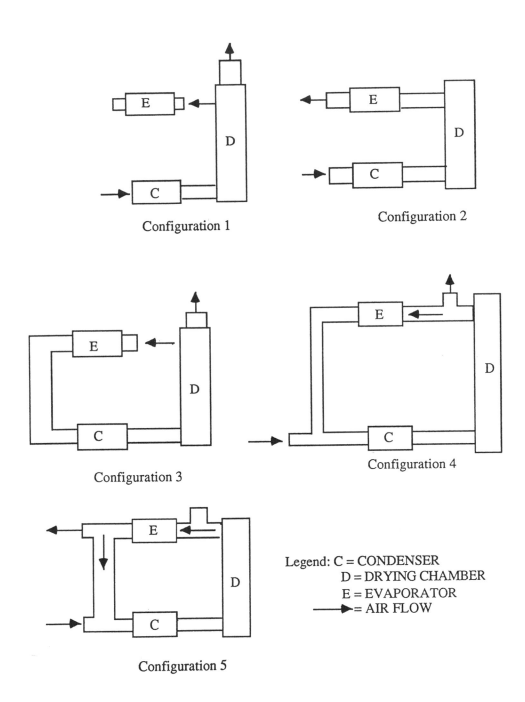

Configuration 1

Configuration 2

Configuration 3

Configuration 4

Configuration 5

Legend: C = CONDENSER
D = DRYING CHAMBER
E = EVAPORATOR
⟶ = AIR FLOW

Fig. 1.
Basic Configurations of Heat Pump Drying System

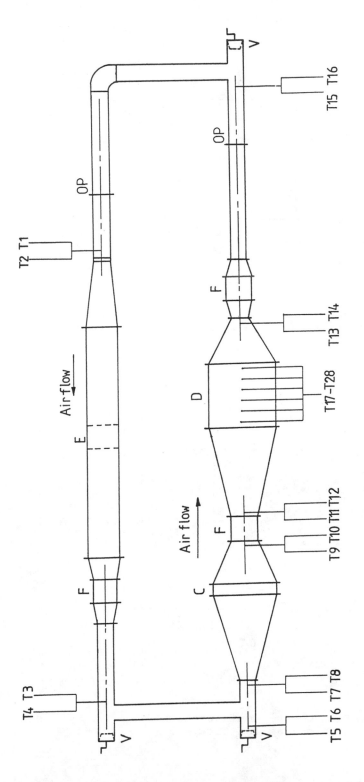

Schematic of experimental rig illustrating air duct
and position or thermocouples for measuring air condition

Fig.2.
Experimental Drying Rig

Description of main components of air duct
C=Condenser, D=Drying chamber, E=Evaporator, F=Fan, Op=Orifice plate, v= Cone valve, T1-T16=Thermocouples
measuring air condition, odd-numbered ones are dry-bulb thermocouples, even-numbered ones are for wet-bulb,
T17-T28 = Wet-and dry-bulb thermocouples for measuring air condition in the grain bed.

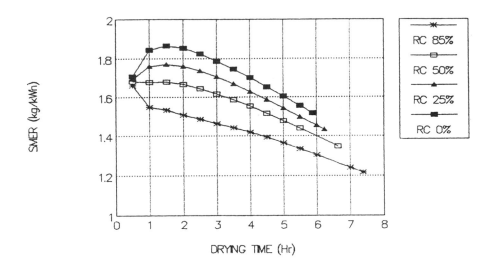

Fig. 3.
Effect of varying air recirculation rate on
Specific Moisture Extraction Rate (Configuration 4)

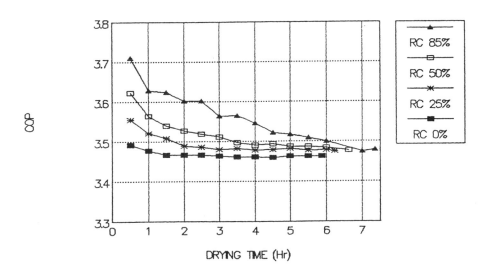

Fig. 4.
Heat Pump COP on Configuration 4
Conditions as for Figure 3

A LARGE SCALE FISH DRYER

H. Farzaad

Iranian Research Organization
for Science and Technology
P.O.Box 15815/3538,Tehran,Iran

ABSTRACT

Warm and dry moving air is needed for drying of many kinds of food, as fruit, vegetable, meat and fish. To accelerate that process; forced air by fans; heaters; electric power and rotating machineries are used. (Imre,L. and colleagues 1986). Natural circulation solar crop dryer is also known (Fleming,1986).

We intend in this paper to introduce a solar powered forced convective controlable warm air Fish-dryer, using the draught of a tall chimney. This way is especially useful in desolate location on long coasts of the continents having adequate solar energy and fish-catch; but lacking power-lines, oil and modern facilities. We present and report this new design and installation of the full scale experimental facility in I.R.O.S.T. in Tehran and the results during the first year of operation.

KEYWORDS

Dryer; fish; greenhouse; chimney-draught.

INTRODUCTION

Different types of solar dryer have been developed in many countries, mainly for agricultural products in order to increase their preservation time and qualities. Of all known methods of fish-preservation; drying is regarded as the easiest and most practical method for the third world countries. The total product-volume is some hundred thousand metric tons per year, which is achieved only by "open-to-sun" drying. This method is associated with some severe drawbacks as follows: salted fish in open air is exposed to unexpected rains, windborn dirt and dust, infestation by insects, rodents and other animals, hence a high percentage of fish is spoiled quickly and gets toxic and useless.

As the solar-dryer in this work is covered by glass and equipped with mosquito-net-doors, the above mentioned losses to the product are practically minimal. So these quality products will sell at higher prices on the world market. Therefore the higher costs are justified.

SYSTEM DESCRIPTION

The solar forced circulation fish-dryer was designed and developed during (1986-1989). General proportions of this facility are illustrated in Fig. 1.

The dryer is a turtle-shaped greenhouse, ending in a tall chimney; directed northwards in northern hemisphere. The outer skin of this housing is made from glass-sheets, secured with rubber-liner in aluminum heavy-profile structure. Fig. 2.

The greenhouse space is partly devided by two long glass walls, forming three channels and one quarter-global space. This space, facing the south-sun, works as an air heater and mixer. The two corridors are equipped with a sliding and a mosquito-net door to the north, acting as the air intake and preheating channels; the eastern corridor in the morning and the western in the afternoon. These corridors also serve as access for loading and unloading fish-carts to the middle corridor, "the dryer". A row of coarse wire-mesh nets from corrosive-resistance material is hanging inside the dryer on frames. These nets are the carrier of the vertically hanging light salted fish, split fish or fish fillets. Fig. 3. Adequate space between the fish pieces allows the passage of the warm air from the mixer to the chimney.

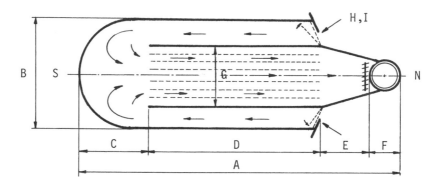

PLAN and SPECIFICATIONS

	DIMENSIONS	METRES
A	TOTAL LENGTH	29.00
B	TOTAL WIDTH	10.40
C	DIAMETER	5.20
D	GLASS WALL LENGTH	18.00
E	VENTURI LENGTH	4.00
F	CHIMNEY DIAMETER	1.80
G	DRYER	4.90
H	DOOR OPENING	0.90
I	DOOR HEIGHT	1.80
TOTAL GLASS SURFACE		260m^2
TOTAL CHIMNEY HEIGHT		27.00 m

Fig. 1.

The solar radiation heats the air inside the greenhouse, and the warm air masses moving from the mixer through the fish material to the chimney will take the moisture out and fulfill the drying process. Optionally, nets could be fixed horizontally, on which spreading of small fishes is possible. Dried small fishes are the raw material for an important product, namely the fish-mill. These nets are also suitable for drying _ agricultural goods, and that is another wide-spread use of this kind of dryer.

CONTROL-MECHANISM OF THE DRYER

Battery-operated sensors and control-boxes allow monitoring the temperature, humidity and air-flow inside the dryer.

The two doors and the chimney's damper are dually controlled day and night to pass the adequate volume of the warm air for the best drying results.

In order to prevent the reverse-flow of the cold air inside the chimney, the damper will close automatically until the cold condition is over. Insulation of the chimney is to be considered in some regions and ambient conditions to maintain the uniformity of the air flow.

To recharge the batteries, which are also eventually used for emergency-lights; solar cells are recommended. Alternatively, a small wind turbine located at the bottom of the chimney, using the remarkable winds, can charge the batteries adequately. (W. Haaf and co-workers, 1983).

EXPERIMENT

450 kg of cleaned silvery croakers; after 10 hours soaking in 25% salt water were hung on nets in the dryer. (June 23. 1990 at 20 pm.). Initial moisture content was 73%. The chimney damper was open and the doors 1/4 open. Automatic control was not available. After 40 hours, the moisture contents of fishes showed 35%; and this was a success. Fig. 4. and Fig. 5. Fishes were ready to be loaded on trucks and transported away.

CHIMNEY HEIGHT

By making some simplifying assumptions, the pressure drop in the chimney can be estimated as follows:

$$\Delta P = \frac{1}{2g} \, \rho v^2 \qquad (1)$$

where ρ is density at the site and is a function of ambient temperature and site elevation.

$$\rho = \frac{T_0 \, \rho_0}{T_a} \left| 1 - kS \right|^{\frac{\rho_0}{P_0 k}} \qquad (2)$$

where ; k = constant = 2.26×10^{-5}

$T_0 = 288^0 K$

$\rho_0 = 1.226$ kg m^{-3}

$P_0 = 1.332$ kg m^{-2}

$T_a = t_a + 273^0 K$

s = elevation above sea level

Height of the chimney is obtained as:

$$H = \Delta P / \Delta \rho \qquad (3)$$

$$\Delta \rho = \rho \ \frac{\Delta T_a}{T_a + \Delta T_a} \qquad (4)$$

$$\Delta T_a = Q \ / \ m_a \ c_p \qquad (5)$$

$$Q = I \ A \ \eta \qquad (6)$$

where: I = Insolation $W \ m^{-2}$

A = Collector area m^2

η = Collector efficiency

m_a is the mass flow rate of air in the the dryer, and can be written as:

$$m_a = \frac{\pi}{4} \ D^2 \rho \ V \qquad (7)$$

In order the velocity in the dryer on the fish to be 1 m sec^{-1}, the air velocity in the chimney should be 6 m sec^{-1} and therefore the chimney height will be ∼40 m.

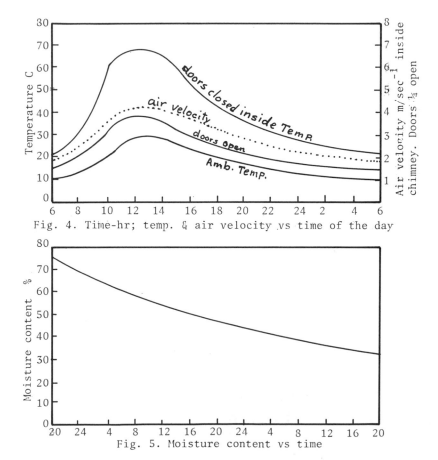

Fig. 4. Time-hr; temp. & air velocity vs time of the day

Fig. 5. Moisture content vs time

Fig. 2. A turtle-shaped Greenhouse

Fig. 3. Inside the Greenhouse

CONCLUSION

Experiments show that the dryer operates successfully for fish drying during all the seasons in the Persian Gulf area. This dryer can also be used as an agricultural dryer in most parts of Iran.

ACKNOWLEDGEMENT

I would like to acknowledge that the need for a big solar fish dryer was introduced to me by Mr. Keshavarz and Mr. Raufi from Fisherman Coop. of Chabahar harbour; (March 1986). Their initial financial help was afterwards generously continued by IROST. I would like also acknowledge the close assistance of my good collegues Dr. E. Azad and Mr. J. Mahallati.

REFERENCES

Imre, L.; Farkas, I.; Gemes, L. (1986). Construction, Simulation and Control of a complex and Integrated Agricultural Solar Drying System. Proceeding of DRYING '86 vol. 2, edited by Arun S. Mujumdar; Hemisphere Publ. Corp. pp. 678-684.

Fleming, P.D. ; and B. Norton, and co-workers. A Large Scale Facility for Experimental Studies of Natural-Circulation Solar-Energy Tropical Crop Dryer. Proceeding of DRYING '86 vol.2, pp. 685-693. Hemisphere Publ. Corp.

Haaf, W. and K. Friedrich, G. MAYR, J. Schlaich. Solar Chimneys, (1983). Int.J. Solar Energy: vol.2. pp. 3-20

THE RATIONAL SELECTION OF APERTURE AREA FOR MIXED MODE SOLAR DRYER

Li, Zongnan   Chen, Chaoxiong

Guangzhou Institute of Energy Conversion
Chinese Academy of Sciences
Guangzhou, China, 510070

ABSTRACT

This paper analyses the physical process of the collector-greenhouse mixed mode solar dryer. A set of equations to describe the dynamic equilibrium processes of the mixed mode solar dryer are established and solved by iterative method. The effect of the proportion of the aperture area of collector to greenhouse, R, on temperature of greenhouse, dehydration of material, and drying period etc. is analysed. And the rational selection of the value of R is concluded in the paper.

KEYWORDS

Mixed mode solar dryer; solar air heater; greenhouse; proportion of aperture area; dehydration rate.

INTRODUCTION

Mixed mode solar dryer is composed of the greenhouse combined with solar air heater. Fig.1 shows the schematic of this kind of dryer.

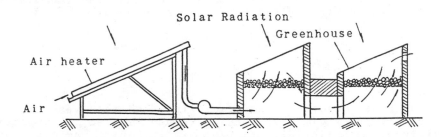

Fig.1. Schematic of mixed mode solar dryer

In mixed mode dryer, the drying material is heated up by both the hot air, preheated in solar heater, and the solar radiation received directly in the drying chamber. The bigger the solar heater area is, the more the solar radiation can be received, the higher the air temperature and the shorter the drying period will be. But they are not in linear relationship. What is the appropriate area of air heater for a certain aperture area of greenhouse? It needs to be studied to find the rational proportion of the aperture area of solar heater to greenhouse.

MATHEMATICAL MODEL

I. Hypotheses to establish model

For simplifying the model, three hypotheses are taken:
1. The temperature and humidity of air and material are uniform.
2. The relationship between drying rate and moisture content of material is taken as linear, i.e. take a small part of the curve as linear.
3. The greenhouse is taken as a solar collector. Then, the thermal performance of greenhouse can be treated as solar collector and can be described by efficiency factor, heat loss coefficient and effective transmittance-absorptance product. The energy collected by unit aperture area of greenhouse can be expressed:

$$Q/A = (\tau \alpha)_e \cdot I - U(T_o - T_a) = F'[(\tau \alpha)_e \cdot I - U(T_r - T_a)] \qquad (1)$$

For solar collector, the temperature of collector plate T_p is always higher than that of flowing air T_r. But in greenhouse, the average temperature T_o and T_r are determined by the drying process. According to the experimental data, T_o and T_r have no big difference. Thus, take $T_o = T_r$, $F' = 1$ for establishing the model.

II. The equations to describe the model

During drying process, the material, greenhouse and solar air heater in dynamic, heat and moisture equilibrium at any moment. A set of energy and mass balance equations is established, based on a unit aperture area of greenhouse.

1. Moisture balance of greenhouse:

The change of the moisture of air in greenhouse is affected by mass transfer between drying material and air, amount of air replaced and moisture discharged from greenhouse, i.e.:

$$\frac{-\rho V_o}{A} \cdot \frac{dX_r}{d\tau} = \frac{A'hj}{A}(X_w - X_r) - \frac{\rho G}{A}(X_a - X_r) \qquad (2)$$

(2) Heat balance of greenhouse

The greenhouse absorbs solar radiation and receives energy from air heater. Part of the energy obtained is used to evaporate the moisture of the material, part of it is lost in ambient, and the rest is exhausted to change the enthalpy of material, greenhouse itself and the air inside the greenhouse. i.e.:

$$(\tau \alpha)_e I - U(T_r - T_a) + \frac{\rho GC_p}{A}(T_{ri} - T_r) - \frac{A'hj}{A}R_J(X_w - X_r)$$

$$= [\frac{\rho V_oCp}{A} + \frac{(MCp)_o}{A} + \frac{(MCp)_1}{A}]\frac{dT_r}{d\tau} \qquad (3)$$

3. Heat balance of solar air heater
Part of the solar energy absorbed by solar air heater is lost in the ambient and the rest is input to greenhouse, i.e.:

$$\int_o^R F'[(\tau \alpha)_{ec}I-U_c(T_{f1}-T_a)]dR=\frac{\rho GCp}{A}(T_{f1}-T_a) \qquad (4)$$

4. The expression of drying characteristics of material
Use Xw, the moisture content of wet air inside the mass transfer boundary layer of material, to express the drying characteristics of material[1]. When the drying rate $dw/d\tau$ is big, the value of Xw is also high, the expression can be described as follows[1]:

$$Xw=Xwi(Tf)[1-k1(Wc1-W)] \qquad (5)$$

5. The calculation of U and $(\tau \alpha)_e$
The calculation of U and $(\tau \alpha)_e$ of greenhouse is similar to that of U_c and $(\tau \alpha)_{ec}$ of collector.

The total heat loss of greenhouse U is the sum of heat loss through the top of the cover and the heat conduct loss through the walls, Uw:

$$U=\frac{1}{[1/(hi+hr)]+(1/ho)}+Uw \qquad (6)$$

$(\tau \alpha)e$ is approximately equal to $(\tau \alpha)ec$.

For calculation, take $(\tau \alpha)=0.488$
Equation (1) to (6) are connected each other. In order to solve the equations, change the differential to algebraic equations and use Gauss-Siedel iterative method[2].

SIMULATION ANALYSIS

1. Reliability of simulation
In order to check the reliability of the simulation, the experiments were conducted in a mixed mode solar dryer with changeable aperture area. The drying material used was water soaked cubic wood with the initial moisture content of 95% (dry basis) and the loading was 33.3kg per m² of aperture area of greenhouse. The maximum error of greenhouse temperature was 9.8% and that of dehydration amount was 1.7%. When the initial moisture content was 65%, the maximum error of greenhouse temperature was 13.4%, and that of dehydration amount was 9.17%. The comparison data, under forced ventilation, and at different R and solar radiation are listed in Table 1.

The results indicate that the simulation analysis is reliable.

2. The effect of different proportion of aperture area on the temperature of greenhouse.

Fig.2 shows the variation tendency of greenhouse temperature with drying time at different proportion of aperture area.

TABLE 1. Comparison Data

Proportion of Area R	1		2		3	
Solar radiation Kwh/m²	5		3.93		4.2	
	A	B	A	B	A	B
Initial moisture %	0.610	0.610	0.551	0.551	0.638	0.638
Final moisture %	0.194	0.155	0.086	0.078	0.060	0.051
Dehydration kg	3.50	3.82	3.90	3.97	4.85	5.30
Error of dehydration %	9.14		1.79		9.27	

Where: A--experimental value; B--calculated value

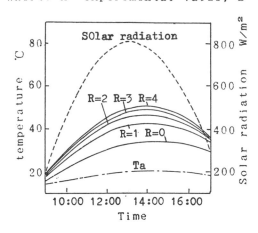

Fig.2. R versus temperature
of greenhouse

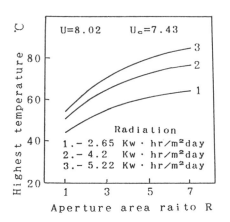

Fig.3. R versus highest
temperature of greenhouse

From Fig.2, it can be seen that, increasing the aperture area of air heater makes the temperature of greenhouse raised by one fold. However, the raising rate decreases with the area increasing. It is beneficial to select the value of R equal to 1-2. When R>2-3, the benefit of increasing area is not obvious.

Fig.3 shows that the highest temperature of greenhouse can reach at different solar radiation and proportion of aperture area. It can be seen that, the more the solar radiation is, the more benefit can be obtained from increasing the area of air heater. At the situation of less solar radiation, there is no considerable effect of increasing air heater area on raising the temperature of greenhouse. The bigger the R value is, the less the temperature can be raised.

3. The effect of different proportion of aperture area on dehydration amount of material.

The Variation of dehydration amount per unit aperture area of greenhouse with the different proportion of aperture area is shown in Fig.4. With the increasing of R, the increment of dehydration decreases. When the increment is expressed in percentage, the changing trend is very remarkable. When R=1, the dehydration amount increases by 47.3% more than that of greenhouse. When R increases from 3 to 4, the increment of dehydration is only 8.6%. It is clear that R<3 is beneficial. However, the profit obtained from R>3 is limited.

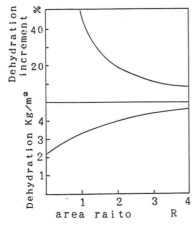

Fig.4. R versus dehydration

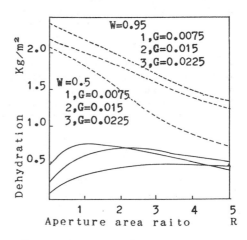

Fig.5. At different moisture content

When dehydration amount is expressed in unit aperture area (including greenhouse and air heater), the changing trend of which with R at different initial moisture content of material and different air flow rate is shown in Fig.5. For high moisture content material, the dehydration amount per unit aperture area drops with the R going up. On the contrary, for low moisture content material, the dehydration amount goes up with R increasing. However, if R keeps on increasing to a certain value, the dehydration amount falls down. It indicates that there is an optimum value of proportion of aperture area. In accordance with the different air flow rate, the optimum value of R is among 1 to 3.

From the point of view of shortening the drying period, it is worth while to increase the area of air heater. Fig.6 shows that at the same loading and air flow rate, comparing with greenhouse dryer, When the moisture content of material is dried from 95% (dry basis) to 50%, the drying period reduces by 45% at R=1, and reduces by 60% at R=2. If the R continues to increase, there is no obvious effect on decreasing the drying period.

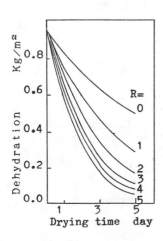

Fig.6. Moisture content change with drying time

CONCLUSION

1. The simulation analysis for mixed mode solar dryer is suitable to describe the actual drying process of this kind of dryer. The calculation values accord with the experimental data.

2. The rational proportion of aperture area of solar air heater to that of greenhouse R is selected to be 1 to 3. When the solar radiation and/or air flow rate is higher, the bigger value of R should be selected. Otherwise, it is proper to choose lower value of R.

3. With the aperture area of solar air heater enlarging, the temperature in the drying chamber of greenhouse is raised, which makes the dehydration of drying material increased and the drying period decreased. However, when R>3, the benefit obtained from increasing the air heater area is not obvious.

REFERENCES

[1] Li, Zongnan & Chen, Chaoxiong, Dehydration Characteristic of Materials, Presented at CSES Biannual conference (1989)(in Chinese)
[2] S.V. Patankar, Numerical Heat Transfer and Fluid Flow, Hemispere, New York, 1980

NOMENCLATURE

A   Aperture area of greenhouse (m^2)
Ac   Aperture area of solar air heater (collector) (m^2)
A'   Mass transfer area of material (m^2)
G   Volumetric flow rate of air (m^3/s)
F'   Greenhouse efficiency factor
F_c' Collector efficiency factor
hi   Heat transfer coefficient in the cover of greenhouse ($W/m^2°C$)
hr   Radiation heat transfer coefficient ($W/m^2°C$)
ho   Heat transfer coefficient outside the greenhouse cover($W/m^2°C$)
hj   Mass transfer coefficient from material (Kg/m^2)
I   Solar insolation (W/m^2)
K_1 Drying characteristic coefficient of material
R   Proportion of the aperture area =Ac/A
Ri   Latent heat of vapour (KJ/kg)
Ta   Ambient temperature (°C)
Tf   Air temperature inside the greenhouse (°C)
Tfi   Inlet air temperature of the greenhouse (°C)
To   Average temperature inside the greenhouse (°C)
U   Over all heat loss coefficient from greenhouse system ($W/m^2°C$)
Uc   Heat loss coefficient from solar air heater to ambient ($W/m^2°C$)
Uw   Heat loss coefficient from walls to ambient ($W/m^2°C$)
V   Greenhouse volume (m^3)
W   Dry basis moisture content of material (Kg/kg) %
Wc_1 Characteristic moisture content (%)
Xw   Moisture of air in mass transfer boundary layer of material %
Xwi   Initial moisture content of material
Xf   Moisture of air inside the greenhouse %
Xa   Moisture of ambient air (%)
($\tau \alpha$)e Effective transmittance-absorptance product

MATHEMATICAL, TECHNIC AND ECONOMIC ANALYSIS OF A
SOLAR DRYER FOR TOBACCO IN COLOMBIA

C. Sánchez*, E. López*, N. Arias and A. Arias**

*Solar Energy Area, Instituto de Asuntos Nucleares,
IAN, Bogotá, Colombia. A.A. 8595.
**Physics Department, Universidad Popular del Cesar,
Colombia.

ABSTRACT

This article presents the equations that describe the performance of a solar
dryer, designed to meet the requirements for   tobacco curing. Both the
theoretical and experimental   .   temperatures within the kiln are
compared, as well as the product's humidities. An economic analysis was done
in order to show the feasibility of a dryer with a capacity of 1000 kg. of
fresh-cut tobacco.

KEYWORDS

Solar dryer, curing of tobacco, solar dryer simulation, economic solar dryer
analysis, solar dryer performance.

INTRODUCTION

Tobacco, as a product, has high water content in its leaf and its midrib,
almost 80%   water. During the curing process, 65% of the water has to be
evaporated in order to transform the freshcut green leaf into a finished
product that has both a pleasant aroma and a brown color.

The drying has to be carefully done taking care of the temperature, as well
as the relative humidity within the tobacco dryer volume in each one of the
three stages of the process (Akehurst, 1978) Fig. 1.

The evaporation of a big amount of water (almost 800 g. out of 1000 g. of
fresh cut green leaves) requires burning about a kilogram of coal per kilogram
of dry tobacco.

The temperatures required for drying, at least in the two first stages, are
less than 50°C; these temperatures are easily reached with solar dryers
permitting not only significant savings of fuels maintaining the drying quality,
but contributing   to safeguarding the environment.

The purpose of this study is to present a mathematical simulation of the solar
dryer, its experimental evaluation and the economic analysis of the solar
kiln for Virginia type tobacco.

A timber structure covered with plastic was designed and constructed for the
experimental part of the study Fig. 2. The effective area of the absorber

plate (cooper) has 2 m², the a usuable drying volume of 2 m³, a height of 3 m, and a cross section of 2.13 m². The dryer was installed near a tobacco production zone in the city of Valledupar, which is located longitude 73°16' East, latitude 10°27' North, altitude 200 m. about sea level and has an average annual temperature of 32°C, average solar radiation of 5500 Wh./m² daily, and an average relative humidity of 40%.

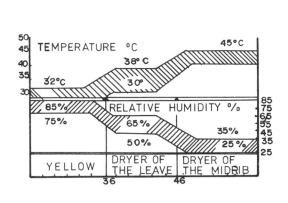

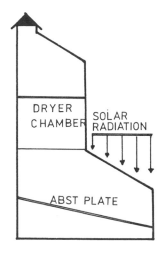

Fig.1. Phases of tobacco curing   Fig.2. Schematic of a solar dryer

MODEL

The purpose of the model is to theoretically predict the dryer's performance, the temperature of absorbing plate, the air's temperature and the speed of the air within the drying chamber.

The equation system proposed below are the result of analysis and review of some models (Zahedand, 1989; Taylor, 1984) adjusted to the conditions required for tobacco and the geometry forsee for the kiln.

$$v_a = \frac{1.7 \, \delta \, U^3 b^2}{(V \, L_p R_{eLp})^{0.5}} \tag{1}$$

$$\frac{\partial Ta}{\partial t} = \frac{HpAp}{\rho \, aC_{va}Va} \, (\, Tp - Ta \,) - \frac{Cpa}{Cva} \, v_a \, \frac{\partial Ta}{\partial X} \qquad X \leqq L \tag{2}$$

$$\frac{\partial Tp}{\partial t} = \frac{Qab \, Aef}{mp \, Cp} - \frac{Hp \, Ap}{\rho aCvaVa} \, (\, 2Tp - (Ta - Tb)) \tag{3}$$

$$\theta \, (t) = \beta e^{-kt} \tag{4}$$

To solve equations (2) and (3), we are using the finite difference technique, obtaining:

$$Ta_{,t}^{X} + \Delta t = C \, T_{p_1 t} - C_2 T_{a,t}^{X+\Delta x} + C_3 T_{a,t}^{X} \tag{5}$$

$$T_{p,t+\Delta t} = C_4 - C_1 \, (T_{a,t}^{\delta} + T_{b,t}^{\delta}) + C_5 T_{p,t} \tag{6}$$

where:

$$C_1 = \frac{Hp\ Ap}{\rho a Cva Va}\ \Delta t \qquad\qquad C_2 = \frac{Cpa}{Cva}\ va\ \frac{\Delta t}{\Delta x}$$

$$C_3 = 1 + C_2 - C_1 \qquad\qquad C_4 = \frac{Qab\ Aef}{mpCp}\ \Delta t \qquad\qquad (7)$$

$$C_5 = 1 - 2C_1$$

With the results from equation (1), we determine the speed of the air in the drying chamber. Employing equations (5) and (6), we calculate the plate's and chamber air's temperatures.

From the equation (Walton, 1986):

$$\theta(t) = \sum_{n=0}^{\infty} \frac{2}{(\lambda_{n}1)^2}\ e^{-(D^1/1^2)\ (\lambda_{n}1)^2 t}$$

$$\lambda_{n}1 = (2n+1)\ \pi/2 \qquad n = 0,1,2,3$$

and considering the first term of the summatory, it was obtained: $\beta = 0.81$ and $k = 0.0234$

In order to obtain a fine quality tobacco, it is necessary to carry out the drying within well-established humidity ranges for getting optimal drying conditions, the psychrometric equations are used (Trnsys, 1981).

RESULTS AND ANALYSIS

From the initial plate and air temperature employing a computer program for IBM-AT and adjusting Δx, Δt, a unique solution is obtained for equations (5) and (6). To calculate the plate temperature, estimation from the climatic data evaluated previously $I(t) = 785\ \sin(3.14t/43200)\ w/m^2$, $Ta(t) = 21\ \sin(3.14t/57600)°C$, Hp is evaluated employing the methodology of the limit layer (Hollman, 1984).

Utilizing: $Cva = 716.8\ J/kg°C$; $Cpa = 1004\ J/kg°C$; $Ap = 3.14\ m^2$; $Aef = 2.04\ m^2$; $Cp = 381\ J/kg°C$. And running the program, the results were obtained and are summarized in Table 1; and the speed in each stage of the drying: $va1 = 0.054\ m/s$; $va2 = 0.008\ m/s$; $va3 = 0.007\ m/s$.

The first experimental stage was realized in the laboratory simulating the conditions obtained theoretically getting a dryer-fine-quality tobacco in 135 hours. Taking these results, design and the construction of the prototype dryer was made, where the second stage of the experiment was executed.

TABLE 1   Calculated Temperatures

Time (h)	Temperature Plate (°C)	Temperature chamber (°C)
9.45	46.1	36.4
10.30	63.9	40.5
11.15	70.8	43.1
11.55	75.0	43.8
12.05	78.1	44.7
14.45	72.5	48.2

Table 2 shows the typical results for a complete drying cycle in the laboratory as in the dryer. Taking into account that in the first phase of drying, 35 kg. of water are evaporated, in the second 15 kg. and the third 9 kg; we obtain: $va1 = 0.049$ m/s; $va2 = 0.0076$ m/s; $va3 = 0.007$ m/s.

Fig. 3 shows the theoretical and experimental plate and air temperatures. It can be seen in Fig. 4 the theoretical and experimental drying curves. From the figures can be observed a good agreement, the same observations can be made when the theoretical speedies are compared with the experimental ones.

TABLE 2   Experimental Values

Time h.	O V E N Temp. °C	Hmd. %	Θ	S O L A R D R Y E R Temp. °C	Hmd %	Θ
5	30	70	0.90	45	70	0.95
9	30	70	0.77	49	65	
36	30	70	0.40	51	67	
53	38	50	0.26	51	59	
60	38	50	0.22	52	58	
73	38	50	0.170	56	45	
113	40	23	0.062	59	43	
135	40	25	0.024	60	28	0.030

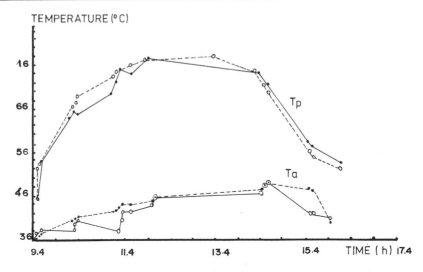

Fig.3. Calculated (----) and experimental (——) temperatures

ECONOMIC ANALYSIS

The analysis has been for a dryer of 50 m² considering the minimum payback period (np), and the net life cycle (LCS) using discounted auxiliary fuel cost (Hawlader, 1987):

$$Cs = \frac{Qs\ Cf}{Crf(i'', n_p)} \qquad LCS = \frac{Qs\ Cf}{Crf(i'',n)} - Cs \qquad (8)$$

$$Crf\ (i'',n) = 1/i''\ [\ 1-(1/(1+i''))^n]^{-1} \qquad (9)$$

Where: Cs is the cost of the investment per square meter of the solar dryer minus the investment made for a conventional dryer, considering that the conventional one dries a volume five times as much. QsCf represents the total energy cost replaced for a year. For the LCS analysis it was assumed that the dryer is in operation 100, 120, 150 and 200 days yearly and n = 10 years, Fig. 5.

In Table 3 are the values used in the Cs and LCS analysis (the costs refer to a colombian context, but are in american dollars).

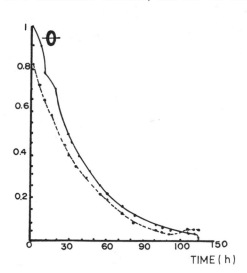 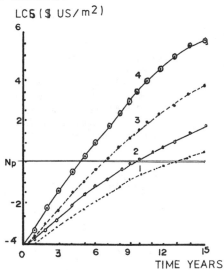

Fig.4. Calculated (----) and experimen-   Fig.5. LCS for 100(1), 120(2), 150(3),
tal (——) drying curves   and 200(4) days

TABLE 3   Economic Parameters

i	0.32	Qs	5.34 MJ/m^2
j	0.25	Cf^2	0.019 $/MJ
e	0.27	Cs	76 $/m^2

CONCLUSIONS

The agreements obtained between the theoretical and experimental results for temperatures, drying curves and air speeds permit to affirm that the model describes,in a right way, the performance of the dryer design.

The proposed model and solution method can be used to simulate the dryer's performance and the drying of products for similar driers and other agricultural products.

The economic analysis indicates that these kind of dryers are an economically viable alternative that has the added advantage that it can be substituted for highly contaminative fuels.

ACKNOWLEDGEMENTS

This study resulted from a project financed by ICFES, IAN and the Universidad Popular del Cesar.

NOMENCLATURE

Aef   Effective plate area m^2
Ap   Plate area m^2
Cf   Fuel cost
Cpa, Cva   Specific heat of air J/kg°C
Cp   Specific heat of the plate J/kg°C
e   Fuel scalation rate
Hp   Heat transfer coefficient w/m^2°C
i   Discount rate
I   Intensity of Solar radiation w/m^2
l   Average leaf thickness m
L   Average height from leaves to the plate m
Lp   Plate length m
mp   Plate mass kg
n   Life cycle system years
np   Minimum pay back period years
Qab   Solar radiation absorbed by the plate w/m^2
Qs   Solar heat gain MJ/m^2
Re   Reynold's number
t   Time s
T   Environment temperature °C
Ta   Air temperature °C
Tb   Air temperature below the plate °C
Tp   Plate temperature °C
U   Speed of air in free flow m/s
va   Speed of the air in the chamber m/s
Va   Volume of the limit layer m^3
δ Thickness of the limit layer m
υ Viscosity m^2/s
ρ_a Air density kg/m^3
θ Moisture ratio db

REFERENCES

Akehurst, B.C. (1978). El Tabaco. Labor Barcelona, España.

Hawlader, M.N.A., Chandatilleke. KC.N.G., and Kelvin K.N. (1987). Energy Conversion Mgmt., 27, 2, 197-204.

Holman, J.P. (1984). Transferencia de Calor. Continental, S.A., México.

Solar Energy Laboratory, University of Wisconsin Madison (1981). Report 38-II TRNSYS.

Walton, L.R., and Casada, M.E. (1984). American Soc. Agr., 29, 271-275.

Zahedand, A., and Elsayed, M.N. (1989). Solar and Wind. Tech., 241-248.

IN-STORAGE SOLAR GRAIN DRYER USING REST-PERIODS

W. Radajewski   D. Gaydon

Agricultural Engineering Section, Queensland Department of Primary Industries,
PO Box 102, Toowoomba Qld 4350, Australia.

ABSTRACT

A low-cost in-storage grain drying-aeration system based on a simple modification of presently used farm storage silos is examined. The rest-period and reversed-airflow principles are used so the size of the solar collector and the grain moisture differential could be keep low. The design and operation of a suitable solar-collector is presented.

KEYWORDS

Solar drying; in-storage drying; drying with rest periods; low-cost drying; stage-loaded drying; grain aeration.

INTRODUCTION

Overall grain losses are directly related to the moisture content of the crop at the time of harvest. The dollar value of these losses depends on the weather conditions, location the crop is grown, farm management, type and variety of crop etc.. The magnitude of these losses cannot be established before they occur and may change considerably from year to year. However, the general trend of the losses is always the same and they are lowest when crop is harvested at maximum harvestable moisture content following full maturation (Fig. 1). The cost of drying always increases with increased moisture content (Fig. 1).

The highest return from the crop would result if the crop is harvested at the annual optimum moisture content. Such a moisture content cannot be defined accurately for each year since the magnitude of annual losses cannot be predicted. Therefore, an average optimum moisture content for a long period is defined using historical data. Using this method the annual return from the crop is lower than it would be otherwise if the annual optimum moisture content were used. One of the potential ways of overcoming this problem is to use a drying system in which an increase in the initial moisture content would have a minimal effect on the overall costs of drying.

At present, artificial drying is performed on farms in expensive continuous-flow, recirculating-batch or field-bin dryers. The air used for drying is heated by the combustion of diesel or gas fuel. After drying, the grain is often temporarily stored in silos so the farmer has more time available for harvesting, and the grain is delivered to the grain depots at a later date. During the period of storage, the grain is artificially ventilated to prevent it from being spoiled. Such a strategy of harvesting, drying and storage is expensive and therefore has a limited application.

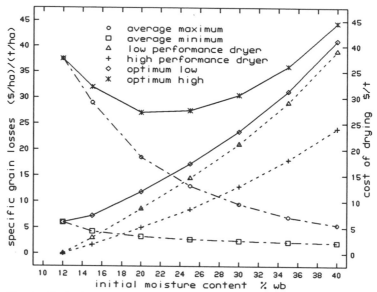

Fig. 1. Grain losses and drying cost in conventional dryer for different moisture content at harvest time.

DRYING CONCEPT

Low-cost drying can be achieved if the specific costs (dollars per tonne) of the energy and the capital investment on drying and heating equipment are low. Solar energy as a source of heat and drying in storage can meet the above requirement and has been used in the USA, Canada and other countries.

Drying in storage can be successfully performed when the ambient air temperature and relative humidity are low. A small temperature rise in the drying air and a low rate of air flow are used, so the grain is dried during a number of days without danger of deterioration. However, in hot and humid conditions the use of a small temperature rise above ambient is not adequate to produce safe drying conditions (Banks, 1981). Drying grain in deep layers using low air flow rates causes the over-drying of grain in the bottom layers and under-drying of the top layers. This could result in spoilage of wet layers of grain and could lead to insect infestation of the grain when in storage. To reduce this risk when drying in deep layers, a high rate of air flow must be used so that the drying zone is much wider, and uniform drying of the whole grain bed takes place. However, this increases the air pressure drop through the grain and results in high capital and running costs. To achieve low-cost drying and good grain quality, a stage-loaded drying system using reverse airflow and the principle of rest-periods is considered.

Principle of Rest Periods

It was shown by Harnoy and Radajewski (1982) that periodical stopping of air during the drying with rest-periods has no affect on grain quality and such a system of drying can be successfully used. During this method of drying, the grain is exposed to the hot air for a period of time t_b and then a 'rest-period' is introduced by stopping the hot air supply to the drier for a time t_r. Alternatively instead of rest-periods, a period of aeration can be applied to the grain (Fig. 2).

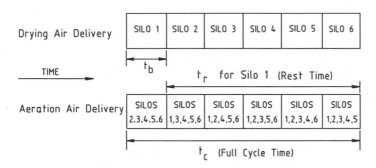

Fig. 2. Logic of rest-periods application.

The time of the full cycle, t_c, is the total of basic time of drying and rest time. The ratio of the full cycle-time, t_c, to the drying-time, t_b, was defined as the drying ratio (R). If R = 1 drying without rest-periods is applied. The most practical way of applying this method of drying is to make the number of silos (silos) in the drying system equal to R and for each of them to hold a similar amount of grain. The system will be operated as follows.

When the process of drying begins, only the grain in first silo is exposed to the hot air. After time t_b, the air supply to the first silo is diverted to the second silo. This operation is repeated for each silo in turn. The first cycle is completed when the air-supply has been closed to the last silo and opened again to the first one. If the total time of drying the first silo from the initial moisture content to the final moisture content with the ratio R is t_t, the second silo in the system will be dried after the time $t_t + t_b$ and the third silo will be dried after the time $t_t + 2t_b$, and so on. The last silo will be dried after a time which represents the total time of drying for all the silos.

$$t_T = t_t + t_b \, (R-1) \tag{1}$$

Stage-Loaded Silo with Reversed Airflow

To overcome problems of high moisture differentials and high power requirements encountered when drying in silos, a stage-loaded (rather than batch-loaded) with reversed airflow philosophy is adopted (Fig. 3). The grain is loaded to silo in layers. The depth of each layer depends on the rate at which grain is coming from the field and in the extreme situation can be equal to the full height of silo. The air is introduced alternately from the top and bottom using valve, V, until grain is dried to the required moisture content. At the time when rest-periods are applied or during the storage period the valve, V, can be set to the aeration mode. During aeration mode, an airflow to the silo is restricted by the valve, V. The incoming day's harvest is split between silos in the system, resulting in a relatively thin layer to be dried in each silo. As the harvest proceeds and the total depth of grain increases, the rate of air-flow to each silo decreases.

SOLAR HEATING SYSTEM

The use of a standard flat-plate solar collector for air heating in conjunction with a stage-loaded in-storage grain dryer presents certain problems. Firstly, since the airflow rate through the dryer varies with grain depth, the airflow rate through the solar collector will be inconsistent and each flow rate will have an optimal collector size to maximise the returns in kWh/dollar. Practically, the best design to suit all airflow rates must be chosen. Since the most convenient airflow space from a

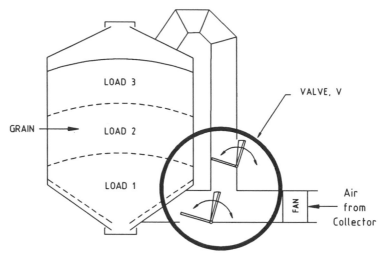

Fig. 3. Cross section of silo showing valve, (V), for reversing airflow and for aeration

construction point of view is defined by the thickness of rafters in the farm-shed roof (Fig. 4), the length and width of the collector only can be optimised. Also, in many farm situations the optimal collector dimensions cannot be accommodated due to practical restrictions placed by farm shed size etc. Both these lead to decreased collector efficiency and, in some cases, highly sub-optimal ratio of energy to overall cost. Figure 5 shows how this problem can be considerably reduced by introducing the rest-periods drying method. The total amount of air used and therefore the size of solar collector and fan is determined by the specific rate of airflow and the volume of grain dried at any time. The size of solar collector in the rest-period drying method can be six times smaller than in the conventional in-storage drying system. This gives much better opportunities to fit optimum dimension collector onto the farm shed roof and therefore achieve higher performance. When reverse airflow is used, a lower specific rate of drying can be applied than in the conventional method of drying without increasing the grain moisture differential. As a result of higher performance achieved by reverse airflow further, a decrease of collector size and therefore cost is possible. An additional advantage of a smaller solar collector is that when fitted in the center of the shed, part of the solar energy absorbed by the remaining portion of roof can be used for drying without additional cost except black painting (Fig. 4).

SIMULATION AND RESULT

A simulation model has been developed so the performance and economics of solar drying in a multi-silo system could be compared with the conventional methods of drying. The overall time of grain drying from initial to final moisture content was obtained using a modified fixed-bed drying model (Brooker and Bakker-Arkema, 1974). The equilibrium moisture content was defined using an equation from Nellist and Dumont, (1978). The pressure drop through the grain was calculated using the method shown by Mathies and Paterson, (1974).

The collector size was determined by optimising length and width using a constant airflow space (Fig. 4) for the average flow-rate encountered by the collector. The total energy used for running a

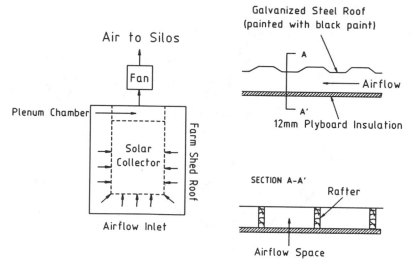

Fig. 4. Design of solar collector used with drying system.

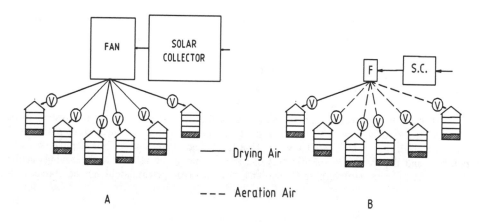

Fig. 5. Multi-silo solar drying system; A - conventional; B - rest-period.

fan was defined using the time of drying defined by Equation 1 and the average power used for air pumping. Since solar energy is used alone for heating, the total energy used is that for air pumping. The objective function was defined as the ratio of net energy derived from solar collector per annum, E_c (kWh/a), divided by the annual running cost of collector, C ($/a). For calculation of the overall costs and optimisation of solar collector, a method outlined by Radajewski, Jolly and Abawi, (1987) and Radajewski and Gaydon (1989) was used.

Simulation of the solar drying system as described above shows that the overall cost of drying in this system can be reduced to such a degree that the overall return from the crop is the highest when drying starts at maximum harvestable moisture content.

Experiments conducted on farm in a full scale (50 tonne) in storage solar drying system have shown that solar energy without auxiliary heating can be used for crop drying from high moisture content.

Ventilation of grain during low solar energy hours and during night time is adequate to prevent grain from being spoiled.

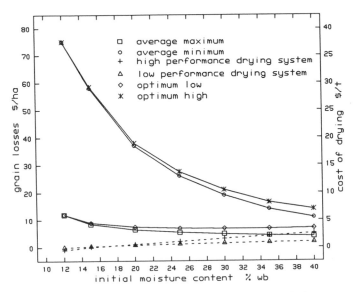

Fig. 6. Grain losses and drying cost in silo rest period drying system for different moisture content at harvest time.

CONCLUSIONS

The use of solar energy for air heating combined with an in-storage grain drying system using rest-period and reversed airflow principles is a most suitable and effective method of reducing the overall cost of drying. The overall savings, to a large degree, depend on the capital cost of installation and therefore the specific cost of collector has to be related to the savings achieved. With the present price structure in Queensland, the collector price should not be above 18 $/m².

REFERENCES

Banks, P.J. (1981). Grain and seed drying. Introduction and principles. *Proceeding of the Australian Development Assistance Course on Preservation of Stored Cereals*, 476-484.

Brooker, D.B., F. M. Bakker-Arkema, and C.W. Hall (1974). *Drying Cereal Grains*. The AVI Publishing Company Inc., Westport.

Duffie, J.A., and W. Beckman (1974). *Solar Energy Thermal Processes*. John Wiley and Sons, New York.

Harnoy, A., and W. Radajewski (1982). Optimization of grain drying with rest-periods. *J. of Agric. Eng. Res., 27*, 291-307.

Matthies, H.J., and H. Paterson (1974). New data for calculating the resistance to airflow of stored granule materials. *Trans. of Am. Soc. of Engrs, 17*, 1144-1149.

Nellist, M.E., and S. Dumont (1978). Desorption isotherms for wheat. *National Institute of Agricultural Engineering*, Silsoe, Dep. Note DN/CDV/983/06010.

Radajewski, W., P. Jolly, and G. Y. Abawi (1987). Optimisation of solar grain drying in a continuous flow dryer. *J. of Agric. Eng. Res., 38(2),*127-144.

Radajewski W., and D. Gaydon (1989). In-storage solar crop-dryer. *Proc. ANZSES,* 60/1-7

DRYING DATA FOR CARNAUBA LEAVES

Pio Caetano Lobo*   and   Julio Wilson Ribeiro**

*Universidade Federal da Paraiba, Centro de Tecnologia-DTM, 58050 Joao Pessoa, Pb, BRAZIL
**Universidade Federal do Ceara, Centro de Tecnologia-DEM&P, 70455, Fortaleza, Ce. BRAZIL

ABSTRACT

Experimental carnauba leaf drying data are obtained. Equilibrium moisture isotherms are determined at air temperatures between 28 and 70 C and relative humidities from 25 to 75 %. Drying curves obtained for overhead direct radiant heating by incandescent lamps in a cabinet dryer tilted at 0, 30, 60 and 90°, provide preliminary base data for checking the analogy between heat and mass transfer as a source of drying rate data.

KEYWORDS

Carnauba, equilibrium moisture content, drying, mass transfer analogy, cabinet dryer, dryer simulation, free convection drying

NOMENCLATURE

C = vapour concentration in gas (kg/m^3)
c_p = specific heat of gas at constant pressure (Jg^{-1}C^{-1})
D = vapour diffusivity in gas (m^2/s)
g = acceleration due to gravity   (m/s^2)
Gr = Grashof number = $\rho^2 g\beta(t_w - t_0)L^3/\mu^2$
h = heat transfer coefficient (W m^{-2}C^{-1})
h_D = mass transfer coefficient (m/s) = m"$_w$/(C$_w$-C$_g$)
h_l = latent enthalpy of water (J/g)
H = radiant flux incident on dryer (W/m^2)
k = thermal conductivity (W m^{-1}C^{-1})
L = characteristic dryer length (m)
$m"$ = vapour mass flux (kg/m^2)
Nu = Nusselt number = L/k
Pr = Prandtl number = $c_p\mu$/k
Ra = Rayleigh number = Gr.Pr
Sc = Schmidt number = μ/(oD)
Sh = Sherwood number = h_DL/D
t = air temperature (C)
α_i = cover glass absorptivity to incident radiation
β,ρ = air thermal volumetric expansion coefficient (C^{-1}), density (kg/m^{-3})
μ = air absolute viscosity (kg m^{-1}s^{-1})
$(\tau\alpha)_e$ = leaf-cover glass transmissivity-absorptivity product

subscript a = ambient   b = back and side to ambient
           e = cover glass to ambient   g = cover glass
           i = absorber to cover glass   w = value near leaf surface
a dot above a symbol (e.g. "ṁ") indicates time rate

INTRODUCTION

Carnauba ("Brazilian") wax, is extracted from the fan-shaped leaf of the carnauba palm, with a radius of 300 to 500 mm. Wax content increases till the apex angle is around 90°, when moisture content is about 55%. Leaf drying promotes wax release. A covered dryer, designed for temperatures below 50C to avoid leaf embrittlement (Lobo, 1987), could improve on open-air sun drying, prone to deterioration and contamination. Since dryer design data are lacking, equilibrium moisture isotherms are obtained as well as drying rate data to compare with values derived through the analogy with mass transfer.

EXPERIMENTAL APPARATUS

Drying chambers were built to the design in Fig. 1 (Ribeiro, 1985; Lobo and Ribeiro, 1987). Five were maintained at constant temperature by electronically controlled 60 W tungsten filament lamps. Three were unheated. Sulphuric acid solutions controlled air relative humidity. Wet and dry bulb temperatures were read to within 0.25 C through side windows.

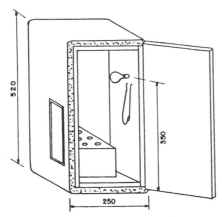

Fig. 1. Drying chamber

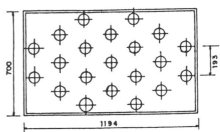

Fig. 2. Distribution of lamps in the radiant panel

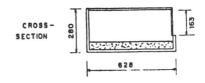

Fig. 3. Dryer model views

The radiant panel, sketched in Fig. 2, is lined with brilliant aluminium foil, and holds twenty-one 100 W tungsten filament spot lamps. It is adjustable in height and tilt, 1.2 m long x 0.7 m wide. The dryer, sketched in Fig. 3, is 1.2 m long x 0.6 m wide x 0.28 m high, with 10 mm plywood base and walls and a 4 mm cover glass. A galvanized iron absorber sheet painted black is back insulated with 50 mm of expanded polystyrene. The dryer could be tilted at 0, 30, 60 or 90°. Horizontal, it could be coupled to an air heater (with the same optical aperture) tilted 15°.

Leaf and air temperatures, measured by thermistor sensors, are read on a multichannel electronic thermometer with 0.1 C resolution and 2 C uncertainty.

The sensors, 4 mm in diameter and 10 mm long are each fitted with aluminium foil radiation shields. Ambient relative humidity is measured by a psychrometer with a resolution of 1 %. and ambient air temperature by a mercury in glass thermometer graduated to 1 C. Ambient relative humidity was measured with 2 % estimated uncertainty by a wet and dry bulb hygrometer.

For further details see Ribeiro (1985), Lobo and Ribeiro (1987), Lobo (1988).

EXPERIMENTAL PROCEDURE

3.1. Equilibrium Moisture Content Isotherms
30 to 50g of 1 to 2 cm square green leaf samples of young carnauba palms from the drying chambers were weighed periodically on a precision balance with a 10 ug resolution. Chamber relative humidity was measured by a psychrometer. The concentration of 1.5 dm^3 of sulphuric acid solution varied less than 2% during the process. Equilibrium was assumed when successive weighings varied less than 0,3% of initial mass. Dry mass, determined after heating in an oven, was subtracted from equilibrium mass to yield equilibrium moisture content.

3.2. Incident Radiant flux
Radiant flux on a horizontal surface equivalent to the dryer area, 770 mm below the filaments of the horizontal radiant panel, measured by a Belfort pyranometer with accuracy estimated at 10%, was used in the simulation. For the tilted dryer the cosine law was applied, with allowance made for shadows. Vertical dryer absorbing area for the simulation was taken as outlet slot width times average leaf corrugation height.

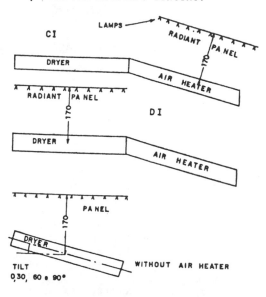

3.3. Drying Curves
Six experiments were performed, with configurations sketched in Fig. 4. The radiant panel is centred 770 mm from the dryer when horizontal or parallel to, and 770 mm from the collector. Uncoupled, the dryer was tilted at 0, 30, 60 and 90° with the panel horizontal; and with the collector coupled at zero tilt.

Fig. 4. Test configurations

Difficulties in transporting leaves from field to laboratory led to a single leaf being dried just upstream of the dryer transverse centreline, apex downstream, in each of three runs that comprised an experiment. Initial leaf thickness varied from about 0.34 mm at the base to about 0.28 mm at the tip; leaf mass from 100 to 130 g. During drying, leaves were weighed on an electronic balance with a resolution of 100 mg, at intervals corresponding to moisture evaporation of 100 g/m^2 of leaf area (plan) subject to a minimum of 50 minutes, ten times the duration of the weighing process, to limit time scale error to 10%. Equilibrium was assumed when successive weighings varied less than 0.3% of initial mass.

SIMULATION MODEL

Leaf physical and thermal properties were considered uniform and constant (though the leaf shrinks by 75% during drying). Total leaf coverage of the absorbing surface and uniform leaf and cover glass temperatures are assumed.

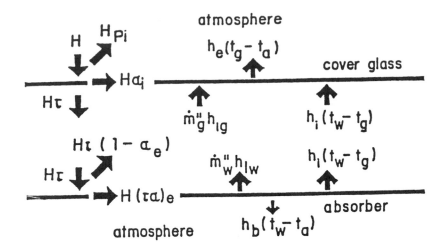

Fig. 5. Simulation model energy transfer rates

The First Law of Thermodynamics for the dryer of Fig. 5 (no collector) is:

$$H(\tau\alpha)_e = h_i(t_w-t_g) + h_b(t_w-t_a) + \dot{m}''_w h_{lw} \tag{1}$$

$$H\alpha_i + \dot{m}''_g h_{lg} + h_i(t_w-t_g) = h_e(t_g-t_a) \tag{2}$$

Heat transfer coefficients in Eqs. (1) and (2) are determined by empirical relations (Lobo and Ribeiro, 1987; Lobo, 1988). Since the constant drying rate period is predominant and the Lewis number (=Sc/Pr) near unity, heat transfer relations for flat plate collector air spaces, that are roughly similar systems, were converted (using the analogy) into corresponding mass transfer relations by a simple change of notation, substituting temperatures by concentrations, Prantl number by Schmidt number and Nusselt number by Sherwood number (Lobo, 1987, 1988). Since many terms are nonlinear in t_g and t_w, Eqs. (4) and (5) are solved by iteration.

Four simulation models are run:
(i) heat transfer only to the cover glass, at ambient temperature. on which no moisture recondenses; back and side losses are neglected.
(ii) heat transfer between cover, absorber and ambient; cover glass absorptivity for incident radiation 0.06; no condensation on the cover; absorber back and side heat loss coefficient 1.4 $Wm^{-2}C^{-1}$.

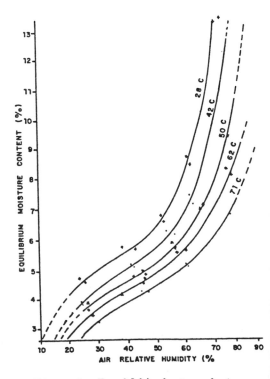

Fig. 6. Equilibrium moisture content isotherms

(iii) as in model (ii) except mass transfer flux is 30% that from Eq.(12).
(iv) as in model (ii) except all moisture is recondensed on cover glass.

RESULTS AND OBSERVATIONS

Figure 6 shows Equilibrium Moisture Content Isotherms in air at temperatures from 28 to 70C and relative humidities from 20 to 80%. For a relative humidity of 45% at 50 C, leaf equilibrium moisture content is about 4.3%, sufficient to release wax. Leaf physical properties and drying data appear in Table 1. Dryer performance, which improves slightly on coupling the collector, worsens when the panel irradiates the dryer, perhaps due to higher resistance. Predicted leaf moisture loss rates are much higher than measured at tilts of 30 and 60°,

TABLE 1. Leaf Physical Data, Measurements and Predictions

CHARACTERISTIC		Units	EXPT. 1	EXPT. 2	EXPT. 3	EXPT. 4
Dryer tilt from horizontal		o	0	30	60	90
Average radiant flux		W/m^2	746	604	366	98
Ambient relative humidity		%	74	74	76	77
Ambient temperature		C	28.8	28.7	28.3	28.3
Mean leaf dry weight		g	45.0	42.3	40.7	36.6
Average leaf area		dm^2	3.65	3.35	3.35	2.80
Dryer air temperature	inlet	C	43	38	33	34
	exit	C	54	46	45	45
	mean	C	49	42	39	39
Leaf temp.	Measured	C	68.8	55.4	37.6	37.9
	Predicted model (i)	C	62.7	59.8	54.3	41.1
	model (ii)	C	63.5	60.6	55.3	41.6
	model(iii)	C	76.9	72.7	63.6	43.1
	model (iv)	C	49.7	46.6	40.5	32.1
Dryer film temp.	Measured	C	49	42	39	39
	Predicted model (i)	C	45.7	44.3	41.3	34.7
	model (ii)	C	54.1	52.2	49.3	40.0
	model(iii)	C	60.8	58.2	53.5	40.7
	model (iv)	C	40.4	38.4	34.8	30.2
Mois-ture flux	Measured	$g/(m^2h)$	190.9	50.7	19.4	24.1
	Predicted model (i)	$g/(m^2h)$	412.1	330.1	196.1	55.5
	model (ii)	$g/(m^2h)$	304.4	232.1	113.3	11.0
	model(iii)	$g/(m^2h)$	249.9	183.4	80.2	5.5
	model (iv)	$g/(m^2h)$	153.3	110.6	50.3	7.5

They are reduced with finite glass thermal conductivity and glass-to-air heat transfer coefficient. They are lowest for model (iv). Measured leaf temperatures drop faster with tilt than predicted. Model (ii) agrees best with measurement for the horizontal dryer, model (i) for 30 and 90° tilts and model (iv) for the 60° tilt. Leaf temperature measurement uncertainty may exceed the 3% estimated for the sensor, due to the small fraction of probe surface in contact with the leaf. Accurate leaf temperature prediction can avoid exceeding the threshold for leaf embrittlement. The influence of changes in the assumed mass transfer coefficient on leaf drying rate and temperature predictions is not uniform.

CONCLUSIONS

Equilibrium moisture content curves confirm solar drying is adequate for wax removal. For the cabinet dryer, mass transfer predictions and experimental determinations are of the same order of magnitude, but discrepancies are significant at tilts of 30 and 60°. Preliminary measurements indicate that total coverage of the absorber may greatly reduce discrepancies. One possible cause is that evaporation cools the leaf below the bare absorber surface, and the resulting convective flow might keep moister air nearer the leaf, decreasing moisture gradient and hence mass transfer rate. If confirmed, the result contradicts the asssumption that, with partial coverage, the hotter absorber quickens drying. Also, solar flat plate collector heat transfer correlations employed exclude the efficiency factor and relate to totally enclosed air spaces with aspect ratios (length to height) over 20, as against an open air space and aspect ratio of 5.6 for the dryer model. Another factor may be uneven leaf heating due to shading by the corrugations. Some published data (Ozoe, 1982; Hollands, 1976) suggest that with tilt, free convection patterns may change.Further model improvement seems worthwhile, since adequate mathematical models would be a quick and economical tool for dryer design.

ACKNOWLEDGEMENTS

The first author is a research fellow of the National Council for Scientific and Technological Development - CNPq (Proc. N° 30.1755/76-EM). The experiments were carried out in the Ceara State Technology Nucleus (NUTEC), Fortaleza,Ce.

REFERENCES

Hollands, K.G.T. and others (1976). Free convection across inclined air layers. Journal of Heat Transfer, 98, n 2, 189.
Lobo, P.C. and J.W. Ribeiro (1987). Experiments on the radiant drying of carnauba leaves. In D.Y. Goswami (Ed.), Solar Engineering - 1987, Vol. 2. ASME, New York. pp 556-562.
Lobo, P.C. (1988). Estimation of mass transfer coefficients from heat transfer data for the prediction of leaf drying rates. In R.K. Shah, E.N. Ganic and K.T. Yang (Eds.), Experimental Heat Transfer, Fluid Mechanics and Thermodynamics 1988. Elsevier, New York. pp 217-223.
Ozoe, H. and others (1982). Laminar natural convection in an inclined rectangular box with the lower surface half heated and half insulated. ASME Paper 82-HT- 72.
Plumb, O.A. and C.C. Wang (1982). Convective mass transfer from partially wetted surfaces. ASME paper 82-HT-59.
Ribeiro, J.W. (1985). Analise da Secagem da Palha de Carnauba com Aquecimento Radiativo (Analysis of Radiant Heat Carnauba Leaf Drying). Master's Thesis, CPGEM, Centro de Tecnolgia, UFPb, Joao Pessoa, Pb, Brazil. In Portuguese.

2.31 Solar Drying II

MODELLING OF FREE CONVECTION AIR FLOW IN A SOLAR DRYER

M. Mahr and J. Blumenberg

Institute C for Thermodynamics, Technical University of Munich,
Augustenstr. 77 Rgb, 8000 München 2, Germany

ABSTRACT

A new kind of solar dryer with free convection air flow was designed and investigated. A dryer consists of an air collector and the drying chamber. An inclination of the collector or a connected chimney make the free convection flow of the heated air possible. For a semicylindrical air collector with double or single foil cover a theoretical model describing the relations between the absorbed solar radiation, the air temperature and the mass flow rate was developed. Based on this model, a computer program calculates the temperature and the flow rate of the air as a function of the weather conditions, the size of the dryer and the properties of the materials.

KEYWORDS

Solar dryer; agricultural products; developing countries; free convection; theoretical model.

INTRODUCTION

Solar drying of agricultural products seems to be a successful way to reduce post harvest losses especially in developing countries. On the occasion of a project supported by the German Federal Ministry of Research and Development (BMFT) two versions of a new kind of solar dryer with free convection air flow were designed and investigated in Germany and Spain. A dryer is basically composed of an air collector and a drying chamber. The collector consists of a semicylindrical horticultural greenhouse covered by one or two transparent foils. A black foil is fixed on the ground inside the tunnel. The foils absorb solar radiation, convective and conductive heat transfer cause an increase in the air temperature and a reduction of the air density inside the greenhouse. A free convection air flow occurs in case of a difference in altitude between the inlet and outlet of the air, which is caused either by an inclination of the collector (type 1: slope-dryer) or by a connected chimney (type 2: chimney-dryer)(Fig. 1). The most important factors concerning convective drying processes are the velocity, temperature and relative humidity of the air entering the drying chamber. A model was developed to describe the relations between the condition of the air and the weather conditions, the size of the dryer and the properties of the materials.

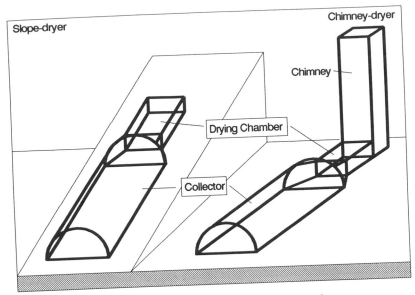

Figure 1: The slope-dryer and the chimney-dryer

THE THEORETICAL MODEL

The Energy Balances

Experiments by Luboschik and colleagues (1989) proved that the movement of the air inside the collector has always to be considered as a turbulent flow with Reynolds number > 10000. This fact leads to a one-dimensional quasi-stationary model of the air flow assuming that the temperature of the foils and the air are a function of the variable x in flow direction. The simplified energy balance of the air includes the temperature gradient between the relevant foils and the air as well as the convective heat transfer coefficient and the mass flow rate. The collector (length l) and the chimney (height h) are divided up into segments of length dx and height dh respectively. The increase in air temperature and the according decrease in density are calculated for every segment as a function of an estimated mass flow and the temperature of the foils and the chimney wall. This requires separate energy balances for each foil and the chimney. The model considers the influences of both diffuse and beam solar radiation as well as convective and radiative heat transfer.

Solar Radiation

Three different types of shortwave solar radiation must be considered:

1. beam radiation;

2. diffuse sky radiation;

3. diffuse radiation reflected by the ground.

The flux per unit area of the diffuse and beam radiation on a horizontal plane outside the tunnel must be calculated or measured. The properties of absorptance, reflectance and transmittance of the foils depend on the angle between the direction of the beam and the normal of the relevant foil area. First, the vector between the location and the sun has to be expres-

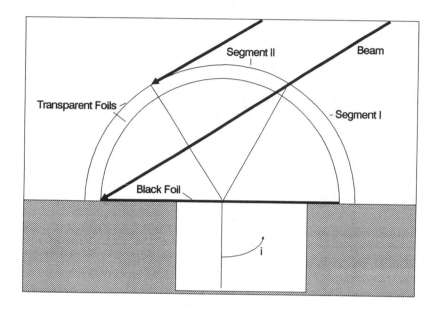

Figure 2: The crossection of the collector

sed as a function of the solar altitude angle and the solar azimuth angle, then the vectors perpendicular to any surface element of the transparent foils are calculated as a function of the angle i according to Fohr and Figueiredo (1987) (Fig. 2). The vector perpendicular to the black plane can be determined by taking $i = \pi$. The angle between the beam vector and a surface vector can be calculated according to the laws of vector analysis. Duffie and Beckman (1980) give a model for the calculation of the optical properties of the foils.

For the investigation of the solar flux per unit area absorbed by each foil the following assumptions are made according to Fohr and Figueiredo (1987):

- the solar flux is taken as diffuse from the first reflection on;

- the consideration of two reflections is sufficient.

An integration over the segments I and II (Fig. 2) hit by beam radiation leads to the values of the beam solar flux per unit area absorbed by the foils. In case of diffuse radiation an integration is not required, because the optical properties are considered equal to those for beam radiation with a constant angle of 60°.

Radiative Heat Transfer

A complete energy balance of the foils requires the investigation of the radiative heat transfer. Five sources emit radiation within the infrared range according to their temperature:

- the black foil;

- the inner transparent foil;

- the outer transparent foil;

- the sky;

- the ground.

For each foil the loss and gain of radiative energy has to be investigated in detailed balances, because the transparent foils show transmittance also in the infrared range. It is assumed that the relevant optical properties do not depend on the wavelength of the radiation in the infrared range. The properties are also calculated according to the model by Duffie and Beckman (1980). Fifteen equations describe the radiative heat transfer between the radiators and the three foils.

Calculation of the Mass Flow

After the calculation of the air temperature inside the collector as a function of an estimated mass flow, it is necessary to calculate the exact mass flow using the theorem of momentum. A system considering both an inclined collector and a chimney is investigated.

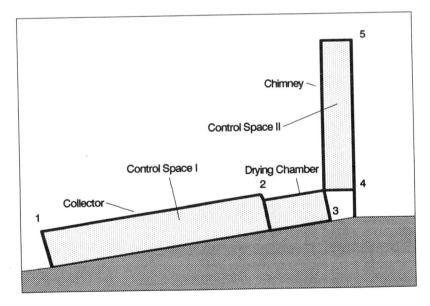

Figure 3: The two separate control spaces

The first control space contains the air inside the collector and the drying chamber, the second control space contains the air inside the chimney (Fig. 3). This model leads to two separate one-dimensional equations of momentum which must be connected. The external forces of pressure, gravity, friction are responsible for the alteration of the mass flow and the velocity respectively. Considering the conservation of mass, two equations for the calculation of the mass flow can be derived, they depend on:

- the external pressure at the locations 1, 2 and 5 (compare Fig. 3);

- the sizes of the components;

- the inclination angle;

- the roughness of the surfaces;

- the pressure drop caused by a loaded drying chamber;

- the air density inside and outside the system.

These two equations are connected by the assumption, that the static pressure at the end of the drying chamber (3) is equal to the static pressure bottom of the chimney (4). The resulting equation is valid for the slope-dryer (chimney: height= 0, diameter $\rightarrow \infty$) and the chimney-dryer (collector: inclination = 0) as well as for combinations of the two. A computer code for the calculation of the condition of the air inside the system was developed. The program works with 144 timesteps per day of 10 minutes duration each. First, the temperature and density of the air inside the collector is calculated as a function of an estimated value for the mass flow. This value is compared with the the result of the equation of momentum, iterative calculations lead to the correct relation between foil temperatures, air temperature and the air flow rate.

RESULTS

The figures 4 and 5 show typical simulation results (-) of the volume flow rate and the temperature of the air entering the drying chamber in comparison with measured values (.). In this case the drying chamber was unloaded. Differences between the results of the experiment and the simulation are mainly caused by the influence of the wind that was not included in the simulation model.

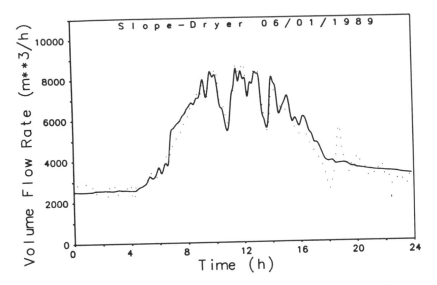

Figure 4: Typical plot of volume flow rate during one day

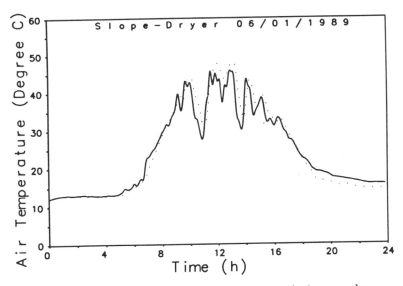

Figure 5: Typical plot of air temperature during one day

REFERENCES

Duffie, J. A., and A. Beckmann (1980). Solar Engineering of Thermal Processes. John Wiley and Sons, New York.

Fohr, J.-P., and A.R. Figueiredo (1987). Agricultural Solar Air Collectors: Design and Performances. Solar Energy, Vol.38, No.5, pp. 311-321.

Luboschik, U., P. Schalajda, M. Reuss, H. Schulz, J. Blumenberg, and M. Mahr (1989). Investigation of a Solar Dryer. ISES Solar World Congress, Kobe/Japan.

EFFECTS OF SEASONAL WEATHER VARIATIONS ON THE MEASURED PERFORMANCE OF A NATURAL-CIRCULATION SOLAR-ENERGY TROPICAL CROP DRYER

O V Ekechukwu
Energy Research Centre, University of Nigeria
Nsukka, NIGERIA
and
B Norton
PROBE, centre for Performance Research On the
Built Environment, Department of Building
& Environmental Engineering, University of Ulster
Newtownabbey, BT37 0QB, NORTHERN IRELAND.

KEYWORDS

Solar drying; tropical climate; measurements.

ABSTRACT

A comprehensive study of the measured transient performance of a large-scale integral-type natural-circulation solar-energy crop dryer suitable for use in the tropics was carried out over the entire duration of weather conditions encountered in a typical tropical environment.

EXPERIMENTAL OBSERVATION

Experimental Facility

The experimental facility consisted of a 5m long by 2.3m wide by 2.7m tall greenhouse type natural-circulation solar-energy dryer with a 5m tall solar chimney. Taut polyethylene cladding totally enclosed the drying chamber. A galvanised steel framework was insulated thermally from the polyethylene cladding by insulant tape to prevent undesirable localised heating of the cladding which would shorten severely its useful life. The crops were placed on metal mesh trays which were hung one beneath another from horizontal galvanised steel struts. Both banks of individual trays were suspended from load cells. Thus moisture loss variations with time during drying could be measured for both the dryer as a whole and for particular locations. Signals from all the sensors were recorded by a microcomputer-based data-acquisition and storage system. Details of the dryer and its instrumentation have been described by Ekechukwu (1987).

A total of ten experimental tests were undertaken under varying dryer configurations and over the complete range of tropical weather conditions (between December 1985 and November 1986).

The main applied parameters measured included; the total and diffuse solar radiation, ambient temperatures and relative humidities, precipitation and wind speed. The drying conditions monitored included; the drying air temperatures and relative humidities and the air flow rates. Also monitored were the moisture loss and crop temperatures. Sensors were located within the dryer as to give information concerning specific locations and the general trend of the prevailing conditions. All the tests were carried out with cassava chips as the drying medium.

Data Reduction

The main performance characteristics evaluated from the measured data were the drying rates (i.e. moisture content variation with time), daily drying efficiencies and the nocturnal crop moisture reabsorption. The instantaneous crop moisture contents (on dry basis) were computed from the measured instantaneous weights of the crop according to the following relation;

$$M_t = [(M_o + 1)W_o/W_t] - 1 \qquad (1)$$

The initial moisture content M_o was calculated from the following;

$$M_o = (W_o - W_d)/W_o \qquad (2)$$

The drying efficiency defined as the ratio of the useful energy (i.e. energy used to evaporate moisture from the crop) to the energy input (i.e. insolation received over the area of the dryer's horizontal projection) is represented as

$$\eta_d = Q_u/Q_{in} = W_v L/IA \qquad (3)$$

The total mass of moisture evaporated from the crop W_v is given by,

$$W_v = (M_t - M_f)W_d \qquad (4)$$

while W_d, the dry mass of the crop is defined as,

$$W_d = W_o/(M_o + 1) \qquad (5)$$

Due to the relatively slow thermal response of the crop mass to instantaneous variations in insolation, the integrated efficiency (i.e. efficiency over a defined period of time) is a more meaningful parameter than an instantaneous drying efficiency. Accordingly, we have computed from the test data the daily drying efficiency.

Combining equations 3-5, the daily drying efficiency (usually expressed as a percentage) is then given as,

$$\eta_d = \frac{[(M_t - M_f)W_o/(M_o + 1)]L}{IA} \, 100 \qquad (6)$$

where the insolation I, is the total insolation received over a horizontal surface for the day considered. Since $_d$ is a function of the quantity of crop being dried, to isolate the effect of the crop mass, the drying efficiencies were normalised against the total initial weight of crop to obtain the corresponding "normalised" drying efficiencies, defined as,

$$\eta_n = \eta_d/W_o \qquad (7)$$

The nocturnal moisture reabsorption has been defined as the percentage increase in moisture content over the night period of the value at sunset of the preceding day, given by,

$$R_n = [\Delta M/M_{ss}]100 \qquad (8)$$

Positive values of R_n indicate moisture reabsorption, while negative values indicate further moisture loss.

Drying Conditions

The frequency distribution of the drying air temperatures for the tests during both seasons are shown in Fig. 1. The results illustrate only a slight predominance of higher drying air temperatures (above 40°C during the dry season compared with the wet season. The corresponding relative humidity frequency distribution histogram for both season tests (Fig. 2) show markedly a more pronounced predominance of lower drying. air relative humidities (below 35%) during the dry season compared with the wet season (predominantly above 50%). Day-time drying air relative humidity levels of below 15% obtainable during the dry season are well below levels required to facilitate quick drying to safe storage moisture content. Frequent rains during the wet season and inclement weather generally, resulted in sporadic changes in the moisture carrying capability of the drying air. Drying conditions during the dry season were generally more predictable.

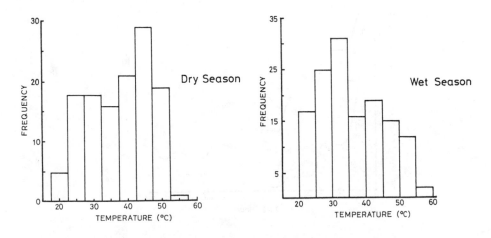

FIG. 1   Frequency distribution of Drying Air Temperatures during the Dry and Wet Seasons

2450

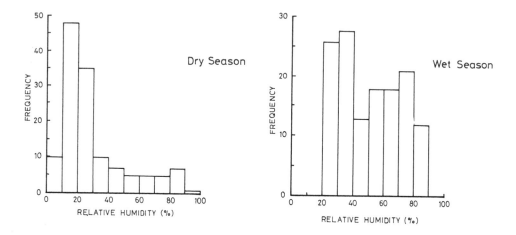

FIG. 2　Frequency Distribution of Drying Air Relative
Humidities during the Dry and Wet Seasons

Drying Progress

The mean drying curves over various locations within the dryer
obtained for both seasons tests are shown in Fig. 3, with the
horizontal broken lines indicating the safe storage moisture
content level (approximately 20% dry basis).　The results show
that for the dry season, crop moisture content was reduced to the
safe storage level within 17 hours of insolation corresponding to
approximately 2 days of drying.　Nocturnal moisture reabsorption
occurred during the third night of drying at a moisture content
of below 5%, much lower than the safe storage level.　The rapid
drying rate is attributable mainly to the prevailing low drying
air relative humidity.　In contrast, the crop moisture content
(at around 35%) was well above the safe storage level after 6
days of drying during the wet season.

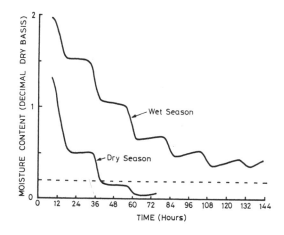

FIG. 3　Typical Mean Drying Curves for the Dry and Wet Seasons

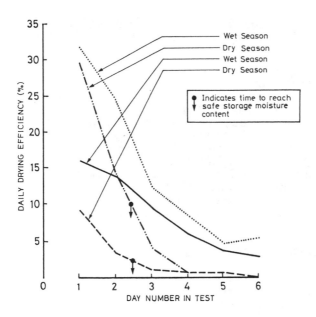

FIG. 4 Variation of Daily Drying Efficiencies with Time

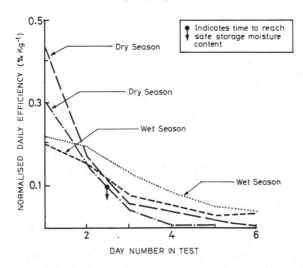

FIG. 5 Variation of Normalised Daily Drying Efficiencies
with Time

Drying Efficiencies

Typical variations in the daily drying efficiencies are shown in Fig. 4. The dry season tests showed a more rapid decrease in the daily efficiencies with time compared with the wet season. The inherent implication of a more rapid decrease in the daily drying efficiency is a faster drying rate (i.e. faster rate of moisture depletion in the crop, thus reducing W_v in equation 3). The trend is the same for the "normalised" efficiency curves shown in Fig. 5 where the possible effect of the loading density has been isolated.

CONCLUSION

The performance of the integral-type natural-circulation solar-energy dryer studies, was found to be affected largely by seasonal weather variations. Drying conditions during the dry season were fairly constant. Variable cloud overcast, frequent rain and high relative humidity affected adversely the drying conditions during the wet season. The overall performance of the dryer was thus better during the dry season, requiring less than 2 days of drying to reduce the moisture content of cassava chips to lower than the desired safe storage level.

NOMENCLATURE

A	Area of dryer's horizontal surface (m^2)
I	Insolation (Wm^2)
L	Latent heat of vapourization (Jkg^{-1})
M_o	Initial moisture content (decimal, dry basis)
M_f	Final moisture content at sunset (decimal, dry basis)
M_{ss}	Moisture content at sunset (decimal, dry basis)
M_t	Instantaneous moisture content (decimal, dry basis)
ΔM	Change in crop moisture content during night-time (decimal, dry basis)
Q_{in}	Energy input (J)
Q_u	Useful heat gain, i.e. energy used to evaporate moisture (J)
R_n	Nocturnal moisture reabsorption (%)
W_o	Initial crop weight (kg)
W_d	Dry weight of crop (kg)
W_t	Instantaneous crop weight (kg)
W_v	Weight of moisture evaporated (kg)
η_d	Drying efficiency (%)
η_n	Normalised drying efficiency ($\%kg^{-1}$)

ACKNOWLEDGEMENTS

We acknowledge grants from the Commission of European Communities Brussels, Belgium and the British Council, London, England, for this study. We acknowledge most warmly the initial contributions of Dr. Paul Fleming. Facilities provided by the University of Nigeria, Nsukka, Nigeria, Cranfield Institute of Technology, UK and the International Centre for Theoretical Physics, Trieste, Italy, during the experimental tests, data analysis and preparation of the manuscript, respectively, and the paid study leave to Dr. Ekechukwu from the University of Nigeria are also acknowledged.

REFERENCE

O V Ekechukwu (1987), Experimental Studies of Integral-Type Natural-Circulation Solar Energy Tropical Crop Dryers, Ph.D. Thesis, Cranfield Institute of Technology UK.

ARCEL SIMULATION MODEL TO PERFORM THE THERMAL BEHAVIOUR
OF A STORAGE IN THE SOIL BELOW A GREENHOUSE

F.Parrini*, S. Vitale*, A. Biondo**
and R. Lo Cicero**

* ENEL (Italian National Electricity Board) - CRTN
Via Monfalcone 15, 20132 Milan, Italy
** Conphoebus
95030 Pantano D'Arci - Catania, Italy

ABSTRACT

ENEL (Italian Electricity Board), for a few years, has carried out a series
of experimental and theoretical studies on the greenhouse management, focusing
the attention on the energy saving and utilization of low temperature water
discharged by thermoelectric power plants.
In this paper will be presented the new version of the mathematical model
ARCEL that has been improved in order to describe the thermal behaviour of the
greenhouse whenever the trapped heat, giving rise to overheating, can be
stored in the ground below the greenhouse. The "greenhouse system" so arranged
can then store the heat, not immediately necessary, in the ground by recirculating the heated air inside pipes buried in the ground. The goal of this
system is: to store heat during sunny days in order to use it either during
night time or when required. The numerical model includes a set of energy
balance equations, assembled by means of implicit finite difference method, in
order to represent the simultaneous energy and mass transfer. The mathematical
model has been implemented into Fortran 77 computer language.

KEYWORDS
Greenhouse management; numerical solution; mathematical model; energy saving;
pipes buried; waste heat.

INTRODUCTION

The simulation code ARCEL has been mainly developed in order to study the
utilization of low grade temperature water discharged by electric power
plants. In general the waste heat may produce thermal pollution on streams or
lakes where biological equilibrium is strongly influenced by thermal levels.
One of the more attractive ways of solving this problem is the use of the
waste water in agriculture to supply thermal requirements of protected
cultivations.
The main energy input of a greenhouse is represented by the solar radiation
striking on the external walls and represents the fundamental contribution to
the energy balance (Garzoli and Blackwell, 1973).
In most cases the internal air temperature swings widely around a mean value
because of the low thermal capacitance of the whole system, requiring, as a
consequence, the intervention of an external heating/cooling system to level
the temperature variations. ARCEL provides the possibility to simulate the use

of low enthalpy heat, coming from any device able to ensure such contribution. In other words it is possible to simulate heaters using water discharged by closed cycle wet tower thermoelectric power plant. In this case low temperature water (22-25°C) is used to heat the greenhouse. To perform the cooling of the greenhouse two ways can be selected: the intervention of a cooling system (energy consuming) and the recirculation of the greenhouse air into pipes buried in the ground. The former system had been already included in the previous version of ARCEL (Alabiso and others, 1989), the latter represents the new implementation which characterises the present version.

The main function of this system is, then, to store energy during overheating periods and give back the stored heat to the greenhouse when internal air temperature goes down a set-point limit. The resultant thermal capacitance of the greenhouse is higher than in the traditional configuration (Tiwari and Dhiman, 1985; Walker, 1963).

THEORY AND MODELLING METHODS

The mathematical model developed to describe the greenhouse system, as said above, is based upon the discretition of the physical system into a number of nodes associated to control volumes.

Both plane and tunnel walls are double-layer polyethylene air inflated; shelters can be used in order to reduce the thermal losses during night-time, their operation can be related to a solar radiation threshold or to a time schedule user defined. The thermal storage is modelled as a soil volume having a face corresponding to the greenhouse floor and a depth defined by the user accordingly to the design requirements. The side walls of the soil thermal storage are considered adiabatic. The pipes are modelled as a network covering the entire greenhouse floor as shown in fig. 1, their deepness can be modified at user request (Karlekar and Desmond, 1977).

The heat fluxes are considered one-dimensional in the greenhouse components and two-dimensional in the soil representation. Sensible heat equations are developed for each component and simultaneously solved at each time step to predict the temperatures which are related to the control operations of equipment such as ventilation system, heating/cooling plant. The mathematical model utilises the finite implicit difference technique in order to transform the

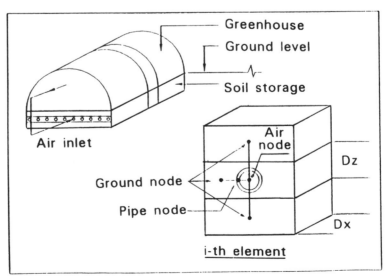

Fig. 1. Greenhouse system and node scheme

basic energy balance equations into a set of algebraic equations, which describes a quasi -steady state by radiative, convective and conductive terms.

The model was designed in order to simulate different greenhouse configurations, auxiliary heat and control devices, and a pipes net adapted to the design needs. In particular the discrete representation of the soil and pipes can be changed accordingly to the accuracy requested.

The soil can be divided into ten layers (maximum, in vertical) and each layer is divided into three nodes (in horizontal). The layer containing the pipes will be divided into ten nodes in the flux direction (see Fig. 1). As shown, the soil nodal representation is two-dimensional and flux equations are written accordingly to such representation. The finest discretization allows the user to represent the whole system as a 101 nodes network maximum.

The typical discretized heat balance equation for the generic j-node, at time t, is given by (Kreith, 1974; Incropera and Dewitt, 1985):

$$Q_j + \sum_{i=1}^{n} H_{i,i}(T_i-T_j) = \rho c \Delta x / \Delta t (T_j-T_j') \tag{1}$$

where Q_j is the heat flux received or exchanged by he node j, $H_{i,j}$ is the generalized heat transfer coefficient between the j node and the i node, T_i and T_j are the associated temperatures, ρ is the density, c the specific heat, ΔV the control volume, Δt the time step, n the number of nodes interacting with the node j and T_j' is the temperature at the time $t' = t - \Delta t$.

The air flow injected into the pipes can be extracted from the greenhouse itself or by other source, if any exists.

SENSITIVITY AND CONSISTENCY ANALYSIS

At the present time the new release of ARCEL hasn't been yet submitted to a validation stage; a complete set of experimental data is going to be acquired. Greenhouse systems with and without pipes buried in the ground have been built in Catania at the Test Site of Conphoebus in order to determine their thermal performance.

In this paper some results of a set of sensitivity and consistency analysis runs exercises will be given.

The parameter set identified to perform a sensitivity analysis is:
- Horizontal step of the pipes in the greenhouse floor (56, 85, 170 cm);
- Depth of the pipes buried in the ground (15, 45, 60 cm);
- Air flow rate in the pipes (1 to 10 greenhouse volumes per hour);
- Pipes fan ON/OFF.

The consistency analysis has been performed changing the input weather files selecting two climatic areas among possible italian sites: Catania and Rome. Aim of this exercise is to check the code prediction under various climatic boundary conditions. The key parameters, to be taken into account, are the daily energy input due to pipes and the mean temperature of the greenhouse, in a free floating operation, to check the different thermal behaviour in the selected sites.

The greenhouse type, simulated in these exercises, is a tunnel one. The walls are double layer polyethylene air inflated (Failla and Cascone, 1987), the dimensions are: 8.5 m width and 30 m lenght. The pipes buried in the ground are concrete made and have the following dimensions: diameter 15 cm, length 30 m. The ground has been considered as wet soil. The air change rate due to natural infiltration is 0.1 volumes per hour.

RESULTS AND DISCUSSION

In this section the results of some runs will be given.
The exercises have been carried out utilizing the test reference years' meteo-data files for Catania and Rome. The data set contains the solar global and diffuse radiation, wind speed, ambient temperature and relative humidity on hourly base. The first test consists in the evaluation of the energy extracted from the ground storage depending on the variation of key parameters: number of pipes buried in the ground and air flow rate inside pipes. The number of pipes have been: 5, 10 and 15 corresponding to a pipes surface to greenhouse floor surface ratio of 0.27, 0.55 and 0.83 respectively. In Fig. 2 the behaviour of the daily total extracted energy is shown for february and march (60 days).
Two exercises have been carried out considering the greenhouse in free floating regime (ten pipes) for Catania and Rome, in Figs. 3-4 the mean daily internal temperature is plotted over a 60 days period, from december to january.
The last result presented gives the heating load behavior for ten days of february with the pipes fan ON and OFF, see Fig. 5; a heating reduction load, about 25%, can be seen.

REFERENCES

Alabiso M., F. Parrini, S Vitale, A. Biondo, R. Licata, and R. Lo Cicero (1989). Greenhouse simulation model 'ARCEL' applied to a double-layer inflated polyethylene tunnel using low temperatures waste heat, Acta Horti-culturae, 245, 356-362.

Failla A., and G. Cascone (1987). Efficienza energetica di apprestamenti co-perti con doppio film plastico, Colture protette, 2

Garzoli K., and J. Blackwell (1973). The response of a glasshouse to high solar radiation and ambient temperature, J. Agricol. Eng. Res., 18

Incropera, F.P., and D.P. Dewitt (1985). Fundamentals of Heat and Mass Transfer, J. Wiley & Sons.

Karlekar B.V., and R.M. Desmond (1977). Engineering heat transfer, West Publ. Co.

Kreith F. (1974). Principi di trasmissione del calore. Ed. Liguori, Napoli.

Tiwari G.N., and N K. Dhiman (1985). Periodic theory of greenhouse, Energy Convers. Mgmt 25 (2).

Walker G.N. (1965). Predicting temperatures in ventilated greenhouses. Trans. of the ASAI.

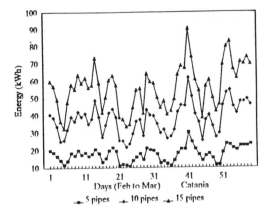

Fig. 2. Energy extracted from the ground with 5, 10 and 15 pipes.

Fig. 3. Greenhouse temperature in free floating regime, for Catania

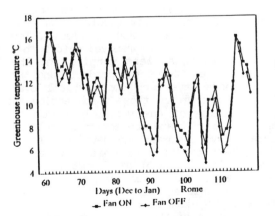
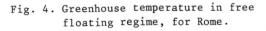

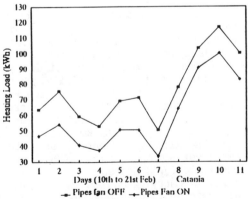

Fig. 4. Greenhouse temperature in free floating regime, for Rome.

Fig. 5. Heating load for Catania, pipes Fan ON/OFF.

THEORETICAL ANALYSIS OF A SOLAR TIMBER DRYING SYSTEM

H.P. Garg[*], Jai Prakash[**] and D.S. Hrishikesan[*]

[*]Centre for Energy Studies, Indian Institute of Technology
Hauz Khas, NEW DELHI - 110 016, India
[**]Department of Physics, Ramjas College, Delhi University
DELHI - 110 007, India

ABSTRACT

A mathematical model of a solar timber drying system is developed. Making use of this model and empirical relations for drying red oak the thermal behaviour and the drying characterstics of the kiln is simulated.

KEY WORDS

Solar timber drying system; moisture content; mathematical model; finite difference technique.

INTRODUCTION

Timber is extensively used for the production of furniture and other wooden fittings in the construction of buildings. But, only a few species like oak, teak,etc.,can be directly used for the manufacture of wood products. The moisture contained in the other species of timber is to be removed before they are used for the manufacture of the wood products, lest warping and the shrinkage of wood would result in the inferior quality of the products and the reduction in their durability. Various methods have been employed all over the world for drying of wood. The kilns widely used are those which employ the conventional fuels for the controlled heating of the wood. In order to investigate the possibility of reducing the use of these conventional fuels in the timber drying systems, various workers have designed and developed kilns working with solar energy and studied the viabilitly of this technology (Martinez, 1984; Taylor, 1985; Zahed,1989). Sharma (1980) at Forest Research Institute (FRI) Dehradun, India has developed an indigenous design of a solar kiln. By carrying out an experimental study of this system which is fabricated at their Institute they have collected the drying data of different species of wood. But, a comprehensive theoretical evaluation of the system, for studying and optimizing its design and working parameters is not successfully carried out. The authors, in this paper, have attempted to develop a mathematical model of the FRI design kiln. By writing the energy balance equations at the different nodes of the system, its thermal behaviour, while it is loaded to its capacity,is predicted. The drying mechanism of wood is taken into account by employing empirical relaions, formulated by Simpson (1980), for the drying of red oak lumber of specified thickness. Finite difference technique is employed to solve the differential equations for the energy balance at the different components of the system. The model is numerically simulated by using the meteorological data for Delhi.

DESCRIPTION OF THE SYSTEM

The design details of the kiln is given in Fig. 1. It is a solar heated compartment kiln of the side mounted external fan type with reversible air circulation. The kiln measures externally 5.76m long x 3.66 m wide x 2.28 m high at the south wall and 3.48 m at the north wall. The difference in the heights at the south and the north walls is for providing a tilt angle of 0.9 times the latitude of the location (which is 30°N in the case of Dehradun), for the roof of the kiln. The long axis of the kiln is oriented east-west. The north wall of the kiln is sheathed with 9 mm thick plywood and all other walls are covered with double layer of 3.0 mm thick clear transparent glass, separated by an air gap of about 37 mm by means of wooden spacer strips. A vertical partition of 9 mm plywood 1.9 m high and 3.46 m long spans the entire length of the kiln and extends from the floor to the false ceiling on its north side. A 0.61 m side plenum gap is thus provided between the fan partition and the north wall. The plywood partition bears a 100 cm dia hole for the fan which is mounted at mid-height of the partition. A propeller fan of reversible type, 0.9 m in diameter and having 12 blades, is mounted in fan housing with shafts supported cross-wise to the length of the stack horizontally, in bearing mounted on one angle iron pedestal grouted into the concrete platform. The shafts are taken out through the plywood north wall and driven at 550 RPM by reversible electric motor. A corrugated black absorber(measuring 5.45m x 2.14m) is placed horizontally in front of the fan-mounted partition wall inside the kiln for the absorption of solar radiation and thus to enhance the energy input in the system. Baffles are provided on the partition wall both above and below the level of the absorber plate. By keeping a horizontal wooden plank behind the partition wall at the same level as that of the absorber plate, the chamber on the northern side of the partition wall can be divided into an upper and a lower chamber completely isolated from each other. By properly adjusting these baffles, the horizontal plank and the openings on the northern wall, the kiln can be made to function in the single pass mode or in recirculation mode.

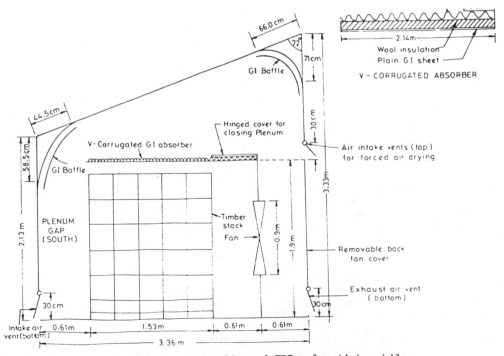

Fig. 1. Schematics of FRI solar timber kiln.

The timber to be dried is cut into long pieces and stacked with the help of spacers in between them. The stacking reaches to a height just below the absorber plate. The system is also provided with a humidifier to spray water inside the kiln to increase the humidity of the environment lest cracking or warping of the wood might take place.

THEORETICAL MODEL OF THE SYSTEM

The system is considered to be made of two different chambers, one below the absorber plate where the timber is stacked and the other above. Heat energy balance equations for different components are written separately in these two chambers as follows:-

For the Upper Chamber

For the plate

$$M_p C_p \frac{dT_p}{dt} = \alpha_p \tau_g S_t - h_{pr2}(T_p - T_{r2}) - h_{pra}(T_p - T_a) - h_{pr1}(T_p - T_{r1})$$
$$- h_{prL}(T_p - T_L) \tag{1}$$

For the air

$$M_{r2} C_r \frac{dT_{r2}}{dt} = h_{pr2}(T_p - T_{r2}) - h_{r2a}(T_{r2} - T_a) - \frac{GC_r}{A}(T_{r2} - T_{r1}) \tag{2}$$

For the Lower Chamber

For the air

$$M_{r1} C_r \frac{dT_{r1}}{dt} = h_{pr1}(T_p - T_{r1}) - (A_v/A_G)h_{Lr1}(T_{r1} - T_L) - h_{r1a}(T_{r1} - T_a)$$
$$- \frac{GC_r}{A}(T_{r1} - T_{r2}) - \frac{G_1 C_r}{A}(T_{r2} - T_a) \tag{3}$$

For the timber load

$$M_L C_L \frac{dT_L}{dt} = A_v h_{Lr1}(T_{r1} - T_L) + \alpha_L \tau_g(A_s S_s + A_e S_e + A_w S_w) - A_G U_1(T_L - T_G)$$
$$+ A_G h_{prL}(T_p - T_L) \tag{4}$$

The drying rate of any material is directly proportional to the average moisture content of the system (Simpson, 1980). Thus,

$$\frac{dW}{dt} = k\,W \tag{5}$$

where W = average MC at time and k is a constant of proportionality. The time required to dry wood of given specifications can be written as:

$$= (-1^{1.52}/b\theta)\ln\left\{(W - W_e)/(W_o - W_e)\right\} \tag{6}$$

where, θ = drying time (days);   l = board thickness (in);
        b = empirical coefficient dependent on temperature and RH
        W = average MC at time θ (%); W_e = EMC conditions in kiln (%);
        W_o = initial MC (%)

b in eq. (6) is given by the relation,

$$b = 0.0575 + 0.00142p \tag{7}$$

where p = vapour pressure of water(mm Hg), which can be related to temperature,

$$p = \exp(20.41 - 5132/T) \tag{8}$$

where T is in degrees Kelvin.

Equation (6) can be put in a more useful form to estimate the moisture content at each step in a kiln schedule. Thus,

$$W = W_e + (W_o - W_e) \exp(-b\theta/l^{1.52}) \tag{9}$$

Equations (7-9) are used for calculating the change in moisture content of the wood with time.

NUMERICAL CALCULATIONS AND RESULTS

In order to study the thermal behaviour of the kiln and the drying characteristics of wood, the differential equations (1-4) are written in their different forms and are simultaneously solved using the finite difference forward time step marching technique. The solar radiation and the ambient temperature for a typical summer day in Delhi are used for the numerical calculations. The variation of solar flux incident on different surface of the wooden stacking with time is given in Fig. 2. The temperatures at different

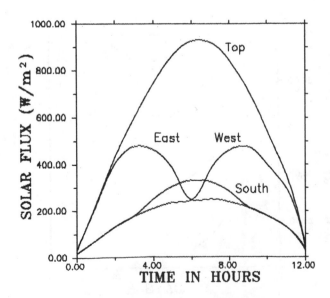

Fig. 2. Variation of incident solar flux on different surfaces
        of the timber stacked in the kiln.

2462

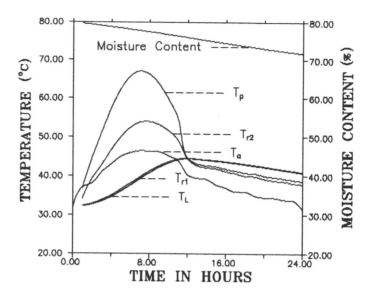

Fig. 3. Variation of temperatures at different nodes in the kiln, moisture content of timber and ambient temperature with time.

nodes, thus calculated, are employed in eqs.(7-9) for calculating the variation in the percentage moisture content of wood. The calculations are carried out for a wooden stacking of dimensions 5.0 m x 2.0 m x 1.0 m. In Fig. 3, the variations of the ambient temperature (T_a), the plate temperature (T_p), the load temperature (T_L), the temperature of air in the lower chamber of

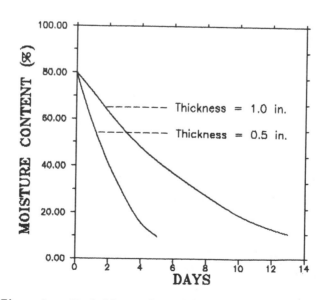

Fig. 4. Variation of moisture content of wood with time for two different thicknesses of the planks.

the kiln (T_{r1}), the temperature of air in upper chamber (T_{r2}) and the percentage moisture content are plotted against time. The moisture content reduces from 80% to nearly 70% in the very first day. The calculations are carried out for evaluating the possiblity of drying wood from an inital moisture content of 80% to a final value of 12%. Here, it assumed that the meteorological condition of a particular day is repeated over a number of days for the whole period of the drying. The variation of the moisture content with time for two thicknesses of the wooden planks are plotted in Fig. 4. It is seen that the required moisture content, mentioned above, is reached within a period of 13 days for wooden planks of thickness 1 inch and the same final moisture content is reached in 5 days for wooden planks of thickness 0.5 inch.

NOMENCLATURE

C_L	specific heat of timber (J/kg K)
C_p	specific heat of absorber plate (J/kg K)
C_r	specific heat of air (J/kg K)
G	mass flow rate of air inside the kiln (kg/s)
G_1	rate of infiltration of air inside the kiln from ambient (kg/s)
h_{Lr1}	heat transfer coefficient between timber and air in lower chamber (W/m^2 K)
h_{pr1}	heat transfer coefficient between plate and air in lower chamber (W/m^2 K)
h_{pr2}	heat transfer coefficient between plate and air in upper chamber (W/m^2 K)
h_{pra}	heat transfer coefficient between plate and ambient (W/m^2 K)
h_{prL}	radiative heat transfer coefficient between plate and timber (W/m^2 K)
h_{r1a}	heat transfer coefficient between air in lower chamber and ambient (W/m^2 K)
M_L	total mass of timber (kg)
M_p	mass per unit area of absorber plate (kg/m^2)
M_{r1}	mass per unit area of air in lower chamber (kg/m^2)
M_{r2}	mass per unit area of air in upper chamber (kg/m^2)
S_e	instantaneous value of solar flux incident on the eastern surface of timber stacking (W/m^2)
S_s	instantaneous value of solar flux incident on the southern surface of timber stacking (W/m^2)
S_w	instantaneous value of solar flux incident on the western surface of timber stacking (W/m^2)
T_a	ambient temperature (^{o}C)
T_G	ground temperature (^{o}C)
T_L	temperature of timber(^{o}C)
T_p	temperature of absorber plate (^{o}C)
T_{r1}	temperature of air in lower chamber (^{o}C)
T_{r2}	temperature of air in upper chamber(^{o}C)
t	time (s)
U_1	heat transfer coefficient between timber and ground (W/m^2 K)
α_L	absorptance of timber surface
α_p	absorptance of plate
τ_g	transmittance of glass

REFERENCES

Martinez, R., E. Martinez, and F. Paez (1984). Solar & Wind Technology, 1 223-227
Sharma, S. N., P. Nath, and S. P. Badoni (1980). Indian Forest Bulletin, No. 274
Simpson, W. T., and J. L. Tschernitz (1980). Forest Products Journal, 30 23-28
Taylor, K. J., and A. D. Weir (1985). Solar Energy, 34 249-255
Zahed, A. H., and M. M. Elsayed (1989). Solar & Wind Technology, 6 19-27

DESIGN, CONSTRUCTION, AND MONITORING RESULTS
OF TWO MEDIUM SIZED SOLAR DRYERS IN GERMANY AND SPAIN
USING NATURAL CONVECTION

U. Luboschik, P.Schalajda

IST Energietechnik GmbH
Ritterweg 1, D-7842 Kandern-Wollbach

ABSTRACT

Two different types of medium sized solar dryers using natural convection are presented. One type is located on the south facing slope of a hill in Germany. The other type is built at a horizontal site in soutern Spain. The difference in altitude, which is nessesary for the buoyancy of the heated air, is achieved by a chimney at the horizontal plant. The collector of all plants is a greenhouse of semicylindrical shape (r_i = 2.0 m), collector areas are between 171 m² and 323 m². The horizontal plant was monitored with chimney hights of 4.2, 7.2 and 11.7 m. The monitoring results of all plants show, that wind velocity has an important impact on the flow rate. Heat storage effects are very small.
Measured optical data of the used transparent EVA-foil and black PE-foil are reported. A first result of a two dimensional FEM calculation with FIDAP is shown.

KEYWORDS

Solar drying; natural convection; buoyancy; chimney; EVA-foil; transmission; finite element method (FEM) calculation.

INTRODUCTION

The first medium sized solar dryer was built in 1988 near Munich on the south facing slope of a hill (Luboschik and co-workers, 1989). The plant was monitored since October 1988. The function of the medium sized solar dryer has been demonstrated. The aim of the actual project is to develop design methods for natural convection applied to solar dryers.

PLANTS DESCRIPTION

Two types of solar convection dryers were built: One type is located in Hohenbachern near Munich, Germany, the other in Tabernas near Almeria, Spain.

In Hohenbachern a commercially available greenhouse, forming the solar collector, is placed on a south orientated slope of a hill. Its ground is covered with black PE-foil. The air streams into the greenhouse at the lower end, is warmed inside the greenhouse and rises to the upper end, where a dryer is connected to the greenhouse. The difference in altitude is 28 m.

Different collector areas and different greenhouse covers were tested (see table 1).

The first tested dryer was a locker with a net to volume of 3 m^3 (h = 2.0 m, w = 1.5 m, l = 1.0 m) and a capacity of 9 trays, each of 1 m x 1.5 m, allowing a fresh goods' loading capacity of 150 to 200 kg. The heated air may flow through or over the trays. In March 1991 a conveyor belt dryer was installed consisting of two belts, one upon another, each of 5 m x 2 m.

Additional experiments with black PE tissue were accomplished for increasing the thermal efficiency of the system.

The second type of solar convection dryer was built on a horizontal site at the R&D center "Plataforma Solar de Almeria" in Spain 1990. The difference in altitude, which is necessary for the buoyancy of the heated air, is achieved by a chimney (r = 0.9 m). The dryer chamber is situated between collector and chimney.

For the collector the same type of greenhouse is used as in Germany: a metallic frame of semicylindercal shape (r = 2 m). A single EVA foil is used for the transparent cover. The main axis of the collector is from south to north. The ground has a slight slope in the direction of the air flow of 0.7°.

TABLE 1. Some Geometric Data of the Medium Sized Solar Dryers

	Hohenbachern 1	Hohenbachern 2	PSA 1/2/3
latitude	48.40°N	48.40°N	37.10°N
longitude	11.70°E	11.70°E	2.35°W
altitude	470 m	470 m	90 m
collector inclination	tilted 17.6°	tilted 17.6°	horizontal 0.7°
azimuth	4.0°w	4.0°w	0.0°
area	323 m²	203 m²	171 m²
cover foil	EVA-double foil with airbag, 0.180 mm each	EVA, 0.180 mm	EVA, 0.180 mm
absorber foil	PE, 0.500 mm	PE, 0.500 mm	PE, 0.150 mm
volume dryer	2.5 m³	2.5 m³	16.9 m³
total	466 m³	312 m³	336/344/348 m³
flow area collector	5.7 m²	6.0 m²	7.0 m²
dryer	1.5 m²	1.5 m²	3.15 m²
chimney	–	–	2.54 m²
diff. of altitude collector	23.5 m	14.5 m	0.5 m
dryer	4.5 m	4.5 m	0.5 m
chimney	–	–	4.2/7.2/11.7 m
total	28.0 m	19.0 m	5.2/8.2/12.7
monitoring period	10/88 – 2/90	6/90 – 3/91	6/90 – 11/90 (4.2m) 11/90 – 2/91 (7.2m) 2/91 – ... (11.7m)

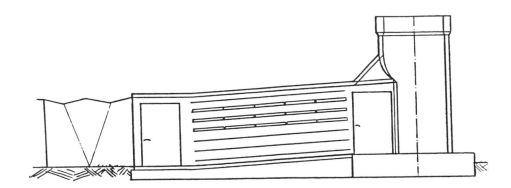

Fig. 1. Dryer and foot of the chimney of the plant PSA 1.

The dryer has a net to volume of 15.8 m^3 (1 = 5.0 m, h = 2.1 m, w = 1.5 m). It is slightly sloped (3.4°) in the direction of the air flow to improve the startup of the convective air flow. The dryer and the foot of the chimney (h = 4.2 m) are made of concrete. The chimney is lengthened by two iron tubes (r_i = 0.9 m) of 3.0 m and 4.5 m.

At the end of the year 1991 another solar dryer in southern Spain will be built with an east/west directed collector/dryer including a heat storage wall. The construction of this plant will be simple and economical.

MONITORING RESULTS

The still running monitoring campaign collecting 10 minutes' mean values was started in October 1988 in Hohenbachern and in June 1990 in Almeria. In Hohenbachern the data of 37 sensors are recorded, in Almeria the plant is monitored with 27 sensors. The weather data are separately recorded by the Plataforma Solar de Almeria. The data are used to validate a simulation model.

Typical temperature rises and flow rates on a clear day at noon were 20 K and 2.0 m^3/s (Hohenbachern 2). In Almeria the temperature rise was only 12 K, but the airflow was up to 6 m^3/s (PSA 1). The differences are due to different flow resistances of the systems.

It was found that not only the thermal buoyancy but also the wind has an important impact on the air flow. For example an airflow of 2.0 m^3/s in Hohenbachern 1 is achieved by an insolation of 500 W/m^2 or by a wind velocity of 8.0 m/s. Fig. 2. and 3. show the dependence of air flow on insolation and wind velocity.

The systems respond quickly to insolation changes. Heat storage effects, e.g. from day to night, are very small. The reason is the small heat transfer from the black ground foil to the ground. The roughness of the ground causes many air spaces between the foil and the ground, reducing the heat transfer.

The thermal efficiency of 25 % (Hohenbachern 1) is not satisfactory. Experiments with PE tissue are accomplished to improve the efficiency.

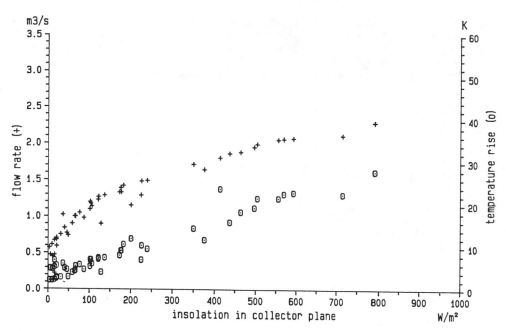

Fig. 2. Flow rate and temperature rise versus insolation in collector plane on a calm day (wind velocity < 0.5 m/s), Hohenbachern 1.

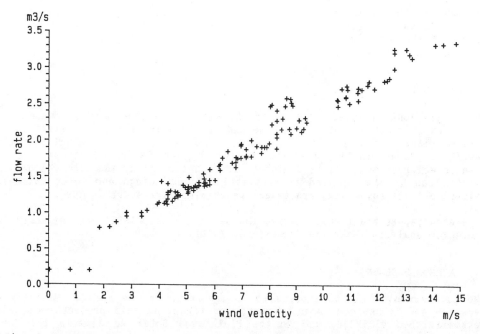

Fig. 3. Flow rate in the night versus wind velocity, Hohenbachern 1.

OPTICAL PROPERTIES OF THE FOILS

The spectral transmission and reflectance of the (clean) foils were measured. For getting τ, ρ and α, a convolution integral was formed with the measured data and AM 1.5 for the solar spectrum, and black body radiation of 323 K for the infrared spectrum. The results of the cover foil ("multieva" sotrafa,s.a.) and the absorber foil used for the plant in Spain are shown in table 2.

TABLE 2   Optical Properties of the Foils used in Almeria

		cover foil EVA, 0.180 mm	absorber foil PE, 0.150 mm
solar	τ	0.89	--
	ρ	0.07	0.04
	α	0.04	0.96
infrared	τ	0.66	--
	ρ	0.06	0.06
	α	0.28	0.94

solar: 0.32 - 2.50 µm, AM 1.5
infrared: 2.50 - 50.0 µm, black body 323 K

FLUID DYNAMICS MODELLING

The flow distribution in the greenhouse was calculated with FIDAP, a finite element methode (FEM) program runing on Cray 2 in Stuttgart. The calculation was two dimensional. The boundary condition was a heated ground with the maximum of temperature in the middle, the cover was on constant temperature. The result was a rising air flow in the middle, dividing in a right and left branch at the top. The air is circulating back along the cover and the ground-foil (see fig. 3.).

OUTLOOK

The evaluation and interpretation of the data is in progress. In summer 1991 drying experiments are planed. At the end of the year 1991 the construction of another plant in southern Spain is projected with a heat storage wall directed east/west. In autumn 1991 a PC program for the calculation of natural convection in solar dryers will be finished. It is anticipated, that the calculation methods can be also applied to ventilation of buildings and solar chimneys. Extensive tests for drying fruits and vegetables will start in 1992.

Separated papers treat solar drying (Müller and Reuss, 1991) and the model for energetic analysis (Mahr and Blumenberg, 1991).

ACKNOWLEDGEMENT

The autors would like to express their gratitude to the Federal Ministry of Research and Technology, Bonn, Gemany, for financing this project under Förderkennzeichen 0338923B, and to the Plataforma Solar de Almeria for local technical support.

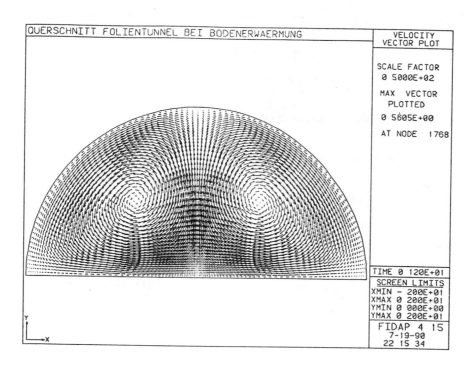

Fig. 4. Velocity vectors in the greenhouse with heated ground, two dimensional calculation.

REFERENCES

Luboschik, U., P. Schalajda, M. Reuß, H. Schulz, J. Blumenberg and M. Mahr (1989). Investigation of a solar convection dryer. Proceedings ISES Congress 1989 in Kobe, Japan.

Mahr, M., and J. Blumenberg (1991). Modelling of free convection air flow in a solar dryer. ISES Congress 1991. Accepted paper.

Müller, K., and M. Reuss (1991). Solar drying - a survey of different techno-lologies and their influence on product quality. ISES Congress 1991. Accepted paper.

THERMAL DESIGN AND PERFORMANCE COMPARISON OF SOLAR TIMBER SEASONING KILNS

K.S. Rao*, D. Singh* and S. Swaroop**

*Gujarat Energy Development Agency,
3rd floor, B.N. Chambers, R.C. Dutt Road, Baroda, India
**Strucon Development Services,
D7, Siddharth nagar Apts., Refinery Road, Baroda, India

ABSTRACT

This paper attempts to present effect of variation in input solar radiation in two different models of solar timber seasoning kilns and importance of thermal design in maximizing kiln's performance. It is found that design developed by Intermediate Techonology Development Group, UK modified suitably is found to absorb 25% to 50% more energy depending on seasons. Seasoning data of three different wood species is also presented. A seasoning time of seven to twelve days was observed. In order to further improve energy collection the role of thermal capacitance, selective absorber and utilization of side walls are considered. In the actual field conditions the seasoning kiln has 82% rate of return & require 72 batches to recover kiln's cost.

KEYWORDS

Solar timber seasoning kiln; Design; Performance; Monitoring; Real time cost; Control Strategy.

INTRODUCTION

With the continued increase in costs of conventional energy, utilization of solar technology is being adopted in increasing number of applications. As a part of furthering this, effectiveness of solar timber seasoning in bringing down the green wood moisture content to 15% are considered. An inherent slower drying rate of solar kilns has shifted the focus from close control over drying rates to more energy collection from source & to its optimum utilization. To bring out the growing improvements in solar designs the text considers a modified version of Forest Research Institute's design (FRIM) of 7.1 m³ capacity and a modified design of Intermediate Technology Group, UK (ITDGM) of 10.0 m³ capacity. These models were closely monitored to bring out the seasoning records.

SYSTEM DESCRIPTION

The salient features of these two models are listed in Table 1 and the schematic diagrams of these are given in Fig. 1. Both the designs can be operated on recirculation mode as well as once through mode with suitable adjustments of air entry & exhausts. Air is circulated by 2 nos of 0.9m dia.2hp axial fans. To condition the air, a disc type humidifier of 40 lits/hr

TABLE 1   Salient Features of Solar Timber Seasoning Kilns

S. No.	Description		FRIM model	ITDGM model
1*	Wood seasoning capacity,	m³	7.1	10.0
2*	Chamber volume,	m³	54.2	97.92
3*	Chamber size, length	m	4.78	6.46
	breadth	m	3.6	6.27
	North high	m	4.0	3.685
	South high	m	2.3	1.15
4*	Glass Area, Roof.	m²	18.55	43.67
	South	m²	11.0	7.43
	East/West	m²	10.5	11.77
5*	Cement Platform	m³	24.0	24.3

evaporation capacity is provided. Both the kilns have twin glazing in all sides except north side with 40 mm air gap inbetween as an insulator. Though both the kilns are equipped wiht V-corrugated abosrber in the ITDGM model the absorber is divided into two portion with the first part placed horizantal and the other part at a slope equal to the roof. This configuration allows the flexbility for seasons.

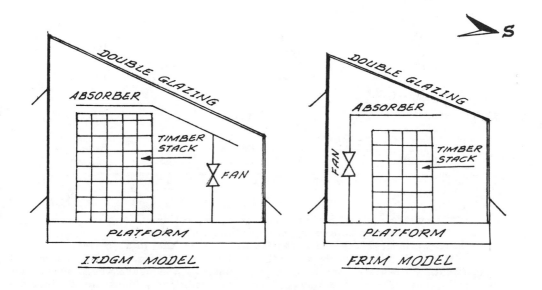

Fig. 1.   Schematic of timber seasoning kilns for 24°N latitude.

THERMAL DESIGN COMPARISON

Solar radiation data required for this analysis is adopted from ASHRAE handbooks. The data has been reworked to suitably predict the thermal performance of the kilns. Daily heat collection by both the kilns from their various surfaces is presented in Table 2. The table also construct the daily heat losses and works out the net heat gain from respective surfaces.

TABLE 2   Comparison of Daily Energy Balance on Walls at 24°N

Walls	Model	December			May		
		Heat In kcal	Heat loss kcal	Heat gain kcal	Heat In kcal	Heat loss kcal	Heat gain kcal
South	FRIM	51050	23350	27700	7160	17500	(-)10340
	ITDGM	34477	15770	18707	4836	11817	(-) 6983
East/	FRIM	17600	18940	(-) 1340	28800	14200	14600
West	ITDGM	19735	21237	(-)1500	32293	15922	16371
Roof	FRIM	92980	38675	54305	109100	30900	78200
	ITDGM	218890	91047	127843	258840	72744	186096
Total	FRIM	179230	99905	79325	173860	76800	97060
	ITDGM	292837	149291	143546	328262	116407	211855

It may be observed that the energy balance is presented for different seasons – Winter & Summer. Table 3 compares the percentage contribution of heat from each surface with respect to total. It also compares the two kilns performance with respect to each surface after correction.

TABLE 3   Contribution of % Heat from Different Walls and a comparison with each other

Models	Month	South	East/West	Roof	Total
Contribution					
FRIM	Dec	34.92	(-)1.74	68.46	100.0
	May	(-)10.65	15.04	80.57	100.0
ITDGM	Dec	13.03	(-)1.045	89.07	100.0
	May	(-) 3.30	7.73	87.84	100.0
Comparison*					
FRIM	Dec	Base	Base	Base	Base
ITDGM		47.95	79.47	167.14	128.48
FRIM	May	Base	Base	Base	Base
ITDGM		47.95	79.84	168.22	154.97

* Ratio of FRIM & ITDGM with ITDGM values reduced by 0.71.

The results of these data can be summarized as follows:
- The ITDGM model collects more energy than FRIM model.
- The side walls of ITDGM kiln contribute less to the total collection.
- In either of these models south surface is ineffective in summer & east/west walls are ineffective in winter. And east/west walls are effective one half of the day. It may be technically possible to insulate a lower 'k' value masonary construction of all side wall with only roof collecting the heat.
- Table 4 estimates the hourly collection rates at different latitutes for FRIM model. Similar could be extended to the other. The energy collection is fairly steady for five hours a day at all latitudes. But the rate of collection is greater at higher latitudes but the collection time is greater at lower latitudes. The maximum kiln temperature is obtained at 1500 hrs. solar time.

TABLE 4   Computed Values of Hourly Heat Collection At
           different latitudes (for FRIM model) in kcal

Solar Time	8°N	24°N	40°N
630	6500	2700	
730	12000	12000	8000
830	16000	18000	16000
930	18000	20500	20500
1030	18500	20700	22000
1130	18000	20250	21500
1230	18000	20400	21700

SEASONING IN ITDGM KILN

A kiln similar to ITDGM model was constructed at a saw mill around 32
kms from Baroda under GEDA subsidy program. The kiln has for accurate
monitoring two dry bulb & wet bulb thermometers. To determine the
moisture content of the wood, two samples were kept alongwith to be
seasoned variety. The samples would be taken out daily in the morning at
8hrs to .check the weight & moisture content and replaced in the same
position. Table 5 gives the seasoning data of different local variety of
wood species.

TABLE   5   Seasoning Data of Different Local Variety of Wood
            Species in ITDGM Model

Wood species	Batch No.	Sample Wt. gms.	Moisture content % initial	%final	Batch qty. m³	Activity Load	Unload
Kapoor	1	1180	46.85	19.45	5.81	29.3.90	6.4.90
Sal	2	870	26.84	17.42	11.47	17.4.90	24.4.90
Marsova	3	192	72.74	11.07	3.46	17.5.90	23.5.90
Kapoor	3	425	37.84	11.6	6.54	17.5.90	23.5.90
Sal	4	2913	38.05	17.25	7.07	28.5.90	5.6.90

The table has recorded a value (last batch) conducted during monsoon
season. The only significant difference found between the batches seasoned
earlier and that of this is in the earlier ones the kiln temperature (dry
bulb) used to raise up from 42°C to 52°C with a constant wet bulb
temperature of 40°C controlled by humidifier. The batch seasoned during
monsoon had a maximum dry bulb temperature of only 49°C from 40°C. Fig.
2 records the daily variation of parameters in the kiln for batch no. 3.
Two different wood species were dried simultaneously. The rate of drying
for moisture contents below 15% is uniform for both where as at higher
values the rates are significantly different.

PERFORMANCE UPGRADATION

As the availability of solar energy is intermittent it is essential to store
the energy for non-sunshine hours to reduce wide temperature fluctuations
in the kiln. Presently only wood and the platform act as the heat storage
components. The thermal capacity could be increased by introducing water
storage in the kiln for humidifier plus storing high specific heat fluids in
the hollow part of the support sturcture. With these the capacity of the
kiln could be increased by 15 to 20%.

2474

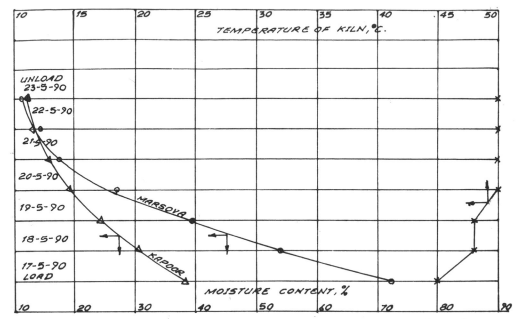

Fig. 2. Drying rate & kiln temperature variation for solar
timber seasoning kiln. (ITDGM model)

Another effective way to control heat collection is by using selectively
coated abosrber to reduce the emissivity lossess.

As discussed earlier the walls could also play a significant role in heat
flow. The side walls contribute nearly 12% of the heat supply to the kiln of
which contribution is from only either south or east/west walls. A suitable
insulating surface with a reflector on top could, under exposure, allow equal
proportion of additional intake energy as well as when in closed position
reduce the heat loss from the walls thus each surface could contribute
positively to heat collection. In case of east/west walls the insulator could
be only for a part as the door & inspection openings may have to be
seperated.

ECONOMICS

Solar timber seasoning kilns could be economically attractive also. These
kilns have a very high rate of return of 82.66% with reduced investment or
45% on the actual cost basis. Table 6   presents a detailed economic analysis
on the basis of actual performance data & the seasoning cost fetched by the
product. The greatest advantage with these kilns are that once the
investment is returned only cost to be serviced is depreciation & manpower
charges. As compared to a conventional kiln the seasoning cost would be far
lower.

CONCLUSIONS

Solar timber seasoning kilns can be very effective in seasoning as compared
to a conventional kiln with a seasoning time of seven to twelve days at a
rate of return of 82.66% over net cost. The thermal design aspects reveal

that ITDGM model has a better collection rate compared to FRIM model. Even this can be further improved with thermal capacity, Selective absorber and better utilization of side walls.

TABLE 6   Economic Analysis of Solar Timber Seasoning kiln
(ITDGM model)

A. Capital cost of kiln.		
Cost of seasoning kiln	:	Rs.167000.00
Subsidy from Government	:	Rs. 78000.00
Net cost to the user	:	Rs. 89000.00
B. Performance of kiln		
No. of batches seasoned	:	9
Quantity of wood seasoned	:	84.1183m^3
Time required for seasoning		93 days
C. Recurring Cost of kiln		
Electricity Expenses	:	Rs. 1200.00
Salary & Wages	:	Rs. 3000.00
Loading & Unloading	:	Rs. 2970.00
Repair & maintenance	:	Rs. 1000.00
Total Recurring expenses	:	Rs. 8170.00
D. Rate of return		
Return from seasoning of		
250m^3/year @ Rs.425/m^3	:	Rs. 107250.00
Annual expenses	:	Rs. 32680.00
Rate of return	:	82.66%

REFERENCES

Singh, D., Bhatia, P.N., and K.S. Rao (1989). Proc. R. Solar World Congress (Kobe)
ASHRAE Hand book (1982). Applications
ASHRAE Hand book (1985). Fundamentals
Swaroop, S., and Sudhir Mohan (1990). Proceedings National Solar Energy Convention (Calcutta)